AF772275

ADVANCES IN OPTICAL AMPLIFIERS

Edited by **Paul Urquhart**

Advances in Optical Amplifiers
Edited by Paul Urquhart

Published by InTech
Janeza Trdine 9, 51000 Rijeka, Croatia

Copyright © 2011 InTech
All chapters are distributed under the Creative Commons Non Commercial Share Alike Attribution 3.0 license, which permits to copy, distribute, transmit, and adapt the work in any medium, so long as the original work is properly cited. After this work has been published by InTech, authors have the right to republish it, in whole or part, in any publication of which they are the author, and to make other personal use of the work. Any republication, referencing or personal use of the work must explicitly identify the original source.

Statements and opinions expressed in the chapters are these of the individual contributors and not necessarily those of the editors or publisher. No responsibility is accepted for the accuracy of information contained in the published articles. The publisher assumes no responsibility for any damage or injury to persons or property arising out of the use of any materials, instructions, methods or ideas contained in the book.

Publishing Process Manager Ana Nikolic

Technical Editor Teodora Smiljanic

Cover Designer InTech Design Team

Additional hard copies can be obtained from orders@intechopen.com

Advances in Optical Amplifiers, Edited by Paul Urquhart
p. cm.
ISBN 978-953-307-186-2

We are IntechOpen, the world's leading publisher of Open Access books

Built by scientists, for scientists

4,000+
Open access books available

116,000+
International authors and editors

120M+
Downloads

151
Countries delivered to

Our authors are among the
Top 1%
most cited scientists

12.2%
Contributors from top 500 universities

CLARIVATE ANALYTICS
BOOK CITATION INDEX
INDEXED

WEB OF SCIENCE™

Selection of our books indexed in the Book Citation Index
in Web of Science™ Core Collection (BKCI)

Interested in publishing with us?
Contact book.department@intechopen.com

Numbers displayed above are based on latest data collected.
For more information visit www.intechopen.com

For EU product safety concerns:
IN TECH d.o.o., Prolaz Marije Krucifikse Kozulić 3, 51000 Rijeka, Croatia,
info@intechopen.com or visit our website at intechopen.com.

Meet the editor

Paul Urquhart is a visiting professor in the Department of Electronic and Electrical Engineering of the Universidad Pública de Navarra (UPNA), Pamplona, Spain, where his research and teaching are on optical communications. He is also a visiting lecturer at the Hochschule Niederrhein, Krefeld, Germany and at ESIEE Paris, France, where he collaborates in research. Dr Urquhart was previously a professor in the Department of Optics at Télécom Bretagne, Brest, France and a senior research engineer at BT Laboratories, Ipswich, UK. He has had a long-standing association with the University of Essex, Colchester, UK, where he has taught optical communications and organised workshops for industry. Throughout his career he has participated in and lead research projects on optical amplifiers, filters and sources, as well as the design and simulation of optically amplified fibre networks for telecommunications and sensing. Paul Urquhart is a FEANI-registered European Engineer (Eur Ing), a member of the Institution of Engineering and Technology (IET) and a chartered engineer.

Contents

Preface

Optical Amplifiers and Their Applications in Networks

Optical amplifiers are essential elements in advanced fibre-based telecommunications networks. They provide the means to counteract the losses caused by the fibre transmission medium, the components placed in the propagation path and the power division at optical splitters. Amplifiers therefore facilitate the high global capacities, long transmission spans and multipoint-to-multipoint connectivity required for operation with growing data volumes. In their absence fibre networks would need many more optical-electrical-optical (O-E-O) converters for the electronic repeating, retiming and reshaping of attenuated and noisy bit streams. The consequences would be transmission at significantly lower data rates, requiring numerous fibres in each cable; more node buildings, often in expensive city centre locations; larger equipment cabinets, occupying valuable floor space; increased total power consumption, with its associated environmental impact and, very importantly, higher operating costs to be passed on to the customer. For these reasons optical amplifier technologies have been key enablers en route to ubiquitous information availability.

All-optical amplification has found application in all categories of fibre network, whether they be single modulated wavelength or multi-channel operation through the use of wavelength division multiplexing (WDM). When incorporated in the tree topology of a passive optical network (PON) for fibre to the home (FTTH), a single amplifier module allows around one thousand customers to be served from one head end. Such "long reach PONs" offer considerable cost savings.

Amplifiers in metropolitan area networks tend to be housed within node buildings that are interconnected by WDM "self-healing" fibre rings. They enable operation with increased inter-node spans and ensure that the channel powers are sufficient for wavelength routeing at the nodes by optical add-drop multiplexers. Wide area terrestrial networks, which are ring or mesh topologies, range in scope from the interlinking of a few towns to major trans-continental trunk routes. Operation is commonly with several dozen WDM channels, each at a data rate of 10 Gbit/s or above. Wavelength routeing, by optical cross-connects, is desirable but it is possible only if the signal powers and optical signal-to-noise ratios are maintained at high values throughout the transmission path by re-amplification at suitable locations. Owing to the demands of electrical power feeds, the amplifiers for terrestrial operation preferably reside in node buildings but this is not always possible in larger networks and reliable external electrical powering is required. Innovations in remote optical pumping and distributed amplification are promising in this context.

A specialised but important application of optical amplifiers is in "repeaterless" subsea transmission for festoon, island-to-mainland and island-to-island links with spans of up to a few hundred kilometres. Costly submerged repeaters and their associated electrical power supplies can often be eliminated by using distributed amplification and remote optical pumping and confining all discrete amplifiers to the terminal buildings.

The longest span optical telecommunications networks traverse the world's oceans. Their amplifiers are housed within repeaters, which are normally spaced every 40-60 km over total transmission distances of up to 10 000 km. The ocean-bed is not readily accessible and reliability is vital to minimise the number of expensive and time-consuming repairs. Trans-oceanic systems are often designed for twenty-five year working lives, indicating the faith that network operators now have in optical amplifier technology.

The applications described here easily justify the substantial investment in amplifier research and development that has taken place over the past three decades. However, what is now particularly impressive is that it is not a complete list of uses. Other domains include the incorporation of amplification in computer interconnects. These can range from fibre-based local area networks (LANs) with star or ring topologies to serve a building or campus to multi-branched optical back-planes within supercomputers. Another growing field is in bus, ring and star based fibre networks for sensors of many types. The know-how developed for telecommunications engineering thus has numerous potential applications in, for example, the structural monitoring of buildings, bridges and dams, site perimeter security, industrial process control, pollution detection, and human safety in the mining, aviation, nuclear power, oil extraction and chemical processing industries.

The view presented so far is of the amplifier as a gain element, in which attenuated input signals pass through an appropriate photonic medium to emerge with significantly enhanced powers. However, research, especially in semiconductor media, has concentrated on other amplifier functionalities. When one or more high intensity signals traverse a suitable semiconductor optical amplifier (SOA) they experience various nonlinear effects. The most important are self- and cross-phase modulation, sum and difference frequency generation, four-wave mixing and cross-gain and cross-polarisation modulation. These phenomena, often in combination with advanced waveguide-based interferometers, provide alternative device possibilities. Examples include: (i) wavelength converters, (ii) all-optical logic elements, (iii) photonic space switches, (iv) optical regenerators to repeat, retime and re-shape corrupted optical bit streams, (v) time domain demultiplexers for very high data rate signals consisting of picosecond pulses and (vi) optical clock-recovery modules for use at the receiver to overcome high frequency jitter.

To take one example, wavelength converters, which offer greatest potential in wide area and metropolitan terrestrial networks, allow a channel to be transferred to another carrier wavelength without O-E-O regeneration. This is a particularly desirable functionality in high capacity networks that must be reconfigured by wavelength rerouteing. Wavelength converters enable economies on the total number of channels and they avoid contention, where two different data streams with the same carrier wavelength converge on a common switch. By performing their function in the optical domain, wavelength converting semiconductor amplifiers can thus allow a marked reduction in the number of large racks of electronic equipment in node buildings, with an associated saving of floor space, energy consumption and cost.

The Main Amplifier Types

A global overview of the main types of optical amplifier for telecommunications is presented in Fig. 1 and it illustrates the diversity of the subject. Many categorisations are possible. In the one shown, the devices have two main configurations, both of which profit from the power confinement offered by waveguiding, which is nearly always in the fundamental mode. They are: fibre based and planar optical waveguide based. All of the fibre amplifiers and the erbium doped planar waveguide amplifiers are optically pumped by launching light from one or more lasers into the waveguide along with the propagating signal(s). In contrast, semiconductor amplifiers provide the population inversion necessary to establish gain by the direct injection of electrical current into the active region via attached electrodes.

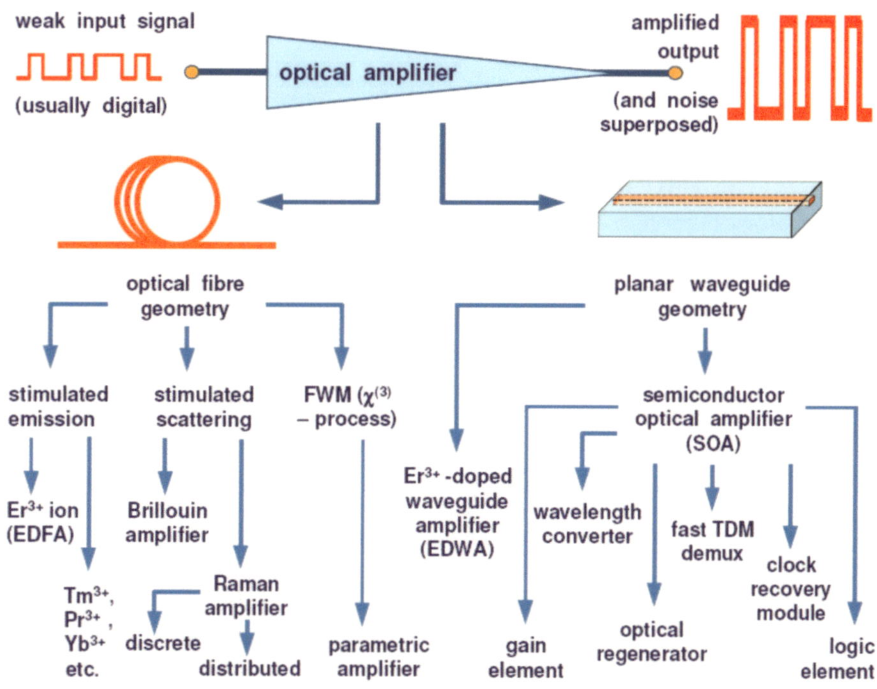

Fig. 1. The main types of optical amplifiers for telecommunications, in which the functions of semiconductor optical amplifiers are a sample of those available.

Fibre optical amplifiers are cylindrical dielectric waveguides with a narrow diameter core of relatively high refractive index, surrounded by a lower index cladding. Pumping is by a comparatively high power laser, normally with a shorter wavelength than the signals. The pump light can be launched to travel in the same or the opposite direction to the signals. As marked in Fig. 1, there are three main groups of operating mechanisms: (i) stimulated emission between resonant states, known as energy levels, provided by rare earth ions in the core, (ii) non-resonant processes called stimulated light scattering and (iii) parametric interactions between guided waves, using the ◎(3) nonlinear process of four wave mixing in the fibre. An important practical difference between them is that the amplifiers based on stimulated emission depend on the presence of a special ionic dopant in the fibre core, whereas those based on stimulated scattering and parametric interactions do not.

The main type of rare earth doped fibre amplifier incorporates the trivalent erbium ion and is therefore known as the erbium doped fibre amplifier (EDFA). It can be optimised to provide gain in the 1530-1565 nm or 1570-1610 nm bands and it is the most widely commercialised amplifier for telecommunications. As products, EDFAs are commonly targeted at specific markets, such as compact line amplifiers for cost-sensitive metropolitan networks with only a few WDM channels or power amplifiers for very high capacity trunk systems. Other rare earth doped fibre amplifiers, such as thulium, praseodymium and ytterbium, show great promise for diverse applications in other operating spectra. However, they normally consist of a special glass type, such as one of the fluorides, which imposes particular design, operation, fabrication and device challenges.

Two scattering processes have been used for amplification in fibres: stimulated Brillouin scattering (SBS) and stimulated Raman scattering (SRS). Brillouin amplification, which provides gain within a very narrow frequency band by contra-directional pumping, has found use in distributed optical sensing, signal processing for microwave photonics and laboratory demonstrations with more futuristic domains, such as slow light, in mind. Fibre Raman amplifiers have been commercialised for WDM systems, using either existing telecommunications fibre or dispersion compensating fibre as broad-bandwidth gain media and they are pumped at wavelengths that are available from high power semiconductor lasers. They can provide distributed amplification, which enables low total noise figures, or discrete amplification, where one fibre can simultaneously provide gain and chromatic dispersion compensation. There are many reports of combining erbium and Raman amplifiers, especially for application in high capacity systems.

The nonlinear optical phenomenon of four wave mixing (FWM) occurs in WDM transmission in optical fibres when the propagating wavelengths are in the vicinity of the wavelength of zero dispersion. The effect is normally to be avoided because it causes cross-talk. However, FWM can be beneficial by providing low noise and large bandwidth "parametric" amplification with high peak gains. The gain band can be designed to coincide with any optical communications window by the provision of suitable high power pump lasers and the selection of appropriate pump and zero dispersion wavelengths. The preferred gain medium is "highly nonlinear fibre", which is dispersion shifted and has a narrow core diameter. Fibre parametric amplifiers offer phase-sensitive operation, which is advantageous as it enables optical noise figures below the 3 dB quantum limit for rare earth, Raman, Brillouin and semiconductor amplifiers.

The smallest and potentially lowest cost category of amplifiers with the best wall plug efficiencies use semiconductor gain media. Their materials and planar waveguide format allow integration with other photonic components, such as switches, filters, lasers and electro-absorption modulators, thereby enhancing their anticipated cost and size advantage over other amplifier types. The gain spectra of SOAs can exceed 100 nm and their profiles are parabolic, which are easily spectrally flattened. Moreover, the whole profile can be designed to exhibit a peak at any of the optical fibre transmission wavelengths by appropriate selection of the ratios of their constituent materials. Operation in the 1460 – 1625 nm communications bands is achieved using group III-V InGaAsP/InP (quaternary) systems, mainly with quantum well structures, but other materials, particularly "dilute nitride" GaInNAs semiconductors, are also being explored.

Unfortunately, SOAs have some disadvantages when used merely as gain elements. It is difficult to achieve low loss coupling between (rectangular cross-section and high refractive index) semiconductor waveguides and (circular cross-section and low refractive index) transmission fibres. Such loss, especially at its input interface, increases an amplifier's total noise figure. Moreover, feed-back by facet reflections leads to unwanted ripples in the gain spectrum, which must be suppressed by appropriate device design. SOAs also tend to exhibit polarisation-dependent gain and up to 1 dB difference between the TE and TM waveguide modes is common. SOAs do not make good power amplifiers and multi-channel WDM operation can lead to cross-talk when operating at 10 Gbit/s or above because the conduction band lifetime time is comparable to the bit period; modulation patterns from each channel are imposed on the others. The phenomenon is cross-gain modulation (XGM). However, as explained above, XGM (and other active phenomena in the gain media) can be turned to advantage to enable alternative amplifier functionalities, such as photonic logic elements and wavelength converters. Therein lies the great potential of semiconductor optical amplifiers.

Other developments have included erbium doped waveguide amplifiers. These are glass planar optical configurations that aim to combine the useful gain bands and low noise figure characteristics of EDFAs with the potential for device integration and small package sizes exhibited by SOAs. To date, progress has been difficult; the peak gains and saturation output powers are relatively low, spectral gain flattening is demanding and operation has been confined to the 1530-1565 nm band. Nevertheless, the ability to integrate a gain functionality with glass components, such as couplers and filters, and the possibility of host materials that are not readily fabricated in fibre format are very desirable.

Advances in optical amplifiers for telecommunications have taken place in parallel with developments in laser physics, where the visible and ultra-violet spectra and femto-second pulses are of great interest. It is to be hoped that these lines of investigation will in future converge with telecommunications engineering. Technology is most beneficial to mankind when its practitioners are open to radically new approaches.

Organisation of the Book

The authors with whom I have had the pleasure to collaborate have written chapters that report recent developments in key optical amplifier technologies. They cover a number of themes, which include the fibre based and planar waveguide based designs described above. No book of the current length can encompass the full scope of the subject but I am pleased at the range of topics that we have been able to include. The context for each contribution can be understood by referring to Fig. 1.

The chapters have been grouped according to the following themes: Semiconductor Optical Amplifiers: General Concepts, Semiconductor Optical Amplifiers: Wavelength Converters, Semiconductor Optical Amplifiers: Other Processing Functions, Erbium-Doped Amplifiers and Lasers and Other Amplifier Mechanisms. This categorisation is, of course, imperfect because some chapters are inter-disciplinary. However, not residing within an easily identifiable subject boundary is a positive sign; it is one of the indicators of scientific progress.

I am grateful to Ana Nikolic, publishing process manager at Intech, for her prompt answers to my questions. I wish my collaborators every success in their future research activities.

January 2011.

Paul Urquhart
Pamplona,
Spain

Part 1

Semiconductor Optical Amplifiers: General Concepts

1

Semiconductor Optical Amplifiers

M. Haridim, B.I. Lembrikov, Y. Ben-Ezra
Holon Institute of Technology (HIT),52 Golomb Str., Holon 58102
Israel

1. Introduction

The theoretical and experimental studies of semiconductor optical amplifiers (SOAs) started immediately after the invention of semiconductor lasers as early as in 1962, but practical applications of SOAs began in the 1980s when SOAs have emerged as an important component in many optical fiber communication systems Agrawal (2002). The applications of SOAs in optical communications, switching and signal processing based on their high performance manifested in laboratory prototypes were predicted as early as 1989 Eisenstein (1989). SOA is characterized by extremely strong non-linearity, low power, high operation rate, and small size as compared to erbium doped fiber amplifiers (EDFAs) and Raman optical amplifiers. The cross gain modulation (XGM), cross phase modulation (XPM), four-wave mixing (FWM) phenomena are strongly manifested in SOA Agrawal (2002).

The two main application areas of SOA are the linear in-line amplification in gigabit passive optical networks (GPON), and fast nonlinear all-optical signal processing Freude (2010). In particular, SOAs are among the most promising candidates for all-optical processing devices due to their high-speed capability, low switching energy, compactness, and optical integration compatibility Dong (2008). Hence, besides its use as an in-line optical amplifier, SOA provides a wide range of functional applications including wavelength conversion (WC), regeneration, photonic switching and various all-optical signal processing components.

High-speed WC, all-optical logic operations, and signal regeneration are important operations in all-optical signal processing where SOAs are widely used Agrawal (2002), Ramamurthy (2001), Dong (2008), Sun (2005). WC is essential for optical wavelength division multiplexing (WDM) network operation Ramamurthy (2001). There exist several all-optical techniques for wavelength conversion based on SOAs using XGM and XPM effects between a pulsed signal and a continuous wave (CW) beam at a desired wavelength of the converted signal Agrawal (2002).

All-optical logic gates operation is based on nonlinearities of optical fibers and SOAs. However, optical fibers suffer from disadvantages such as weak nonlinearity, long interaction length, and/or high control energy required in order to achieve a reasonable switching efficiency Sun (2005). SOA, on the other hand, has high nonlinearity, small dimensions, low energy consumption, high operation speed, and can be easily integrated into photonic and electronic systems Sun (2005), Hamié (2002), Kanellos (2007), Dong (2008).

The major challenges of improving optical transmission systems stem mainly from signal-to-noise ratio (SNR) degradation and chromatic dispersion. For this reason, the optical signal reamplification, reshaping, and retiming, i.e. the so-called 3R regeneration, are necessary in order to avoid accumulation of noise, crosstalk and nonlinear distortions and to provide a good signal quality for transmission over any path in all-optical networks Sartorius (2001), Zhu (2007), Leem (2006), Kanellos (2007). Optical regeneration technology is characterized by lower power, much more compact size, and can provide transparency in the needed region of spectrum Zhu (2007). All-optical 3R regeneration is also less complex, provides better performance and uses fewer optoelectronic/electronic components than electrical regeneration Leem (2006), Kanellos (2007). SOA chips and packaged SOAs are used as tunable lasers with ultra-wide tuning range and kHz scan rate for sensing and medical applications Spiekman (2009). The wavelength agility and low power requirements of SOA stimulate their use in telecommunications as a range extender in Fiber to the Home networks and in transmission system optical channels outside the EDFA band Spiekman (2009).

Tunable optical delay lines and continuously controllable phase shifters for microwave signal processing are based on the slow and fast light propagation Dúill (2009). One of the most efficient mechanisms for the slow and fast light propagation is the coherent population oscillations (CPO) in SOAs due to its strongly manifested XPM, XGM and FWM Chang-Hasnain (2006), Dúill (2009).

Recent advances in quantum well (QW) SOAs, quantum dash (Q-dash) SOAs and especially quantum dot (QD) SOAs make them promising candidates for use in evolving optical fiber communication systems since a bulk SOA performance may be substantially improved in terms of threshold current density, saturation power, gain bandwidth, and the temperature dependence of quantum devices Bimberg (1999), Sugawara (2004), Ustinov (2003). The possibility of integrating into Si photonic technology makes QD SOA an especially important candidate for integrated electronic and photonic circuits on the same silicon platform.

The chapter is constructed as follows. The operation principle and the phenomenological theory of bulk SOA is presented in Section 2. Structure, dynamics and peculiarities of QW and QD SOAs are discussed in Section 3. Novel applications of SOAs such as photonic pulse generation, all-optical signal processing, all-optical logics, slow and fast light generation are reviewed in Section 4. Conclusions are presented in Section 5.

2. Theory of bulk SOA

2.1 Structure and operation principle of bulk SOA

In this section, we consider briefly the operation principle and the structure of a bulk SOA Eisenstein (1989), Agrawal (2002), Chang (2005). SOAs amplify incident light by the stimulated emission process using the same mechanism as laser diodes Agrawal (2002). In fact, SOA can be characterized as a semiconductor laser without feedback Agrawal (2002). Optical gain can be realized when the electrically or optically pumped SOA achieves population inversion Agrawal (2002). The optical gain depends on the incident signal frequency ω and the local beam intensity at any point inside the amplifier Agrawal (2002).

Bulk SOA consists of a laser diode with low facet reflectivities. Typically, the device contains several layers Eisenstein (1989), Agrawal (2002). Carriers from an external bias circuit are injected into the active region of SOA where they are confined by layers of materials with higher energy band gap. An optical signal impinging on the active region will induce simulated emission and will be amplified under the condition of population inversion, i.e. when the bias current is sufficiently large Eisenstein (1989). The active region serves as the core of a slab optical waveguide since its refractive index n is larger than the refractive index of the cladding. In a semiconductor laser, the end facets are cleaved perpendicular to the waveguide forming mirrors to provide the feedback necessary for laser oscillations. The SOA structure is shown in Fig.1.

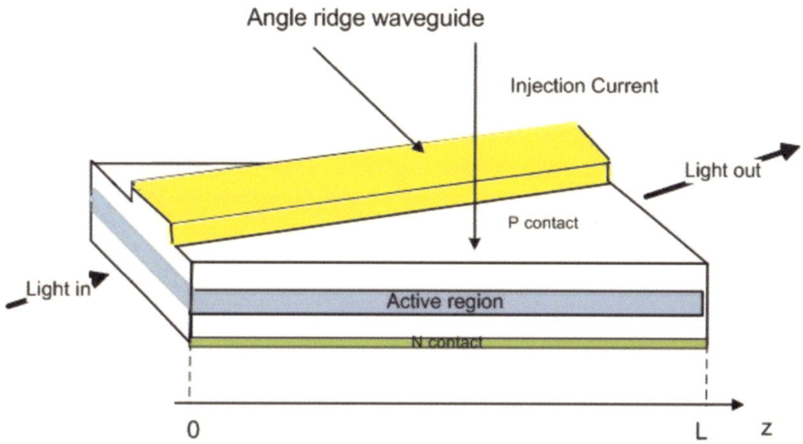

Fig. 1. Structure of SOA

SOA can operate in two different regimes. In the first case called the travelling-Wave (TW) regime, the oscillations are prevented in order to create a single pass gain , Eisenstein (1989), Agrawal (2002). The active medium is described as a homogeneously broadened two-level system. In such a case, the gain coefficient $g(\omega)$ has the form Agrawal (2002)

$$g(\omega) = \frac{g_0}{1 + (\omega - \omega_0)^2 T_2^2 + P/P_s} \tag{1}$$

where g_0 is the peak value of the gain, ω_0 is the transition frequency, P is the optical power of the amplified signal, P_s is the saturation power, and $T_2 \leq 1ps$ is the dipole relaxation time. Under the condition of $P/P_s \ll 1$ corresponding to the unsaturated regime, equation (1) takes the form

$$g(\omega) = \frac{g_0}{1 + (\omega - \omega_0)^2 T_2^2} \tag{2}$$

It is seen from equations (1) and (2) that the frequency dependence of the gain is characterized by a Lorentzian profile with the resonance frequency ω_0. The gain bandwidth $\Delta\omega_g$ defined as the full width at half maximum (FWHM) is given by Agrawal (2002).

$$\Delta\omega_g = \frac{2}{T_2} \tag{3}$$

The local optical power $P(z)$ in a gain medium can be defined as Agrawal (2002).

$$P(z) = P(0)\exp(gz) \tag{4}$$

where $P(0) = P_{in}$ is the input optical power. We can define an amplification factor $G(\omega)$ for a SOA of length L as the ratio between the output optical power $P_{out} = P(L)$ and P_{in} Agrawal (2002).

$$G(\omega) = \frac{P_{out}}{P_{in}} = \frac{P(L)}{P(0)} = \exp(gL) \tag{5}$$

The gain saturation regime is defined by the gain dependence on the optical power according to equation (1). It can be shown that for the large-signal amplification factor the following relationship prevails Agrawal (2002).

$$G = G_0 \exp\left[-\frac{(G-1)\,P_{out}}{GP_s}\right] \tag{6}$$

where G_0 is the unsaturated amplification factor. The main noise in SOA is due to the spontaneous emission. The SNR degradation is characterized by the amplifier noise figure (NF) F_n defined by Agrawal (2002).

$$F_n = \frac{(SNR)_{in}}{(SNR)_{out}} \tag{7}$$

NF can reach $6 - 8dB$. However for optical communication applications it should be as low as possible Agrawal (2002).

In the second regime, called Fabry-Perot (FP) SOA, a semiconductor laser biased slightly below threshold is used as SOA in a FP cavity where the FP cavity is characterized by multiple reflections at the facets with reflectivities $R_{1,2}$ Agrawal (2002). The amplification factor $G_{FP}(\omega)$ for such a FP SOA is given by Agrawal (2002)

$$G_{FP}(\nu) = \frac{(1-R_1)(1-R_2)\,G}{\left(1-G\sqrt{R_1R_2}\right)^2 + 4G\sqrt{R_1R_2}\sin^2\left[\pi(\nu-\nu_m)/\Delta\nu_L\right]} \tag{8}$$

where $\nu = \omega/2\pi$, $\nu_m = mc/2nL$ are the cavity resonance frequencies, c is the free space light velocity, $\Delta\nu_L = c/2nL$ is the longitudinal mode spacing, and $m = 1, 2, 3, ...$ is the mode number. The FP SOA bandwidth $\Delta\nu_A$ at the $3dB$ level of $G_{FP}(\nu)$ is given by Agrawal (2002)

$$\Delta\nu_A = \frac{2\Delta\nu_L}{\pi}\arcsin\left[\frac{1-G\sqrt{R_1R_2}}{\left(4G\sqrt{R_1R_2}\right)^{1/2}}\right] \tag{9}$$

2.2 Rate equations and optical field propagation equations

The efficiency of SOA applications in all-optical integrated circuits for optical signal processing and functional devices is mainly determined by their high gain coefficient and a relatively low saturation power Agrawal (1989), Agrawal (2002), Premaratne (2008). Device engineering and performance optimization require simple and quantitatively accurate models for SOAs providing an adequate description of SOAs peculiarities Premaratne (2008).

The theory of pulse propagation in amplifiers based on a two-level system model is also valid for SOAs Agrawal (1989). The SOA's active region is modeled as an ensemble of noninteracting two-level systems with transition energies extending over the range of the conduction and valence bands Agrawal (1989). If the pulse width $T_p \gtrsim 1ps$ is much larger than the intraband relaxation time $\tau_{in} \sim 0.1ps$ and, on the other hand, $T_p \ll \tau_e \sim 10^{-10}s$ where τ_e is the carrier lifetime, the rate-equation approximation can be applied where the amplification process within the SOA is described by the following rate equations for the carrier density $N(z,t)$ and the optical signal intensity $I(z,t)$ Agrawal (1989), Agrawal (2002), Premaratne (2008).

$$\frac{\partial N(z,t)}{\partial t} = \rho(z) - \frac{N(z,t)}{\tau_e} - g(z,t)\frac{\lambda I(z,t)}{hc} \tag{10}$$

$$\frac{\partial I(z,t)}{\partial z} + \frac{1}{v_g}\frac{\partial I(z,t)}{\partial t} = g(z,t)\,I(z,t) - \alpha I(z,t) \tag{11}$$

where t is the time, z is the distance along the light propagation direction in the SOA active region, $\rho(z)$ is the current-injection density, α is the loss coefficient, λ is the operating wavelength, and h is the Planck's constant. The input pulse intensity profile $I_0(t)$ is given by Premaratne (2008)

$$E_g = A_m \int_{-\infty}^{\infty} I_0(t)\, dt \tag{12}$$

where A_m is the effective mode area of the SOA active region, and E_g is the pulse energy. The gain coefficient $g(z,t)$ is defined as Premaratne (2008).

$$g(z,t) = \Gamma a \left[N(z,t) - N_0\right]$$

where Γ is the mode confinement factor, a is the differential gain coefficient, and N_0 is the carrier density corresponding to the transparency state. By using the coordinate transformations $z, \tau = t - z/v_g$ and introducing the new variable

$$h(z,\tau_n) = \int_0^z g(z,\tau)\, dz \tag{13}$$

where $z \subset [0,L]$, $\tau_n = (\tau v_g)/L$ is normalized time, and L is the length of SOA, equations (10) and (11) can be reduced to the following integro-differential equation describing the gain recovery dynamics for a short optical pulse Premaratne (2008)

$$\frac{\partial h(z,\tau_n)}{\partial \tau_n} = \epsilon\left[h_\rho - h(z,\tau_n)\right] - \beta(\tau_n)\left\{\exp\left[h(z,\tau_n) - \alpha z\right] - 1\right\}$$
$$- \beta(\tau_n)\left[\alpha \int_0^z \exp\left[h(z,\tau_n) - \alpha z\right] dz\right] \tag{14}$$

where $\epsilon = L/(v_g \tau_e)$, $\beta(\tau_n) = \Gamma a \lambda I_0(\tau) L/(v_g h c)$, and

$$h_\rho(z) = \int_0^z \left[\tau_e \Gamma a \rho(z) - \Gamma a N_0\right] dz \tag{15}$$

Equation (14) has been solved analytically and numerically for different situations using a multiple-scales technique Premaratne (2008). In particular, the signal gain $G(z,T,U)$ is given by Premaratne (2008).

$$G(z,T,U)\exp(-\alpha z) = \exp\left\{\left[h(z) - \alpha z\right]\left[1 - \exp(-U)\right]\right\}$$
$$\times \left\{1 - \frac{1 - \exp(-h_I(z) + \alpha z)}{E_\gamma(T)}\right\}^{-\exp(-U)} \tag{16}$$

where $T = \tau_n$ is the pulse time scale, $U = \epsilon \tau_n$ is a slow time scale corresponding to the carrier recovery, and $h_I(z)$ is the initial profile of $h(z,T,U,\epsilon)$,

$$E_\gamma(T) = \exp\left(\int_0^T \frac{\beta(T)\,\tilde{g}(z,T,U)}{\left[\tilde{g}(z,T,U) - \alpha\right]} dT\right) \tag{17}$$

$$\tilde{g}(z,T,U) = \frac{h_\rho(z)}{z}\left[1-\exp(-U)\right] - \frac{\exp(-U)}{z} \times \ln\left[1 - \frac{(1-\exp(-h_I(z)))}{E_\beta(T)}\right] \tag{18}$$

and

$$E_\beta(T) = \exp\left(\int_0^T \beta(T)\,dT\right) \tag{19}$$

The electric field $E(z,t)$ of the pulse is given by Premaratne (2008)

$$E(z,t) = \sqrt{I(z,t)}\exp\left[i\left(\phi(z,t) - \omega t\right)\right]; \phi(z,t) = \frac{\alpha_L}{2}h(z,t) \tag{20}$$

where $\phi(z,t)$ is the phase of the pulse inside SOA and α_L is the linewidth enhancement factor (LEF).

Comparison of the analytical results of the theory developed above with the numerical simulation results for Gaussian picosecond pulses propagating through a SOA of length $L \approx 400\mu m$, active region width and thickness of $2.5\mu m$ and $0.2\mu m$, respectively, $\Gamma = 0.3$, carrier injection rate $1.177 \times 10^{34}s^{-1}m^{-3}$, $n_g = 3.7$, $\alpha = 3000m^{-1}$, $\tau_e = 3 \times 10^{-10}s$, $a = 2.5 \times 10^{-20}m^2$, $N_0 = 1.5 \times 10^{24}m^{-3}$, $\alpha_L = 5$, and $\lambda = 1552.52nm$ shows a good accord Premaratne (2008).

3. Advanced structures

3.1 QW SOA

The structure of QW SOA devices is very similar to that of bulk SOA, except that the active layer thickness of the former is reduced to the order of $10nm$ where quantum effects play an essential role. A thin layer sandwiched between two layers of a wide band gap material form a finite potential well. As a consequence of carriers confinement in the z direction in the thin layer, the conduction and valence bands break up into a series of well-defined sub-bands with a step-like density of states (DOS) function $\rho_{QW}(E)$ which is energy independent. DOS function for electrons has the form Zhao (1999)

$$\rho_{QW}(E) = \sum_{n_e} \frac{m_e}{\pi\hbar^2} H\left[E_e - E_{ez}(n_e)\right] \tag{21}$$

where m_e is the electron effective mass, $\hbar = h/(2\pi)$, E_e is the total electron energy given by

$$E_e = E_{ez}(n_e) + \frac{(\hbar k)^2}{2m_e} \tag{22}$$

$E_{ez}(n_e)$ are the quantized energy levels, $n_e = 1, 2, ...$is the level number, $k = \sqrt{k_x^2 + k_y^2}$ is the electron wave vector corresponding to the motion in the (x,y) directions, and $H[x]$ is the Heaviside function given by

$$H[x] = \begin{cases} 1, & x \geq 0 \\ 0, & x < 0 \end{cases} \tag{23}$$

There is no quantum confinement in the (x, y) plane. The quantized electron energies $E_{ez}(n_e)$ can be evaluated by solving the Schrödinger equation for the case of the confinement in the z direction Zhao (1999)

$$\left[-\frac{\hbar^2}{2m_e}\frac{\partial^2}{\partial z^2} + V(z)\right]\Phi_{n,k}(z) = E_n\Phi_{n,k}(z) \tag{24}$$

where

$$E_n = E_{ez}(n_e) + E_{n0} + \frac{(\hbar k)^2}{2m_e} \tag{25}$$

$V(z)$ is the confinement potential, and the electron wave function $\Psi_n(x, y, z)$ is given by Zhao (1999)

$$\Psi_n(x, y, z) = u_n(x, y, z)\,\Phi_{n,k}(z)\,\frac{1}{\sqrt{L_x L_y}}\exp i\left(k_x x + k_y y\right) \tag{26}$$

Here, $u_n(x, y, z)$ is the periodic Bloch function, L_x and L_y are the SOA dimensions in the (x, y) directions, respectively. In the simplest case of a finite square-well potential V_0, the energy spectrum of QW is determined by the solutions of the following transcendental equations Brennan (1999).

$$k_1 = k_2 \tan\left(\frac{k_2 d_z}{2}\right); k_1 = -k_2 \cot\left(\frac{k_2 d_z}{2}\right) \tag{27}$$

where

$$k_1 = \sqrt{\frac{2m_e(V_0 - E)}{\hbar^2}}; k_2 = \sqrt{\frac{2m_e E}{\hbar^2}}; \tag{28}$$

and d_z is the QW dimension in the z direction.

In this structure stimulated emission occurs only between discrete energy levels. QW-SOAs have superior performance as compared to bulk SOAs in terms of optical bandwidth and maximum output power. In SOAs, the saturation output power is inversely proportional to the differential gain coefficient, which is smaller in QW-SOAs.

The conventional SOA has rather poor carrier and photon confinement. Optical confinement can be enhanced by the implementation of a tapered graded index (GRIN) region on both sides of the well. A multi quantum well SOA (MQW-SOA) can be produced by stacking several layers of wells and barriers, usually created by molecular beam epitaxy (MBE) and/or metal organic vapor deposition (MOCVD) techniques. The optical gain of MQW is higher due to multiple QWs. Further improvement of the SOA performance is achieved by introducing an outer cladding region of a higher energy gap compared to the MQW barrier layers. The optical confinement factor of MQW SOAs increases approximately linearly with the number of QWs.

Many applications of SOA such as optical amplification, optical switching and signal processing require gain and phase shift insensitivity to the polarization. The polarization dependence of the phase shift is due to intrinsic or induced birefringence in SOA, so that the effective refraction index for the transverse electric (TE) mode is different from that of the transverse magnetic (TM) mode. These dependencies stem, in turn, from different quantization levels for heavy-hole (HH) bands, which provide the TE-mode dominant optical gain, and light-hole (LH) bands, which provide the TM-mode dominant optical gain. The difference in the confinement factors for TE and TM modes leads to different TE and TM signal gains.

The polarization sensitivity of MQW–SOAs can be significantly reduced when the active layer is strained, e.g. by introduction of lattice mismatch between the well and adjacent layers. In order to mitigate the polarization sensitivity problems in QW SOAs, several QW structures have been proposed such as low tensile-strained QWs, QWs with tensile barriers, tensile-strained QWs with compressive barriers, alternation of tensile and compressive QWs, and the delta-strained QW where strain is applied only at a shallow and highly strained layer called delta layer, that can yield a polarization-insensitive SOA at $\lambda = 1550nm$. In conventional QWs without delta layers, HH band energy levels are higher than those of LH bands, resulting in more TE gain relative to the TM mode. In delta-strained QWs, the delta layer causes a larger valence band discontinuity for HH bands than LH bands, so that the quantized energy levels of HH bands have a larger shift downward than those for the LH bands that shift upward. The delta layer pushes the LH band upward and the HH band downward such that their order is reversed.

In SOA applications requiring high optical power levels such as XGM, XPM and FWM, additional nonlinear birefringence is induced by the strong optical probe beam in such a way that the weak signal beam will undergo polarization dependent phase modulation. In some applications, however, this phenomenon can be utilized beneficially.

3.2 QW SOA dynamics

The dynamics of a QW heterostructure is usually described by a set of coupled rate equations containing injection current, injected carrier density, and photon density in the active region Zhao (1999). These equations provide an adequate tool for the analysis of XGM, XPM and FWM in QW SOA which is important for applications related to modern optical communications, especially in WDM architectures Reale (2001). Consider the carrier dynamics in a MQW SOA during the propagation of Gaussian pulses with the duration of several ps. The main transport mechanisms across the active region are the exchange of carriers between different QWs, exchange between a QW and separate confinement heterolayers (SCH), and the carrier injection from the SCH. In such a model each QW interacts with the neighboring QWs and with the surrounding SCHs Reale (2001). The rate equations in such a case differ significantly from the bulk SOA equations (10), (11) due to the number M of QWs. These rate equations have the form Reale (2001)

$$\frac{\partial S}{\partial z}+\frac{1}{v_g}\frac{\partial S}{\partial t}=\sum_{i=1}^{M} g\left(N_i\right)S_i-\frac{1}{v_g}\frac{S}{\tau_p}+\frac{1}{v_g}\sum_{i=1}^{M}\beta R_{spon}\left(N_i\right) \tag{29}$$

$$\frac{\partial N_i}{\partial t}=\eta_{inj}\frac{I_{inj}}{qL_i}-R\left(N_i\right)-R_{st,i} \tag{30}$$

where $S(z,t)=\sum_{i=1}^{M} S_i$ is the total photon concentration, S_i is the photon density generated in ith QW, $N_i(z,t)$ is the carrier density in ith QW, $g(N_i)$ is the gain-carrier density relationship, τ_p is the photon lifetime, β is the coupling factor of the spontaneous emission $R_{spon}(N_i)$, η_{inj} is the injection efficiency, I_{inj} is the injection current, $R(N_i)$ is the recombination term accounting for trap-related, spontaneous, and Auger recombinations, L_i is the thickness of the particular layer considered, and $R_{st,i}$ accounts for stimulated recombination. The polarization dependence of the gain for TE and TM modes is neglected. The typical values of the MQW SOA main parameters are: $M = 10$, the QW width $L_w = 48\mathring{A}$, device length $L = 700\mu m$, $\beta = 10^{-4}$, the threshold density $N_{th} = 1.5\times10^{18}cm^{-3}$, $v_g = 8.5\times10^{9}cm/s$, optical confinement

factor 0.02, the losses in the active region $\alpha_l = (v_g \tau_p)^{-1} \sim 10cm^{-1}$, the bias currents $I_{inj} \sim 50 - 150mA$, the gain for the probe light at $\lambda = 1560nm$ is $17 - 21dB$ Reale (2001).
We introduce a delayed time reference $t = t' - z/v_g$ and rewrite equation (29) as

$$\frac{\partial S}{\partial z} = g(N) - \frac{1}{v_g}\frac{S}{\tau_p} + \frac{1}{v_g}\beta R_{spon} \tag{31}$$

which yields a formal solution

$$S_{out} = S_{in} \exp\left[\int_0^L \left(g - \alpha_l + \frac{1}{v_g}\beta R_{spon}\right) dz\right] \tag{32}$$

Equations (30) and (31) have been solved numerically and showed a good accord with the experimental data for picosecond optical pulse propagation through a MQW SOA Reale (2001). Typically, the ultrafast pulse is simultaneously amplified and undergoes shape modification for high bias currents of about $150mA$. The shape evolution is mainly due to the gain saturation in the MQW SOA. As a result, the total gain for a single optical pulse reduces from $20dB$ at the lowest energy to $1dB$ at the highest energy demonstrating strong nonlinearity of MQW SOAs Reale (2001).

3.3 QD structure and energy spectrum

Quantization of electron states in all three dimensions results in creation of a novel physical object - a macroatom, or quantum dot (QD) containing a zero dimensional electron gas. Size quantization is effective when the QD's three dimensions are of the same order of magnitude as the electron de Broglie wavelength which is about several nanometers Ustinov (2003). QD is a nanomaterial confined in all the three dimensions, and for this reason it has unique electronic and optical properties that do not exist in bulk semiconductor material Ohtsu (2008). An electron-hole pair created by light in a QD has discrete energy eigenvalues due to carrier confinement. This phenomenon is called a quantum confinement effect Ohtsu (2008).
There exist different types of QDs based on different technologies that operate in different parts of spectrum such as In(Ga)As QDs grown on GaAs substrates, InAs QDs grown on InP substrates, and colloidal free-standing InAs QDs. QD structures are commonly realized by a self-organized epitaxial growth where QDs are statistically distributed in size and area. A widely used QD fabrication method is a direct synthesis of semiconductor nanostructures based on island formation during strained-layer heteroepitaxy called the Stranski-Krastanow (SK) growth mode Ustinov (2003). Spontaneously growing QDs are said to be self-assembling. SK growth has been investigated intensively for InAs on GaAs, InP on GaInP, and Ge on Si structures Ustinov (2003). The energy shift of the emitted light is determined by the size of QDs that can be adjusted within a certain range by changing the amount of deposited QD material. Evidently, smaller QDs emit photons of shorter wavelengths Ustinov (2003). The main advantages of the SK growth are Ustinov (2003):

1. SK growth permits the preparation of extremely small QDs in a maskless process without lithography and etching which makes it a promising technique to realize QD lasers.
2. A great number of QDs is formed in one simple deposition step.
3. Synthesized QDs are highly uniform in size and composition.
4. QDs can be covered epitaxially by a host material without any crystal or interface defects.

The simplest QD models are described by the spherical boundary conditions for an electron or a hole confined in a spherical QD with radius R, or by the cubic boundary conditions for a parallelepiped QD with side length $L_{x,y,z}$ Ohtsu (2008). In the first case, the electron and hole energy spectra $E_{e,nlm}$ and $E_{h,nlm}$ are respectively given by Ohtsu (2008)

$$E_{e,nlm} = E_g + \frac{\hbar^2}{2m_e}\left(\frac{\alpha_{nl}}{R}\right)^2; E_{h,nlm} = \frac{\hbar^2}{2m_h}\left(\frac{\alpha_{nl}}{R}\right)^2 \tag{33}$$

where

$$n = 1,2,3,...; l = 0,1,2,...n-1; m = 0,\pm 1,\pm 2,..\pm l, \tag{34}$$

E_g is the QD semiconductor material band gap, $m_{e,h}$ are the electron and hole effective mass, respectively, and α_{nl} is the n-th root of the spherical Bessel function of order l. In the second case, the energy eigenvalues $E_{e,nlm}$ and $E_{h,nlm}$ are respectively given by Ohtsu (2008)

$$E_{e,nlm} = E_g + \frac{\hbar^2\pi^2}{2m_e}\left[\left(\frac{n}{L_x}\right)^2 + \left(\frac{l}{L_y}\right)^2 + \left(\frac{m}{L_z}\right)^2\right] \tag{35}$$

$$E_{h,nlm} = \frac{\hbar^2\pi^2}{2m_h}\left[\left(\frac{n}{L_x}\right)^2 + \left(\frac{l}{L_y}\right)^2 + \left(\frac{m}{L_z}\right)^2\right] \tag{36}$$

The DOS function $\rho_{QD}(E)$ for an array of QDs has the form Ustinov (2003)

$$\rho_{QD}(E) = \sum_n \sum_m \sum_l 2n_{QD}\delta\left(E - E_{e,nlm}\right) \tag{37}$$

where $\delta\left(E - E_{e,nlm}\right)$ is the δ-function, and n_{QD} is the surface density of QDs. The active layer with QDs and the structure of QD conduction band are shown in Fig. 2.

Detailed theoretical and experimental investigations of InAs/GaAs and InAs QDs electronic structure taking into account their realistic lens, or pyramidal shape, size, composition profile, and production technique (SK, colloidal) have been carried out Bimberg (1999), Bányai (2005), Ustinov (2003). A system of QDs can be approximated with a three energy level model in the conduction band containing a spin degenerate ground state (GS), fourfold degenerate excited state (ES) with comparatively large energy separations of about $50-70meV$, and a narrow continuum wetting layer (WL). The electron WL is situated $150meV$ above the lowest electron energy level in the conduction band, i.e. GS and has a width of approximately $120meV$. In real cases, QDs vary in size, shape, and local strain which leads to fluctuations in the quantized energy levels and inhomogeneous broadening in the optical transition energy. Gaussian distribution may be used for description of the QD sizes Bányai (2005). The QDs and WL are surrounded by a barrier material which prevents direct coupling between QD layers. The absolute number of states in the WL is much larger than in the QDs. GS and ES in QDs are characterized by homogeneous and inhomogeneous broadening Bányai (2005). The homogeneous broadening caused by scattering of the optically generated electrons and holes with imperfections, impurities, phonons, or through the radiative electron-hole pair recombination Bányai (2005) is about $15meV$ at room temperature.

The eigenspectrum of a single QD fully quantized in three dimensions consists of a discrete set of eigenvalues depending only on the number of atoms in it. Variations of eigenenergies from QD to QD are caused by variations of QD's strain and shape. The finite carrier lifetime results in Lorentzian broadening of a finite width Ustinov (2003). The optical spectrum of

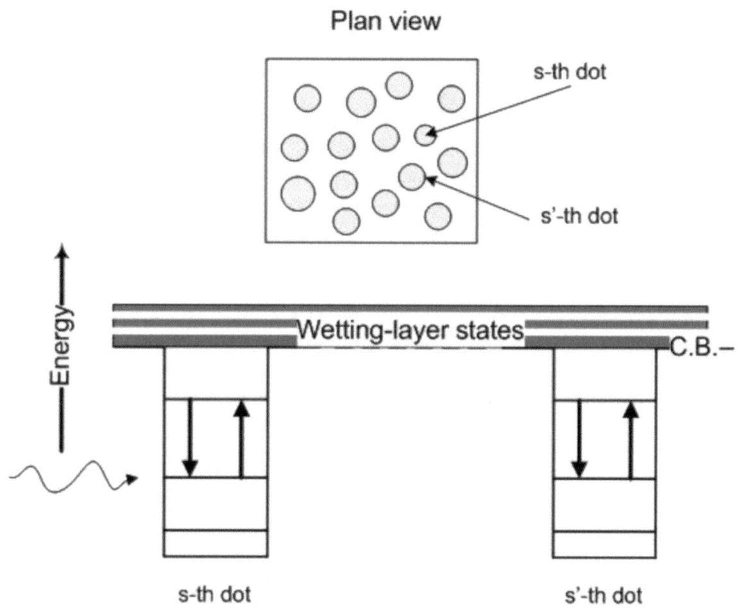

Fig. 2. Plan view of active layer with QDs (above); the QD conduction band structure (below)

QDs consists of a series of transitions between the zero-dimensional electron gas energy states where the selection rules are determined by the form and symmetry of QDs Ustinov (2003). In(Ga)As/GaAs QDs are characterized by emission at wavelengths no longer than $\lambda = 1.35\mu m$, while the InAs/InP QDs have been proposed for emission at the usual telecommunication wavelength $\lambda = 1.55\mu m$ Ustinov (2003).

3.4 Structure and operation of QD SOA

In this section, we will discuss the structure and operation principles of QD SOA. Theory of QD SOA operation based on the electron rate equations and photon propagation equation has been developed in a large number of works Berg (2004), Qasaimeh (2003), Qasaimeh (2004), Ben Ezra (September 2005), Ben Ezra (October 2005), Ben Ezra (2007).

The active region of a QD SOA is a layer including self-assembled InGaAs QDs on a GaAs substrate Sugawara (2004). Typically, the QD density per unit area is about $\left(10^{10} - 10^{11}\right) cm^{-2}$. The bias current is injected into the active layer including QDs, and the input optical signals are amplified via the stimulated emission or processed via the optical nonlinearity by QDs Sugawara (2004). The stimulated radiative transitions occur between GS and the valence band of QDs. A detailed theory of QD SOAs based on the density matrix approach has been developed in the pioneering work of Sugawara (2004) where the linear and nonlinear optical responses of QD SOAs with arbitrary spectral and spatial distribution of quantum dots in active region under the multimode light propagation have been considered. It has been shown theoretically that XGM takes place due to the coherent terms under the condition that the mode separation is comparable to or less than the polarization relaxation rate $|\omega_m - \omega_n| \leq \Gamma_g$ where $\omega_{m,n}$ are the mode frequencies and the relaxation time $\tau = \Gamma_g^{-1} = 130 fs$ Sugawara (2004). XGM is also possible in the case of the incoherent nonlinear polarization, or the so-called incoherent spectral hole burning Sugawara (2004). It has been assumed that XGM occurred only for signals with a detuning limited by

the comparatively small homogeneous broadening, and for this reason the ensemble of QDs should be divided into groups by their resonant frequency of the GS transition between the conduction and valence bands Sugawara (2004).
The phenomenological approach to the QD SOA dynamics is based on the rate equations for the electron densities of GS, ES and for combined WL and barrier serving as a reservoir. It is determined by electrons, because of the much larger effective mass of holes and their smaller state spacing Berg (2004).
In the QD SOA-MZI, optical signals propagate in an active medium with a gain determined by the rate equations for the electron transitions in QD-SOA between WL, GS and ES Qasaimeh (2003), Qasaimeh (2004), Ben Ezra (2008), Ben Ezra (September 2005). Unlike this model Sun (2005), we have taken into account two energy levels in the conduction band: GS and ES. The stimulated and spontaneous radiative transitions occur from GS to the QD valence band level. The system of the rate equations accounts for the following transitions:

1. the fast electron transitions from WL to ES with the relaxation time $\tau_{w2} \sim 3ps$;
2. the fast electron transitions between ES and GS with the relaxation time from ES to GS $\tau_{21} = 0.16ps$ and the relaxation time from GS to ES $\tau_{12} \sim 1.2ps$;
3. the slow electron escape transitions from ES back to WL with the electron escape time $\tau_{2w} \sim 1ns$.

The balance between the WL and ES is determined by the shorter time τ_{w2} of QDs filling. Carriers relax quickly from the ES level to the GS level, while the former serves as a carrier reservoir for the latter Berg (2001). In general case, the radiative relaxation times depend on the bias current. However, it can be shown that for moderate values of the WL carrier density $N_w \sim \left(10^{14} - 10^{15}\right) cm^{-3}$ this dependence can be neglected Berg (2001), Berg (November 2004).
The spontaneous radiative time $\tau_{1R} \gtrsim (0.4 - 0.5)\, ns$ in QDs remains large enough Sakamoto (2000), Qasaimeh (2003), Sugawara (2004), Qasaimeh (2004), Matthews (2005). The carrier dynamics is characterized by slow relaxation processes between WL and ES, and the rapidly varying coherent nonlinear population terms vanish after averaging over the comparatively large relaxation time τ_{w2} ~several ps from the two-dimensional WL to the ES. We have taken into account only incoherent population terms because for XGM between modes with the maximum detuning $\Delta\lambda_{\max} = 30nm$, within the especially important in optical communications conventional band (C-band) of $\lambda = (1530 \div 1565)\, nm$, the condition $\omega_1 - \omega_2 > \Gamma_g^{-1}$ is valid even for the lowest relaxation time from the ES to GS $\tau_{21} = 0.16ps$, and the rapidly varying coherent beating terms are insignificant Sugawara (2004). The direct carrier capture into the GS is neglected due to the fast intradot carrier relaxation and the large energy separation between GS and WL and it is assumed that the charge neutrality condition in GS is valid. The rate equations account for both fast transitions form WL to ES and GS and slow dynamics of the spontaneous transitions and electron escape from ES back to WL Qasaimeh (2003), Qasaimeh (2004), Ben Ezra (2007). They have the form Qasaimeh (2003), Qasaimeh (2004), Ben Ezra (2007).

$$\frac{\partial N_w}{\partial t} = \frac{J}{eL_w} - \frac{N_w (1-h)}{\tau_{w2}} + \frac{N_w h}{\tau_{2w}} - \frac{N_w}{\tau_{wR}}, \tag{38}$$

$$\frac{\partial h}{\partial t} = \frac{N_w L_w (1-h)}{N_Q \tau_{w2}} - \frac{N_w L_w h}{N_Q \tau_{2w}} - \frac{(1-f)\, h}{\tau_{21}} + \frac{f\,(1-h)}{\tau_{12}}, \tag{39}$$

$$\frac{\partial f}{\partial t} = \frac{(1-f)\,h}{\tau_{21}} - \frac{f\,(1-h)}{\tau_{12}} - \frac{f^2}{\tau_{1R}}$$
$$- \frac{g_p L}{N_Q}\,(2f-1)\,S_p \frac{c}{\sqrt{\varepsilon_r}} - \frac{g_s L}{N_Q}\,(2f-1)\,S_s \frac{c}{\sqrt{\varepsilon_r}}. \tag{40}$$

Here, S_p, S_s are the CW pump and on-off-keying (OOK) modulated signal wave photon densities, respectively, L is the length of SOA, g_p, g_s are the pump and signal wave modal gains, respectively, f is the electron occupation probability of GS, h is the electron occupation probability of ES, e is the electron charge, τ_{2w} is the electron escape time from the ES to the WL, τ_{wR} is the spontaneous radiative lifetime in WL, τ_{1R} is the spontaneous radiative lifetime in QDs, N_Q is the surface density of QDs, N_w is the electron density in the WL, L_w is the effective thickness of the active layer, τ_{21} is the electron relaxation time from the ES to GS and τ_{12} is the electron relaxation time from the GS to the ES, and ε_r is the SOA material permittivity. The modal gain $g_{p,s}(\omega)$ is given by Uskov (2004)

$$g_{p,s}(\omega) = \frac{2\Gamma N_Q}{a} \int d\omega F(\omega)\,\sigma(\omega_0)\,(2f-1) \tag{41}$$

where the number l of QD layers is assumed to be $l = 1$, the confinement factor Γ is assumed to be the same for both the signal and the pump waves, a is the mean size of QDs, $\sigma(\omega_0)$ is the cross section of interaction of photons of frequency ω_0 with carriers in QD at the transition frequency ω including the homogeneous broadening factor, $F(\omega)$ is the distribution of the transition frequency in the QD ensemble which is assumed to be Gaussian Qasaimeh (2004), Uskov (2004). It is related to the inhomogeneous broadening and it is described by the expression Uskov (2004)

$$F(\omega) = \frac{1}{\Delta\omega\sqrt{\pi}} \exp\left[-\frac{(\omega-\overline{\omega})^2}{(\Delta\omega)^2}\right] \tag{42}$$

where the parameter $\Delta\omega$ is related to the inhomogeneous linewidth $\gamma_{in\,\mathrm{hom}} = 2\sqrt{\ln 2}\Delta\omega$, and $\overline{\omega}$ is the average transition frequency.

In order to describe adequately XGM and XPM in QD SOA we should take into account the interaction of QDs with optical signals. The optical signal propagation in a QD SOA is described by the following truncated equations for the slowly varying CW and pulse signals photon densities and phases $S_{CW,P} = P_{CW,P} / \left(\hbar\omega_{CW,P} (v_g)_{CW,P} A_{eff}\right)$ and $\theta_{CW,P}$ Agrawal (1989).

$$\frac{\partial S_{CW,P}(z,\tau)}{\partial z} = (g_{CW,P} - \alpha_{int})\,S_{CW,P}(z,\tau) \tag{43}$$

$$\frac{\partial \theta_{CW,P}}{\partial z} = -\frac{\alpha}{2} g_{CW,P} \tag{44}$$

Here $P_{CW,P}$ are the CW and pulse signal optical powers, respectively, A_{eff} is the QD SOA effective cross-section, $\omega_{CW,P}$, $(v_g)_{CW,P}$ are the CW and pulse signal group angular frequencies and velocities, respectively, $g_{CW,P}$ are the active medium (SOA) gains at the corresponding optical frequencies, and α_{int} is the absorption coefficient of the SOA material. For the pulse propagation analysis, we replace the variables (z,t) with the retarded frame variables $\left(z, \tau = t \mp z/v_g\right)$. For optical pulses with a duration $T \gtrsim 10ps$ the optical radiation of the pulse fills the entire active region of a QD SOA of length $L \lesssim 1mm$ and the propagation effects can be neglected Gehrig (2002). Hence, in our case the photon densities

$$S_{CW,P}(z,\tau) = (S_{CW,P}(\tau))_{in} \exp\left[\int_0^z (g_{CW,P} - \alpha_{int})\, dz'\right] \tag{45}$$

can be averaged over the QD SOA length L which yields

$$S_{CW,P}(\tau) = \frac{1}{L}(S_{CW,P}(\tau))_{in} \int_0^L dz \exp\left[\int_0^z (g_{CW,P} - \alpha_{int})\, dz'\right] \tag{46}$$

Solution of equation (44) yields for the phases which should be inserted into MZI equation (53)

$$\theta_{CW,P}(\tau) = -(\alpha/2) \int_0^L dz g_{CW,P}. \tag{47}$$

The time-dependent variations of the carrier distributions in the QDs and WL result in strong phase changes (44) during the light propagation in the QD SOA Gehrig (2002). System of equations (38)-(40) with the average pump and signal photon densities (46) and phases (47) constitutes a complete set of equations describing XGM and XPM in QD SOA related by the LEF α as it is seen from equations (43), (44) and (47). The possibility of XGM in QD SOAs due to the connections between different QDs through WL at detunings between a signal and a pumping larger than the homogeneous broadening has been thoroughly investigated theoretically Ben Ezra (2007).
The advantages of QD SOAs as compared to bulk SOAs are the ultrafast gain recovery of about a few picoseconds, broadband gain, low NF, high saturation output power and high FWM efficiency Akiyama (2007). For instance, distortion free output power of $23dBm$ has been realized which is the highest among all the SOAs Akiyama (2007). A gain of $> 25dB$, NF of $< 5dB$ and output saturation power of $> 20dBm$ can be obtained simultaneously in the wavelength range of $90nm$ Akiyama (2007).

4. Recent advances in SOA applications

4.1 All-optical pulse generation

Ultra wideband (UWB) communication is a fast emerging technology that offers new opportunities such as high data rates, low equipment cost, low power, precise positioning capability and extremely low signal interference. A contiguous bandwidth of $7.5GHz$ is available in the frequency interval of $(3.1 - 10.6)\, GHz$ at an extremely low maximum power output of $-41.3dBm/MHz$ limited by the regulations of Federal Communication Commission (FCC) Ghawami (2005). Impulse radio (IR) UWB communication technique is a carrier free modulation using very narrow radio frequency (RF) pulses generated by UWB pulse generators Yao (2007). However, high data rate UWB systems are limited to distances less than $10m$ due the constraints on allowed emission levels Yao (2007), Ran (2009). In order to increase IR UWB transmission distances, a new concept based on UWB technologies and the optical fiber technology has been proposed that is called UWB radio over optical fibre (UROOF) Ran (2009). The IR UWB signals of several GHz are superimposed on the optical continuous wave (CW) carrier and transmitted transparently over an optical fiber Ran (2009), Yao (2007). The

UROOF technology permits to avoid the high cost additional electronic components required for signal processing and enables the integration of all RF and optical transmitter/receiver components on a single chip.
In order to distribute UWB signals via optical fibers, it is desirable to generate these signals directly in the optical domain. The advantages of the all-optical methods are following: decreasing of interference between electrical devices, low loss and light weight of optical fibers Lin (2005), Yao (2007), Wang (2006).
Typically, Gaussian waveforms are used in UWB communications due to their simplicity, achievability, and almost uniform distribution over their frequency spectrum Yao (2007), Ghawami (2005). The basic Gaussian pulse y_{g1}, a Gaussian monocycle y_{g2} and a Gaussian doublet y_{g3} are given by Ghawami (2005).

$$y_{g1} = K_1 \exp\left(-\frac{t^2}{\tau^2}\right); \tag{48}$$

$$y_{g2} = K_2 \left(-\frac{2t}{\tau^2}\right) \exp\left(-\frac{t^2}{\tau^2}\right); \; y_{g3} = K_3 \left(-\frac{2}{\tau^2}\right)\left(1-\frac{2t^2}{\tau^2}\right) \exp\left(-\frac{t^2}{\tau^2}\right) \tag{49}$$

where τ is the time-scaling factor, and $K_{1,2,3}$ are the normalization constants:

$$K_1 = \sqrt{\frac{E_1}{\tau\sqrt{\pi/2}}}; \; K_2 = \sqrt{\frac{\tau E_2}{\sqrt{\pi/2}}}; \; K_3 = \tau\sqrt{\frac{\tau E_3}{3\sqrt{\pi/2}}} \tag{50}$$

There exist three main optical IR UWB generation techniques Yao (2007)

1. UWB pulse generation based on phase-modulation-to-intensity-modulation (PM-IM) conversion.
2. UWB pulse generation based on a photonic microwave delay line using SOA.
3. UWB pulse generation based on optical spectral shaping and dispersion-induced frequency-to-time mapping. All-optical methods of UWB pulse generation are based on nonlinear optical processes in SOA such as XPM and XGM.

We concentrate on the all-optical methods of UWB pulse generation based on XPM and XGM in SOA. Consider first the method based on XPM. A probe CW signal generated by CW laser diode and a light wave modulated by the Mach-Zehnder modulator (MZM) are simultaneously fed into SOA, the probe signal will undergo both XGM and XPM, and the phase Φ_c of the output signal varies approximately proportionally to Gaussian pulse train power $P_s(t)$ Dong (2009)

$$\Phi_c = KP_s(t) + \Phi_0 \tag{51}$$

where K is the proportionality constant and Φ_0 is the initial phase. The chirp $\Delta\nu_c(t)$ of the probe signal is the first order derivative of the phase given by Dong (2009)

$$\Delta\nu_c(t) = -\frac{1}{2\pi}\frac{d\Phi_c}{dt} = -\frac{K}{2\pi}\frac{dP_s(t)}{dt} \tag{52}$$

The chirp (52) is a monocycle, according to definition (49). Its value may be positive or negative. UWB doublet pulses can be obtained by combining positive and negative monocycles with a proper delay Dong (2009). The shortages of the proposed method are the necessity for complicated electronic circuit for generation short electric Gaussian pulses, the use of an electro-optic phase modulator (EOM), the need for a comparatively long singlemode fiber (SMF), and a comparatively low operation rate and high bias currents of bulk SOAs. Recently, the theory of a novel all-optical method of the IR UWB pulse generation has been proposed Ben Ezra (2008). QD SOA can be inserted into one arm of an integrated Mach-Zehnder interferometer (MZI) which results in an intensity dependent optical signal interference at the output of MZI Ben Ezra (2008). The IR UWB pulse generation process is based both on XPM and XGM in QD SOA characterized by an extremely high optical nonlinearity, low bias current, and high operation rate Sugawara (2004). Unlike other proposed all-optical methods, we need no optical fibers, FBG and EOM substantially reducing the cost and complexity of the IR UWB generator. The IR UWB signals generated by the proposed QD SOA based MZI structure have the form of the Gaussian doublet. The shape of the signal and its spectrum can be tailor-made for different applications by changing the QD SOA bias current and optical power. The diagram of the MZI with QD SOA is shown in Fig. 3.

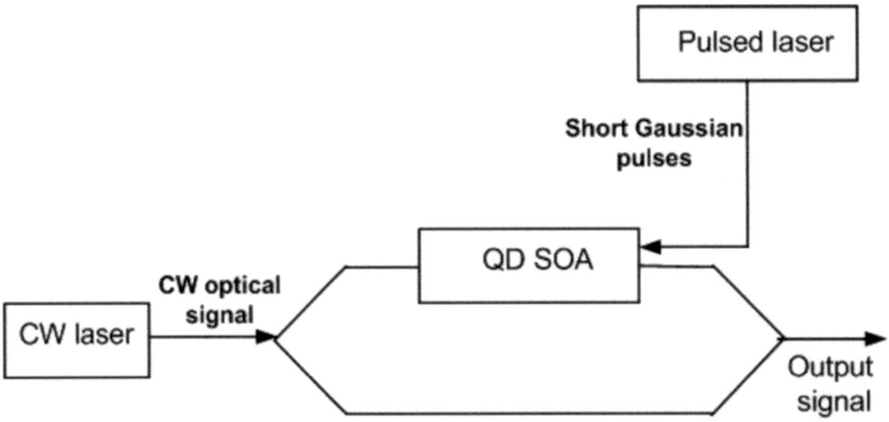

Fig. 3. MZI with QD SOA in the upper arm

The pulsed laser produces a train of short Gaussian pulses counter-propagating with respect to the input CW optical signal. The CW signal propagating through the upper arm of MZI transforms into the Gaussian pulse at the output of the MZI due to XPM and XGM with the train of Gaussian pulses. The optical signal in the linear lower arm of MZI remains CW, and the phase shift $\phi_2 = const$ in the lower arm of MZI is constant. Both these pulses interfere at the output of MZI, and the output pulse shape is defined by the power dependent phase difference $\Delta\phi(t) = \phi_1(t) - \phi_2(t)$ where $\phi_{1,2}(t)$ are the phase shifts in the upper and lower arms of MZI, respectively. The MZI output optical power P_{out} is given by Wang (2004).

$$P_{out} = \frac{P_0}{4}\left[G_1(t) + G_2(t) - 2\sqrt{G_1(t)\,G_2(t)}\cos\Delta\phi(t)\right] \tag{53}$$

where $G_{1,2}(t) = \exp(g_{1,2}L_{1,2})$, $g_{1,2}$, $L_{1,2}$ are the amplification factors of the upper and lower arms of MZI, the time-dependent gain, the SOA gain, and the active medium length, respectively. The relation between the MZI phase shift and its amplification factor is given by $\Delta\phi(t) = -(\alpha_L/2)\ln G_1(t)$. The shape of the output pulse is determined by the time dependence of $G_1(t)$ both directly and through $\Delta\phi(t)$ according to equation (53) resulting in a Gaussian doublet under certain conditions determined by the QD SOA dynamics.

4.2 All-optical signal processing

Recently, theoretical model of an ultra-fast all-optical signal processor based on the QD SOA-MZI where XOR operation, WC, and 3R signal regeneration can be simultaneously carried out by AO-XOR logic gates for bit rates up to $(100-200)\ Gb/s$ depending on the value of the bias current $I \sim (30-50)\ mA$ has been proposed. Ben Ezra (2009). The structure of the proposed processor is shown in Fig. 4.

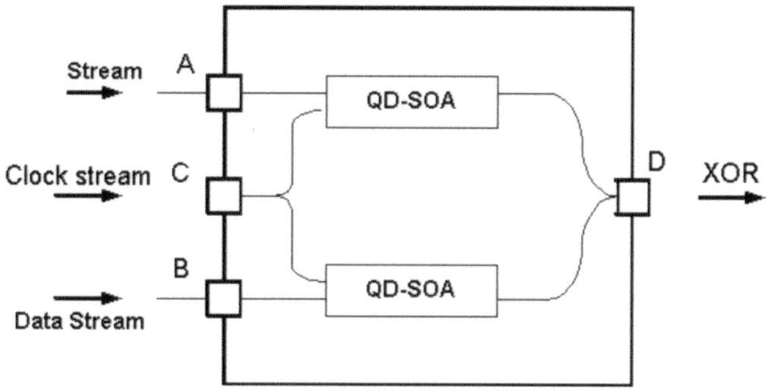

Fig. 4. The structure of the ultra-fast all-optical signal processor based on QD SOA-MZI

The theoretical analysis of the proposed ultra-fast QD SOA-MZI processor is based on combination of the MZI model with the QD-SOA nonlinear characteristics and the dynamics. At the output of MZI, the CW optical signals from the two QD SOAs interfere giving the output intensity are determined by equation (53) with the CW or the clock stream optical signal power P_{in} instead of P_0 Sun (2005), Wang (2004). When the control signals A and/or B are fed into the two SOAs they modulate the gain of the SOAs and give rise to the phase modulation of the co-propagating CW signal due to LEF α_L Agrawal (2001), Agrawal (2002), Newell (1999). LEF values may vary in a large interval from the experimentally measured value of LEF $\alpha_L = 0.1$ in InAs QD lasers near the gain saturation regime Newell (1999) up to the giant values of LEF as high as $\alpha_L = 60$ measured in InAs/InGaAs QD lasers Dagens (2005). However, such limiting cases can be achieved for specific electronic band structure Newell (1999), Dagens (2005), Sun (2004). The typical values of LEF in QD lasers are $\alpha_L \approx (2-7)$ Sun (2005). Detailed measurements of the LEF dependence on injection current, photon energy, and temperature in QD SOAs have also been carried out Schneider (2004). For low-injection currents, the LEF of the dot GS transition is between 0.4 and 1 increasing up to about 10 with the increase of the carrier density at room temperature Schneider (2004). The phase shift at the QD SOA-MZI output is given by Wang (2004)

$$\phi_1(t) - \phi_2(t) = -\frac{\alpha_L}{2} \ln\left(\frac{G_1(t)}{G_2(t)}\right) \tag{54}$$

It is seen from equation (54) that the phase shift $\phi_1(t) - \phi_2(t)$ is determined by both LEF and the gain. For the typical values of LEF $\alpha_L \approx (2-7)$, gain $g_{1,2} = 11.5 cm^{-1}$, $L_{1,2} = 1500 \mu m$ the phase shift of about π is feasible.

4.3 All-optical logics

Consider an AO-XOR gate based on integrated SOA-MZI which consists of a symmetrical MZI where one QD SOA is located in each arm of the interferometer Sun (2005). Two optical control beams A and B at the same wavelength λ are inserted into ports A and B of MZI separately. A third signal, which represents a clock stream of continuous series of unit pulses is split into two equal parts and injected into the two SOAs. The detuning $\Delta\omega$ between the signals A, B and the third signal should be less than the homogeneous broadening of QDs spectrum. In such a case the ultrafast operation occurs. In the opposite case of a sufficiently large detuning comparable with the inhomogeneous broadening, XGM in a QD SOA is also possible due to the interaction of QDs groups with essentially different resonance frequencies through WL for optical pulse bit rates up to $10Gb/s$ Ben Ezra (September 2005). When $A = B = 0$, the signal at port C traveling through the two arms of the SOA acquires a phase difference of π when it recombines at the output port D, and the output is "0" due to the destructive interference. When $A = 1$, $B = 0$, the signal traveling through the arm with signal A acquires a phase change due to XPM between the pulse train A and the signal. The signal traveling through the lower arm does not have this additional phase change which results in an output "1" Sun (2005). The same result occurs when $A = 0$, $B = 1$ Sun (2005). When $A = 1$ and $B = 1$ the phase changes for the signal traveling through both arms are equal, and the output is "0".

4.4 Wavelength conversion

An ideal wavelength convertor (WC) should have the following properties: transparency to bit rates and signal formats, fast setup time of output wavelength, conversion to both shorter and longer wavelengths, moderate input power levels, possibility for no conversion regime, insensitivity to input signal polarization, low-chirp output signal with high extinction ratio and large signal-to-noise ratio (SNR), and simple implementation Ramamurthy (2001). Most of these requirements can be met by using SOA. The XGM method using SOAs is especially attractive due to its simple realization scheme for WC Agrawal (2001). However, the main disadvantages of this method are substantial phase distortions due to frequency chirping, degradation due to spontaneous emission, and a relatively low extinction ratio Agrawal (2001). These parameters may be improved by using QD-SOAs instead of bulk SOAs due to pattern-effect-free high-speed WC of optical signals by XGM, a low threshold current density, a high material gain, high saturation power, broad gain bandwidth, and a weak temperature dependence as compared to bulk and MQW devices Ustinov (2003). We combine the advantages of QD-SOAs as a nonlinear component and MZI as a system whose output signal can be easily controlled. In the situation where one of the propagating signals A or B is absent, the CW signal with the desired output wavelength is split asymmetrically to each arm of MZI and interferes at the output either constructively or destructively with the intensity modulated input signal at another wavelength. The state of interference depends on the relative phase difference between the two MZI arms which is determined by the SOAs. In such a case the QD SOA-MZI operates as an amplifier of the remaining propagating signal. Then, the operation with the output "1" may be characterized as a kind of WC due to XGM between the input signal A or B and the clock stream signal. The possibility of the pattern-effect-free WC by XGM in QD SOAs has been demonstrated experimentally at the wavelength of $1.3\mu m$ Sugawara (2004).

4.5 3R regeneration

Short optical pulses propagating in optical fibers are distorted due to the fiber losses caused by material absorption, Rayleigh scattering, fiber bending, and broadening caused by the material dispersion, waveguide dispersion, polarization-mode dispersion, intermodal dispersion Agrawal (2001), Agrawal (2002). 3R regeneration is essential for successful logic operations because of the ultra-fast data signal distortions. 3R regeneration requires an optical clock and a suitable architecture of the regenerator in order to perform a clocked decision function Sartorius (2001). In such a case, the shape of the regenerated pulses is defined by the shape of the clock pulses Sartorius (2001).

The proposed QD SOA-MZI ultra-fast all-optical processor can successfully solve three problems of 3R regeneration. Indeed, the efficient pattern–effect free optical signal re-amplification may be carried out in each arm by QD-SOAs. WC based on an all-optical logic gate provides the re-shaping since noise cannot close the gate, and only the data signals have enough power to close the gate Sartorius (2001). The re-timing in QD-SOA-MZI based processor is provided by the optical clock which is also essential for the re-shaping Sartorius (2001). Hence, if the CW signal is replaced with the clock stream, the 3R regeneration can be carried out simultaneously with logic operations. The analysis shows that for strongly distorted data signals a separate processor is needed providing 3R regeneration before the data signal input to the logic gate.

4.6 Slow light propagation in SOA

One of the challenges of the optoelectronic technology is the ability to store an optical signal in optical format. Such an ability can significantly improve the routing process by reducing the routing delay, introducing data transparency for secure communications, and reducing the power and size of electronic routers Chang-Hasnain (2006). A controllable optical delay line can function as an optical buffer where the storage is proportional to variability of the light group velocity v_g defined as Chang-Hasnain (2006)

$$v_g = \frac{\partial \omega}{\partial k} = \frac{c - \omega \frac{\partial n(\omega,k)}{\partial \omega}}{n(\omega,k) + \omega \frac{\partial n(\omega,k)}{\partial k}} \tag{55}$$

Here $n(\omega,k)$ is the real part of the refractive index, and k is waveguide (WG) propagation constant. The signal velocity can be identified as the light group velocity v_g for the signals used in the optical communications where the signal bandwidth $(1-100)\,GHz$ is much less compared to the carrier frequency of about $193GHz$ Chang-Hasnain (2006). It is seen from equation (55) that the group velocity v_g can be essentially reduced for a large positive WG dispersion $\partial n/\partial k$ and/or material dispersion $\partial n/\partial \omega$ Chang-Hasnain (2006). Such a phenomenon is called a slow light (SL) propagation Chang-Hasnain (2006), Dúill (2009), Chen (2008). The WG dispersion can be realized by using gratings, periodic resonant cavities, or photonic crystals Chang-Hasnain (2006). The material dispersion can be achieved by gain or absorption spectral change. For instance, an absorption dip leads to a variation of the refractive index spectrum with a positive slope in the same frequency range, due to the Kramers-Kronig dispersion relation, which results in the SL propagation Chang-Hasnain (2006). The slowdown factor S is given by Chang-Hasnain (2006).

$$S = \frac{c}{v_g} = \frac{n(\omega,k) + \frac{\omega}{c}\frac{\partial n(\omega,k)}{\partial k}}{1 - \frac{\omega}{c}\frac{\partial n(\omega,k)}{\partial \omega}} \tag{56}$$

Large material dispersion necessary for SL phenomenon can be obtained by using different nonlinear optical effects such as electromagnetically induced transparency, FWM, stimulated Brillouin scattering, stimulated Raman scattering, coherent population oscillations (CPO) Chang-Hasnain (2006), Dúill (2009), Chen (2008). A sinusoidally modulated pump propagating in a SOA induces XGM, XPM and FWM which results in amplitude and phase changes. The sinusoidal envelope of the detected total field at SOA output exhibits a nonlinear phase change that defines the slowdown factor S controllable via the SOA gain Dúill (2009). It has been experimentally demonstrated that light velocity control by CPO can be realized in bulk, QW and QD SOAs Chen (2008). The nanosecond radiative lifetime in SOAs corresponds to a GHz bandwidth and is suitable for practical applications Chang-Hasnain (2006).

QW SOA is modelled as a two-level system. In such a system, a pump laser and a probe laser of frequencies ν_p nd ν_s, respectively create coherent beating of carriers changing the absorption and refractive index spectra Chang-Hasnain (2006). The sharp absorption dip caused by CPO induced by the pump and probe was centered at zero detuning. For the pump and probe intensities of 1 and $0.09kW/cm^2$, respectively, a slowdown factor $S = 31200$ and a group velocity $v_g = 9600m/s$ at zero detuning have been demonstrated Chang-Hasnain (2006).

QD SOAs characterized by discrete electronic levels, efficient confinement of electrons and holes, and temperature stability have been used for room temperature observation of CPO based SL Chang-Hasnain (2006). SL effects have been observed in QD SOA under reverse bias, or under a small forward bias current below the transparency level behaving as an absorptive WG Chang-Hasnain (2006).

5. Conclusions

We reviewed the structure, operation principles, dynamics and performance characteristics of bulk, QW and QD SOAs. The latest experimental and theoretical results concerning the SOAs applications in modern communication systems clearly show that SOAs in general and especially QW and QD SOAs are the most promising candidates for all-optical pulse generation, WC, all-optical logics, and even SL generation. These applications are due to SOA's extremely high nonlinearity which results in efficient XGM, XPM and FWM processes. In particular, QD SOAs are characterized by extremely low bias currents, low power level, tunable radiation wavelength, temperature stability and compatibility with the integrated Si photonics systems.

6. References

Agrawal, G.P. & Olsson, N.A. (1989). Self-phase modulation and spectral broadening of optical pulses in semiconductor laser amplifiers. *IEEE Journal of Quantum Electronics*, Vol. 25, No.11, (November 1989) 2297-2306, ISSN 0018-9197

Agrawal, G.P. (2001). *Applications of Nonlinear Fiber Optics*. Academic Press, ISBN 0-12-045144-1, New York

Agrawal, G.P. (2002). *Fiber-Optic Communication Systems*. Wiley, ISBN 0-471-21571-6, New York

Akiyama, T., Sugawara, M. & Arakawa, Y. (2007). Quantum-Dot semicondyctor optical amplifiers. *Proceedings of the IEEE*, Vol. 95, No. 9 (September 2007) 1757-1766, ISSN 0018-9219

Bányai, L. & Koch, S. W. (2005). *Semiconductor Quantum Dots* (Second Edition). World Scientific, ISBN 9810213905, London,

Ben-Ezra, Y.; Haridim, M. & Lembrikov, B. I. (2005). Theoretical analysis of gain-recovery time and chirp in QD-SOA. *IEEE Photonics Technology Letters*, Vol. 17, No. 9, (September 2005) 1803-1805, ISSN 1041-1135

Ben-Ezra, Y.; Lembrikov, B. I. & Haridim, M. (2005). Acceleration of gain recovery and dynamics of electrons in QD-SOA. *IEEE Journal of Quantum Electronics*, Vol. 41, No. 10, (October 2005) 1268-1273, ISSN 0018-9197

Ben-Ezra, Y.; Lembrikov, B. I. & Haridim, M. (2007). Specific features of XGM in QD-SOA. *IEEE Journal of Quantum Electronics*, Vol.43, No. 8, (August 2007) 730-737, ISSN 0018-9197

Ben Ezra, Y.; Haridim, M.; Lembrikov, B.I. & Ran, M. (2008). Proposal for All-optical Generation of Ultra Wideband Impulse Radio Signals in Mach-Zehnder Interferometer with Quantum Dot Optical Amplifier. *IEEE Photonics Technology Letters*, Vol. 20, No. 7 (April 2008) 484-486, ISSN 1041-1135

Ben Ezra, Y.; Lembrikov, B.I. & Haridim, M. (2009). Ultra-Fast All-Optical Processor Based on Quantum Dot Semiconductor Optical Amplifiers (QD-SOA). *IEEE Journal of Quantum Electronics*, Vol. 45, No.1 (January 2009) 34-41, ISSN 0018-9197

Berg, T.W.; Bischoff, S.; Magnusdottir, I. & Mørk, J. (2001). Ultrafast gain recovery and modulation limitations in self-assembled quantum-dot devices. *IEEE Photonics Technology Letters*, Vol. 13, No. 6 (June 2001) 541-543, ISSN 1041-1135

Berg, T.W.; Mørk, J. & Hvam, J.M. (2004). Gain dynamics and saturation in semiconductor quantum dot amplifiers. *New Journal of Physics*, Vol. 6, No. 178, (2004) 1-23, ISSN 1367-2630

Berg, T.W. & Mørk, J. (2004). Saturation and Noise Properties of Quantum-Dot Optical Amplifiers. *IEEE J. of Quantum Electronics*, Vol. 40, No. 11, (November 2004) 1527-1539, ISSN 0018-9197

Bimberg, D.; Grundmann, M. & Ledentsov, N. N. (1999). *Quantum Dot Heterostructures*. John Wiley, ISBN 047 1973882, New York

Brennan, K.F. (1999). *The Physics of Semiconductors with applications to optoelectronic devices.* Cambridge University Press, ISBN 0 521 59662 9, New York

Chang, W.S.C. (2005). *Principles of Lasers and Optics.* Cambridge University Press, ISBN 0-511-08061-1, Cambridge

Chang-Hasnain, C.J. & Chuang, S.L. (2006). Slow and fast light in semiconductor quantum-well and quantum-dot devices. *Journal of Lightwave Technology*, Vol. 24, No. 12, (December 2006) 4642-4654, ISSN 0733-8724

Chen, H.; Zhu, G.; Wang, Q.; Jaques, J.; Leuthold, J.; Picirilli, A.B. & Dutta, N.K. (2002). All-optical logic XOR using differential scheme and Mach-Zender interferometer. *Electronic Letters*, Vol. 38, No. 21, (October 2002) 1271-1273, ISSN 0013-5194

Chen, Y., Xue, W., Öhman, F. & Mørk, J. (2008). Theory of optical-filtering enhanced slow and fast light effects in semiconductor optical waveguides. *Journal of Lightwave Technology*, Vol. 26, No. 23, (December 2008) 3734-3743, ISSN 0733-8724

Dagens, B.; Markus, A.; Chen, J.X.; Provost, J.-G.; Make, D.; de Gouezigou, O.; Landreau, J.; Fiore, A. & Thedrez, B. (2005). Giant linewidth enhancement factor and purely frequency modulated emission from quantum dot laser. *Electronics Letters*, Vol. 41, No. 6, (17th March 2005) 323-324, ISSN 0013-5194

Dong, J.; Zhang, X.; Fu, S.; Xu, J.; Shum, P. & Huang, D. (2008) Ultrafast all-optical signal processing based on single semiconductor optical amplifier. *IEEE Journal of Selected Topics in Quantum Electronics*, Vol. 14, No. 3 (May/June 2008) 770-778, ISSN 1077-260X

Dong, J.; Zhang, X. & Huang (2009). All-optical ultra-wideband pulse generation based on semiconductor optical amplifiers. *Frontiers of Optoelectronics in China*, Vol. 2, No.1, (March 2009) 40-49, ISSN 1674-4128

Dúill, Seán Ó; O'Dowd, R. F. & Eisenstein, G. (2009) On the role of high-order coherent population oscillations in slow and fast light propagation using semiconductor optical amplifiers. *IEEE Journal of Selected Topics in Quantum Electronics*, Vol. 15, No. 3 (May/June 2009) 578-584, ISSN 1077-260X

Eisenstein, G. (1989). Semiconductor optical amplifiers. *IEEE Circuits and Devices Magazine*, Vol. 5, No. 4 (July 1989) 25-30, ISSN 8755-3996

Freude, W. & al. (2010). Linear and nonlinear semiconductor optical amplifiers. *Proceedings of 12th International Conference on Transparent Optical Networks* (ICTON 2010), We.D4.1, 1-4, ISBN 978-1-4244-7799-9, Munich, Germany June 27-July 1, 2010 IEEE

Ghavami, M.; Michael, L.B. & Kohno, R. (2005). *Ultra Wideband Signals and Systems in Communication Engineering*, Wiley, ISBN-10 0-470-86571-5(H/B), Chichester, England

Gehrig, E. & O. Hess, O. (2002). Mesoscopic spatiotemporal theory for quantum-dot lasers. *Phys. Rev. A*, Vol. 65, No. 3, (March 2002) 033804-1-16, ISSN 1050-2947

Hamié, A.; Sharaiha, A.; Guégan, M. & Pucel, B. (2002). All-Optical logic NOR Gate Using Two-Cascaded Semiconductor Optical Amplifiers. *IEEE Photonics Technology Letters*, Vol. 14, No. 10, (October 2002) 1439-1441, ISSN 1041-1135

Joergensen, C. ; Danielsen, S.L.; Durhuus, T.; Mikkelsen, B. ; Stubkjaer, K.E.; Vodjdani, N.; Ratovelomanana, F.; Enard, A.; Glastre, G.; Rondi, D. & Blondeau, R. (1996). Wavelength conversion by optimized monolithic integrated Mach-Zender interferometer, *IEEE Photonics Technology Letters*, Vol. 8, No. 4, (April 1996) 521-523, ISSN 1041-1135

Kanellos, G.T.; Petrantonakis, D.; Tsiokos, D.; Bakopoulos, P.; Zakynthinos, P.; Pleros, N.; Apostolopoulos, D.; Maxwell, G.; Poustie, A. & Avramopoulos, H. (2007). All-optical 3R burst-mode reception at 40 Gb/s using four integrated MZI switches. *Journal of Lightwave Technology*, Vol. 25, No. 1, (January 2007) 184-192, ISSN 0733-8724

Kim, J.; Laemmlin, M.; Meuer, C.; Bimberg, D. & Eisenstein, G. (2009). Theoretical and experimental study of high-speed small-signal cross-gain modulation of quantum-dot semiconductor optical amplifiers. *IEEE J. of Quantum Electronics*, Vol. 45, No. 3, (March 2009) 240-248, ISSN 0018-9197

Leem, Y.A.; Kim, D. C.; Sim, E.; Kim, S.-B.; Ko, H.; Park, K.H.; Yee, D.-S.; Oh, J. O.; Lee, S. H. & Jeon, M. Y. (2006). The characterization of all-optical 3R regeneration based on InP-related semiconductor optical devices, *IEEE J. of selected topics in Quantum Electronics*, Vol. 12, No. 4, (July/August 2006) 726-735, ISSN 1077-260X

Lin, W.-P. & Chen, J.-Y. (2005). Implementation of a new ultrawide-band impulse system, *IEEE Photonics Technology Letters*, Vol. 17, No. 11, (November 2005) 2418-2420, ISSN 1041-1135

Matthews, D.K. ; Summers, H.D.; Smowton, P.M.; Blood, P.; Rees, P. & Hopkinson, M. (2005). Dynamics of the wetting-layer-quantum-dot interaction in GaAs self-assembled systems, *IEEE Journal of Quantum Electronics*, Vol. 41, No. 3, (March 2005), 344-350, ISSN 0018-9197

Mukherjee, B. & Zang, H. (2001). Introduction. Survey of State-of-the-Art, In: *Optical WDM Networks. Principles and Practice*, Sivalingam, K.M. & Subramaniam, S. (Ed.), 3-24, Kluwer, ISBN 0-7923-7825-3, Boston

Newell, T.C.; Bossert, D.J.; Stinz, A.; Fuchs, A. & Malloy, K.J. (1999). Gain and linewidth enhancement factor in InAs quantum-dot laser diodes, *IEEE Photonics Technology Letters*, Vol. 11, (November 1999), 1527-1529, ISSN 1041-1135

Ohtsu, N.; Kobayashi, K.; Kawazoe, T.; Yatsui, T. & Naruse, N. (2008). *Principles of Nanophotonics*, CRC Press, ISBN-13 978-1-58488-972-4, London

Premaratne, M., Nešić, D. & Agrawal, G.P. (2008) Pulse amplification and gain recovery in semiconductor optical amplifiers: a systematic analytical approach. *Journal of Lightwave Technology*, Vol. 26, No. 12, (June 2008) 1653-1660, ISSN 0733-8724

Qasaimeh, O. (2003). Optical gain and saturation characteristics quantum-dot semiconductor optical amplifiers. *IEEE J. of Quantum Electronics*, Vol. 39, No. 6, (June 2003) 793-798, ISSN 0018-9197

Qasaimeh, O. (2004). Characteristics of cross-gain (XG) wavelength conversion in quantum dot semiconductor optical amplifiers. *IEEE Photonics Technology Letters*, Vol. 16, No. 2, (February 2004) 542-544, ISSN 1041-1135

Ramamurthy, B. (2001). Swithches, wavelength routers, and wavelength converters. In: *Optical WDM Networks. Principles and Practice*, Sivalingam, K.M. & Subramaniam, S. (Ed.), 51-75, Kluwer, ISBN 0-7923-7825-3, Boston

Ran, M; Ben Ezra, Y. & Lembrikov B.I. (2009). Ultra-wideband Radio-over-optical-fibre Technologies, *In Short-Range Wireless Communications*, Kraemer, R. & Katz, M. D. (Eds.), 271-327, Wiley, ISBN 978-0-470-69995-9 (H/B), Chichester, England

Reale, A., Di Carlo, A. & Lugli, P. (2001). Gain dynamics in traveling-wave semiconductor optical amplifiers. *IEEE Journal of Selected Topics in Quantum Electronics*, Vol. 7, No. 2 (March/April 2001) 293-299, ISSN 1077-260X

Sakamoto, A. & Sugawara, M. (2000). Theoretical calculation of lasing spectra of quantum-dot lasers: effect of homogeneous broadening of optical gain, *IEEE Photonics Technology Letters*, Vol. 12, No. 2, (February 2000) 107-109, ISSN 1041-1135

Sartorius, B. (2001). 3R regeneration for all-optical networks, *Proceedings of 3rd International Conference on Transparent Optical Networks* (ICTON 2001), Th. A.4, pp. 333-337, ISBN 10 0780370961, Krakow, Poland, June 18-21, IEEE, Krakow

Schneider, S.; Borri, P.; Langbein, W.; Woggon, U.; Sellin, R.L.; Ouyang, D. & Bimberg, D. (2004). Linewidth enhancement factor in InGaAs quantum-dot amplifiers, *IEEE of Quantum Electronics*, Vol. 40, No. 10, (October 2004) 1423-1429, ISSN 0018-9197

Spiekman, L. (2009). Economics and markets of semiconductor optical amplifiers, *Proceedings of 11th International Conference on Transparent Optical Networks* (ICTON 2009), Mo.B4.1, 1, ISBN 978-1-4244-4826-5, S. Miguel, Azores, Portugal, June 28-July 2, 2009 IEEE

Sugawara, M.; T. Akiyama, T.; N. Hatori, N. ; Y. Nakata, Y.; Ebe, H. & H. Ishikava, H. (2002). Quantum-dot semiconductor optical amplifiers for high-bit-rate signal processing up to 160 Gbs^{-1} and a new scheme of 3R regenerators, *Meas. Sci. Technol.*, Vol. 13, (2002), 1683-1691, ISSN 0957-0233

Sugawara, M.; Ebe, H.; Hatori, N.; Ishida, M.; Arakawa, Y.; Akiyama, T.; Otsubo, K. & Nakata, Y. (2004) Theory of optical signal amplification and processing by quantum-dot semiconductor optical amplifiers. *Phys. Rev.B*, Vol. 69, No. 23 (June 2004) 235332-1-39, ISSN 1098-0121

Sun, G.; Khurgin, J.B. & Soref, R.A. (2004). Design of quantum-dot lasers with an indirect bandgap short-period superlattice for reducing the linewidth enhancement factor, *IEEE Photonics Technology Letters*, Vol.16, No. 10 (October 2004), 2203-2205, ISSN 1041-1135

Sun, H.; Wang, Q.; Dong, H. & Dutta, N.K. (2005). XOR performance of a quantum dot semiconductor optical amplifier based Mach-Zender interferometer, *Optics Express*, Vol. 13, No. 6, (March 2005) pp.1892-1899, ISSN 10944087

Uskov, A.V. ; Berg, T.W. & Mørk, J. (2004). Theory of pulse-train amplification without patterning effects in quantum-dot semiconductor optical amplifiers. *IEEE J. of Quantum Electronics*, Vol. 40, No. 3, (March 2004) 306-320, ISSN 0018-9197

Ustinov, V.M.; Zhukov, A.E.; Egorov, A. Yu. & Maleev, N. A. (2003). *Quantum Dot Lasers*, Oxford University Press, ISBN 0 19 852679 2, Oxford

Wada, O. (2007). Femtosecond all-optical devices for ultrafast communication and signal processing, In: *Microwave Photonics*, Lee, C. H. (Ed), 31-75, CRC Press, ISBN-10: 0-8493-3924-3

Wang, Q.; Zhu, G.; Chen, H. ; Jaques, J. ; Leuthold, J.; Picirilli, A.B. & Dutta, N.K. (2004). Study of all-optical XOR using Mach-Zehnder interferometer and differential scheme. *IEEE J. of Quant. Electr.*, Vol. 40, No. 6, (June 2004) 703-710, ISSN 0018-9197

Wang, Q. & Yao, J. (2006). UWB doublet generation using nonlinearly-biased electro-optic intensity modulator, *Electronic Letters*, Vol. 42, No. 22, (October 2006)1304-1305, ISSN 0013-5194

Yao, J.; Zeng, F. & Wang, Q. (2007). Photonic generation of ultrawideband signals. *Journal of Lightwave Technology*, Vol. 25, No. 11, (November 2007) 3219-3235, ISSN 0733-8724

Zhao, B. & Yariv, A. (1999). Quantum well semiconductor lasers. In: *Semiconductor Lasers I. Fundamentals*, Kapon, E. (Ed.), 1-121, Academic Press, ISBN 0-12-397630-8, San Diego, USA

Zhu, Z.; Funabashi, M.; Zhong Pan; Paraschis, L.; Harris, D. & Ben Yoo, S.J. (2007). High-performance optical 3R regeneration for scalable fiber transmission system applications. *Journal of Lightwave Technology*, Vol. 25, No. 2, (January 2007) 504-511, ISSN 0733-8724

2

Semiconductor Optical Amplifier Nonlinearities and Their Applications for Next Generation of Optical Networks

Youssef Said and Houria Rezig
Sys'Com Laboratory, National Engineering School of Tunis (ENIT)
Tunisia

1. Introduction

Semiconductor optical amplifiers (SOAs) have attracted a lot of interest because of their application potential in the field of optical communications. Their use has been envisaged in different applications in the access, core and metropolitan networks. Particularly, they have been envisioned for all-optical signal processing tasks at very high bit rates that cannot be handled by electronics, such as wavelength conversion, signal regeneration, optical switching as well as logic operations. To implement such all-optical processing features, the phenomena mostly used are: cross gain modulation (XGM), cross phase modulation (XPM), four-wave mixing (FWM) and cross polarization modulation (XPolM).

The aim of the present work is to present a qualitative and an exhaustive study of the nonlinear effects in the SOA structure and their applications to achieve important functions for next generation of optical networks. These phenomena are exploited in high speed optical communication networks to assure high speed devices and various applications, such as: wavelength converters in WDM networks, all-optical switches, optical logic gates, etc. Particularly, we focus on analyzing the impact of variation of intrinsic and extrinsic parameters of the SOA on the polarization rotation effect in the structure. This nonlinear behavior is investigated referring to numerical simulations using a numerical model that we developed based on the Coupled Mode Theory (CMT) and the formalism of Stokes. Consequently, it is shown that the azimuth and the ellipticity parameters of the output signal undergo changes according to injection conditions, i.e. by varying the operating wavelength, the input polarization state, the bias current, the confinement factor and obviously the SOA length, which plays an important role in the gain dynamics of the structure. We will show that the obtained results by the developed model are consistent with those obtained following the experimental measurements that have been carried out in free space.

In addition, an investigation of the impact of nonlinear effects on the SOA behavior in linear operating and saturation regimes will be reported. Their exploitation feasibility for applications in high bit rate optical networks are therefore discussed. Hence, the impact of variation of the SOA parameters on the saturation phenomena is analyzed by our numerical simulations. It was shown that high saturation power feature, which is particularly required in wavelength division multiplexing (WDM) applications to avoid crosstalk arising from gain saturation effects, can be achieved by choosing moderate values of the operating parameters. Moreover, we will address one of the essential processes to consider in SOAs analysis, which is the noise. Particularly, we numerically simulate the impact of noise effects on the SOA behavior by measuring the gain, the optical signal to noise ratio and the noise figure. Although its gain dynamics provide very attractive features of high speed optical signal processing, we show that the noise is important in SOAs and can limit the performance of the structure. In order to remedy this, we show that using high bias current at moderate input signal power is recommended.

We report and characterize the impact of the nonlinear polarization rotation on the behavior of a wavelength converter based on XGM effect in a SOA at 40 Gbit/s. Moreover, we investigate and evaluate its performance as function of the intrinsic and extrinsic SOA parameters, such as the bias current, the signal format, the input signal power and its polarization state that determine the magnitude of the polarization rotation by measuring the ellipticity and the azimuth. Also, the impact of noise effects on the structure behavior is investigated through determining the noise figure. In particular, we focus on the performance of an improved wavelength conversion system via the analysis of quality factor and bit error rate referring to numerical simulation.

In this chapter, we deal either with the investigation of the SOA nonlinearities; particularly those are related to the polarization rotation, to exploit them to assure important optical functions for high bit rate optical networks. The dependence of SOA on the polarization of the light is an intrinsic feature which can lead to the deterioration of its performance. As a system, it is very inconvenient because of the impossibility to control the light polarization state, which evolves in a random way during the distribution in the optical fiber communication networks. For that reason, the technological efforts of the designers were essentially deployed in the minimization of the residual polarimetric anisotropy of the SOAs, through the development of almost insensitive polarization structures. On the other hand, various current studies have exploited the polarization concept to assure and optimize some very interesting optical functions for the future generation of the optical networks, as the wavelength converters, the optical regenerators and the optical logical gates. In this frame, many studies have demonstrated, by exploiting the nonlinear polarization rotation, the feasibility of the implementation of optical logical gates, wavelength converters and 2R optical regenerators.

2. Semiconductor optical amplifier: Concept and state of the art

2.1 SOA architecture

A semiconductor optical amplifier (SOA) is an optoelectronic component, which is characterized by a unidirectional or bidirectional access. Its basic structure, represented in figure 1, is slightly different from that of the laser diode. Indeed, there will be creation of the following effects: the inversion of population due to the electric current injection, the spontaneous and stimulated emission, the non-radiative recombination. Contrary to semiconductor lasers, there are no mirrors in their extremities but an antireflection coating, angled or window facet structures have been adopted to reduce light reflections into the circuit. SOAs manufacturing is generally made with III-V alloys, such as the gallium arsenide (GaAs), indium phosphide (InP) and various combinations of these elements according to the required band gap and the characteristics of the crystal lattice. In particular case for use around 1,55 μm, the couple InGaAsP and InP is usually used for the active layer and the substratum, respectively.

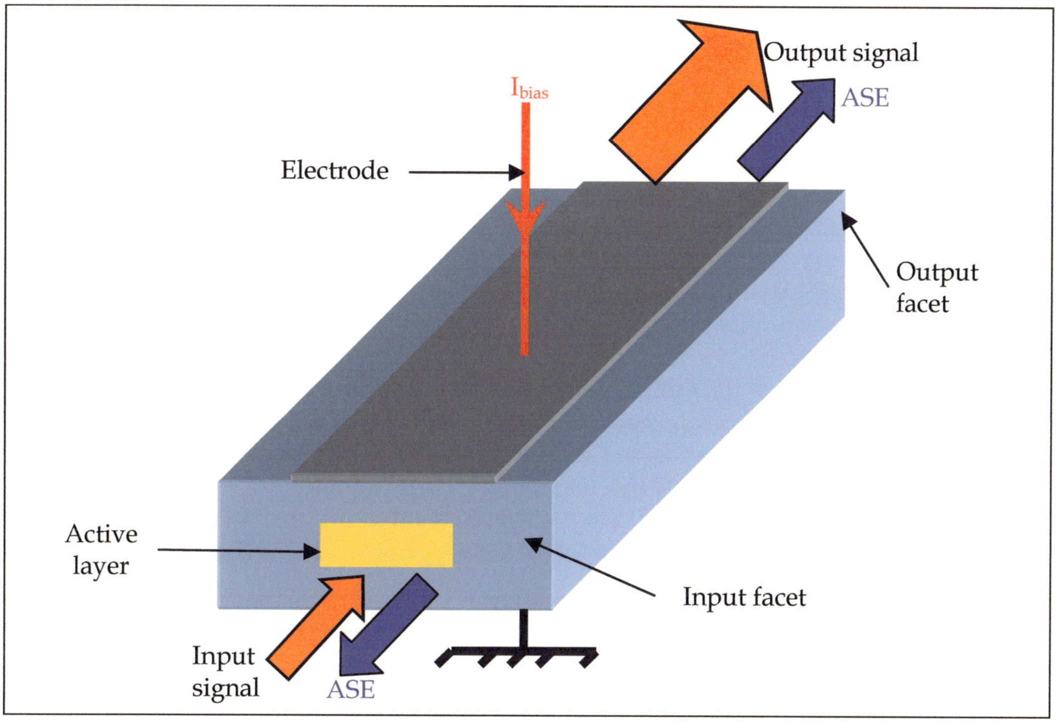

Fig. 1. SOA architecture.

Typical physical features of the SOA structure, used in simulations, are listed in Table 1.

Symbol	Description	Value
I_{bias}	Injection current	200 mA
η_{in}	Input coupling loss	3 dB
η_{out}	Output coupling loss	3 dB
R_1	Input facet reflectivity	5e-005
R_2	Output facet reflectivity	5e-005
L	Active layer length	500 μm
W	Active layer width	2.5 μm
d	Active layer height	0.2 μm
Γ	Optical confinement factor	30%
v_g	Group velocity	75 000 000 m/s
n_r	Active refractive index	3.7

Table 1. SOA parameters used in simulation.

2.2 SOA structure characteristics

The SOA has proven to be a versatile and multifunctional device that will be a key building block for next generation of optical networks. The parameters of importance, used to characterize SOAs, are:

- the gain bandwidth,

- the gain saturation,
- the noise figure,
- the polarisation independence,
- the conversion efficiencies,
- the input dynamic range,
- the extinction ratio/crosstalk,
- the tuning speed,
- the wavelength of operation.

The evolution of the SOA output power as function of the wavelength for various values of the input power is represented in figure 2. It shows that when the wavelength increases, the output power decreases. So, we can notice that when the input power injected into the SOA increases, the maximum of the output power will be moved towards the high wavelengths, which is due to the decrease of the carriers' density. For example, for an input power of - 18 dBm, the maximal output power is 5,41 dBm for a wavelength equal to 1520 nm; but for an injected power equal to -5 dBm, the maximum of the output power is 8,29 dBm and corresponds to a wavelength of 1540 nm. Whereas for an input power equal to 5 dBm, the maximal value of the output power is 8,95 dBm for a wavelength of 1550 nm.

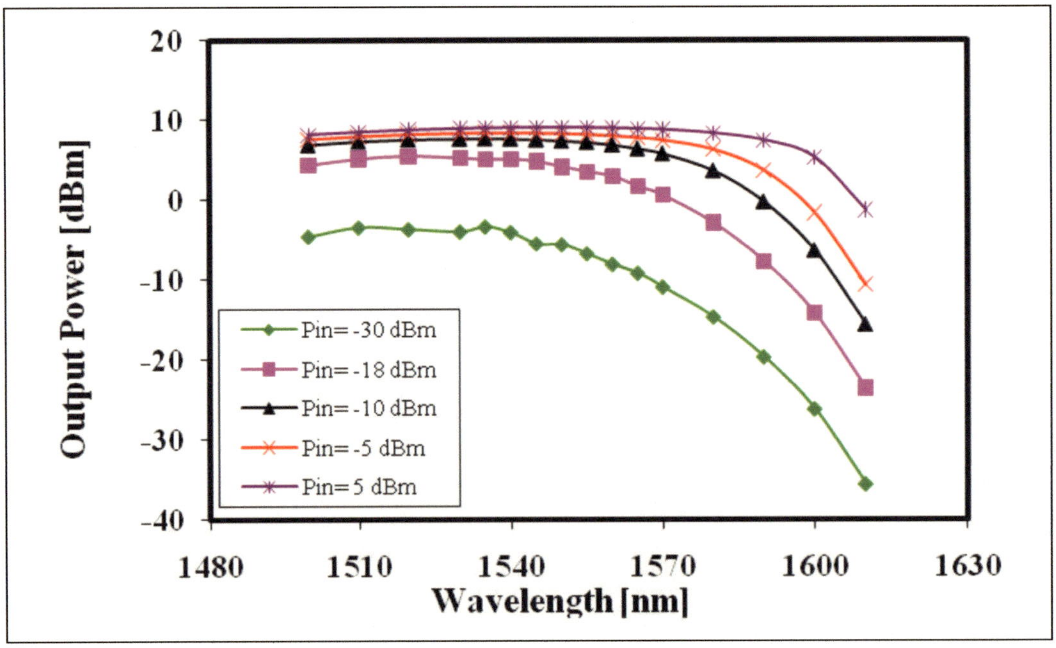

Fig. 2. SOA output power versus the wavelength of operation for different input powers.

A wide optical bandwidth is a desirable feature for a SOA, so that it can amplify a wide range of signal wavelengths. In order to analyze the impact of the injection condition on this parameter, we represent simulation results of the SOA gain as function of the wavelength of the signal for different input powers. Referring to figure 3, we note that wavelength variations and the injected power have a significant impact on the gain bandwidth evolution. However, we can notice according to the obtained curves, which are drawn for a bias current of 200 mA, that when the input power increases, the gain maximum (known as the peak of the gain) is moved towards the high wavelengths, which is due to the decrease of the carriers' density. For an input power of -30 dBm, the gain peak is 26,6 dB at a wavelength equal to 1510 nm, but for an injected power of -10 dBm, the gain maximum is 17,65 dBm for a wavelength of 1535 nm. On the other hand, for a high input power of 5 dBm, for example, the gain peak, having a value of 3,96 dB, is reached for a signal wavelength of 1550 nm, which is higher than the wavelength corresponding to the last case.

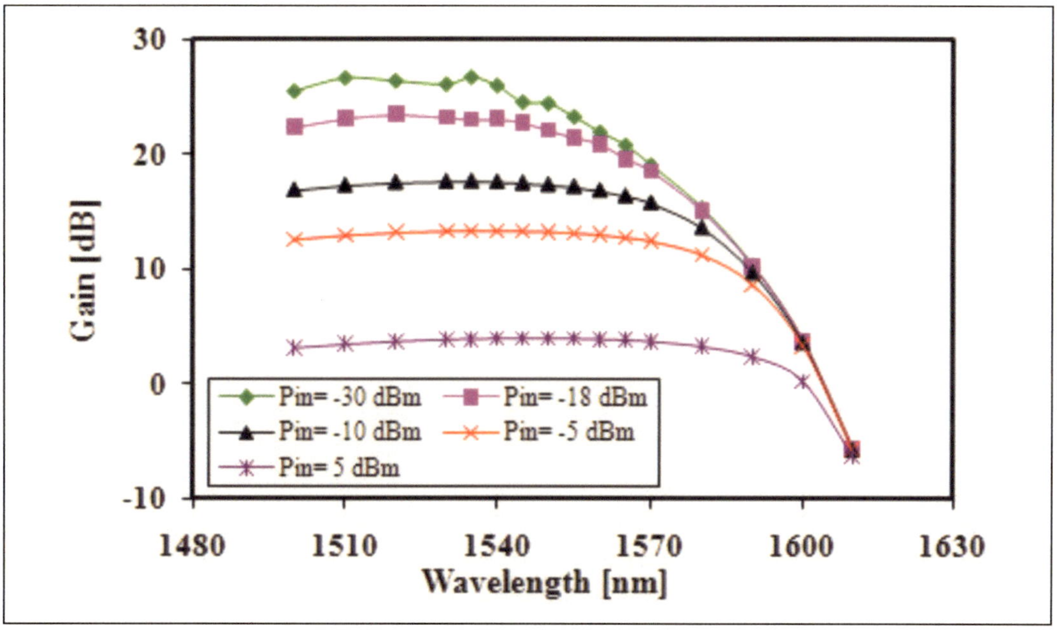

Fig. 3. Gain spectrum as a function of the injected input power.

In order to look for the conditions which correspond to an improvement of the SOA functioning, we have analyzed the influence of the intrinsic parameters on the SOA performance by representing, in figure 4, the gain and the noise figure as function of the bimolecular recombination coefficient (B). We notice that the increase of the B coefficient entails a diminution of the gain and consequently an increase of the noise figure. These results are justified by the fact that when the B coefficient increases, there will be an increase of the carriers' losses that are caused by the radiative and non-radiative recombination processes and consequently the carriers' density decreases, which involves a gain decrease. In that case, the maximal value of the gain is 26,16 dB, which corresponds to a minimum of noise figure of 5,27 dB, a B coefficient equal to 9.10^{-16} $m^3.s^{-1}$ and an input power P_{in}= -30 dBm.

2.3 Noise effects in a SOA structure

One of major processes to consider in the SOA analysis is the amplified spontaneous emission (ASE) noise, because it strongly affects the structure performance. It is also crucial in determining the bit error rate (BER) of the transmission system within which the amplifier resides. The injected signal and the ASE noise interact nonlinearly as they propagate along the SOA structure. Then, the interaction correlates different spectral components of the noise. Consequently, we can distinguish three types of noise, which are:

- The shot noise.
- The signal-spontaneous beat noise.
- The spontaneous-spontaneous beat noise.

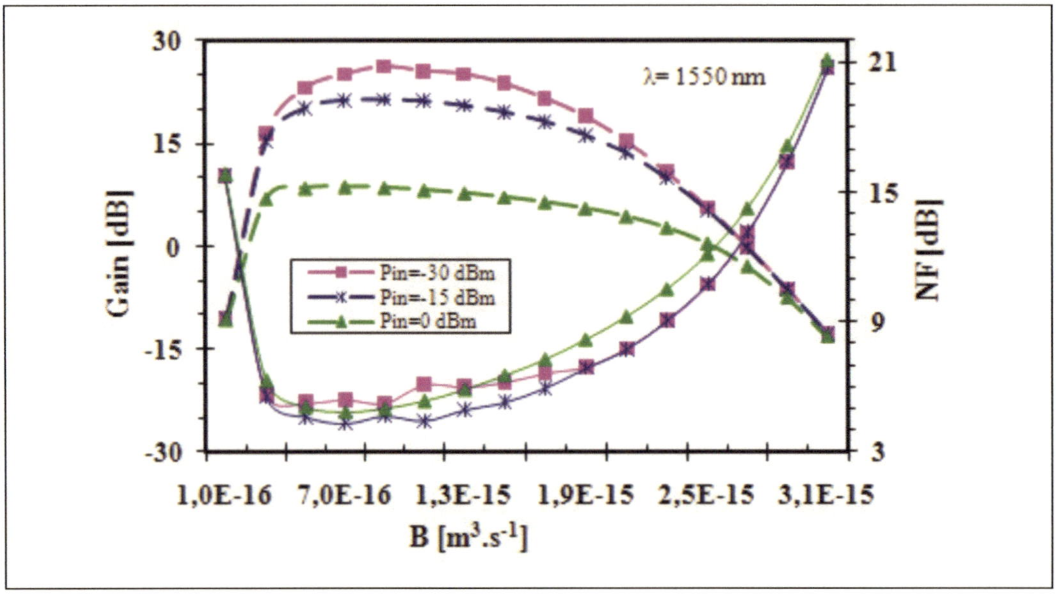

Fig. 4. Evolution of the gain and the noise as a function of the bimolecular recombination coefficient (B) and the SOA injected power.

The power of the ASE noise generated internally within the SOA is given by:

$$P_{ASE} = 2.n_{sp}.h.\nu.(G-1).B_0 \tag{1}$$

Where:
G: is the gain at the optical frequency ν,
h: represents the Planck's constant,
B_0: is the optical bandwidth of a filter within which P_{ASE} is determined,
n_{sp}: refers to the population inversion factor (sometimes called the spontaneous emission factor).
For an ideal amplifier, n_{sp} is equal to 1, corresponding to a complete inversion of the medium. However, in the usual case, the population inversion is partial and so $n_{sp} > 1$.
The shot noise results in the detection of the received total optical power due to the signal and the power of the ASE noise. It is given by the following equation:

$$N_{shot} = 2.e^2.B_e\left(\frac{G.P_{in}}{h.\nu} + n_{sp}.B_0.(G-1)\right) \tag{2}$$

Where B_e is the electrical bandwidth of the photo-detector.
The noise contribution due to the signal exists as well there is no optical amplifier; this later simply modifies the signal power, then the shot noise power related to this introduces a supplementary shot noise that is associated to the detection of the ASE.

The two intrinsic components related to beat noise are produced when optical signals and ASE coexist together. The first type of beat noise which is the signal-spontaneous beat noise occurs between optical signals and ASE having frequency close to that of the optical signals. It is given by the following equation:

$$N_{s-sp} = 4.\frac{e^2}{h.\nu}.B_e.P_{in}.n_{sp}.G.(G-1) \tag{3}$$

The second type, which is the spontaneous-spontaneous beat noise, occurs between ASEs. It is expressed as follows:

$$N_{sp-sp} = e^2.(2B_0 - B_e).B_e.n_{sp}^2.(G-1)^2 \tag{4}$$

The signal-spontaneous beat noise is preponderant for a strong input signal, whereas the spontaneous-spontaneous beat noise is dominating when there is an injection of a small input power. Compared to the shot noise and the signal-spontaneous beat noise, the spontaneous-spontaneous beat noise can be significantly minimized by placing an optical filter having a bandwidth B_0 after the amplifier.

A convenient way to quantify and characterize the noise and describe its influence on the SOA performance is in terms of Noise Figure (NF) parameter. It represents the amount of degradation in the signal to noise ratio caused by amplification process, and it is defined as the ratio between the optical signal to noise ratio (OSNR) of the signal at the input and output of SOA:

$$NF = \frac{OSNR_{in}}{OSNR_{out}} \tag{5}$$

The OSNR of the input signal is given by the following equation (Koga & Matsumoto, 1991):

$$OSNR_{in} = \frac{P_{in}}{2.h.\nu.B_e} \tag{6}$$

The OSNR of the input signal is proportional to the optical power of the input signal, or more specifically to the input number of photons per unit time ($P_{in}/h\nu$). Whereas, the OSNR of the output signal is defined by:

$$OSNR_{out} = \left(\frac{e.P_{in}.G}{h.\nu}\right)^2.\frac{1}{N_{shot} + N_{s-sp} + N_{sp-sp}} \tag{7}$$

Accordingly, by substituting equations (2), (3), (4), (6) and (7) into (5), the noise figure can be written as follows:

$$NF = \frac{1}{G} + 2.n_{sp}.\frac{G-1}{G} + \frac{h.\nu.B_0.n_{sp}.P_{in}.(G-1)}{P_{out}^2} + \frac{h.\nu.(2B_0 - B_e).n_{sp}^2.P_{in}.(G-1)^2}{2.P_{out}^2} \tag{8}$$

In practical case, the last two terms can be neglected because the ASE power is weak compared with the signal power; otherwise the spontaneous-spontaneous beat noise can be minimized by placing an optical filter at the output. So the noise figure can be rewritten as (Simon et al., 1989):

$$NF \approx \frac{1}{G} + 2.n_{sp}.\frac{G-1}{G} \tag{9}$$

Since spontaneous emission factor (n_{sp}) is always greater than 1, the minimum value of NF is obtained for n_{sp}=1. So, for large value of gain (G>>1), the noise figure of an ideal optical amplifier is 3dB. This is considered as the lowest NF that can be achieved. This implies that every time an optical signal is amplified, the signal to noise ratio is reduced to the half.
The NF can be expressed as function of the power of ASE noise, which is given by (1), as follows:

$$NF \approx \frac{1}{G} + 2.\frac{P_{ASE}}{h.\nu.G.B_0} \tag{10}$$

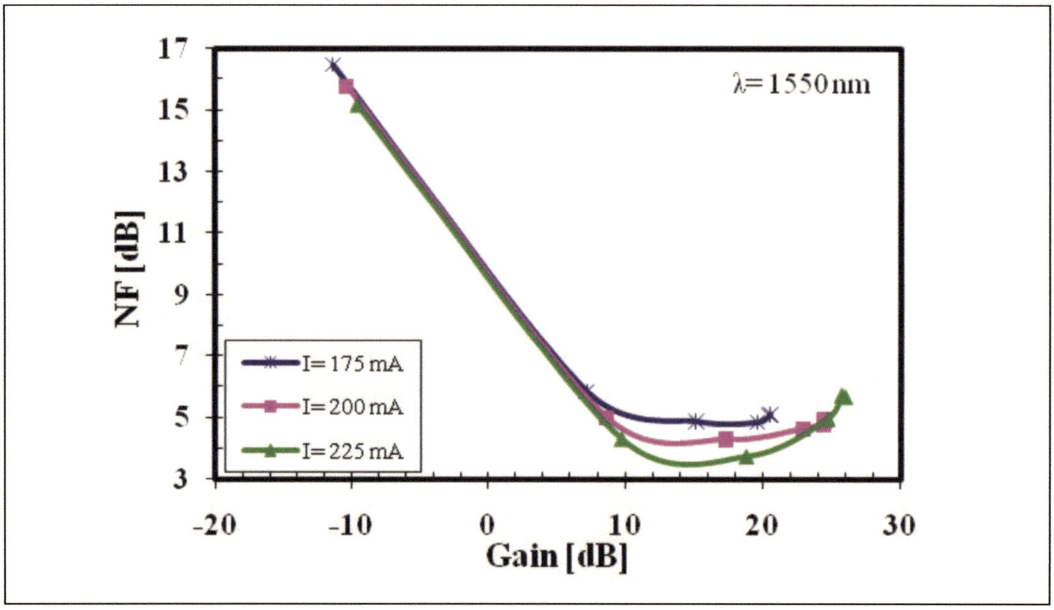

Fig. 5. Evolution of the noise figure as a function of the SOA gain for different bias current values.

The NF is represented as function of the gain in figure 5. This result is very significant because it allows us to choose the characteristics of the SOA in order to obtain the highest value of the gain for a minimum noise figure. So, we can notice that a low gain corresponds to a high value of NF; whereas to have the possible maximum of the gain while satisfying the criterion of low noise, it is necessary to choose the highest bias current possible.

2.4 Linear and saturation operating regimes in a SOA structure

A SOA amplifies input light through stimulated emission by electrically pumping the amplifier to achieve population inversion. It should have large enough gain for such application. The gain is dependent on different parameters, such as the injected current, the

device length, the wavelength and the input power levels. The SOA gain decreases as the input power is increased.
This gain saturation of the SOA is caused not only by the depletion of carrier density owing to stimulated emission, but also by the main intraband processes, such as spectral hole burning (SHB) and carrier heating (CH). However, when the SOA is operated with pulses shorter than a few picoseconds, intraband effects become important.
The origin of gain saturation lies in the power dependence of the gain coefficient where the population inversion due to injection current pumping is reduced with the stimulated emission induced by the input signal.
The saturation power parameter of the SOAs is of practical interest. It is a key parameter of the amplifier, which influences both the linear and non-linear properties. It is defined as the optical power at which the gain drops by 3 dB from the small signal value. That is to say, it is the optical power which reduces the modal gain to half of the unsaturated gain.
The saturation output power of the SOA is given by (Connelly, 2002):

$$P_{sat} = \frac{A}{\Gamma}.I_s \tag{11}$$

Where:

$$I_s = \frac{h.\nu}{a_N.\tau_s} \tag{12}$$

A: denotes the active region cross-section area,
Γ: represents the optical confinement factor coefficient,
a_N: symbolizes the differential modal gain,
τ_s: makes reference to spontaneous carrier lifetime.
High saturation output power is a desirable SOA characteristic, particularly for power booster and multi-channel applications. Referring to equation (11), the saturation output power can be improved by increasing the saturation output intensity (I_s) or reducing the optical confinement factor. The former case can be achieved either by reducing the differential modal gain and/or the spontaneous carrier lifetime. Since the last parameter (τ_s) is inversely proportional to carrier density, operation at a high bias current leads to an increase in the saturation output power. Nevertheless, when the carrier density increases, the amplifier gain also increases, making resonance effects more significant.
As the saturation output power depends inversely on the optical confinement factor, the single pass gain can be maintained by reducing this coefficient or by increasing the amplifier length. This process is not always necessary for the reason that the peak material gain coefficient shifts to shorter wavelengths as the carrier density is increasing.
When the average output power is at least 6 dB less than the output saturation power, non-linear effects are not observed and the SOA is in the linear regime. This linear operating regime, which is closely related to the output saturation power, is defined as the output power of an SOA where the non-linear effects do not affect the input multi-channel signal.
Gain saturation effects introduce undesirable distortion to the output signal. So, an ideal SOA should have very high saturation output power to achieve good linearity and to maximize its dynamic range with minimum distortion. Moreover, high saturation output power is desired for using SOAs especially in wavelength division multiplexing (WDM)

systems. Figure 6 shows that the highest value of the saturation output power, which corresponds to very fast dynamics of the carriers' density, is obtained when a strong bias current is used. Furthermore, we can notice that a high value of the bias current can engender a high gain with a high saturation output power. On the other hand, a low bias current corresponds to a less strong gain with a less high saturation output power, but the saturation input power is stronger.

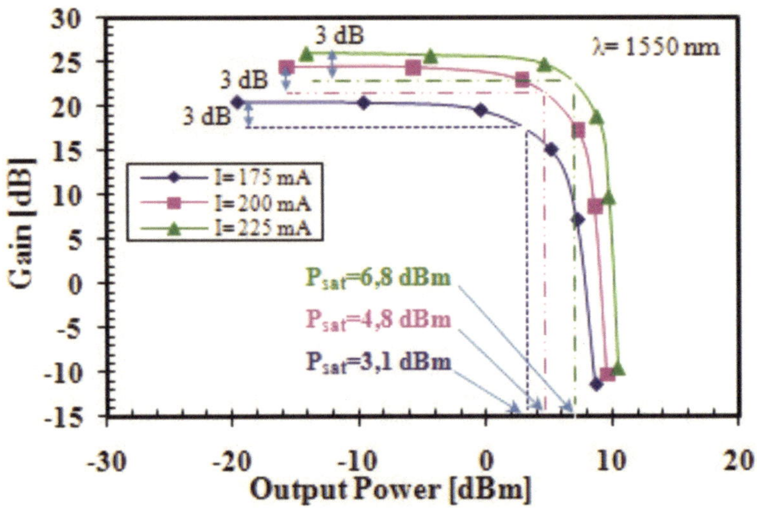

Fig. 6. Evolution of the gain as a function of the output power and the SOA bias current.

Because of the SOA's amplification and nonlinear characteristics, SOAs or integrated SOAs with other optical components can be exploited to assure various applications for high bit rate network systems. Moreover, large switching matrices comprised of SOA gates can be constructed to take advantage of the SOA gain to reduce insertion losses, to overcome electronic bottlenecks in switching and routing. The fast response speed can also be utilized effectively to perform packet switching.

3. SOA nonlinearities

SOAs are showing great promise for use in evolving optical networks and they are becoming a key technology for the next generation optical networks. They have been exploited in many functional applications, including switching (Kawaguchi, 2005), wavelength conversion (Liu et al., 2007), power equalization (Gopalakrishnapillai et al., 2005), 3R regeneration (Bramerie et al., 2004), logic operations (Berrettini et al., 2006), etc., thanks to their nonlinear effects, which are the subject of the current section. The effects are: the self gain modulation (SGM), the self phase modulation (SPM), the self induced nonlinear polarization modulation (SPR), the cross-gain modulation (XGM), the cross-phase modulation (XPM), the four-wave mixing (FWM) and the cross-polarization modulation (XPolM). These functions, where there is no conversion of optical signal to an electrical one, are very useful in transparent optical networks.

In SOA operational regime, there is a variation of the total density of the carriers and their distributions. This variation engenders intraband and interband transitions. The interband transition changes the carrier density but does not affect the carrier distribution. It is produced by the stimulated emission, the spontaneous emission and the non-radiative recombination. The modification of the total density of the carriers comes along with the modification of the carriers in the same band. The intraband transitions, such as spectral hole burning (SHB) and carrier heating (CH) are at the origin of the fast dynamics of the SOAs. They change the carrier distribution in the conduction band.
The main nonlinear effects involved in the SOAs, having for origin carriers dynamics and caused mainly by the change of the carriers density induced by input signals, are the following ones:

3.1 Self Gain Modulation

The self gain modulation (SGM) is an effect which corresponds to the modulation of the gain induced by the variation of the input signal power. It can be used to conceive a compensator of signal distortion.

3.2 Self Phase Modulation

The self phase modulation (SPM) is a nonlinear effect that implies the phase modulation of the SOA output signal caused by the refractive index variation induced by the variation of the input signal power.

3.3 Self induced nonlinear Polarization Rotation

The self induced nonlinear polarization rotation (SPR) translates the self rotation of the polarization state of the SOA output signal with regard to input one.

3.4 Cross-Gain Modulation

The cross-gain modulation (XGM) is a nonlinear effect, which is similar to the SGM. It implies the modulation of the gain induced by an optical signal (known as a control or pump signal), which affects the gain of a probe signal propagating simultaneously in the SOA. The XGM can take place in a SOA with a co-propagative or counter-propagative configuration.

3.5 Cross-Phase Modulation

The cross-phase modulation (XPM) is a nonlinear effect, which is similar to the SPM. It corresponds to the change of the refractive index induced by an optical signal (known as a control or pump signal), which affects the phase of another optical signal (probe) propagating at the same time in the SOA structure.

3.6 Four Wave Mixing

The four wave mixing (FWM) is a parametric process, which is at the origin of the production of new frequencies. It can be explained by the beating between two or several optical signals having different wavelengths propagating in the SOA structure, which generates signals having new optical frequencies.
The FWM effect in SOAs has been shown to be a promising method for wavelength conversion. It is attractive since it is independent of modulation format, capable of

dispersion compensation and ultra fast. So, wavelength conversion based on FWM effect offers strict transparency, including modulation-format and bit-rate transparency, and it is capable of multi-wavelength conversions. However, it has low conversion efficiency and needs careful control of the polarization of the input lights (Politi et al., 2006). The main drawbacks of wavelength conversion based on FWM are polarization sensitivity and the frequency-shift dependent conversion efficiency.

3.7 Cross-Polarization Modulation

The cross-polarization modulation (XPolM) effect in a SOA structure, which has been subject of many investigations in recent years, is a nonlinear effect similar to the SPR. It denotes the polarization rotation of a beam propagating in a SOA affected by the polarization and the power of a relatively strong control beam, introduced simultaneously into the amplifier. When two signals are injected in the SOA, an additional birefringence and gain compression affects the SOA. The two signals affect one another by producing different phase and gain compression on the transverse electric (TE) and transverse magnetic (TM) components (because the gain saturation of the TE and TM modes is different). This results in a rotation of the polarization state for each signal. The SOA bias current, and the input signal power are among the parameters that determine the magnitude of the polarization rotation. As a result, the XPolM effect in SOA is then directly related to the TE/TM mode discrepancy of XPM and XGM.

The nonlinear polarization rotation that occurs in the SOA is demonstrated to perform very interesting functionalities in optical networks. However, it is exploited in optical gating, in wavelength conversion, in regeneration and in all-optical switching configurations that are required for wavelength routing in high-speed optical time-division multiplexing networks.

4. Modelling of polarization rotation in SOAs using the Coupled Mode Theory

4.1 Analysis of the polarization rotation in SOA with application of Stokes parameters

A convenient method to describe the state of polarization is in terms of Stokes parameters. They provide a very useful description of the polarization state of an electromagnetic wave. Moreover, they characterize the time-averaged electric-field intensity and the distribution of polarization among three orthogonal polarization directions on the Poincaré sphere. They are used in this work to analyze the polarization change at the SOA output with relation to its state at the input for various length of the active region. They are noted as (S_0, S_1, S_2, S_3) and defined as (Flossmann et al., 2006):

$$\begin{pmatrix} S_0 \\ S_1 \\ S_2 \\ S_3 \end{pmatrix} = \begin{pmatrix} A_{TE}^{\ 2} + A_{TM}^{\ 2} \\ A_{TE}^{\ 2} - A_{TM}^{\ 2} \\ 2A_{TE}.A_{TM}.\cos(\phi_{TM} - \phi_{TE}) \\ 2A_{TE}.A_{TM}.\sin(\phi_{TM} - \phi_{TE}) \end{pmatrix} \tag{13}$$

Where:

S_0 is a parameter that translates the total intensity.

S_1 refers to the intensity difference between the horizontal polarization and the vertical polarization.

S_2 makes reference to the difference between intensities transmitted by the axes (45°, 135°).

S_3 is a parameter that expresses the difference between intensities transmitted for the left and right circular polarizations.
ϕ_{TE} and ϕ_{TM} denote the phase shift for the TE and TM modes, respectively.
The normalized Stokes parameters that can be measured at the SOA structure output by using a polarization analyzer are given by:

$$s_i = \frac{S_i}{S_0} \quad \text{with i} \in \{1, 2, 3\} \tag{14}$$

The phase shift variation can be written as follows:

$$\Delta\phi = \phi_{TM} - \phi_{TE} = \arctan\left(\frac{s_3}{s_2}\right) \tag{15}$$

The relationship of the normalized Stokes parameters to the orientation (azimuth) and the ellipticity angles, ψ and χ, associated with the Poincaré Sphere is shown in the following equations (Guo & Connelly, 2005):

$$\begin{pmatrix} s_1 \\ s_2 \\ s_3 \end{pmatrix} = \begin{pmatrix} \cos(2\psi).\cos(2\chi) \\ \sin(2\psi).\cos(2\chi) \\ \sin(2\chi) \end{pmatrix} \tag{16}$$

Therefore, the polarization change at the SOA output can be analyzed and evaluated by the azimuth and the ellipticity that can be expressed as function of normalized Stokes parameters:

$$\begin{cases} \psi = \dfrac{1}{2}\arctan\left(\dfrac{s_2}{s_1}\right) \\ \chi = \dfrac{1}{2}\arcsin(s_3) \end{cases} \tag{17}$$

4.2 Concept of the proposed model

In this model, which is based on the coupled mode theory (CMT), we assume that the optical field is propagating in the z-direction of the SOA structure and it is decomposed into TE and TM component. In addition, the TE/TM gain coefficients are supposed, in a saturated SOA, to be not constant along the amplifier length and then can be written as the following forms (Connelly, 2002):

$$\begin{cases} g_{TE}(z) = \Gamma_{TE} \cdot g_m(z) - \alpha_{TE} \\ g_{TM}(z) = \Gamma_{TM} \cdot g_m(z) - \alpha_{TM} \end{cases} \tag{18}$$

Where g_{TE} and g_{TM} are the gain coefficients, Γ_{TE} and Γ_{TM} denote the confinement factors, α_{TE} and α_{TM} symbolize the efficient losses, respectively for TE and TM modes. g_m designates the gain material coefficient.
To estimate the polarization sensitivity of a saturated amplifier, the material intensity gain coefficient is assumed to be saturated by the light intensity as the following equation (Gustavsson, 1993):

$$g_m(z) = \frac{g_{m,0}}{1 + \left(|A_{TE}(z)|^2 + |A_{TM}(z)|^2 \right).E_s^{-2}} \tag{19}$$

Referring to the coupled mode equations developed in (Gustavsson, 1993) that take into account the coupling between the TE and TM modes, the evolution of the electromagnetic field envelope in the SOA active region can be written under the following equations:

$$\begin{cases} \dfrac{\partial A_{TE}}{\partial z} = \dfrac{1}{2} g_{TE}(z).A_{TE}(z) + C_{cpl,1}.A_{TM}(z).e^{-j.\Delta\beta.z} \\ \dfrac{\partial A_{TM}}{\partial z} = \dfrac{1}{2} g_{TM}(z).A_{TM}(z) - C_{cpl,2}.A_{TE}(z).e^{j.\Delta\beta.z} \end{cases} \tag{20}$$

Where

$\Delta\beta$ represents the difference between the propagation constants β_{TE} and β_{TM} for TE and TM modes, respectively.

$C_{cpl,i}$ (with i={1,2}) denotes the coupling coefficient given by the following equation:

$$C_{cpl,i}(z) = \kappa_i - \kappa_i.\left(1 + \frac{\left(|A_{TE}(z)|^2 + |A_{TM}(z)|^2 \right)}{|E_s|^2} \right)^{-1} - \kappa_i.\left(1 + \frac{\left(|A_{TE}(z)|^2 + |A_{TM}(z)|^2 \right)}{|E_s|^2} \right)^{-2} \tag{21}$$

With: "κ_i" is a constant.

Hence, the evolution of the electromagnetic field envelope in the active region of the SOA can also be written under the following matrix form:

$$\frac{\partial}{\partial z}\begin{pmatrix} A_{TE}(z) \\ A_{TM}(z) \end{pmatrix} = j.\begin{pmatrix} m_{11} & m_{12} \\ m_{21} & m_{22} \end{pmatrix}.\begin{pmatrix} A_{TE}(z) \\ A_{TM}(z) \end{pmatrix} \tag{22}$$

Where

$$\begin{cases} m_{11} = -\dfrac{j}{2}.(\Gamma_{TE}.g_m(z) - \alpha_{TE}) \\ m_{12} = -j.C_{cpl1}.e^{-j.\Delta\beta.z} \\ m_{21} = j.C_{cpl2}.e^{j.\Delta\beta.z} \\ m_{22} = -\dfrac{j}{2}.(\Gamma_{TM}.g_m(z) - \alpha_{TM}) \end{cases} \tag{23}$$

The solution of the set of differential equations (20) is not available in analytical form. Then, for calculating the electromagnetic field envelope in the SOA structure, it is primordial to use a numerical method; that is the object of the next section.

4.3 Numerical method formulation

The numerical method adopted to calculate electromagnetic field envelope of SOA, is based on a numerical integration approach of the differential equations in the z-direction. Firstly, the equation (22) is reformulated as:

$$\frac{\partial A(z)}{\partial z} = j.M(A(z)).A(z) \tag{24}$$

Where

$$A(z)=\begin{pmatrix} A_{TE}(z) \\ A_{TM}(z) \end{pmatrix} \tag{25}$$

$$M\left(A(z)\right)=\begin{pmatrix} m_{11} & m_{12} \\ m_{21} & m_{22} \end{pmatrix} \tag{26}$$

The initial solution of equation (24) at a position $z=z_0+\Delta z$ is given by:

$$A_0(z_0+\Delta z)=A(z_0).\exp\left(-j.M(z_0).\Delta z\right) \tag{27}$$

For the fact that the matrix M is not constant in the interval Δz, it is necessary to apply a correction to the initial solution. The correction term is expressed as:

$$A_c(z_0+\Delta z)=\frac{j.\Delta z}{2}\left[M(z_0+\Delta z)-M(z_0)\right].\overline{A_0} \tag{28}$$

Where $\overline{A_0}$ is the average value of A(z) in the interval $[z_0, z_0+\Delta z]$.
Then, the final solution will be written as the following form:

$$A(z_0+\Delta z)=A_0(z_0+\Delta z)+A_c(z_0+\Delta z) \tag{29}$$

For the numerical implementation of the described method, the exponential term in equation (27) is developed as a finite summation of Taylor series terms as:

$$A_0(z_0+\Delta z)=\sum_{p=0}^{q}\frac{1}{p!}.\left(-j.\Delta z.M(z_0)\right)^p.A(z_0)+O^{q+1}(\Delta z) \tag{30}$$

In which $O^{q+1}(\Delta z)$ denotes that the remaining error is order (q+1) in Δz.
Finally, in order to reduce the calculation time, it is worthwhile to calculate the electromagnetic field envelope of SOA recursively as the following form:

$$\frac{1}{p!}.\left[-j.\Delta z.M(A(z_0))\right]^p.A(z_0)=\frac{\Delta z}{p!}.\left[-j.M(A(z_0))\right]*\frac{\Delta z^{p-1}}{(p-1)!}.\left[-j.M(A(z_0))\right]^{p-1}.A(z_0) \tag{31}$$

4.4 Numerical simulation results and their experimental validation

In the implementation of the developed model, which is proposed in the section 4.3, using the theoretical background described above, the Taylor series is evaluated up to the twentieth order. In order to validate the results obtained by this approach, we have performed an experiment which was done in free-space. Its setup consists of a commercial InGaAsP/InP SOA structure, which is positioned so that their TE and TM axes correspond to the horizontal and vertical axes of the laboratory referential, respectively. The laboratory refers to the RESO lab in the National Engineering School of Brest, France. Light emitted from the SOA was collected and collimated with a microscope objective, then passed through a quarter-wave plate and a linear polarizer acting as an analyzer, before being recollected with a fibred collimator, connected to an optical spectrum analyzer with an optical band-pass filter, having a bandwidth of 0.07 nm in order to reject the amplified spontaneous emission. The passing axis of the linear polarizer, when set vertically, coincided with the TM axis in the sample and defined a reference direction from which the orientation angle θ of the fast axis of the quarter-wave plate was estimated. This orientation could be modified, as the quarter-wave plate was mounted on a rotation stage whose movements were accurately determined by a computer-controlled step motor.

The presence of an injected optical signal affects the carrier density and includes strong modifications of the birefringence and dichroism experienced by the signal itself in the SOA active zone. Consequently, the input optical signal experiences a modification of its polarization state due to the intrinsic birefringence and residual differential gain of the active region. In the linear operating regime, the SOA output polarization remains nearly independent of the input power. However, within the saturation regime, a self-induced nonlinear rotation of polarization takes place and depends upon input power, because of carrier density variations which modify induced birefringence and residual differential gain. This causes fast variations of the state of polarization of the output signal, both in terms of azimuth and ellipticity parameters, which are analyzed in this section by varying the SOA injection conditions.

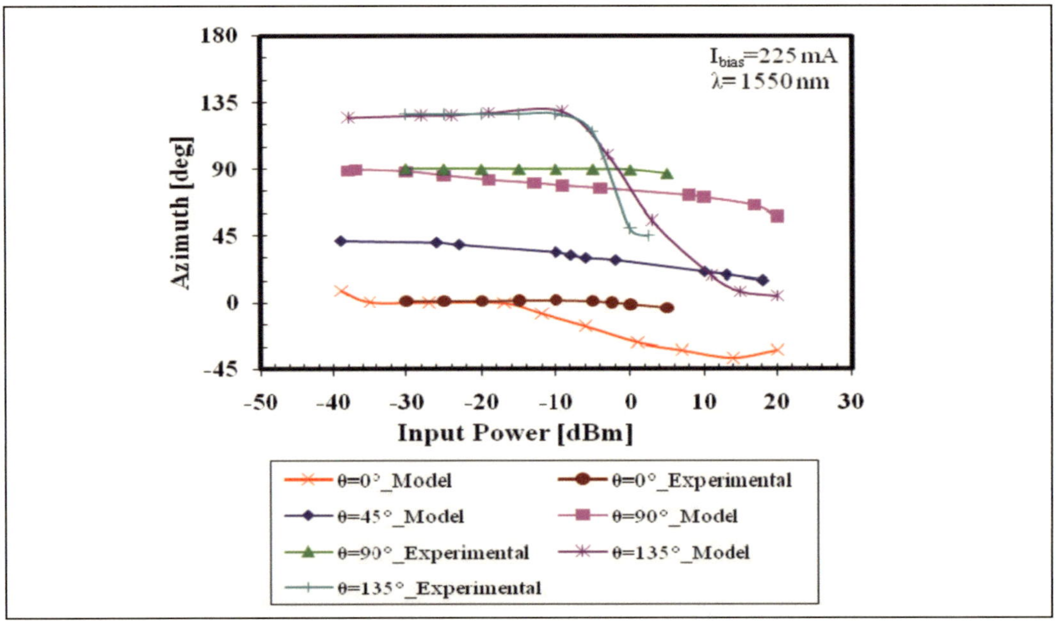

Fig. 7. Evolution of the azimuth of the output state of polarization of the SOA as a function of the injected input power for a bias current of 225 mA.

Figure 7 depicts the evolution of the azimuth at the output as a function of the input signal power initially injected at an angle θ. When θ = 0° or θ = 90°, the azimuth undergoes a small variation. Whereas, when θ = 45° or θ = 135° that corresponds to the injection with identical TE/TM powers, a significant change of the polarization state is shown when the input signal power becomes high, which corresponds to the saturation regime, contrarily for low values that refer to the linear operating regime. However, we can notice that the results obtained are in good agreement with the experimental measurements.

The ellipticity, which is shown in figure 8, experiences a slight increase when there is an augmentation of the injected power for an angle θ equal to 0° or 90°. However, for the case of injecting the input power with θ = 45° or θ = 135°, the ellipticity parameter experiences significant variations.

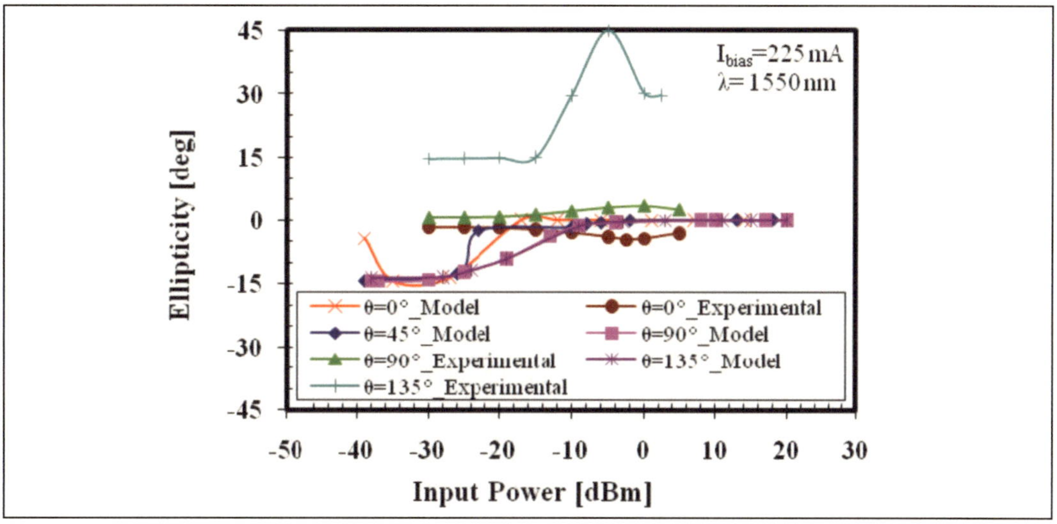

Fig. 8. Evolution of the ellipticity of the output state of polarization of the SOA versus the input signal power for a bias current of 225 mA.

According to the results presented in figure 9, we can notice that the phase shift is almost constant when the input signal power is very low. It decreases rapidly by augmenting the injected power. This behavior is explained by the diminution of carrier density due to the stimulated emission as the input power is increased. Also, it mirrors the variation of birefringence induced by effective refractive index variations with carrier densities. Moreover, this change of ΔΦ is significant only in the gain band; i.e. when the injected input power is high and thus corresponds to the operation at the saturation regime.

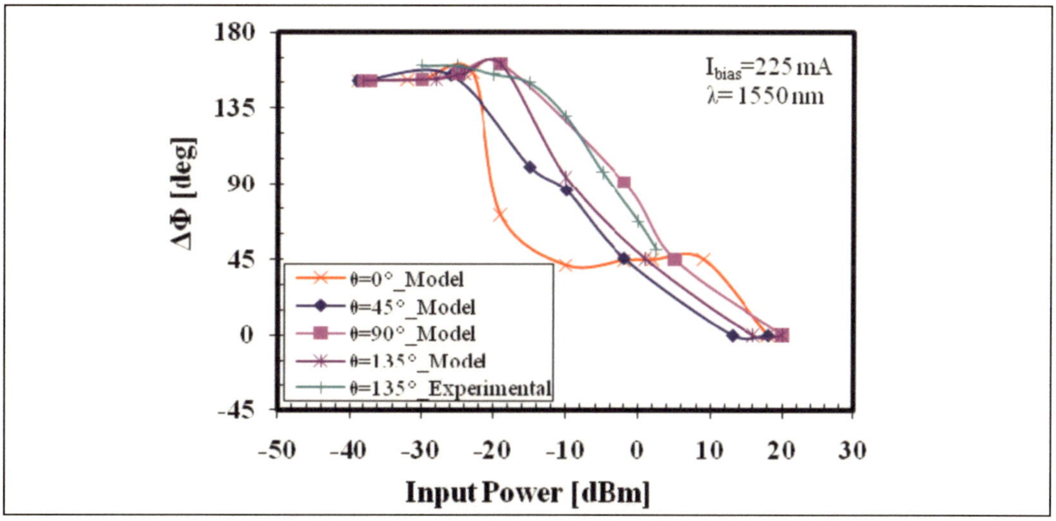

Fig. 9. Variation of the phase shift of the optical signal at the SOA output as a function of the injected input power for a bias current of 225 mA.

5. Application of SOA nonlinearities to achieve wavelength conversion

All-optical wavelength converters are considered key components in future WDM optical networks due to their main advantage that consists of increasing the flexibility and the capacity of these networks and facilitating WDM network management. Moreover, they form an essential part of the wavelength routing switch that is employed in the all-optical buffering concept. All-optical wavelength conversion can be realized by using fiber nonlinearities or nonlinearities in semiconductor devices.

In the last few years, a considerable attention has been focusing on SOAs and their potential use in optical communication systems. Especially, SOAs have generated more and more interest when optical signal processing is involved. Subsequently, they are exploited to achieve wavelength conversion at high bit rates, which is a very important function in conjunction with WDM systems. This reason makes them also very useful in wavelength routers, which manage wavelength paths through optical networks based on complex meshes, rather than point-to-point architectures (Wei et al., 2005).

Many studies have paid more attention on SOA performance for implementing and configuring wavelength conversion sub-systems. Hence, several optical wavelength converters based on SOA nonlinearities have been proposed and discussed, such as XGM (Tzanakaki & O'Mahony, 2000), XPM (Matsumoto et al., 2006), FWM (Contestabile et al., 2004), and XPolM (Wei et al., 2005). Each configuration has its advantages and disadvantages and thus its framework of application in optical communication networks.

Optical wavelength converters based on SOA nonlinearities, which are fundamental components in today's photonic networks, offer advantages in terms of integration potential, power consumption, and optical power efficiency. However, the major limitation of SOA-based wavelength converters is the slow SOA recovery, causing unwanted pattern effects in the converted signal and limiting the maximum operation speed of the wavelength converters. It has already been theoretically and experimentally clarified that the increase in electrical pumping power, confinement factor and the device interaction length effectively improve the speed performance (Joergensen et al., 1997). For improving the SOA-based wavelength converters, some techniques are proposed, such as: Fiber Bragg grating at 100 Gbit/s (Ellis et al., 1998), interferometric configuration at 168 Gbit/s (Nakamura et al., 2001), two cascaded SOAs at 170.4 Gbit/s (Manning et al., 2006) and optical filtering at 320 Gbit/s (Liu et al., 2007).

In this section, we evaluate the influence of SOA parameters and the signal format (non return-to-zero "NRZ" or return-to-zero "RZ") on the behavior of the structure used in wavelength conversion configuration and we analyze the performance dependence on several critical operation parameters of the SOA structure.

5.1 Concept of wavelength conversion based on SOA nonlinear effects

All-optical wavelength conversion refers to the operation that consists of the transfer of the information carried in one wavelength channel to another wavelength channel in the optical domain. It is a key requirement for optical networks because it has to be used to extend the degree of freedom to the wavelength domain. Moreover, All-optical wavelength conversion is also indispensable in future optical packet switching (OPS) networks to optimize the network performance metrics, such as packet loss rate and packet delay (Danielsen et al., 1998). Also, it is very useful in the implementation of switches in WDM networks. In addition, it is crucial to lower the access blocking probability and therefore to increase the utilization efficiency of the network resources in wavelength routed optical networks.

While a significant part of network design, routing and wavelength assignment depends on the availability and performance of wavelength converters; and as many techniques have been explored and discussed in this context, all-optical wavelength converters based on SOA structures have attracted a lot of interest thanks to their attractive features, such as the small size, the fast carrier dynamics, the multifunctional aspect and the high potential of integration. The main features of a wavelength converter include its transparency to bit rate and signal format, operation at moderate optical power levels, low electrical power consumption, small frequency chirp, cascadability of multiple stages of converters, and signal reshaping.

When a RZ pump (the data signal) at wavelength λ_1 and a continuous wave (CW) probe signal at wavelength λ_2 are injected into an SOA, the pump modulates the carrier density in its active region and hence its gain and refractive index. This leads to a change in the amplitude and phase of the CW probe signal. In the case of XGM, the output probe signal from the SOA carries the inverted modulation of the RZ input data signal.

The XPM is used to obtain an output probe signal with non-inverted modulation, whereby the phase modulation of the probe signal is converted to amplitude modulation by an interferometer. Particularly, in the wavelength conversion based on the XGM scheme, a strong input signal is needed to saturate the SOA gain and thereby to modulate the CW signal carrying the new wavelength. While the XGM effect is accompanied by large chirp and a low extinction ratio, and limited by the relatively slow carrier recovery time within the SOA structure, impressive wavelength conversion of up to 40 Gbit/s and with some degradation even up to 100 Gbit/s (Ellis et al., 1998), has been demonstrated.

To overcome the XGM disadvantages, SOAs have been integrated in interferometric configurations, where the intensity modulation of the input signal is transferred into a phase modulation of the CW signal and exploited for switching. These XPM schemes enable wavelength conversion with lower signal powers, reduced chirp, enhanced extinction ratios and ultra fast switching transients that are not limited by the carrier recovery time. Subsequently, wavelength conversion based on the XPM effect with excellent signal quality up to 100 Gbit/s, has been demonstrated (Leuthold et al., 2000) by using a fully integrated and packaged SOA delayed interference configuration that comprises a monolithically integrated delay loop, phase shifter and tunable coupler.

The FWM effect in SOAs has been shown to be a promising method for wavelength conversion. It is attractive since it is independent of modulation format, capable of dispersion compensation and ultra fast. So, wavelength conversion based on FWM offers strict transparency, including modulation-format and bit-rate transparency, and it is capable of multi-wavelength conversions. However, it has a low conversion efficiency and needs careful control of the polarization of the input lights (Politi et al., 2006). The main drawbacks of wavelength conversion based on FWM are polarization sensitivity and the frequency-shift dependent conversion efficiency.

Wavelength conversion based on XPolM is another promising approach. It uses the optically induced birefringence and dichroism in an SOA and it has great potential to offer wavelength conversion with a high extinction ratio.

The influence of the nonlinear polarization rotation and the intrinsic and extrinsic SOA parameters on the performance of a wavelength converter based on XGM effect is the subject of the next section.

5.2 Impact of polarization rotation on the performance of wavelength conversion based on XGM at 40 Gbit/s

Gain saturation of the SOA structure induces nonlinear polarization rotation that can be used to realize wavelength converters (Liu et al., 2003). Depending on the system configuration, inverted and non-inverted polarity output can be achieved. Recently, a remarkable wavelength conversion at 40 Gb/s with multi-casting functionality based on nonlinear polarization rotation has been demonstrated (Contestabile et al., 2005). The proposed wavelength converter based on XGM effect in a wideband traveling wave SOA (TW-SOA) at 40 Gbit/s, is presented in figure 10.

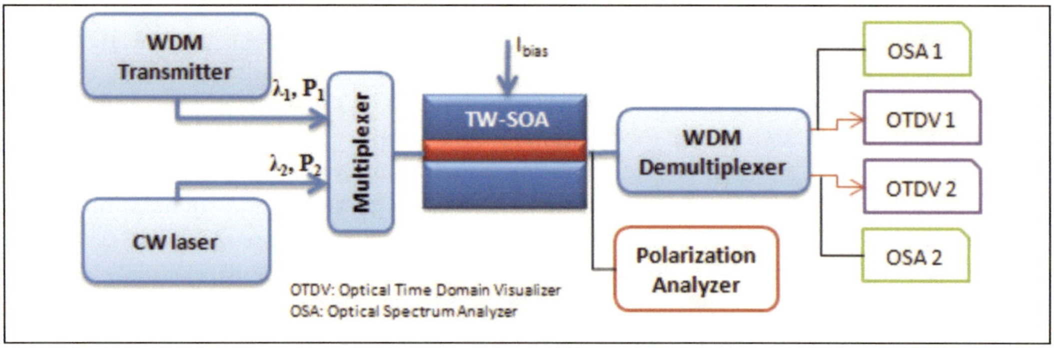

Fig. 10. Schematic of the wavelength converter configuration.

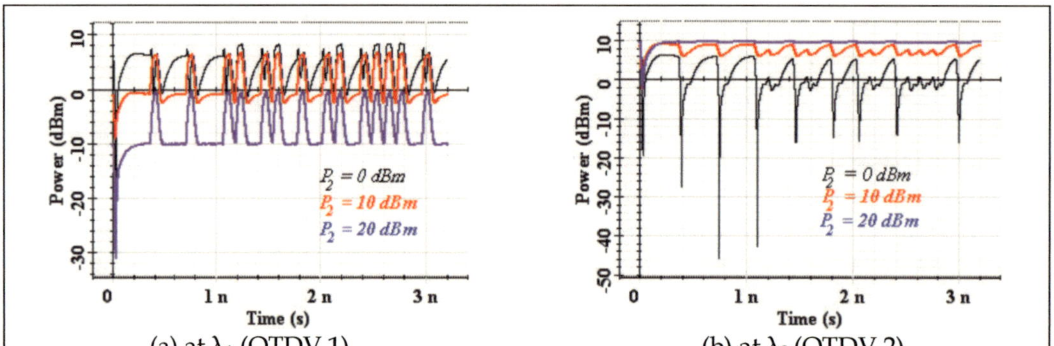

(a) at λ_1 (OTDV 1) (b) at λ_2 (OTDV 2)

Fig. 11. Evolution of the output signal by varying the CW input power for an RZ format signal.

An input signal obtained from a WDM transmitter, called the pump, at the wavelength λ_1= 1554 nm and a CW signal, called the probe light, at the desired output wavelength λ_2=1550 nm are multiplexed and launched co-directionally in the wideband TW-SOA. The pump wave modulates the carrier density and consequently the gain of the SOA. The modulated gain modulates the probe light, so that the output probe light, which is known as the converted signal, contains the information of the input signal, and achieve wavelength conversion (from λ_1 to λ_2).

By varying the CW input power and the input format signal, we visualized the output signal power by using the OTDV1 and OTDV2, as illustrated in figures 11 and 12. So, we can notice that a strong input signal is needed to saturate the SOA gain and thereby to modulate the CW signal, as shown in figures 11b and 12b. Also, this is accompanied by a modulation inversion of the output signal, which is considered one among the drawbacks of the wavelength conversion using XGM.

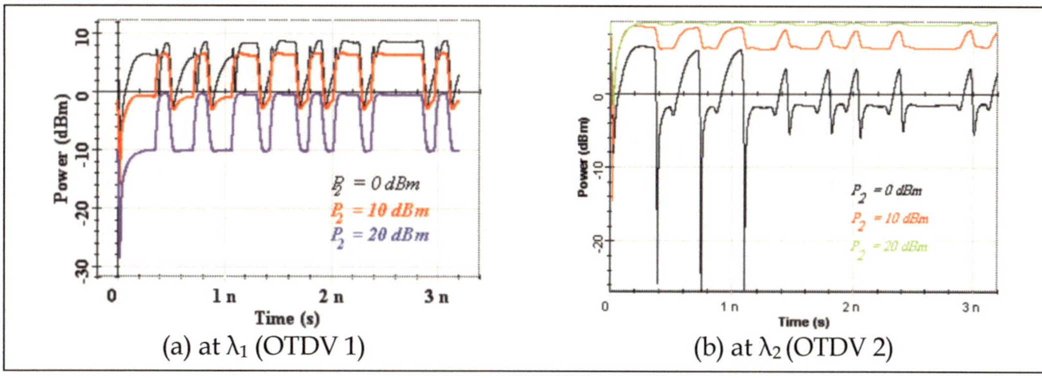

Fig. 12. Evolution of the output signal power as a function of the CW power for an NRZ format signal.

Birefringent effects are induced when the pump is coupled into the structure, owing to the TE/TM asymmetry of the confinement factors, the carriers' distributions, the induced nonlinear refractive indices and the absorption coefficients of the SOA. Consequently, the linear input polarization is changed and becomes elliptical at the output as the input power is increased. Thus, the azimuth and ellipticity vary at the SOA output, as shown in figure 13a. A significant change of the polarization state is shown when the CW input power is high, contrarily for low values that correspond to a linear operating regime. Moreover, this polarization rotation varies not only with the pump power but also as a function of the RZ/NRZ signal format and the optical confinement factor.

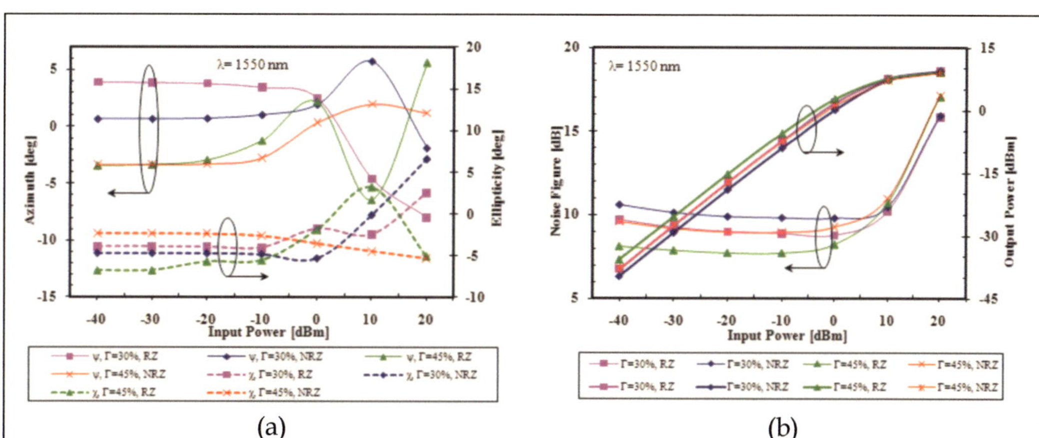

Fig. 13. Evolution of the azimuth, the ellipticity, the noise figure and the output power as a function of the input signal power, the signal format and the optical confinement factor.

The transfer function, illustrated in figure 13b, shows that the linear operating regime is exhibited when the input power is low; then the RZ signal format with a high optical confinement factor is the privileged case. The saturation regime occurs as we are increasing the input powers, which corresponds to a gain saturation that can cause significant signal distortion at the output of the wavelength converter. Consequently, in the proposed wavelength converter scheme, we can use a band-pass filter just after the SOA, centered on λ_2 to suppress the spontaneous noise and to extract only the converted signal containing the information of the input signal. Moreover, the discussed wavelength converter configuration can be used to interface access-metro systems with the core network by achieving wavelength conversion of 1310 to 1550 nm since multi-Gbit/s 1310 nm transmission technology is commonly used in access and metro networks and the long-haul core network is centered on 1550 nm window.

In order to analyze the wavelength converter performance in detail, we adopt a wavelength conversion scheme based on an RZ configuration. The used SOA has a bias current I= 150mA and is connected to a receiver composed of a Bessel optical filter centered on λ_2, a photo-detector PIN, a low pass Bessel filter and a Bit-Error-Rate (BER) analyzer. The default order of the Bessel optical filter was set to 4 in the subsequent simulations.

By varying the input power, the maximum value for the Q-factor, the minimum value for the BER, the eye extinction ratio and the eye opening factor versus decision instant are shown in figures 14 and 15.

The results obtained demonstrate that the optimal point corresponds to an input power equal to -39 dBm. The BER analyzer eye diagram for this case is represented in figure 16. As for the order of the Bessel low pass filter at the receiver, it has been also studied to observe its effects on performance of the system. It appears from figure 16, that the change of the filter order "m" has a slight variation on the performance of the simulated system.

So, we can conclude that high-speed wavelength conversion seems to be one of the most important functionalities required to assure more flexibility in the next generation optical networks, since wavelength converters, which are the key elements in future WDM networks, can reduce wavelength blocking and offer data regeneration.

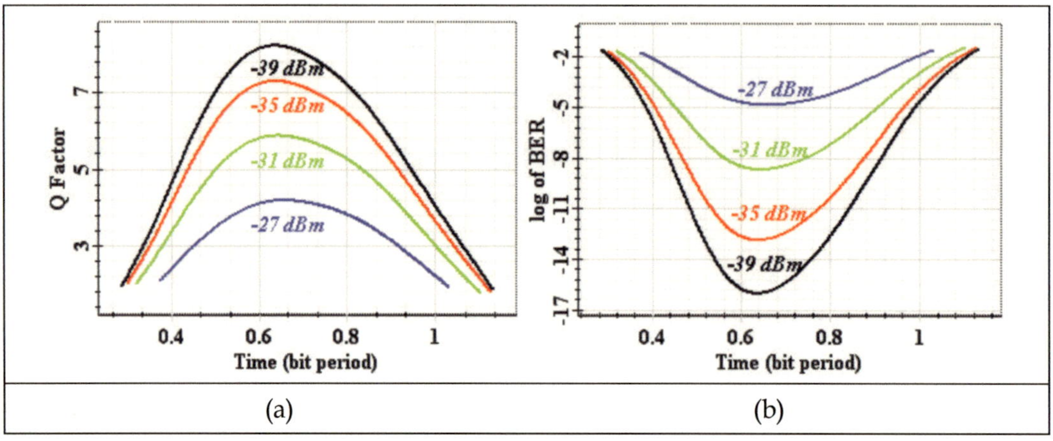

Fig. 14. Evolution of the Q-factor and the BER for different values of the input power.

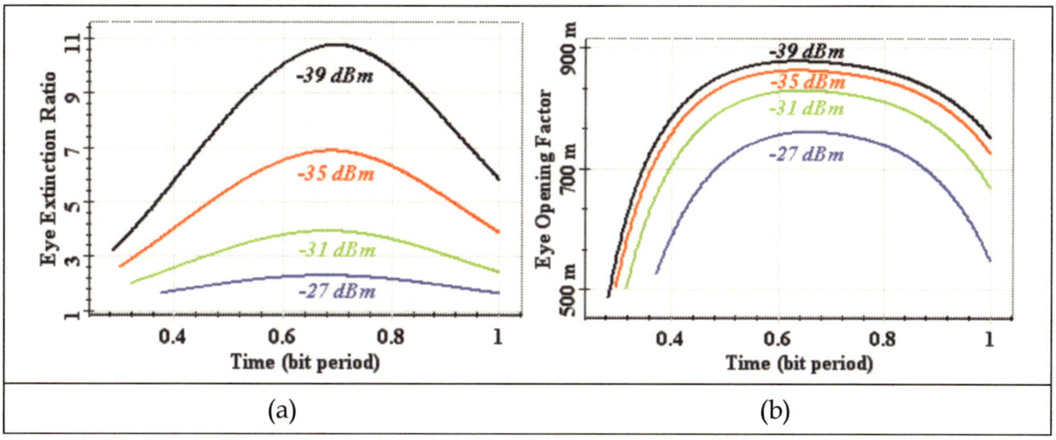

Fig. 15. Evolution of the eye extinction ratio and the eye opening factor for different values of the input power.

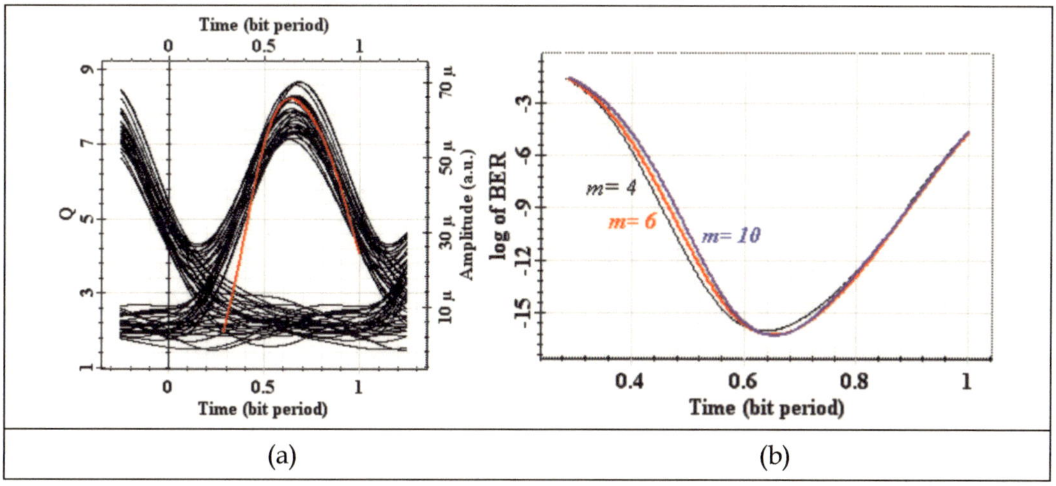

Fig. 16. Evolution the eye diagram, the Q factor and the BER by changing the order of the Bessel low pass filter "m" for an input power equal to -39 dBm.

6. Conclusion

In this chapter, an investigation of SOA nonlinearities and their applications for future optical networks are presented and discussed. We have shown that intrinsic and extrinsic SOA parameters, such as the bias current, the active region length, etc. play an important role in the SOA gain dynamics. As results, high saturation output power, which is especially preferred in WDM systems, can be achieved by increasing the bias current or by using short SOAs. An accurate choice of these parameters is very important for the determination of the best device operation conditions to achieve the desired functionality based on SOAs and exploiting their linear or saturation operating regime in a variety of different applications for all-optical signal processing and in long-haul optical transmissions.

We have also analyzed the impact of SOA parameter variations on the polarization rotation effect, which is investigated referring to a numerical model that we developed based on the Coupled Mode Theory and the formalism of Stokes. Subsequently, it is shown that the azimuth and the ellipticity parameters undergo changes according to injection conditions. Our model agrees well with available experimental measurements that have been carried out in free space and also reveals the conditions for the validity of previous simpler approaches.

7. References

Berrettini, G.; Simi, A.; Malacarne, A.; Bogoni, A. & Poti, L. (2006). Ultrafast integrable and reconfigurable XNOR, AND, NOR, and NOT photonic logic gate. *IEEE Photon. Technol. Lett.,* Vol. 18, No. 8, (Apr. 2006), (917-919), ISSN 1041-1135.

Bramerie, L.; Gay, M.; Girault, G.; Roncin, V.; Feve, S. & Simon, J.C. (2004). Performance of a Polarization Insensitive 3R Optical Regenerator Based on a new SOA-NOLM Architecture. *Proceedings of ECOC,* paper We2.5.2, Stockholm, 2004.

Connelly, M.J. (2002). *Semiconductor Optical Amplifier,* Kluwer Academic Publishers, ISBN 0-7923-7657-9, Boston, London.

Contestabile, G.; Presi, M. & Ciaramella, E. (2004). Multiple wavelength conversion for WDM multicasting by FWM in an SOA. *IEEE Photon. Technol. Lett.,* Vol. 16, No. 7, (Jul. 2004), (1775–1777), ISSN 1041-1135.

Contestabile, G.; Calabretta, N.; Presi, M. & Ciaramella, E. (2005). Single and multicast wavelength conversion at 40 Gb/s by means of fast nonlinear polarization switching in an SOA. *IEEE Photon. Technol. Lett.,* Vol. 17, No. 12, (Dec. 2005), (2652–2654), ISSN 1041-1135.

Danielsen, S.; Hansen, P. & Stubkjaer, K.E. (1998). Wavelength conversion in optical packet switching. *J. Lightwave Technology,* Vol. 16, No. 12, (Dec. 1998), (2095–2108), ISSN 0733-8724.

Ellis, A.D.; Kelly, A.E.; Nesset, D; Pitcher, D.; Moodie, D.G. & Kashyap, R. (1998). Error free 100 Gb/s wavelength conversion using grating assisted cross gain modulation in 2 mm long semiconductor amplifier. *Electronics Lett.,* Vol. 34, No. 20, (Oct. 1998), (1958–1959), ISSN 0013-5194.

Flossmann, F.; Schwarz, U.T.; Maier, M. & Dennis, M.R. (2006). Stokes parameters in the unfolding of an optical vortex through a birefringent crystal. *J. Optics Express,* Vol. 14, No. 23, (Nov. 2006), (11402-11411).

Gopalakrishnapillai, B.S.; Premaratne, M.; Nirmalathas, A. & Lim, C. (2005). Power Equalization using polarization rotation in semiconductor optical amplifiers. *IEEE Photon. Technol. Lett,.* Vol. 17, No. 8, (Aug. 2005), (1695-1697), ISSN 1041-1135.

Guo, L.Q. & Connelly, M.J. (2005). Signal-Induced Birefringence and Dichroism in a Tensile-Strained Bulk semiconductor optical amplifier and its Application to Wavelength Conversion. *J. Lightwave Technology,* Vol. 23, No. 12, (Dec. 2005), (4037-4045), ISSN 0733-8724.

Gustavsson, M. (1993). Analysis of Polarization Independent Optical Amplifiers and Filters Based on Polarization Rotation in Periodically Asymmetric Waveguides. *IEEE J. Quantum Electronics,* Vol. 29, No. 4, (Apr. 1993), (1168-1178), ISSN 0018-9197.

Joergensen, C.; Danielsen, S.L.; Stubkjaer, K.E.; Schilling. M.; Daub, K. & Doussiere, P. (1997). All-optical wavelength conversion at bit rates above 10 Gb/s using semiconductor optical amplifiers. *IEEE J. Select. Top. Quantum Electron.,* Vol. 3, No. 5, (Oct. 1997), (1168–1180), ISSN 1077-260X.

Kawaguchi, H. (2005). All-Optical Switching of Picosecond Pulses by Four-Wave Mixing in a Semiconductor Optical Amplifier, In: *IPAP Books 2: Photonics Based on Wavelength Integration and Manipulation,* (271–282), Japan.

Koga, M. & Matsumoto, T. (1991). High-gain polarization-insensitive optical amplifier consisting of two serial semiconductor laser amplifiers. *J. Lightwave Technology,* Vol. 9, No. 3, (Feb. 1991), (284–290), ISSN 0733-8724.

Leuthold, J.; Joyner, C.H.; Mikkelsen, B.; Raybon, G.; Pleumeekers, J.L.; Miller, B.I.; Dreyer, K. & Burrus, C.A. (2000). 100Gbit/s all-optical wavelength conversion with integrated SOA delayed-interference configuration. *Electronics Lett.,* Vol. 36, No. 13, (Jun 2000), (1129–1130), ISSN 0013- 5194.

Liu, Y.; Hill, M.T.; Tangdiongga, E.; de Waardt, H.; Galabretta, N. & Khoe, G.D. (2003). Wavelength conversion using nonlinear polarization rotation in a single semiconductor optical amplifier. *IEEE Photon. Technol. Lett,.* Vol. 15, (2003), (90–92), ISSN 1041-1135.

Liu, Y.; Tangdiongga, E.; Li, Z.; de Waardt, H.; Koonen, A.M.J.; Khoe, G.D.; Xuewen, S.; Bennion, I. & Dorren, H.J.S. (2007). Error-free 320-Gb/s all-optical wavelength conversion using a single semiconductor optical amplifier. *IEEE J. Lightwave Technology,* Vol. 25, No. 1, (Jan. 2007), (103–108), ISSN 0733-8724.

Manning, R.J.; Yang, X; Webb, R.P.; Giller, R.; Garcia Gunning, F.C. & Ellis, A.D. (2006). The "turbo-switch" a novel technique to increase the high-speed response of SOAs for wavelength conversion. *Proceedings of OFC/NFOEC,* paper OWS8, Anaheim, USA, Mar. 2006, Optical Society of America.

Matsumoto, A.; Nishimura, K.; Utaka, K. & Usami, M. (2006). Operational design on high-speed semiconductor optical amplifier with assist light for application to wavelength converters using cross-phase modulation. *IEEE J. Quantum Electronics,* Vol. 42, No. 3, (Mar. 2006), (313-323), ISSN 0018-9197.

Nakamura, S.; Ueno, Y. & Tajima, K. (2001). 168-Gb/s all-optical wavelength conversion with a symmetric Mach-Zehnder-type switch. *IEEE Photon. Technol. Lett.,* Vol. 13, No. 10, (Oct. 2001), (1091–1093), ISSN 1041- 1135.

Politi, C.; Klonidis, D. & O'Mahony, M.J. (2006). Dynamic behavior of wavelength converters based on FWM in SOAs. *IEEE J. Quantum Electronics,* Vol. 42, No. 2, (Feb. 2006), (108-125), ISSN 0018-9197.

Simon, J.C.; Doussiere, P.; Pophillat, L. & Fernier, B. (1989). Gain and noise characteristics of a 1.5 μm near-travelling-wave semiconductor laser amplifier. *J Electronics Letters,* Vol. 25, No. 7, (1989), (434–436), ISSN 0013- 5194.

Tzanakaki, A. & O'Mahony, M.J. (2000). Analysis of tunable wavelength converters based on cross-gain modulation in semiconductor optical amplifiers operating in the counter propagating mode, *Proceedings of IEE Optoelectronics,* Vol. 147, No. 1, pp. 49–55, Feb. 2000, Institution of Electrical Engineers.

Wei, C.C.; Huang, M.F. & Chen, J. (2005). Enhancing the frequency response of cross-polarization wavelength conversion. *IEEE Photon. Technol. Lett.,* Vol. 17, No. 8, (Aug. 2005), (1683-1685), ISSN 1041-1135.

3

A Frequency Domain Systems Theory Perspective for Semiconductor Optical Amplifier - Mach Zehnder Interferometer Circuitry in Routing and Signal Processing Applications

George T. Kanellos[1], Maria Spyropoulou[2], Konstantinos Vyrsokinos[1], Amalia Miliou[2] and Nikos Pleros[2]

[1]*Informatics and Telematics Institute, Center for Research and Technology Hellas*
[2]*Department of Informatics, Aristotle University of Thessaloniki*
Greece

1. Introduction

1.1 SOAs as nonlinear elements in Mach-Zehnder Interferometers

Although SOAs have been initially introduced as integrated modules mainly for optical amplification purposes, they have been widely used in all optical signal processing applications, like all-optical switching and wavelength conversion, utilizing the exhibited nonlinearities such as gain saturation, cross-gain (XGM) and cross-phase modulation (XPM). These nonlinear effects that present the most severe problem and limit the usefulness of SOAs as optical amplifiers in lightwave systems can be proven attractive in optically transparent networks. The origin of the nonlinearities lies in the SOA gain saturation and in its correlation with the phase of the propagating wave, since the carrier density changes induced by the input signals are affecting not only the gain but also the refractive index in the active region of the SOA. The carrier density dynamics within the SOA are very fast (picosecond scale) and thus the gain responds in tune with the fluctuations in the input power on a bit by bit basis even for optical data at 10 or 40 Gb/s bit-rates (Ramaswami & Sivarajan, 2002).

If more than one signal is injected into the SOA, their nonlinear interaction will lead to XPM between the signals. However, in order to take advantage of the XPM phenomenon and create functional devices, the SOAs have to be placed in an interferometric configuration such as a Mach-Zehnder Interferometer (MZI) that converts phase changes in the signals to intensity variations at its output exploiting interference effects. Semiconductor Optical Amplifier-based Mach-Zehnder Interferometers (SOA-MZIs) have been widely used in the past years as all-optical high speed switches for signal conditioning and signal processing, mainly due to their low switching power requirements and their potential for integration (Maxwell, 2006). Using this type of switch, a set of processing operations ranging from demultiplexing (Duelk et al.,1999) to regeneration (Ueno et al., 2001) and wavelength conversion (Nielsen et al. 2003) to optical sampling (Fischer et al. 2001) and optical flip-flops (Hill et al. 2001, Pleros et al. 2009) has been demonstrated, highlighting multi-functionality as an additional advantage of SOA-MZI devices. Within the same frame, SOA-MZI devices have proven very efficient in dealing with packet-formatted optical traffic allowing for their exploitation in several routing/processing demonstrations for optical packet or burst switched applications, performing successfully in challenging and demanding functionalities like packet envelope detection (Stampoulidis et al, 2007), packet clock recovery (Kanellos et al, 2007a), label/payload separation (Ramos et al, 2005), burst-mode reception (Kanellos et al, 2007a, 2007b) and contention resolution (Stampoulidis et al, 2007).A brief description of the most important SOA-MZI signal processing applications and their principle of operation is provided in the following paragraphs.

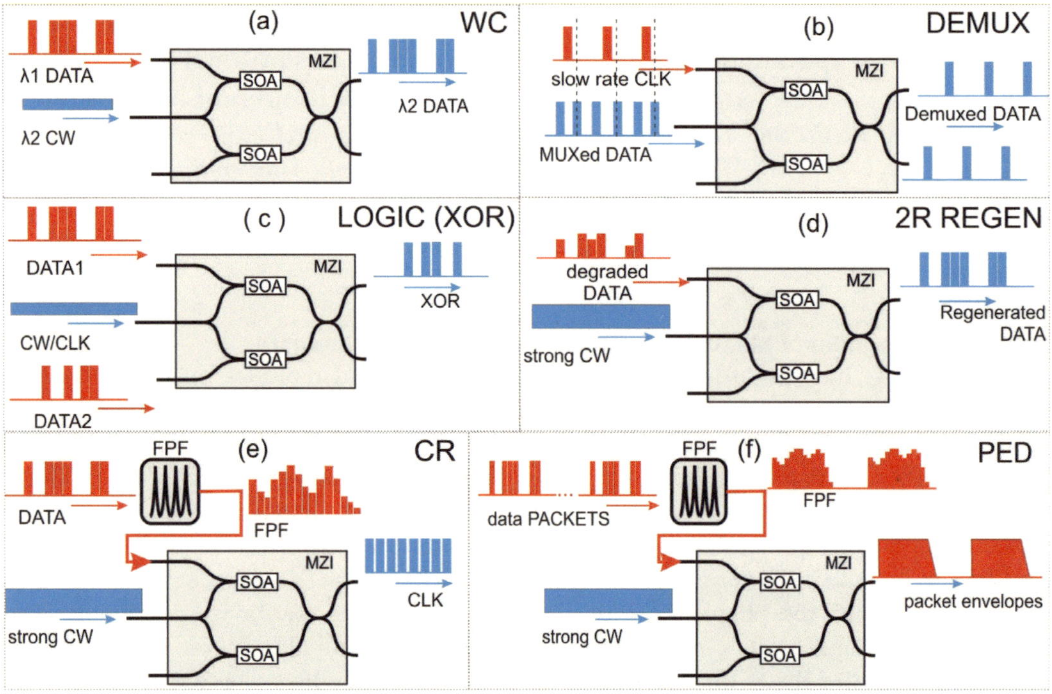

Fig. 1. Single MZI basic functionalities: a) wavelength conversion b) demultiplexing c) Boolean logic (XOR) operation d) 2R regeneration e) Clock recovery (CR) f) Packet envelope detection (PED)

1.1.1 Wavelength converters

An important class of application area for SOA-MZIs is wavelength conversion both for RZ and NRZ data formats, offering also 2R regenerative characteristics to the wavelength converted signal as a result of their nonlinear transfer function. Several schemes were developed during the decade of the 1990s (Durhuus et al. 1994), and many others have been proposed since then ((Stubkjaer, 2000; Wolfson et al. 2000; Leuthold, J. 2001; Nakamura et al. 2001; M. Masanovic et al. 2003; Apostolopoulos et al. 2009a).

Figure 1(a) depicts the standard WC layout employing a single control signal that is inserted into one of the two MZI arms causing a gain and phase variation only to one of the two CW signal components. This configuration offers the advantage of reduced complexity but is liable to result in pulse broadening and significant patterning effects due to the unbalanced gain saturation in the two SOAs, severely limiting the operational speed of the device especially when NRZ pulses are used. In the case of RZ signal formats, unequal SOA gain-induced speed restrictions can be overcome by the well-known "push-pull" architecture that employs two identical control signal entering the two SOAs with a differential time delay. In the case of NRZ data format, these effects are partially compensated in the bidirectional push-pull scheme, which employs two identical control pulses travelling in opposite directions through the two MZI branches, while further improvement is achieved by the recently proposed Differentially-biased NRZ wavelength conversion scheme that provides enhanced 2R regenerative characteristics (Apostolopoulos et al. 2009a,b).

1.1.2 All-optical logic gates

The successful employment of SOA-MZIs in all-optical Boolean logic configurations has been the main reason for referring to SOA-MZIs as the all-optical version of the electronic transistor. All-optical logic gates based on SOA-MZI structures using cross phase modulation have demonstrated several interesting merits, i.e. high extinction ratio, regenerative capability, high speed of operation, and low chirp in addition to low energy requirement and integration capability. Particular attention has been paid to the all-optical XOR gate that forms a key technology for implementing primary systems for binary address and header recognition, binary addition and counting, pattern matching, decision and comparison, generation of pseudorandom binary sequences, encryption and coding. This gate has been demonstrated at 40 Gb/s (Webb et al. 2003) using SOA-MZI schemes. Figure 1 (c) presents the principle of Boolean XOR operation. Moreover, Kim et al. proposed and experimentally demonstrated all-optical multiple logic gates with XOR, NOR, OR, and NAND functions using SOA-MZI structures that enable simultaneous operations of various logic functions with high ER at high speed (Kim et al, 2005).

1.1.3 All- Optical 2R/3R regeneration

All-optical regeneration is employed at the input of an optical node in order to relieve the incoming data traffic from the accumulated signal quality distortions and to restore a high-quality signal directly in the optical domain prior continuing its route through the network. 2R regeneration generic layout comprises a SOA-MZI interferometer configured in wavelength conversion operation and powered with a strong CW signal. The saturated SOAs in combination with the interferometric transfer function of the gate exhibit a highly non-linear step-like response and the configuration operates as a power limiter (Pleros et al. 2004), forcing unequal power level pulses to equalization. SOA-MZIs have been usually utilized as the nonlinear regenerating elements in several optical 2R regeneration experiments up to 40Gb/s (Apostolopoulos et al,2009a). The block diagram of this setup is shown in Fig. 1(d). All-optical 3R regeneration combines the 2R regeneration module, acting as a decision element, with a clock recovery unit, an important subsystem that produces high-quality data-rate clock pulses. The decision element is used for imprinting the incoming data logical information onto the "fresh" clock signal. The clock recovery process has been demonstrated to perform with different length 40Gb/s asynchronous packets, using a low-Q FPF filter with a highly saturated SOA-MZI gate (Kanellos et al. 2007a). The block diagram setup of the clock recovery subsystem is shown in Fig. 1(e).

1.1.4 DEMUX and Add/Drop multiplexer

Demultiplexing and add/drop multiplexing have been among the first applications areas of SOA-MZI devices, successfully providing the required wavelength and/or data-rate adaptation at the node's front-end. The use of SOA-MZI configurations has initially been demonstrated for demultiplexing purposes from 40 to 10 Gbit/s (Duelk et al.,1999). Figure 1(b) presents the experimental set-up used for the demultiplexing operation.

1.1.5 Burst-Mode Receiver

Burst-mode reception (BMR) is a highly challenging yet necessary functionality on the way to optical packet and burst-mode switched networks, as it has to be capable of adapting to and handling arriving packets with different phase alignment and optical power levels, ensuring at the same time successful regeneration at the intermediate network nodes or error-free reception at the end-user terminals. SOA-MZI-based designs have been already presented in several 2R and 3R setups to simplify the BMR circuit design, whereas the interconnection of four cascaded SOA-MZI gates has led also to the first BMR architecture demonstrated at 40Gb/s (Kanellos et al. 2007a). Each one of the four SOA-MZI modules in this BMR setup provides a different functional task, namely wavelength conversion, power level equalization, clock recovery and finally regeneration or reception.

1.1.6 Optical RAM

Buffering and Random Access Memory (RAM) functionality have been the main weakness of photonic technologies compared to electronics, mainly due to the neutral charge of photon particles that impedes them to mimic the storage behavior of electrons. The first all-optical static RAM cell with true random access read/write functionality has been only recently feasible by exploiting a SOA-MZI-based optical flip-flop and two optically controlled SOA-based ON/OFF switches (Pleros et al. 2009), providing a proof-of-principle solution towards high-speed all-optical RAM circuitry. The optical flip-flop serves as the single-bit memory element utilizing the wavelength dimension and the coupling mechanism between the two SOA-MZIs for determining the memory content, whereas the two SOAs operate as XGM ON/OFF switches controlling access to the flip-flop configuration.

1.1.7 Contention resolution

Packet envelope detection is performed by extracting the envelope of the incoming optical packets (Figure 1(f)). All-optical PED has been demonstrated using an integrated SOA-MZI (Stampoulidis et al. 2007). In these experiments the PED circuit consists of a passive filter in combination with an SOA-MZI gate operated as a low-bandwidth 2R regenerator. The PED circuit generates a packet envelope, indicating the presence of a packet at the specific timeslot. The same experimental work has demonstrated contention resolution in the wavelength domain, using the PED signal to wavelength-convert the deflected packets (Stampoulidis et al. 2007).

1.2 Progress in fabrication and integration: technology overview

The large variety of signal processing and routing applications demonstrated during the last decade by means of SOA-MZI-based circuitry has been mainly a result of the remarkable progress in monolithic and hybrid photonic integration. This has allowed for increased integration densities at high-speed channel rates, offering at the same time the potential for

multiple SOA-MZI interconnection even in cascaded stages (Zakynthinos et al.2007, Apostolopoulos et al. 2009b), lower cost, smaller footprints and lower power consumption.
The silica-on-silicon hybrid integration platform developed by the Center for Integrated Photonics (CIP, U.K.) has been proven a technique of great potential, enabling flip-chip bonding of pre-fabricated InP and InGaAsP components, including SOAs and modulators, on silicon boards with low loss waveguides (Maxwell et al. 2006). This technique relies on the design and development of a planar silica waveguide acting as a motherboard, which is capable of hosting active and passive devices, similar to the electronic printed circuit board used in electronics. The active elements of the device are independently developed on precision-machined silicon submounts called "daughterboards." In the case of SOA-MZI development, the daughterboards are designed to host monolithic SOA chips and provide all suitable alignment stops. The daughterboard consists of a double SOA array and is flip chipped onto the motherboard. This platform has led also to the implementation of multielement photonic integrated circuits, paving the way towards true all-optical systems on-chip that can yield both packaging and fiber pig-tailing cost reduction while retaining cost effectiveness through a unified integration platform for a variety of all-optical devices. Toward this milestone, SOA_MZI regenerators integrated on the same chip with bandpass filtering elements have been shown to perform successfully in WDM applications (Maxwell et al. 2006), while high-level system applications have been demonstrated by making use of the first quadruple arrays of hybridly integrated SOA-MZI gates (Stampoulidis et al. 2008).
Monolithic integration has also witnessed significant progress, presenting Photonic Integrated Circuits (PICs) that incorporate several active and passive components, being capable of meeting different performance requirements on a single chip. A butt-joint growth-based integration platform was explored to incorporate both high- and low-confinement active regions in the same device. To this end, monolithically widely tunable all-optical differential SOA-MZI wavelength converter operating at 40Gb/s have been implemented (Lal et al. 2006), whereas the device functionality was extended by incorporating an electrical modulation stage yielding a monolithic Packet Forwarding Chip (PFC). These structures enabled the successful realization of three major high-rate packet switching functions in a single monolithic device, allowing for simultaneous tunability, all-optical wavelength conversion, and optical label encoding. Monolithic integration holds also the record number of more than 200 passive and active elements integrated on the same functional chip, leading to the first 8x8 InP monolithic tunable optical router capable of operating at 40-Gb/s (Nicholes et al. 2010).

1.3 Modeling and Theory approaches so far

Despite the significant progress in SOA-MZI-based applications and the maturity reached in this technology during the last years, its theoretical toolkit is still missing a holistic frequency domain analysis that will be capable of yielding a common basis for the qualitative understanding of all SOA-MZI enabled nonlinear functionalities. Although single SOAs have been extensively investigated using both time- and frequency-domain theoretical methods (Davies, 1995, Mørk et al.1999), SOA-MZI theory has mainly relied on time-domain simulation-based approaches (Melo et al. 2007) employing customized frequency-domain analytical methods only for specific applications (Kanellos et al 2007b). However, it is well-known from system's theory that a unified frequency-domain argumentation of the experimentally proven multifunctional potential of SOA-MZIs can allow for simple analytical procedures in the performance analysis of complex SOA-MZI-based setups, leading also to optimized configurations. The modulation bandwidth enhancement of the SOA-based Delayed Interferometer Signal Converter (DISC) through chirp filtering (Nielsen et al 2004) and its associated subsequent applications are timely examples for the research strength that can be unleashed by a solid theoretical frequency domain model. This chapter aims to introduce a frequency-domain description of the SOA-MZI transfer function providing a solid system's theory platform for the analysis and performance evaluation of SOA-MZI circuitry. The following sections have been structured so as to introduce the mathematical framework required for the SOA-MZI transfer function extraction, to present its application in several SOA-MZI architectural designs and, finally, to use it in the analysis and evaluation of well-known complex SOA-MZI-based devices exploiting well-established system's theory procedures.

2. Theory development

2.1 SOA-MZI all-optical wavelength conversion generic layout

SOA-MZI all optical wavelength conversion is the fundamental operation for the interferometric devices, as it represents either a stand-alone key network application or an essential sub-system functionality to be employed in larger and more complicated processing systems. The principle of operation of the SOA-MZI AOWCs relies on splitting the injected CW input signal into two spatial components that propagate through the two MZI branches and are forced to interfere at the output coupler after experiencing the induced SOA carrier density changes and the associated cross-gain (XGM) and cross-phase modulation (XPM) phenomena imposed by the control signals. In this way, the optical data signal serving as the Control (CTR) signal into the SOA-MZI is imprinted onto the CW input beam, resulting to a replica of the inserted data sequence carried by a new wavelength and emerging at the Switched-port (S-port) of the MZI.

Fig. 2(a-c) illustrates all possible configurations of SOA-MZI-based wavelength converters and their principle of operation, with their classification being determined by the type of signals injected into the SOA-MZI. Fig. 2(a-c) illustrate the operation principle for each corresponding case. Fig. 2(d-f) show the amplitude and phase of the upper (x) and lower (y) cw components with the red solid line and the dotted lines, respectively, after passing through the SOA devices at the impulse of an NRZ pulse.

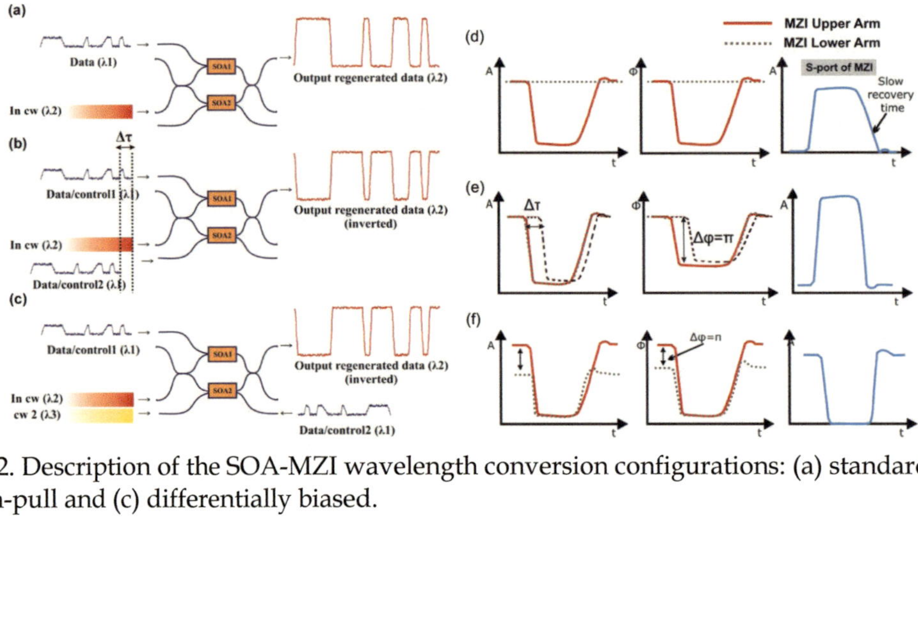

Fig. 2. Description of the SOA-MZI wavelength conversion configurations: (a) standard, (b) push-pull and (c) differentially biased.

To this end, Fig. 2(a) illustrates the simplest configuration called the standard SOA-MZI scheme, employing a strong continuous wave (CW) signal at λ_1 commonly inserted into both SOA-MZI branches after passing through the input 3 dB coupler. A weak data signal at λ_2 is inserted into the upper branch as the control (CTR_x) and is responsible for altering the gain and phase dynamics of the upper branch CW signal by stimulating the cross-gain (XGM) and cross-phase modulation (XPM) phenomena on the upper SOA. Finally, the two CW components traveling along the upper and lower SOA-MZI arms interfere constructively at the output 3 dB coupler, yielding the wavelength converted signal at the switched port. As shown in Fig.2(a), only the amplitude and phase of the upper CW component are affected (red line) due to the carrier modulation imposed by the input control pulse (CTR_x), whereas the amplitude and phase of the lower CW component remain constant (dotted line) and they are only subject to the waveguide propagation loss. The result of the interference between the two output CW components is the wavelength converted output signal shown at the top right corner, suffering though from slow rise and fall times due to the slow gain recovery time of the SOA. The second SOA-MZI wavelength conversion scheme is illustrated in Fig.2(b) and is called the PUSH PULL configuration. In this scheme, the control signal is split through a coupler and the two parts of the signal are fed as the control signals CTR_x and CTR_y of the upper and lower branches of the SOA-MZI after experiencing a differential time delay $\Delta\tau$. In this way, the gains of the two SOAs are suppressed at different time instances separated by Δt, while the optical powers of the controls signals may be adjusted properly in order to induce a differential phase shift close to π rad (Fig.2.b) for the time fraction that the optical pulses of the two control signals are not overlapping in time. The output wavelength converted signal is dominated by this differential phase shift and appears at the switched port of the SOA-MZI. The width of the output optical pulse is strongly dependent on the differential time delay $\Delta\tau$, while the rise and fall times are shorter compared to the standard operation since the symmetrical gain saturation at the upper and lower SOA-MZI arms cancels out the slow recovery time of the SOAs. The switching speed of the push-pull SOA-MZI configuration will increase as $\Delta\tau$ reduces and the switching window shrinks, enabling in this way high operational speeds. However, the requirement for $\Delta\tau$ delay between the two control signals enforces the availability of empty time durations within a single bit-slot, rendering this scheme suitable only for RZ optical pulses.

The differentially biased scheme illustrated in Fig. 2(c) employs again two identical control signals entering the respective MZI arms, enabling again for their individual tuning of their power levels prior injected into the SOAs. However, the two control signals are now coinciding in time, whereas an additional CW′ signal at λ_3 is inserted into the lower branch co-propagating in the respective SOA with the lower cw component. The additional CW′ is responsible for controlling independently the induced differential gain saturation at the upper and the lower SOA-MZI arms so as to yield an exact differential phase shift of π rad between the two CW input signal components at the absence of any control pulse, while exhibiting the same gain modulation when a control pulse is injected. Therefore, almost perfect interference conditions are met generating a high quality wavelength converted output signal. Again, interference conditions yield the wavelength converted output signal inverted at the switched port.

As successful WC relies on producing a replica of the original data signal onto the wavelength of the CW input signal, the spectral content of the original signal has to be transferred to the SOA-MZI output without experiencing any frequency dependent alterations. To this end, the transfer function of an ideal WC device should follow the profile of a low-pass filtering element with a flat frequency response and with its -3dB cut-off frequency denoting the maximum bit-rate that can be supported, as shown in Fig.3. Fig.3 depicts the ideal frequency domain transfer function that should apply to the single SOA-MZI when WC operation is targeted.

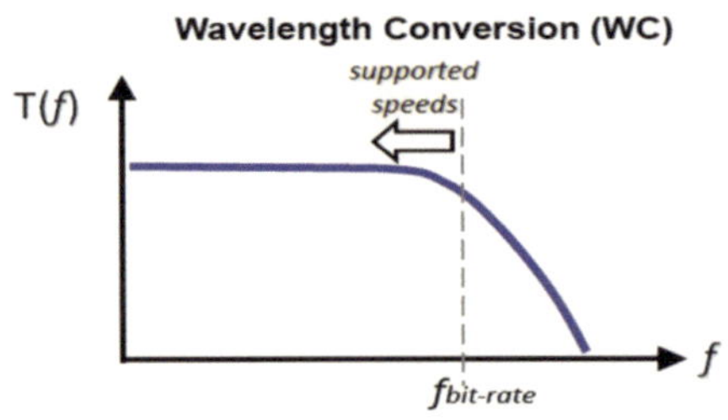

Fig. 3. Ideal frequency domain transfer function for the SOA-MZI based wavelength conversion operation

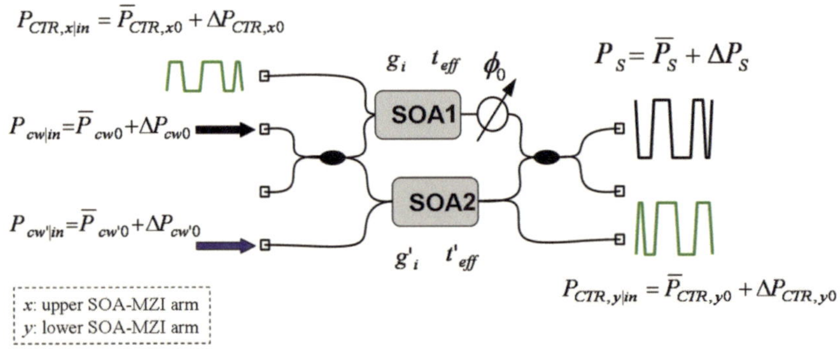

Fig. 4. Generic layout of SOA-MZI all-optical wavelength converter

In order to perform a theoretical frequency domain analysis of the SOA-MZI under wavelength conversion operation, a generic layout of the SOA-MZI AOWC has to be employed that will be capable at expressing all three sub-cases. Fig.4 depicts a generic layout of SOA-MZI AOWC, depicting the input and output signals that may be employed in all schemes. As such, a cw signal with λ_1 wavelength is inserted in the input port of the SOA-MZI while two counter propagating control signals of λ_2 wavelength are considered at the input of upper and lower branches of the SOA-MZI respectively. Finally, an additional cw′ signal of λ_3 wavelength is considered at the input of the lower SOA-MZI branch. Note that each of the possible SOA-MZI AOWC configurations may be retrieved when eliminating the unnecessary inputs. For example, the standard SOA-MZI scheme is retrieved when considering $P_{cw'}=0$ and $P_{CTR,y}=0$, while PUSH-PULL SOA-MZI scheme is retrieved when considering $P_{cw'}=0$. To this end, the derivation of a mathematical expression that describes the frequency response of the SOA-MZI AOWC generic layout will serve in easily extracting the frequency domain transfer function for each of the SOA-MZI schemes, simply by incorporating in the final expression the input signal assumptions described. Finally, one important assumption is that the lower branch control signal is a copy of the upper branch control signal with different power levels and a differential time delay.This indicates that $P_{CTR,y}$ should be expressed as $P_{CTR,y}= mP_{CTR,x} \exp(-j\omega\tau)$.

2.2 Frequency domain analysis theory development

In this section, we theoretically derive an analytical expression for the frequency domain transfer function for the SOA-MZI AOWC generic layout by employing simple approximations for modeling the SOA carrier dynamics based on first order perturbation theory. This will allow to gain qualitative insight in the efficiency of the wavelength conversion functionality versus operational frequency while allowing for the investigation of key device parameters on the wavelength conversion performance such as the material gain and the effective carrier lifetime of the SOAs.

Our approach relies on small-signal analysis and first-order perturbation theory approximations for describing the response of both the SOA devices and the interferometer, following the principles employed in (Davies et al., 1999; Marcenac et al 1997;Nielsen et al 2004). As both the SOA and the MZI configuration are governed by strongly nonlinear coupled equations, small-signal analysis and perturbation theory are employed in order to reduce them into linear expressions so as to enable their subsequent frequency domain treatment towards extracting their transfer function. The employment of first-order perturbation theory tools allows for the description of every time-dependent physical quantity as a sum of a direct current (dc) term and a perturbing term, which in our case is considered to oscillate at a frequency ω. To this end, every time-dependent variable X(t) incorporated in the equations governing the SOA-MZI layout is approximated as

$$X(t)=X_{dc}+\Delta X\cdot e^{j\omega t} \tag{1a}$$

By applying this formalism in a nonlinear equation, separating the dc- from the perturbing-equation terms and neglecting all higher order perturbing terms, two distinct linear equations are obtained: one employing the dc components and one employing all oscillating expressions. This simplifies their solution procedure, leading to linear relationships for the final system's expressions

As shown in Fig. 4, the power of every signal i follows the formalism $P_i = \overline{P_i} + \Delta P_i$, where $\overline{P_i}$ represents a steady state term and ΔP_i represents a small perturbation superimposed on it. The SOA-MZI CW input signal employs only a dc power component $\overline{P}_{cw}$ that is split into two equal parts prior entering the SOAs, so that the power expressions of the CW signal components at the x- and y-branch SOA inputs equal:

$$\overline{P}_{cw,x0} = \overline{P}_{cw,y0} = \overline{P}_{cw} / 2$$

The two spatial CW signal components are forced to interfere at the output 3dB coupler of the SOA-MZI after propagating through the x- and y- MZI arms and the respective SOA1 and SOA2 modules. This interference results to an optical power P_S emerging at the S-port of the MZI that is provided by the well-known expression (Pleros et al 2004):

$$P_s = \frac{1}{4} \cdot \left(P_{cw,x} + P_{cw,y} - 2\sqrt{P_{cw,x} \cdot P_{cw,y}} \cdot \cos\left(\phi_0 + \phi_{cw,x} - \phi_{cw,y}\right)\right) \tag{1b}$$

, where $P_{cw,x}$ and $P_{cw,y}$ stand for the power levels and $\phi_{cw,x}$, $\phi_{cw,y}$ denote the respective phase values of the x- and y-branch CW signal constituents when interfering at the MZI output 3dB coupler. By writing the power and phase quantities again as the sum of a dc and a perturbing signal component at frequency ω, i.e.

$$P_{cw,x} = \overline{P}_{cw,x} + \Delta P_{cw,x}, \tag{2a}$$

$$P_{cw,y} = \overline{P}_{cw,y} + \Delta P_{cw,y}, \tag{2b}$$

$$\phi_{cw,x} = \overline{\phi}_{cw,x} + \Delta\phi_{cw,x}, \tag{2c}$$

$$\phi_{cw,y} = \overline{\phi}_{cw,y} + \Delta\phi_{cw,y} \tag{2d}$$

and expanding equation (1) into a first-order Taylor series around [$\Delta P_{cw,x}$, $\Delta P_{cw,y}$, $\Delta\phi_{cw,x}$, $\Delta\phi_{cw,y}$]=[0,0,0,0], the P_S output power can be expressed as:

$$P_s = \frac{1}{4}\cdot\left\{\begin{aligned} &\overline{P}_{cw,x} + \overline{P}_{cw,y} - 2\cdot\sqrt{\overline{P}_{cw,x}\cdot\overline{P}_{cw,y}}\cdot\cos\left(\phi_0 + \overline{\phi}_{cw,x} - \overline{\phi}_{cw,y}\right) + \\ &+\left[1 - \frac{\overline{P}_{cw,y}}{\sqrt{\overline{P}_{cw,x}\cdot\overline{P}_{cw,y}}}\cdot\cos\left(\phi_0 + \overline{\phi}_{cw,x} - \overline{\phi}_{cw,y}\right)\right]\cdot\Delta P_{cw,x} + \\ &+\left[1 - \frac{\overline{P}_{cw,x}}{\sqrt{\overline{P}_{cw,x}\cdot\overline{P}_{cw,y}}}\cdot\cos\left(\phi_0 + \overline{\phi}_{cw,x} - \overline{\phi}_{cw,y}\right)\right]\cdot\Delta P_{cw,y} + \\ &+2\cdot\sqrt{\overline{P}_{cw,x}\cdot\overline{P}_{cw,y}}\cdot\sin\left(\phi_0 + \overline{\phi}_{cw,x} - \overline{\phi}_{cw,y}\right)\cdot\left(\Delta\phi_{cw,x} - \Delta\phi_{cw,y}\right) \end{aligned}\right\} \tag{3}$$

The steady-state term $\overline{P}_s$ and the perturbing term ΔP_s of the switched output power can now be easily identified as:

$$\overline{P}_s = \frac{1}{4}\cdot\left(\overline{P}_{cw,x} + \overline{P}_{cw,y} - 2\cdot\sqrt{\overline{P}_{cw,x}\cdot\overline{P}_{cw,y}}\cdot\cos\left(\phi_0 + \overline{\phi}_{cw,x} - \overline{\phi}_{cw,y}\right)\right) \tag{4}$$

and

$$\Delta P_s = \frac{1}{4}\cdot\left\{\begin{aligned} &\left[1 - \frac{\overline{P}_{cw,y}}{\sqrt{\overline{P}_{cw,x}\cdot\overline{P}_{cw,y}}}\cdot\cos\left(\phi_0 + \overline{\phi}_{cw,x} - \overline{\phi}_{cw,y}\right)\right]\cdot\Delta P_{cw,x} + \\ &+\left[1 - \frac{\overline{P}_{cw,x}}{\sqrt{\overline{P}_{cw,x}\cdot\overline{P}_{cw,y}}}\cdot\cos\left(\phi_0 + \overline{\phi}_{cw,x} - \overline{\phi}_{cw,y}\right)\right]\cdot\Delta P_{cw,y} + \\ &+2\cdot\sqrt{\overline{P}_{cw,x}\cdot\overline{P}_{cw,y}}\cdot\sin\left(\phi_0 + \overline{\phi}_{cw,x} - \overline{\phi}_{cw,y}\right)\cdot\left(\Delta\phi_{cw,x} - \Delta\phi_{cw,y}\right) \end{aligned}\right\} \tag{5}$$

The theoretical analysis carried out so far applies to any type of MZI interferometers even when nonlinear media different than SOAs are employed in the MZI arms. Moreover, the wavelength conversion efficiency η_{WC} of the SOA-MZI is obtained by the ratio of the acquired over the initial perturbation, given in equation (6) below.

$$\eta_{WC} = \frac{\Delta P_s}{\Delta P_{CTR,x}} \tag{6}$$

Our goal is to express the perturbing component of the output ΔP_s as a linear function of the perturbing component of the input $\Delta P_{CTR.x}$, so as to obtain an expression of the wavelength conversion efficiency η_{WC} that is independent from the input signal. However, the exact solutions of (4) and (5) can only be derived if the nonlinear medium characteristics and their impact onto the propagating signal are taken into consideration in order to allow for the analytic expressions of both the steady-state and the perturbing power and phase terms at the two x- and y- branch CW signal components. This implies that analytic power and phase expressions in the case of the SOA-MZI can be only obtained by taking into account the interaction of the propagating beams with the respective SOA1 and SOA2 devices.
The SOA carrier dynamics are described by a rate equation model that relates the power (7) and the phase (8) of every input signal i with the carrier density N,

$$\frac{\partial P_i}{\partial z} = \left[\Gamma \cdot \alpha_i \cdot \left(N - N_{t,i}\right) - a_{\text{int}}\right] \cdot P_i \tag{7}$$

$$\frac{\partial \phi_i}{\partial z} = -\frac{1}{2} \cdot \alpha_f \cdot \Gamma \cdot \alpha_i \cdot N \tag{8}$$

where Γ is the confinement factor, a_i is the gain parameter at wavelength i, N is the carrier density and $N_{t,i}$ is the carrier density at transparency, a_f is the linewidth enhancement factor and a_{int} is the internal waveguide loss of the semiconductor. Following the same approach as we did for describing the optical signal power, both the phase information of the signals and the carrier density of the SOA device are expressed as the sum of a steady state term and a small varying term, respectively. Separating the steady state from the perturbation terms, in equations (7) and (8), generates a new set of equations (9)-(12) describing the evolution of the steady state terms $\overline{P_i}$ and $\overline{\phi_i}$ and the evolution of the perturbation terms ΔP_i and $\Delta\phi_i$ for the power and phase, along the propagation axis z.

$$\frac{\partial \overline{P_i}}{\partial z} = \left[\Gamma \cdot \alpha_i \cdot \left(\overline{N} - N_{t,i}\right) - a_{\text{int}}\right] \cdot \overline{P_i} \tag{9}$$

$$\frac{\partial \Delta P_i}{\partial z} = \left[\Gamma \cdot \alpha_i \cdot \left(\overline{N} - N_{t,i}\right) - a_{\text{int}}\right] \cdot \Delta P_i + \Gamma \cdot \alpha_i \cdot \Delta n \cdot \overline{P_i} \tag{10}$$

$$\frac{\partial \overline{\phi_i}}{\partial z} = -\frac{1}{2} \cdot \alpha_f \cdot \Gamma \cdot \alpha_i \cdot \left(\overline{N} - N_{t,i}\right) \tag{11}$$

$$\frac{\partial \Delta\phi_i}{\partial z} = -\frac{\alpha_f}{2} \cdot \Gamma \cdot \alpha_i \cdot \Delta n \tag{12}$$

Δn describes the SOA carrier density change as given in (Davies et al 1999)

$$\Delta n = \frac{-\sum_i \left(N - N_{t,i}\right) \cdot \frac{\Delta P_i}{E_{sat,i}}}{1/\tau_c + \sum_i 1/\tau_{s,i} + j \cdot \omega} \tag{13}$$

where $E_{sat,i}$ is the saturation energy of the SOA at wavelength λ_i and ω is the cyclic frequency. The effective carrier lifetime of the SOA denoted by t_{eff} is given by $1/t_{eff} = 1/\tau_c + \sum_i 1/\tau_{s,i}$ where, τ_c is the carrier lifetime, $\tau_{s,i}$ is the contribution of the signal i to the effective carrier lifetime and $\tau_{s,i} = E_{sat,i}/\overline{P}_i$ is the stimulated carrier recombination time.

Analytical solution of equations (9)-(12) yields the power and phase information of the signals at the output of every SOA-MZI branch. Throughout the analysis, it has been assumed that the input perturbation imposed by the control signals $\Delta P_{CTR,x0}$ and $\Delta P_{CTR,y0}$ is responsible for modulating the carrier density of the SOAs and that the input continuous wave signals cw and cw′ have no initial perturbation (i.e. $\Delta P_{cw0} = \Delta P_{cw'0} = 0$). It has been further assumed that, the input cw components traveling the upper and lower SOA-MZI branches are responsible for driving the SOAs into deep saturation and dominate their carrier dynamics (i.e. $\overline{P}_{CTR,x0} = \overline{P}_{CTR,y0} = 0$).

At the upper SOA-MZI branch of the SOA-MZI the $CTR_{,x}$ signal imposes a perturbation on the carrier density of the upper SOA described by equation (14),

$$\Delta n = \frac{-\left(\overline{N} - N_{t,cw,x}\right)\cdot\frac{\Delta P_{cw,x}}{E_{sat}} - \left(\overline{N} - N_{t,CTR,x}\right)\cdot\frac{\Delta P_{CTR,x}}{E_{sat}}}{1/t_{eff} + j\cdot\omega} \tag{14}$$

where the effective carrier lifetime of the upper SOA is given by $1/t_{eff} = 1/\tau_c + 1/\tau_{s,cw,x}$ assuming that the contribution of the control signal to the effective carrier lifetime is zero. The power of the upper cw component is obtained solving equations (9),(10) and its phase is obtained solving equations (11),(12) using (14). Therefore, for the steady state terms and the perturbation terms of the power and phase we get,

$$\overline{P}_{cw,x} = \overline{P}_{cw,x0}\cdot\exp\left[\left(g - a_{\text{int}}\right)\cdot z\right] = \frac{\overline{P}_{cw,0}}{2}\cdot\exp\left[\left(g - a_{\text{int}}\right)\cdot z\right] \tag{15}$$

$$\Delta P_{cw,x} = \Delta P_{CTR,x}\cdot\left[\exp\left(\frac{-g\cdot z}{\tau_{s,cw}\cdot\left(1/t_{eff} + j\cdot\omega\right)}\right) - 1\right] \tag{16}$$

$$\overline{\phi}_{cw,x} = -\frac{\alpha_f}{2}\cdot g\cdot z \tag{17}$$

$$\Delta\phi_{cw,x}(z) = -\frac{\alpha_f}{2}\cdot\frac{\Delta P_{CTR,x}}{\left(g - \alpha_{\text{int}}\right) + \frac{(-g)}{\tau_{s,cw}\cdot\left(1/t_{eff} + j\cdot\omega\right)}}\cdot\frac{(-g)}{E_{sat}\cdot\left(1/t_{eff} + j\cdot\omega\right)}\cdot\exp\left[\frac{-g\cdot z}{\tau_{s,cw}\cdot\left(1/t_{eff} + j\cdot\omega\right)}\right] \tag{18}$$

where, $P_{cw,0}$ is the power of the cw signal in the input of the interferometer (Fig.4), g is the modal gain given by $g_i = \Gamma\cdot\alpha_i\cdot\left(N - N_{t,i}\right)$ (Davies et al 1999), where i is denoting the signal wavelength.

In the same way, the optical power of the CTR_x signal at the output of the SOA is obtained by solving (10), taking into account that $\overline{P}_{CTR,x0} = 0$.

$$\Delta P_{CTR,x} = \Delta P_{CTR,x0} \cdot \exp\left[\left(g - a_{\text{int}}\right) \cdot z\right] \tag{19}$$

where $\Delta P_{CTR,x0}$ is the initial perturbation imposed on the CTR_x signal.
At the lower SOA-MZI branch, the additional cw' signal employed is responsible for driving the lower SOA device into a slightly different saturation level than the upper SOA and thus the respective gain parameters are no longer equal (i.e. $g' \neq g$). Furthermore, the SOA dimensions are assumed to be infinitely small and the direction of propagation is not taken into account in this model. Finally, the effective carrier lifetime of the lower SOA device comprises also by the additional contribution of the CW' signal ($1/t'_{eff} = 1/\tau_c + 1/\tau_{s,cw,y} + 1/\tau_{s,cw'}$). At the lower SOA-MZI branch, the CW' signal is additionally contributing to the carrier density change of the lower SOA, now given by equation (20).

$$\Delta n' = \frac{-\left(\overline{N}_0 - N_{t,cw,y}\right) \cdot \dfrac{\Delta P_{cw,y}}{E_{sat}} - \left(\overline{N}_0 - N_{t,cw',y}\right) \cdot \dfrac{\Delta P_{cw',y}}{E_{sat}} - \left(\overline{N}_0 - N_{t,CTR,y}\right) \cdot \dfrac{\Delta P_{CTR,y}}{E_{sat}}}{1/t'_{eff} + j \cdot \omega} \tag{20}$$

The steady state power and phase evolution of the $CW_{,y}$ component are given by equations (21) and (22).

$$\overline{P}_{cw,y} = \overline{P}_{cw,y0} \cdot \exp\left[\left(g' - a_{\text{int}}\right) \cdot z\right] = \frac{\overline{P}_{cw,0}}{2} \cdot \exp\left[\left(g' - a_{\text{int}}\right) \cdot z\right] \tag{21}$$

$$\overline{P}_{cw,'y} = \overline{P}_{cw',y0} \cdot \exp\left[\left(g' - a_{\text{int}}\right) \cdot z\right]$$

$$\overline{\phi}_{cw,y} = -\frac{\alpha_f}{2} \cdot g' \cdot z \tag{22}$$

$$\overline{\phi}_{cw',y} = -\frac{\alpha_f}{2} \cdot g' \cdot z$$

The respective power perturbation of the $CW_{,y}$ signal is described by equation (23), taking into account equation (24), and it is derived following the same analysis.

$$\Delta P_{cw',y} = \frac{\Delta P_{CTR,y} \cdot \overline{P}_{cw'.y}}{\overline{P}_{cw,y} + \overline{P}_{cw',y}} \cdot \left\{ \exp\left[\frac{-g' \cdot z \cdot \left(\overline{P}_{cw,y} + \overline{P}_{cw',y}\right)}{E_{sat} \cdot \left(1/t'_{eff} + j \cdot \omega\right)}\right] - 1 \right\} \tag{23}$$

$$\begin{gathered} \Delta P_{cw,y} = \Delta P_{CTR,y} \cdot \left\{ \exp\left[\frac{-g' \cdot z \cdot \left(\overline{P}_{cw,y} + \overline{P}_{cw',y}\right)}{E_{sat} \cdot \left(1/t'_{eff} + j \cdot \omega\right)}\right] - 1 \right\} - \Delta P_{cw',y} \rightarrow \\ \Delta P_{cw,y} = \Delta P_{CTR,y} \cdot \left(\exp\left[\frac{-g' \cdot z \cdot \left(\overline{P}_{cw,y} + \overline{P}_{cw',y}\right)}{E_{sat} \cdot \left(1/t'_{eff} + j \cdot \omega\right)}\right] - 1 \right) \left(1 - \frac{\overline{P}_{cw',y}}{\left(\overline{P}_{cw,y} + \overline{P}_{cw',y}\right)} \right) \end{gathered} \tag{24}$$

The spatial evolution of the frequency domain phase variation of the $CW_{,y}$ signal is given

$$\Delta\phi_{cw,y}(z) = -\frac{\alpha_f}{2}\cdot\frac{\Delta P_{CTR,y}}{(g'-\alpha_{\text{int}}) + \dfrac{(-g')\cdot(\bar{P}_{cw,y}+\bar{P}_{cw',y})}{E_{sat}\cdot(1/t'_{eff}+j\cdot\omega)}}\cdot\frac{(-g')}{E_{sat}\cdot(1/t'_{eff}+j\cdot\omega)}\cdot\exp\left[\frac{-g'\cdot z\cdot(\bar{P}_{cw,y}+\bar{P}_{cw',y})}{E_{sat}\cdot(1/t'_{eff}+j\cdot\omega)}\right] \quad (25)$$

To this end, the wavelength conversion efficiency of the differentially biased SOA-MZI scheme can be analytically derived by substituting (15)-(19) and (21)-(25) in equations (5) and (6).

$$\eta_{WC} = \frac{1}{4}\cdot$$

$$\left|\begin{array}{l} \left(1-\sqrt{\dfrac{\exp(g'-a_{\text{int}})z}{\exp(g-a_{\text{int}})z}}\cos(\phi_o-\dfrac{a_f}{2}(g'-g))\right)\left(\exp\left(\dfrac{-g\cdot z}{\tau_{s,cw}\cdot(1/t_{eff}+j\cdot\omega)}\right)-1\right)+ \\ \left(m\exp(j\omega T)\left(1-\sqrt{\dfrac{\exp(g-a_{\text{int}})z}{\exp(g'-a_{\text{int}})z}}\cos(\phi_o-\dfrac{a_f}{2}(g'-g))\right)\cdot\left(\exp\left[\dfrac{-g'\cdot z\cdot(\bar{P}_{cw,y}+\bar{P}_{cw',y})}{E_{sat}\cdot(1/t'_{eff}+j\cdot\omega)}\right]-1\right)\left(1-\dfrac{\bar{P}_{cw',y}}{(\bar{P}_{cw,y}+\bar{P}_{cw',y})}\right)\right)- \\ \dfrac{a_f\cdot(-g)}{\tau_{S,CW}}\sqrt{\dfrac{\exp(g'-a_{\text{int}})z}{\exp(g-a_{\text{int}})z}}\sin(\phi_0-\dfrac{a_f}{2}(g'-g))\cdot \\ \left(\dfrac{\exp\left[\dfrac{-g\cdot z}{\tau_{s,cw}\cdot(1/t_{eff}+j\cdot\omega)}\right]}{(1/t_{eff}+j\cdot\omega)\left((g-\alpha_{\text{int}})+\dfrac{(-g)}{\tau_{s,cw}\cdot(1/t_{eff}+j\cdot\omega)}\right)}-\dfrac{m\exp(j\omega T)\cdot\exp\left[\dfrac{-g'\cdot z\cdot(\bar{P}_{cw,y}+\bar{P}_{cw',y})}{E_{sat}\cdot(1/t'_{eff}+j\cdot\omega)}\right]}{\left((g'-\alpha_{\text{int}})+\dfrac{(-g')\cdot(\bar{P}_{cw,y}+\bar{P}_{cw',y})}{E_{sat}\cdot(1/t'_{eff}+j\cdot\omega)}\right)(1/t'_{eff}+j\cdot\omega)}\right) \end{array}\right| \quad (26)$$

3. SOA-MZI AOWC frequency domain transfer function analysis

3.1 Standard SOA-based MZI scheme

Standard SOA-based MZI scheme is the simplest configuration for performing wavelength conversion. This scheme can be modeled by equation (26) if we consider that at the lower SOA-MZI arm there is no input perturbation, i..e $P_{CTR,y} = 0$ and $\Delta P_{CTR,y} = 0$. This is represented in (26) by setting m=0. Moreover, there is no second CW signal entering the lower SOA-MZI arm, leading to $P_{CW',y}=0$. Thus, the modal gain g' of the lower SOA equals g since the upper and lower SOAs are driven into the same saturation level (i.e. $\bar{P}_{cw,x0} = \bar{P}_{cw,y0} = \bar{P}_{cw0}/2$). By setting in (26) $m=0$, $g=g'$ and $P_{CW',y}=0$ we derive the analytical expression of the wavelength conversion efficiency for the standard SOA-MZI scheme given by

$$\eta_{STD}(z,\omega) = \frac{1}{4}\cdot\left|(1-\cos(\phi_0))\cdot\left\{\exp\left[\frac{-g\cdot z}{\tau_{s,cw}\cdot(1/t_{eff}+j\cdot\omega)}\right]-1\right\}-\frac{\alpha_f\cdot\sin(\phi_0)(-g)\exp\left[\dfrac{-g\cdot z}{\tau_{s,cw}\cdot(1/t_{eff}+j\cdot\omega)}\right]}{\tau_{s,cw}\cdot(1/t_{eff}+j\cdot\omega)\left((g-\alpha_{\text{int}})+\dfrac{(-g)}{\tau_{s,cw}\cdot(1/t_{eff}+j\cdot\omega)}\right)}\right| \quad (27)$$

Equation (27) provides an analytical expression for the frequency-domain transfer function of the SOA-MZI configuration. Fig. 5 depicts the respective normalized S-port conversion efficiency versus frequency for various values of the *gz* factor assuming that t_{eff} is 50psec, *z*=1mm and τ_c=200psec. As shown in Fig.5(a), three different operational regions with respect to the total gain *gz* can be identified for the MZI response. For small *gz* values like 1 and 3, the MZI frequency response resembles that of a low-pass filtering element with a rather low 3-dB bandwidth that increases with *gz*. When *gz* continues to increase reaching *gz*=5, the MZI response extends to higher frequencies but exhibits a resonance peak around 10GHz forming a transfer function shape that is more similar to that of a band-pass filter centered around this resonance, as shown experimentally in (Yan et al 2009). As *gz* exceeds the value of 5 approaching higher total gain values, this resonance peak continues to exist

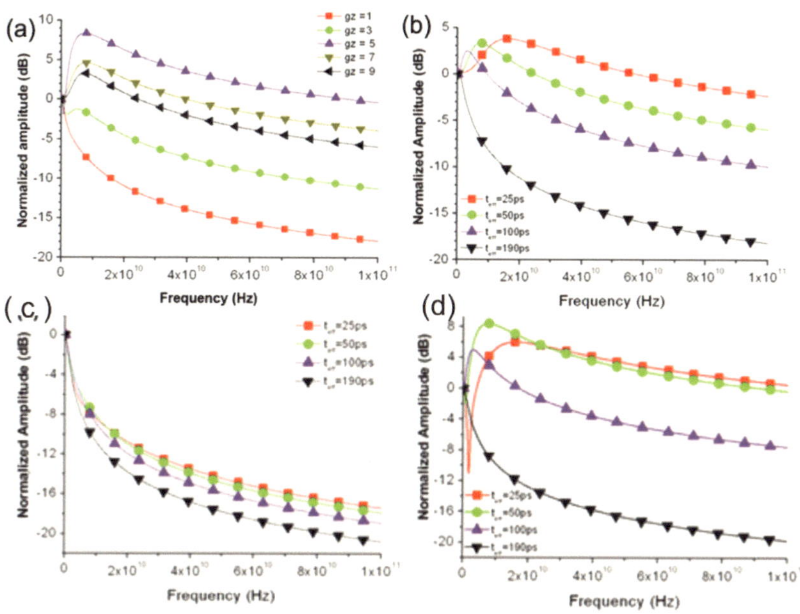

Fig. 5. a) The SOA-MZI frequency response for different total gain values and for an effective SOA gain recovery time of 50psec, (b,c,d) SOA-MZI response with high SOA gain for four different t_{eff} values (25ps,50ps, 100ps, 190ps) (b) gz=9, (c) *gz*=1, (d) *gz*=5.

becoming, however, significantly weaker as *gz* increases. This indicates that the MZI response tends to develop a smoother frequency dependence for frequencies up to 25 GHz with spectral amplitude deviations residing within a 3dB limit. The existence of a resonance peak with varying amplitude depending on the SOA gain conditions has been presented experimentally in (Cao et al 2002).

Fig. 5(b) depicts the SOA-MZI response when high SOA gain operation is utilized, again for the four different t_{eff} values. A low recovery time of 190psec yields just a low-pass filtering characteristic curve with small bandwidth values, but when t_{eff} decreases below 100psec the bandwidth is significantly enhanced extending beyond 40GHz for t_{eff}=25psec. At the same time, the MZI response continues to exhibit a resonance peak around a certain frequency that shifts to higher values for decreasing t_{eff}, however its effect is significantly weaker than in the case of moderate SOA gain levels, having an amplitude that is only 3dB higher compared to the amplitudes of the lower frequency components. This suggests that increased gain values yield a smoother frequency dependence with higher supporting bandwidths, indicating this region as the operational regime in WC applications.

Optimization of the SOA-MZI operational parameters with respect to each separate application can be achieved by proper control of the SOA gain recovery dynamics. The dependence of the SOA-MZI frequency response on the effective SOA carrier recovery time for all three different operational regimes is depicted in Fig.5 (a). Fig. 5(c) shows the transfer function for a gz=1 value and for four different t_{eff} values ranging between 25psec and 190psec. A typical low-pass filter transfer function can be observed with a 3-dB bandwidth being always below 1GHz and decreasing as t_{eff} increases.

Figure 5(d) illustrates the four graphs of the SOA-MZI response obtained for the respective t_{eff} values of 25, 50, 100 and 190 psec and for a SOA gain level of gz=5. In the case of 190 psec recovery times the transfer function is again a low-pass filtering curve, however its shape changes dramatically when t_{eff} values lower than 100psec are used: the MZI response exhibits a resonance peak around a certain frequency that shifts to higher frequency values as t_{eff} is decreasing. This resonance implies that the SOA-MZI circuit starts to behave like a band-pass filtering device suppressing the power of the frequency components lying around the central resonance. This effect is even more pronounced when t_{eff}=25psec, where the spectral content between 0 and 8 GHz is suppressed by more than 3dB compared to the approximately 12GHz resonant frequency, reaching in some spectral areas around the 3GHz region also suppression values up to 15dB.

3.2 Differential time delay ("Push-Pull") configuration

The differential delay configuration relies on the principle that the output modulated signal is the portion of the initial CW that falls within the switching window induced by the non-perfect overlapping of the two control signals. Moreover, the broadness of the switching window is controlled by differential time delay imposed between the two control signals. To this end, the switching window can be set to be very thin, dramatically increasing the capacity for very high frequencies wavelength conversion operations at the expense of reduced noise figure. However, numerous experimental setups have demonstrated successful operation for high bit rate WC employing the differential delay scheme, including 40 Gb/s wavelength conversion (Wolfson et al. 2000, Leuthold et al. 2001, Nakamura 2001).

Frequency domain analysis allows for a thorough investigation of the WC operation in order to reveal its limitations and suggest a systematic optimization procedure. The wavelength conversion efficiency for differential time delay scheme is analytically described by equation (26) by considering the modal gains of the upper (g) and lower (g′) SOAs equal (g=g′), as the absence of a second CW signal entering the lower SOA-MZI branch ($P_{CW,y=0}$) allows SOA carrier density modulation imposed by the CTRy data signal to be copied to the amplitude and phase of the lower cw component. Hence, equal power level of control signals at the upper and lower SOA-MZI arms induce the same gain saturation. Moreover, initial perturbations imposed by the control signals are related through the parameter m taking values within the range (0,1]. Differential time delay between the two SOA-MZI branches is expressed through the perturbation $\Delta P_{CTR,y}$ in the lower branch of the interferometer, as the time delay with respect to the perturbation $\Delta P_{CTR,x}$ of the upper branch is expressed as $\Delta P_{CTR,y0} = m \cdot \Delta P_{CTR,x0} \exp\left[j\omega \Delta T \right]$, where ΔT is the differential time delay.

Substituting these considerations in (26) leads in the following expression for the wavelength conversion efficiency of the differential delay scheme :

$$\eta_{DELAY}(z,\omega)=\frac{1}{4}\left| \begin{array}{l} (m\cdot\exp[j\omega\Delta T]+1)(1-\cos(\varphi_0)\cdot\left\{\exp\left[\frac{-g\cdot z}{\tau_{s,cw}\cdot\left(1/t_{eff}+j\cdot\omega\right)}\right]-1\right\}- \\ a_f\sin(\varphi_0)\cdot(1-m\cdot\exp[j\omega\Delta T])\cdot\dfrac{(-g)\exp\left[\frac{-g\cdot z}{\tau_{s,cw}\cdot\left(1/t_{eff}+j\cdot\omega\right)}\right]}{\tau_{s,cw}\cdot\left(1/t_{eff}+j\cdot\omega\right)\left((g-\alpha_{int})+\frac{(-g)}{\tau_{s,cw}\cdot\left(1/t_{eff}+j\cdot\omega\right)}\right)} \end{array}\right| \qquad (28)$$

Based on (28), the first row of graphs in Figure 6 (a,b,c) depict the normalized frequency response for the differentially biased SOA-MZI AOWC for different time delay values, namely $\Delta\tau$=1ps (a), $\Delta\tau$=20ps (a) and $\Delta\tau$=50ps (c). The t_{eff} for all graphs is 100 ps while the five lines on each graph correspond to different values of modal gain (gz=1,3,5,7,9). Finally, for Fig. 6 (a,b,c) the frequency response is optimized versus the power ratio m by means of flattening the transfer function. The graphs clearly show that the differential delay scheme can significantly broaden the dyanmic range of wavelength conversion operation to very high frequencies even when employing slow SOAs (t_{eff} = 100 ps), as the flat area of the frequency response exceeds the 10 GHz in all graphs. Especially in Fig.6(a) that corresponds to $\Delta\tau$=1ps, an effective bandwidth of at least 100 GHz is revealed for successful wavelength conversion. In fact, the cut-off frequency is inversely proportional to the differential time delay $\Delta\tau$ employed, as this is limited by periodic zero points of the transfer function at frequency values equal to $1/\Delta\tau$.

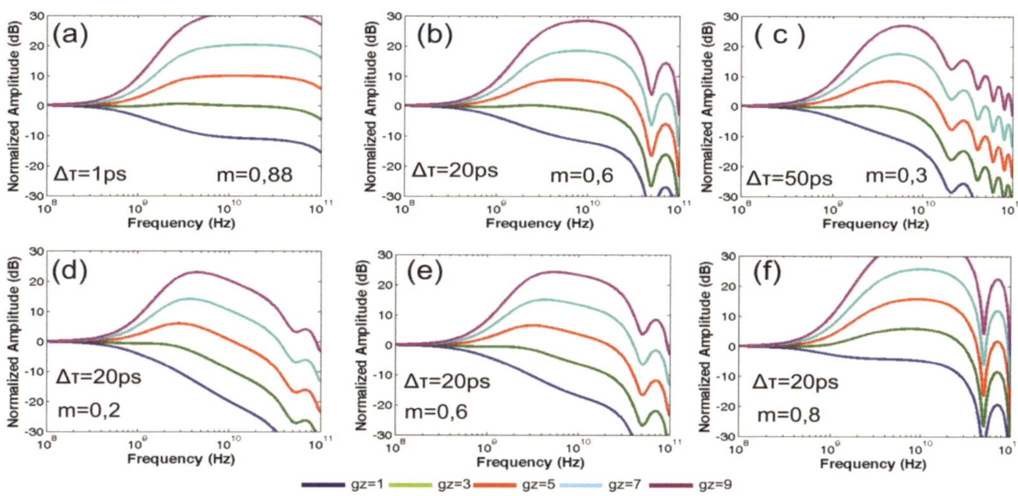

Fig. 6. (a,b,c) normalized frequency response for the differentially biased SOA-MZI AOWC for different time delay values (1ps,20ps,50ps),(d,e,f) normalized frequency response for $\Delta\tau$=20ps and different contorl signal power ratio values.

Frequency response optimization of the differential time delay scheme for a given Δτ is strongly dependent on the parameters gz (gain) and m (control signals power ratio). For the gain parameter, lower values of gain are more tolerant to control signals power ratio variations, meaning that they allow a flat frequency response for a wider range of m and for that reason they are preferable. On the other hand, control signals power ratio m is critical in achieving a flat frequency response, as shown by the graphs in Fig. 6(d,e,b,f).

These graphs depict the frequency response for a differential time delay Δτ=20ps, for values of m equal to m=0,2, m=0,4, m=0,6 and m=0,8 respectively. Higher values of m tend to lift the higher frequencies response while the part until 1GHz remains unaffected. The analysis also shows that higher differential time dealy values require smaller values of m in order to achieve optimum performance. This is evident if we consider the physical mechanism of the differential time delay scheme: as the lower control signal is time delayed compared to the upper control signal, its power has to decrease in order to fit the exponential recovery of the SOA and cancel it out.

3.3 Differentially biased SOA-based MZI scheme

The additional CW′ signal employed in the differentially biased SOA-MZI scheme is responsible for driving the lower SOA device into a slightly different saturation level than upper SOA and thus the respective gain parameters are no longer equal (i.e. $g' \neq g$). Moreover, the power levels of the continuous wave signals are such that a differential phase shift of π rad between the two SOA-MZI branches is achieved, yielding $g' = g - 2\pi / \alpha_f$. The carrier dynamics of the upper SOA device is dominated by the same effective carrier lifetime with the one given for the standard and bidirectional schemes ($1/t_{eff} = 1/\tau_c + 1/\tau_{s,cw,x}$), whereas the effective carrier lifetime of the lower SOA device comprises also by the additional contribution of the cw' signal ($1/t'_{eff} = 1/\tau_c + 1/\tau_{s,cw,y} + 1/\tau_{s,cw'}$). The wavelength conversion efficiency of the differentially biased SOA-MZI scheme is expressed as

$$\eta_{DIF} = \frac{1}{4} \cdot$$

$$\left| \begin{array}{l} \left(1-\sqrt{\frac{\exp(g'-a_{\text{int}})z}{\exp(g-a_{\text{int}})z}}\cos(\varphi_o-\frac{a_f}{2}(g'-g))\right)\left(\exp\left(\frac{-g\cdot z}{\tau_{s,cw}\cdot\left(1/t_{eff}+j\cdot\omega\right)}\right)-1\right)+ \\ \left(m\left(1-\sqrt{\frac{\exp(g-a_{\text{int}})z}{\exp(g'-a_{\text{int}})z}}\cos(\varphi_o-\frac{a_f}{2}(g'-g))\right)\cdot\left(\exp\left[\frac{-g'\cdot z\cdot\left(\bar{P}_{cw,y}+\bar{P}_{cw',y}\right)}{E_{sat}\cdot\left(1/t'_{eff}+j\cdot\omega\right)}\right]-1\right)\left(1-\frac{\bar{P}_{cw',y}}{\left(\bar{P}_{cw,y}+\bar{P}_{cw',y}\right)}\right)\right)- \\ \frac{a_f\cdot(-g)}{\tau_{S,CW}}\sqrt{\frac{\exp(g'-a_{\text{int}})z}{\exp(g-a_{\text{int}})z}}\sin(\varphi_0-\frac{a_f}{2}(g'-g))\cdot \\ \left(\frac{\exp\left[\frac{-g\cdot z}{\tau_{s,cw}\cdot\left(1/t_{eff}+j\cdot\omega\right)}\right]}{\left(1/t_{eff}+j\cdot\omega\right)\left(\left(g-\alpha_{\text{int}}\right)+\frac{(-g)}{\tau_{s,cw}\cdot\left(1/t_{eff}+j\cdot\omega\right)}\right)}-\frac{m\cdot\exp\left[\frac{-g'\cdot z\cdot\left(\bar{P}_{cw,y}+\bar{P}_{cw',y}\right)}{E_{sat}\cdot\left(1/t'_{eff}+j\cdot\omega\right)}\right]}{\left(\left(g'-\alpha_{\text{int}}\right)+\frac{(-g')\cdot\left(\bar{P}_{cw,y}+\bar{P}_{cw',y}\right)}{E_{sat}\cdot\left(1/t'_{eff}+j\cdot\omega\right)}\right)\left(1/t'_{eff}+j\cdot\omega\right)}\right) \end{array} \right| \quad (29)$$

The normalized frequency response for the differentially biased SOA-MZI AOWC, depicted in Fig. 7(a) for t_{eff} 25 ps is flat at low frequencies and starts to decrease above 20 GHz, exhibiting low-pass filter characteristics and resembling the ideal transfer function for WC as described in Fig. 3. In this case, the 3 dB cut-off frequency indicating the speed capabilities for efficient wavelength conversion operation, reaches several tenths of GHz when the gain value increases up to 9. For longer effective carrier lifetime (t_{eff} = 100 ps), the frequency response curves for the differentially biased SOA-MZI schemes shifts towards lower frequencies, limiting the available wavelength conversion bandwidth below 2-3 GHz, as illustrated in Fig. 7(b).

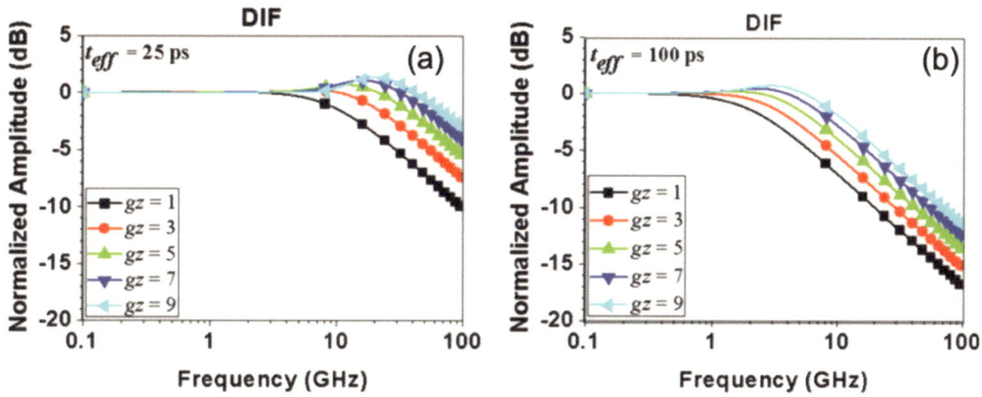

Fig. 7. Normalized frequency response for the differentially biased SOA-MZI AOWC, (a) for t_{eff} 25 ps (b) for t_{eff} 100 ps

3.4 Origin of the frequency domain behavior: XPM/XGM

In order to gain insight into the physical mechanisms dominating the carrier dynamics of the SOA-MZI AOWC and to unveil the key role of the $\Delta P_{CTR,y}$ and *cw'* signals, we treat the generic frequency domain transfer function given by equation (29) as a sum of two terms and we investigate their individual contributions to the overall AOWC frequency response. The first term is governed by the amplitude modulation (*AM*) $\Delta P_{cw,x}$ and $\Delta P_{cw,y}$ acquired by the two interfering cw components at the output of the upper and lower SOA-MZI branch whilst, the second term is governed by the difference of their acquired phase modulation (*PM*) $\left(\Delta\phi_{cw,x} - \Delta\phi_{cw,y}\right)$, respectively, given by equations (30) and (31).

$$\eta_{AM} = \frac{1}{4 \cdot \Delta P_{CTR,x}} \cdot \left\{ A(z,\omega) \cdot \Delta P_{cw,x} + B(z,\omega) \cdot \Delta P_{cw,y} \right\} \tag{30}$$

$$\eta_{PM} = \frac{1}{4 \cdot \Delta P_{CTR,x}} \cdot C(z,\omega) \cdot \left(\Delta\phi_{cw,x} - \Delta\phi_{cw,y} \right) \tag{31}$$

Fig. 8(a) illustrates the individual frequency domain responses of η_{AM} and η_{PM} on the left and middle graphs, and the normalized with respect to the total SOA-MZI AOWC transfer function, for three different values of *m* and three different states of the induced differential phase shift Δφ between the two SOA-MZI branches. On the right, the normalized total

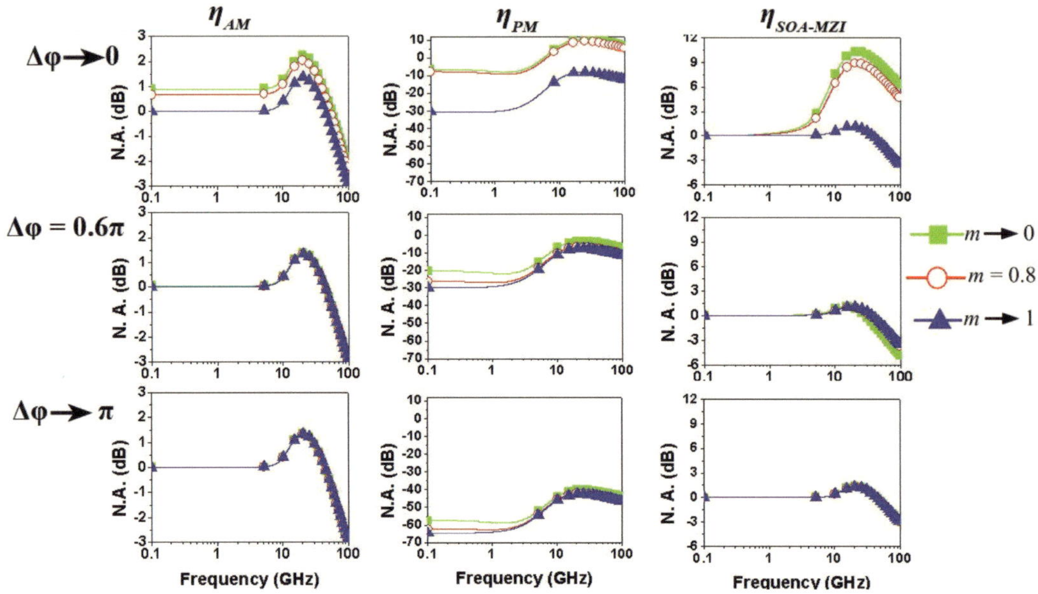

Fig. 8. Normalized amplitude (N.A.) frequency response of η_{AM} (left) and η_{PM} (middle) with respect to the total transfer function (right) for various values of *m* and the differential phase shift Δφ. Results shown for SOA effective carrier lifetime 25 ps and gain 9.

frequency response is illustrated to allow for direct comparison. When both Δφ and *m* are close to 0, the total transfer function resembles a bandpass frequency response that is mainly attributed to the contribution of the η_{PM} term, corresponding to the standard SOA-MZI AOWC operation. At the same time, increasing *m*, which is equivalent to introducing the control signal $\Delta P_{CTR,y}$ at the lower SOA-MZI branch, suppresses η_{PM} by 20 dB and yields an almost flat total response dominated by the η_{AM} term. In the case that the additional CW' signal is introduced yielding a differential phase shift of 0.6π, the η_{PM} contribution is still suppressed for all *m* values, and the total response also follows the characteristic shape of η_{AM}. A strong control signal ($m \rightarrow 1$) at the lower SOA-MZI branch is required in order to shift the total transfer function towards higher frequencies and thus to increase the AOWC operation speed. Finally, when Δφ is very close to π rad corresponding to the differentially biased SOA-MZI AOWC scheme, the SOA-MZI is fully balanced for all *m* values yielding optimum interference of the two cw components.

3.5 Theoretical assessment of SOA-MZI-based System Applications: PED and CR

The theoretical analysis presented in this chapter may expand in order to create a theoretical tool that is able to be applied in the various multi-functional applications whose architecture is based on SOA-MZIs. To this end, this section presents two examples that will assess a theoretical performance analysis in critical SOA-MZI based applications, such as the FPF-assisted SOA-MZI circuit in both Packet envelope detection and and Clock Recovery operations. Fig.9 illustrates the SOA-MZI configurations when performing in two different signal processing functionalities: Fig.9(a) and 9(b) depict a SOA-MZI assisted by a FPF and operating as PED (Stampoulidis et al 2007) and as CR circuit (Kanellos et al 2007), respectively. In both system architectures, the SOA-MZI comprises two SOAs, with SOA1 located at the upper and SOA2 at the lower MZI arm, and is powered by a CW optical beam used as the input signal.

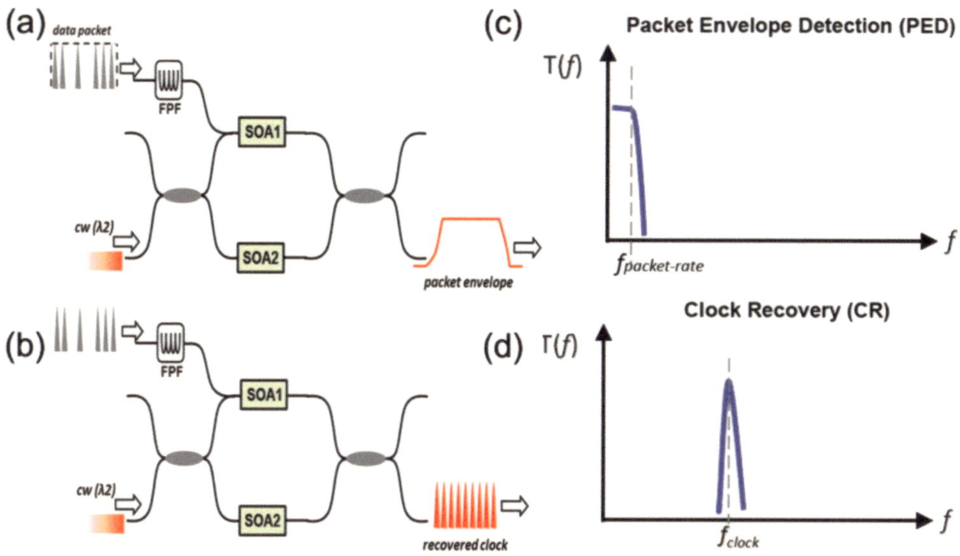

Fig. 9. (a,b) PED and CR setup employing a AOWC MZI and FPF filter, (c,d) ideal transfer functions for PED and CR operation.

When operating as PED or CR circuit, shown in Fig. 9(a) and 9(b), respectively, the optical data signal travels through a low-Finesse FPF prior entering SOA1 as the CTR signal, with the FPF having a Free-Spectral Range (FSR) equal to the data line-rate. As a result, the light signal obtained at the FPF output resembles a strongly amplitude modulated clock pulse signal that has a slightly enhanced time duration with respect to the original input. When PED functionality is targeted in the case of a packet-formatted input data signal, SOA-MZI delivers a single optical pulse at its S-port, with the pulse duration being similar to the duration of the entire original packet. In this application, line-rate information is completely ignored at the MZI output and only the packet-rate information can be identified at the output signal. On the other hand, when CR functionality is performed, the same SOA-MZI configuration yields a power-equalized clock pulse signal at its S-port, almost entirely removing the amplitude modulation of the clock resembling signal being inserted as the CTR signal. In this example, the data information originally carried by the input optical sequence has vanished at the SOA-MZI output and only the clock-rate information is retrieved at the S-port.

This multi-functional time-domain performance of the SOA-MZI-based configurations is inevitably correlated with different frequency domain characteristics for each of the system applications. Fig.9(c,d) depict the ideal frequency domain transfer function that should apply to the single SOA-MZI when PED or CR functions are performed by combining the preceded SOA-MZI WC operation with the subsystem of FPF. In the case of PED functionality shown in Fig. 9(d), all the data- and clock- information carrying frequency components have to be eliminated at the system's output and only the spectral content ranging from dc to the first packet-rate harmonic has to remain unaffected. This suggests that the ideal PED transfer function has again to follow the characteristic curve of a low-pass filter but with a -3dB cut-off frequency not exceeding the value of the second packet-rate harmonic employed in the signal. This can be easily optimized by correctly tuning the SOAs gain, according to the analysis of Figure 5.

Finally, when CR functionality is intended, the complete spectral content of the original data signal except the clock-rate frequency harmonics has to be ignored. This indicates that the ideal CR circuit transfer function has to suppress the entire data spectral content and to exhibit periodic pass-bands only at the multiple clock-rate harmonics or at least at the first clock harmonic, as shown in Fig. 9(d).

PED and CR complete systems transfer function can be easily calculated by multiplying the theoretically obtained SOA-MZI frequency domain transfer function provided by equation (27) with the well known FPF's comb-like transfer function:

$$H_{PED/CR}(\omega) = H_{FPF}(\omega) \cdot H_{SOA-MZI}(\omega)$$

Fig. 10(a) depicts the FPF and the SOA-MZI response curves when gz=1 and t_{eff}=190psec, whereas Fig. 10(b) shows the respective graphs when gz=5 and t_{eff}=25psec. The total transfer function of the FPF-assisted SOA-MZI subsystem obtained through multiplication of the corresponding transfer functions of Fig. 10(a) and 10(b) are shown in Fig. 10(c) and 10(d), respectively. In order to be consistent with the experimental setups reported in (Kanellos et al, 2007), (Satmpoulidis et al 2007), in our theoretical analysis we have considered a 10 Gb/s 2^7-1 Pseudo-Random Bit Sequence (PRBS) optical signal with 10psec Gaussian pulses entering a FPF with a FSR equal to the bitrate and a finesse of 30.

Fig. 10(c) shows clearly that the complete spectral content has been suppressed by more than 15dB compared to the dc frequency component, allowing in this way only the spectral content residing within the bandwidth of the dc-centered resonance peak to emerge at the output. At the same time, Fig. 10(d) illustrates that the total transfer function retains the periodic FPF frequency characteristics suppressing, however, significantly the spectral content lying between dc and the first FPF harmonic. The effect of these transfer function curves on the original 10Gb/s PRBS data signal can be identified by transforming the data sequence into its Fourier-domain representation, multiplying this expression with the respective transfer function and then applying an inverse Fourier transform. The time-domain outcome of this procedure is depicted in Fig. 10(e) and 10(f) for the transfer functions of Fig. 10(c) and 10(d), respectively. Fig.10(e) shows a time-domain signal of almost constant amplitude confirming the effective elimination of the high-rate clock signal harmonics and the generation of the signal's low-rate envelope. This indicates that the proposed FPF-assisted SOA-MZI subsystem can lead to successful PED functionality when packet-formatted traffic is used by simply utilizing a 3-dB dc-peak bandwidth equal to the packet harmonic. Fig. 10(f) illustrates the time-domain signal obtained at the output of the SOA-MZI when the transfer function of Fig.10(d) is employed. As can be clearly identified, the initial PRBS data stream has been converted into a clock resembling signal with reduced amplitude modulation between its clock pulses, confirming the data harmonic suppression properties of the circuit and its effective use as a CR module.

PED and CR applications employ the same experimental setup, i.e. a FPF followed by a SOA-MZI switch that is powered by a CW beam as the input signal. This setup has been reported so far in experimental demonstrations to perform as a PED circuit (Stampoulidis et al 2007) when the SOAs are driven at low dc current values and as a CR circuit (Kanellos et

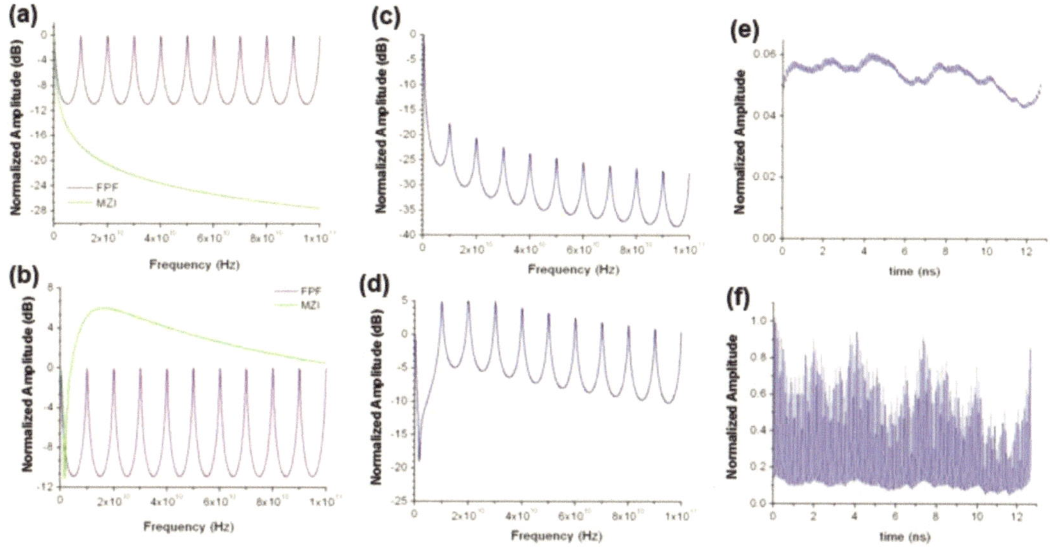

Fig. 10. (a), (b) Individual transfer functions of the FPF and SOA-MZI modules when gz=1, t_{eff}=190psec and gz=5, t_{eff}=25psec, respectively. (c), (d) corresponding total system transfer functions for the FPF-assisted SOA-MZI, and (e), (f) respective time-domain outputs when a 10Gb/s 2^7-1 PRBS input signal is employed.

al 2007) when the SOAs are operated in the saturated regime with moderate gain values and fast recovery speeds. This becomes evident if we consider the three SOA-MZI gain operational regimes discussed in Section 3.1. Low gain values for the SOAs yield the required low-pass filtering response required by a PED circuit that relies on a FPF-assisted SOA-MZI module. At the same time, the resonant behavior of the SOA-MZI at moderate SOA gain levels holds the credentials for supporting CR operation when placed after a FPF element, supporting the clock rate harmonics whilst suppressing the remaining spectral content. Finally, the smoother frequency dependence and the enhanced bandwidth arising when the SOAs are operating in their high gain regime render the circuit more suitable for WC applications, where a replica of the high-rate control signal has to be obtained at a new wavelength.

4. Conclusion

The theoretical frequency-domain transfer function of the SOA-MZI circuit and its qualitative verification even in more complex SOA-MZI-based subsystems allows for its exploitation in several functional applications yielding significant advantages in their theoretical performance analysis. A typical example is the use of cascaded SOA-MZI-based WC stages and its impact in terms of speed limitations and signal degradation due to accumulated jitter and amplitude modulation. This scenario can be now described for *n* cascaded WC stages by means of *n*-times multiplication of the single SOA-MZI transfer function, significantly improving the reliability and accuracy of the results compared to phenomenological transfer function descriptions that have been employed so far. Moreover, the SOA-MZI frequency domain transfer function enables the reliable theoretical proof-of-principle verification of more complex structures and applications that employ SOA-MZIs as sub-elements, as long as the frequency response of the additional sub-modules is known. This can be of significant advantage in the case of novel photonic integrated circuitry where several configurations can be tested theoretically without necessitating the a priori circuit fabrication and its experimental evaluation.

5. References

Apostolopoulos, D.; Vyrsokinos, K.; Zakynthinos, P.; Pleros, N.; Avramopoulos, H. (2009a). An SOA-MZI NRZ Wavelength Conversion Scheme With Enhanced 2R Regeneration Characteristics, *IEEE Photon, Technol. Lett.*, Vol. 21, No. 19, 1363-1365, 1041-1135

Apostolopoulos, D.; Klonidis, D.; Zakynthinos, P.; Vyrsokinos, K.; Pleros, N.; Tomkos, I.; Avramopoulos, H.; (2009b). Cascadability Performance Evaluation of a new NRZ SOA-MZI Wavelength Converter, *IEEE Photon. Technol. Lett.*, Vol. 21, No. 18, 1341-1343, 1041-1135

Cao S.C. and J.C. Cartledge, "Characterization of the chirp and intensity modulation properties of an SOA-MZI wavelength converter"(2002), *J. of Lightwave Technol.*, vol. 20, pp. 689 - 695

Davies D.A.O., "Small-signal analysis of wavelength conversion in semiconductor laser amplifier via gain saturation",(1995) *IEEE Photon. Technol. Lett.*, vol. 7, pp. 617-619

Duelk, M.; Fischer, S.; Gamper, E.; Vogt, W.; Gini, E.; Melchior, H.; Hunziker, W.; Puleo, M.; Girardi, R.; (1999). Full 40 Gbit/s OTDM to WDM conversion: simultaneous four channel 40:10 Gbit/s all-optical demultiplexing and wavelength conversion to individual wavelengths, *Optical Fiber Communication Conference,* San Diego, CA , USA, ISBN: 1-55752-582-X, PD17/1 - PD17/3

Durhuus, T.; Joergensen, C.; Mikkelsen, B.; Pedersen, R.J.S.; and Stubkjaer, K.E. (1994). All Optical Wavelength Conversion by SOA's in Mach-Zehnder Configuration, *IEEE Photon, Technol. Lett.*, Vol. 6, No. 1, 53-55, 1041-1135

Fischer, S.; Bitter, M.; Caraccia, M.; Dülk, M.; Gamper, E.; Vogt, W.; Gini, E.; Melchior, H. and Hunziker, W., (2001). All-optical sampling with a monolithically integrated Mach–Zehnder interferometer gate, Optics Letters, Vol. 26, No. 9, 626–628, 0146-9592

Hill, M.T.; de Waardt, H.; Khoe, G. D. and Dorren, H. J. S. (2001). Fast optical flip-flop by use of Mach-Zehnder interferometers, *Microwave and Optical Technol. Lett.*, Vol. 31, No. 6, 411-415, 1098-2760

Kanellos, G.T.; Petrantonakis, D.; Tsiokos, D.; Bakopoulos, P.; Zakynthinos, P.; Pleros, N.; Apostolopoulos, D.; Maxwell, G.; Poustie, A.; Avramopoulos, H.; (2007a). All-Optical 3R Burst-Mode Reception at 40 Gb/s Using Four Integrated MZI Switches, *IEE/OSA J. Light. Tech.*, Vol. 25, No. 1, 184-192, 0733-8724

Kanellos G. T., et al, (2007b) "40 Gb/s 2R Burst Mode Receiver with a single integrated SOA-MZI switch", *OSA Optics Express*, Vol. 15, No. 8, pp. 5043-5049

Kim, J.Y.; Han S.K.; Lee, S., (2005) All-optical multiple logic gates using parallel SOA-MZI structures, *Lasers and Electro-Optics Society, 2005. LEOS 2005. The 18th Annual Meeting of the IEEE*, ISBN: 0-7803-9217-5, 133 – 134, October 2005, Paper MM1

Lal V., M. Masanovic, D. Wolfson, G. Fish, and D. Blumenthal (2006) "Monolithic Widely Tunable Packet Forwarding Chip in InP for All-Optical Label Switching," in

Integrated Photonics Research and Applications/Nanophotonics, Technical Digest (CD) (Optical Society of America, 2006), paper ITuC3.

Leuthold, J. (2001). Semiconductor Optical Amplifer-Based Devices for All-Optical High-Speed Wavelength Conversion. *Opt. Amplifiers and Their Applications Conf. (OAA'2001), Stresa, Italy, July 2001, paper OWA1*

Marcenac JD and A. Mecozzi, (1997) "Switches and frequency converters based on cross-gain modulation in semiconductor optical amplifiers", *IEEE Photon. Technol. Lett.*, Vol. 9, pp. 749–751

Masanovic, M., Lal, *V.*, Barton, *J.S.*, Skogen, *E.J.*, Coldren, *L.A.*, and Blumenthal, *D.J.* (2003). Monolithically integrated Mach-Zehnder interferometer wavelength converter and widely tunable laser in InP, *IEEE Photon. Technol. Lett.*, vol. 15, No. 8, 1117-1119, 1041-1135

Maxwell, G.; (2006). Low-Cost Hybrid Photonic Integrated Circuits using Passive Alignment Techniques, invited paper MJ2, *IEEE-LEOS Annual Meeting*, Montreal, Canada (2006).

Melo A. Marques de , S. Randel, and K. Petermann,(2007)"Mach–Zehnder Interferometer-Based High-Speed OTDM Add–Drop Multiplexing", *J. of Lightwave Technol.*, vol. 25, no. 4, pp. 1017 - 1026

Nakamura, S.; Ueno, Y.; Tajima, K., (2001). 168-Gb/s all-optical wavelength conversion with a symmetric-Mach-Zehnder-type switch, *IEEE Photon. Technol. Lett.*, Vol. 13, No. 10, 1091-1093, 1041-1135

Nicholes, S.C.; Masanovic, M. L.; Jevremovic, B.; Lively, E.; Coldren, L.A. and Blumenthal, D.J. (2010). An 8x8 InP Monolithic Tunable Optical Router (MOTOR) Packet Forwarding Chip", *IEEE J. of Lightwave Technol.*, vol. 28, 641-650

Nielsen ML and J. (Mork,2004) "Increasing the modulation bandwidth of semiconductor-optical-amplifier-based switches by using optical filtering", *J. Opt. Soc. Am. B*, Vol. 21, pp. 1606-1619

Pleros N., C. Bintjas, G.T.Kanellos, K.Vlachos, H.Avramopoulos, G.Guekos (2004), Recipe for Intensity Modulation Reduction in SOA-Based Interferometric Switches *Journal of Lightwave Technology*, Vol. 22 , No. 12

Pleros N.; Apostolopoulos, D.; Petrantonakis, D.; Stamatiadis, C.; Avramopoulos, H.; (2009). Optical Static RAM Cell, *IEEE Photon, Technol. Lett.*, Vol. 21, No. 2, 73-75, 1041-1135

Ramaswami, R. & Sivarajan, K.N. (2002). *Optical Networks: a Practical Perspective,* R. Adams Editor, Second Ed., Morgan Kaufmann Publisher, ISBN 1-55860-655-6, USA

F. Ramos et al. (2005) IST-LASAGNE:Towards All-Optical Label Swapping Employing Optical Logic Gates and Optical Flip-Flops *Journal of Lightwave Technology*, Vol. 23 , No. 10

Stampoulidis, L.; Kehayas, E.; Apostolopoulos, D.; Bakopoulos, P.; Vyrsokinos, K.; Avramopoulos, H.; (2007). On-the-Fly All-Optical Contention Resolution for NRZ and RZ Data Formats Using Packet Envelope Detection and Integrated Optical Switches, *IEEE Photon, Technol. Lett.*, Vol. 19, No. 8, 538-540, 1041-1135

Stampoulidis, et al (2008). Enabling Tb/s Photonic Routing: Development of Advanced Hybrid Integrated Photonic Devices to Realize High-Speed, All-Optical Packet Switching, *IEEE J. of Sel. Topics in Quantum Electron.*, Vol. 14, No. 3, 849 - 860, 1077-260X

Stubkjaer, K.E. (2000). Semiconductor Optical Amplifier-Based All-Optical Gates for High-Speed Optical Processing. *IEEE J. on Selected Topics in Quantum Electronics,* Vol. 6, No. 6, (November/December 2000), 1428-1435, 1077-260X

Ueno, Y.; Nakamura, S.; Tajima, K. (2001). Penalty-free error-free all-optical data pulse regeneration at 84 Gb/s by using a symmetric-Mach-Zehnder-type semiconductor regenerator, *IEEE Photon. Technol. Lett.*, vol. 13, No. 5, 469-471, 1041-1135

Wang, L.; Zhang, M.; Zhao, Y.; Ye, P. (2004). Performance analysis of the all-optical XOR gate using SOA-MZI with a differential modulation scheme, *Microwave and Opt. Tech. Lett.,* Vol. 40, No. 2, 173-177, 1098-2760

Webb, R.P.; Manning, R.J.; Maxwell, G.D.; Poustie, A.J. (2003). 40 Gbit/s all-optical XOR gate based on hybrid-integrated Mach-Zehnder interferometer, *Electron. Lett.* Vol. 39, No. 1, 79-81, 0013-5194

Wolfson, D.; Kloch, A.; Fjelde, T.; Janz, C.; Dagens, B. and Renaud, M. (2000). 40-Gb/s All-Optical Wavelength Conversion, Regeneration, and Demultiplexing in an SOA-Based All-Active Mach–Zehnder Interferometer, *IEEE Photon, Technol. Lett.*, Vol. 12, No. 3, 332-334, 1041-1135

Yan N., J. del Val Puente, T.G. Silveira, A. Teixeira, A.P.S. Ferreira, E. Tangdiongga, P. Monteiro and A.M.J. Koonen, 2009 "Simulation and experimental characterization of SOA-MZI-based multiwavelength conversion", *IEEE J. of Lightwave Technol.*, vol. 27, No. 2, pp. 117 - 127

Zakynthinos, P.; Kanellos, G.T.; Klonidis, D.; Apostolopoulos, D.; Pleros, N.; Poustie, A.; Maxwell, G.; Tomkos, I.; Avramopoulos, H.; (2007). Cascaded operation of a 2R Burst Mode Regenerator for Optical Burst Switching network transmission, *IEEE Photon. Technol. Lett.*, Vol. 19, No. 22, 1834-1836, 1041-1135

Part 2

Semiconductor Optical Amplifiers: Wavelength Converters

4

Semiconductor Optical Amplifiers and their Application for All Optical Wavelength Conversion

Oded Raz
Eindhoven University of Technology
The Netherlands

1. Introduction

All optical networks and switches are envisioned as a solution to the increasing complexity and power consumption of today's communication networks who rely on optical fibers for the transmission of information but use electronics at the connecting points on the network (nodes) to perform the switching operation. All optical networks, in contrast, will use simple signalling methods to trigger all optical switches to forward the optical data, from one optical fiber to another, without the need to convert the information carried by the optical signal into an electric one. This may save up to 50% of the total power consumption of the switches and will allow for simple scaling of the transmission rates. While all optical networks may offer significant breakthroughs in power consumption and network design, they fall back on one essential aspect, contention resolution. In traditional communication networks and in particular those who carry data (which has long surpassed voice traffic, in bandwidth), the nodes on the network use huge amounts of electronic random access memory (RAM) to store incoming data while waiting for their forwarding to be carried out. The storage of data, also called buffering, is essential in resolving contention which occurs when two incoming streams of data need to be forwarded to the same output port at the same time. In contrast all optical switches, who do not convert the data signals into the electrical domain, cannot use electronic buffers for contention resolution. They can however use the unique properties of light signals which at moderate power levels can propagate along the same transmission media without interference if they have different wavelengths. This means that if two competing light signals need to be switched to the same output port, their successful forwarding can be accomplished by assigning them different wavelength. This can be done completely in the optical domain by means of all optical wavelength conversion.

Large optical networks, require optical amplifiers for signal regeneration, especially so if the signal is not regenerated through optical to electrical to optical conversion. Semiconductor Optical Amplifiers (SOAs) are a simple, small size and low power solution for optical amplification. However, unlike fiber based amplifiers such as EDFAs, they suffer from a larger noise figure, which severely limits their use for long haul optical communication networks. Nevertheless, SOAs have found a broad area of applications in non-linear all optical processing, as they exhibit ultra fast dynamic response and strong non-linearities,

which are essential for the implementation of all optical networks and switches. This means that for a most essential function such as all optical wavelength conversions, SOAs are an excellent solution.
Wavelength conversion based on SOAs has followed several trajectories which will be detailed in the following sections. In section 2 we discuss how data patterns can be copied from one optical carrier to another based on the modulation of gain and phase experienced by an idle optical signal in the presence of a modulated carrier. Section 3 is devoted for the use of Kerr effect based wavelength conversion, and specifically to wavelength conversion based on degenerate four wave mixing (FWM). In section 4 we discuss how the introduction of new types of SOAs based on quantum dot gain material (QDSOA) has lead to advances in all optical wavelength conversion due to their unique properties. We conclude the chapter in section 5 where we point at future research directions and the required advancement in SOA designs which will allow for their large scale adoption in all optical switches.

2. Cross gain and cross phase modulation based convertors

When biased above their transparency current, SOAs may deliver considerable optical gain with a typical operational bandwidth of several tens of nanometers. However, since the gain mechanism is based on injection of carriers, the introduction of modulated optical carriers, and especially of short high peak power pulses such as those used for Opitcal Time Domain Multiplexing systems (OTDM), result in severe modulation of gain bearing majority carriers leading to undesirable cross talk in case multiple channels are introduced into the SOA (Inoue, 1989). The gain of an SOA recovers on three different timescales. Ultrafast gain recovery, driven by carrier-carrier scattering takes place at sub-picoseconds timescale (Mark & Mork, 1992). Furthermore, carrier-phonon interactions contribute to the recovery of the amplifier on a timescale of a few picoseconds (Mark & Mork, 1992). Finally, on a tens of picoseconds to nanosecond timescale, there is a contribution driven by electron-hole interactions. This last recovery mechanism dominates the eventual SOA recovery. Careful design of the active layer in the amplifier, injection efficiency and carrier confinement plays a role in the final recovery time which can vary between several hundreds of picoseconds to as low as 25 pico seconds for specially designed Quantum Well structures (CIP white paper , 2008). During the recovery of gain and carriers from the introduction of an optical pulse, the refractive index of the SOA wave guiding layer is also altered, so that not only the gain but also the phase of the CW signals travelling through the device is modulated. These two phenomena, termed Cross Gain Modulation (XGM) and Cross Phase Modulation (XPM), severely limit the use of SOAs for amplification of optical signals in Wavelength Division Multiplexed (WDM) networks.
Yet, the coupling of amplitude modulation of one optical channel into the amplitude and phase of other optical carriers travelling in the same SOAs has caught the attention of researchers working on all optical networks as a simple manner of duplicating data from one wavelength to another, a process also known as wavelength conversion.
Early attempts to exploit XGM in SOAs were already reported in 1993 (Wiesenfeld et al, 1993) where conversion of Non Return to Zero (NRZ) data signal was achieved at a bit rate of 10Gb/s and a tuning range of 17nm. These were later followed with demonstrations of conversion at increasingly higher bit rates but due to the low peak to average power ratio of NRZ signals (which dominated optical communications until the end of the 1990's) could not exceed 40Gb/s (and even this was only made possible with the use of two SOAs nested in a Mach Zehnder interferometer (Miyazaki et al, 2007).

ODTM systems which are based on short optical pulses interleaved together to achieve an effective data rate in the hundreds of Gb/s was conceived as an alternative to WDM for multiplexing data channels into the optical domain. The large peak to average power ratio associated with this transmission technique means that the carrier depletion effect is much stronger leading to a more pronounced drop in gain. For OTDM signals many methods have been proposed to allow high bit-rate All Optical Wavelength Conversion (AOWC) based on an SOA. Higher bit-rate operation was achieved by employing a fiber Bragg grating (FBG) (Yu et al, 1999), or a waveguide filter (Dong et al, 2000). In (Miyazaki et al, 2007), a switch using a differential Mach–Zehnder interferometer with SOAs in both arms has been introduced. The latter configuration allows the creation of a short switching window (several picoseconds), although the SOA in each arm exhibits a slow recovery. A delayed interferometric wavelength converter, in which only one SOA has been implemented, is presented in (Nakamura et al, 2001). The operation speed of this wavelength converter can reach 160 Gb/s and potentially even 320Gb/s (Liu et al, 2005) and allows also photonic integration (Leuthold et al, 2000). This concept has been analyzed theoretically in (Y. Ueno et al, 2002). The delayed interferometer also acts as an optical filter. Nielsen and Mørk (Nielsen & Mørk, 2004) present a theoretical study that reveals how optical filtering can increase the modulation bandwidth of SOA-based switches. Two separate approaches for filter assisted conversion can be considered, inverted and non-inverted.

Inverted wavelength conversion

In case an inversion stage is added after optical filtering, it is possible to obtain ultra high speed conversion (bit rate >300 Gb/s) by combining XGM and XPM. This can be most easily understood by looking at Fig. 1. The CW optical signal (or CW probe) is filtered by a Guassian shaped filter which is detuned relative to the probe's wavelength (peak of filter is placed at a shorter wavelength - blue shifted).

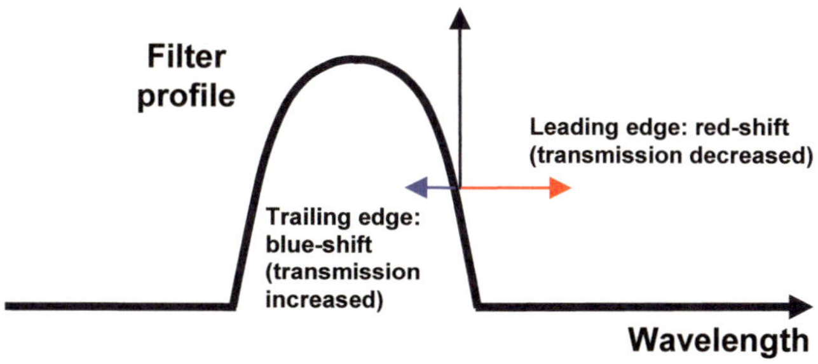

Fig. 1. Operation principle of detuned filtering conversion

As the pump light hits the SOA (leading edge of the pulse), carrier depletion results in a drop of gain as well as a phase change which leads to a wavelength shift to a longer wavelength (red-shift). This means that for the CW probe, on top of the drop in gain, a further drop in power is observed as the signal is further pushed out of the filter's band pass. Once the pump signal has left the SOA, carrier recovery begins, with a steady increase in gain and carrier concentration. The latter is responsible for a blue-shift in the probe's wavelength, which implies that the CW probe is now pushed into the middle of the filter's band, further increasing the output power, and effectively speeding up the eventual recovery of the probe signal. As a result, the net intensity at the filter output is constant although the actual carrier recovery may continue far after the pump pulse has passed the SOA (see Fig. 2).

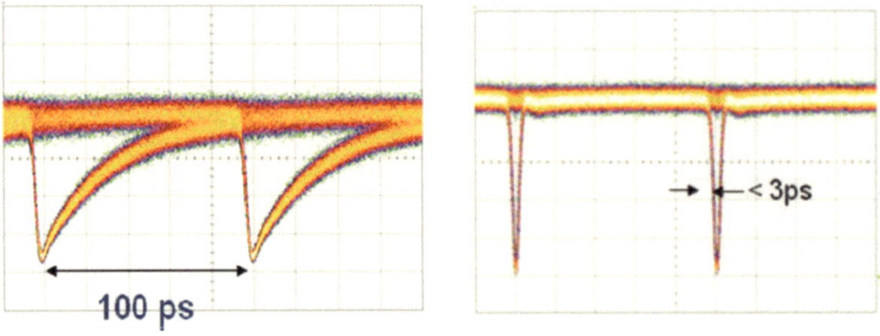

Fig. 2. Effect of filter detuning on probe recovery; (Left) no detuning, (Right) optimum detuning

Using this method, AOWC has been demonstrated at speeds up to and including 320 Gb/s (Y. Liu et al, 2005). The main limitation in extending the technique to even higher bit-rates is that as bit-rate increases the peak to mean power ratio drops, so that patterning effects dominate the performance of the converter and the obtained eye opening of the converted signal degrades. Further limitations of this conversion technique arise from the need to include after the SOA and optical filter, an inversion stage, which essentially suppresses the original CW optical carrier leading to poor optical signal to noise ratio at the output of the complete converter. Typical reported conversion penalties are dependent on the bit rate and might be as high as 10dB for 320Gb/s conversion.

Non-inverted wavelength conversion

For the non inverted conversion, although both XGM and XPM occur with the introduction of a short high power pulse into the SOA, it is mostly the effect of phase modulation that is utilized. As discussed above, during the introduction of a short optical pump pulse into the SOA, the changing levels of carriers leads to changes in refractive index which modulate the phase and frequency of the CW probe. By using a very sharp flat top filter (see Fig. 3), the induced frequency shifts can be converted to amplitude variations, thus having direct rather than inverted relation to the pump signal. Since both red and blue shifting of the probe's wavelength occurs, it is in principal possible to place the sharp filter so that the pass band is

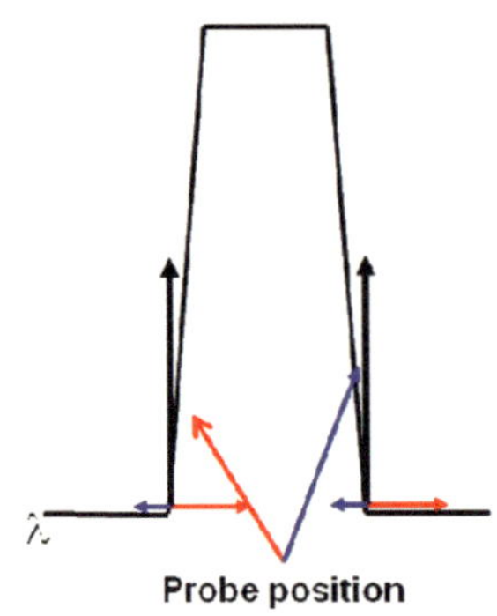

Fig. 3. Operation principle of non-inverted conversion

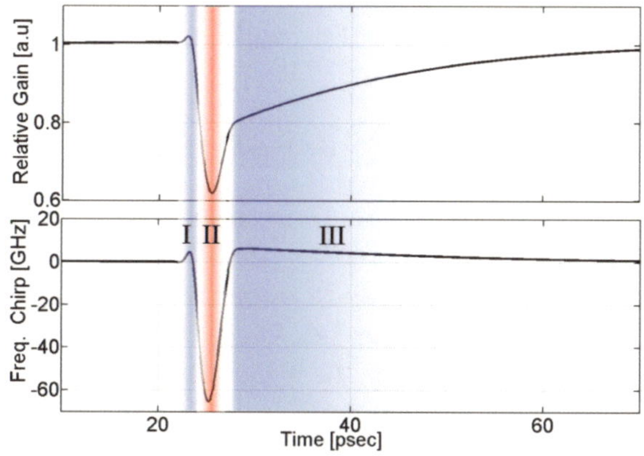

Fig. 4. Gain and frequency shift, experienced by the probe signal

either to the left or the right of the CW probe. While filtering the red component yields a more suitable temporal pulse shape (see section II in Fig. 4, tracing the frequency chirp vs time), the sharp drop in gain implies poorer signal to noise for this option. Alternatively, opting for blue component filtering, a broader pulse is obtained but with improved signal to noise. In the experimental section below demonstration of these two filtering scheme is detailed.

Non-inverted wavelength conversion - simulation and demonstration (Raz et al, 2009)

SOA theory and numerical simulations

The final shape of the time domain pulse is dominated by the duration of the blue/red chirp induced frequency change and the shape of the optical filter used. In order to preserve the original pulse shape one needs the filter's optical bandwidth to be in the order of the spectral width of the original RZ pulses (~5 nm). Another crucial aspect for this kind of WC scheme is the eventual OSNR obtainable as it will determine the penalty incurred. For that purposes it is desired to filter out the CW component without affecting the 1st blue/red modulation side-band as it contains most of the converted pulse energy. In order to fulfill both of the above requirements a special flat top, broad filter with sharp roll off is required (Leuthold et al, 2004). In order to gain a better understanding of the requirements from this sort of filtering technique and its applicability for fast WC we used an SOA band model valid for time responses in the pico-second and sub-picosecond regime (Mork & Mecozzi, 1996; Nielsen et al, 2006; Mark & Mork, 1992; Mork & Mark, 1994). The SOA model includes XGM and XPM effects required to model the wavelength conversion process as well as Two-Photon Absorption (TPA) and Free-Carrier Absorption (FCA) responsible for the Carrier-Heating (CH) and Spectral-Hole Burning (SHB) effects. The equations used for generating the simulation results are detailed in (Mark & Mork, 1992; Mork & Mark, 1994), and are described shortly below:

$$\frac{\partial N}{\partial t} = \frac{I}{eV_c} - \frac{N}{\tau_s} - v_g g S + \frac{\Gamma_2}{\Gamma} v_g \beta_2 S^2 \quad (1)$$

$$\frac{\partial U_i}{\partial t} = (\sigma_i N \hbar \omega_0 - g E_i) v_g S + \frac{\Gamma_2}{\Gamma} v_g \beta_2 E_{2,i} S^2 - \frac{U_i - U_{L,i}}{\tau_{h,i}} \quad (2)$$

$$\frac{\partial S}{\partial z}+\frac{1}{v_g}\frac{\partial S}{\partial t}=(\Gamma g-\alpha_{\text{int}})S-\beta_2 S^2 \tag{3}$$

$$\frac{\partial p}{\partial z}+\frac{1}{v_g'}\frac{\partial p}{\partial t}=(\Gamma g-\alpha_{\text{int}})p-\beta_2 Sp \tag{4}$$

Where N stand for the carrier concentration, U_i the energy densities, and S and p represent the pump and probe photon density. The energy density is computed for both conduction (i=c) and (heavy hole) valence (i=v) band, respectively. $E_{2,i}$ are the carrier energies corresponding to the two-photon transition, i.e., $2\hbar\omega_0 = E_g + E_{2,c} + E_{2,v}$ with $\hbar\omega_0$ being the photon energy and E_g the band-gap energy, β_2 is the TPA coefficient averaged (with weight S^2) over the cross section of the waveguide (σ_i) and Γ_2 is the corresponding confinement factor for the quantum well region. We have $\Gamma_2/\Gamma > 1$ due to the tighter confinement of the square of the intensity profile, as well as the higher value for the TPA coefficient in the lower band-gap well region as compared to the separate confinement and cladding regions (Sheik-Bahae et al, 1991). In (Raz et al, 2009), a more detailed description of the simulation follows but the important results are given below in Fig. 5.

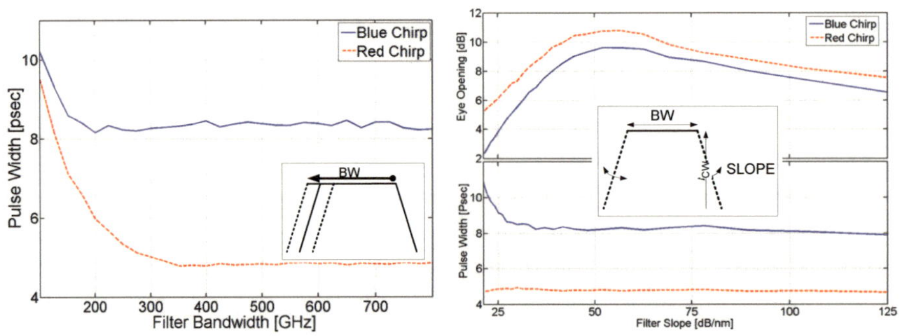

Fig. 5. Simulation results showing the dependence of pulse width on the filter Bandwidth (Left) and slope (Right)

On the left we observe the dependency of final pulse width on the bandwidth of the filter. For the case of blue chirp filtering, the slow response time sets a lower limit (8 ps) on the pulse width which is already apparent for 200 GHz filter bandwidth. However for the case of red chirp filtering the converted signal's pulse width is considerably narrower (<5 ps) and the filter bandwidth at which this value is achieved is almost double (around 400 GHz). Still it is obvious that the fundamental limit for the pulse width lies in the carrier dynamics of the SOA rather than the filter bandwidth. On the right we see how changing the filter's roll-off affects both EO and pulse width. When changing the roll-off the EO goes from a practically closed eye for a roll off lower than 25dB/nm to a maximum value of 10-11 dB for a slope value between 50-60dB/nm. Increasing the roll-off further does not improve EO as it implies sharper spectral slicing which results in ripples in the time domain eye. For EO, the difference between the red and blue filtering is not very pronounced. As for the pulse width, the same values obtained for altering the width are repeated with a minimum required roll-off larger than 30dB/nm. The apparent increase/decrease in pulse width for slopes lower than 25dB/nm is meaningless since for these values the eye is practically closes (or inverted), and only positive EO were computed as explained above.

Experimental demonstration

The experimental set-up used to demonstrate 40 and 80Gb/s direct non-inverted conversion is shown in Fig. 6

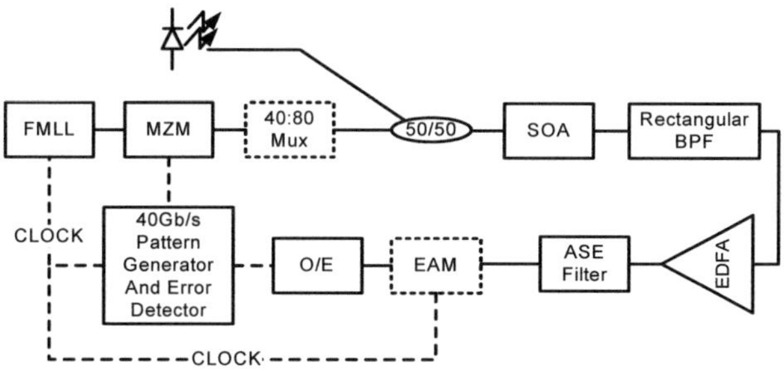

Fig. 6. Experimental set-up

40Gb/s wavelength conversion:

The 40GHz Fiber Mode Locked Laser (FMLL) RZ pulse source, with 2 ps FWHM, is externally modulated by a Mach Zehnder Modulator (MZM) by a 231-1 Pseudo Random Bit Sequence (PRBS) at 40 Gb/s. The pump signal is coupled with the probe signal and launched into the SOA. An SOA similar to the one used in (Liu et al, 2005) was also used for this experiment. The SOA has a measured total recovery time of 56 ps when biased at 400 mA, dominated by a slow blue component. At the output of the SOA the signal is filtered by the special flat top broad band filter with roll-off > 60db/nm and a rejection greater than 50dB of adjacent channels. The signal is then amplified using and Erbium doped fiber amplifiers (EDFA) and filtered again using a standard Gaussian shaped 5 nm filter to remove excess ASE noise. When running the experiment at 80Gb/s, an inter leaver is used after the modulator to go from 40 to 80 Gb/s and a EAM demux is used to gate 40Gb/s tributaries from the 80Gb/s serial data stream for BER estimation. Table 1 summarizes the key parameters for operating the WC for either the blue or red filtered components at 40Gb/s bit rate.

	Red Component Filtering	Blue Component Filtering
Pump Wavelength [nm]	1560	1560
Pump Power [dBm]	1.5	-6.3
Probe Wavelength [nm]	1548.1	1548.1
Probe Power [dBm]	1.5	-2.7
SOA current [mA]	400	262.8
Filter Center Frequency [nm]	1550.968	1545.858
Filter Bandwidth [nm]	4.5	4.31

Table 1. Main operation parameters for both blue and red filtering scenarios

In Fig. 7 the spectra for the wavelength converted signal for both filtering cases as well as the unfiltered spectrum are plotted together. The filtered spectra were taken in both cases after the EDFA so that spectral features on the edges of the filter's band-pass are lost in the ASE noise. Also, the power of the sidebands as it appears in the filtered spectra includes

~20dB of EDFA gain. The non filtered spectra, taken for the case of higher bias current and stronger pump power (green line), has a secondary peak around 1545 nm arising from non linear distortions (Self Phase Modulation) incurred by the original pump signal that are copied to the WC probe through XGM and XPM processes.

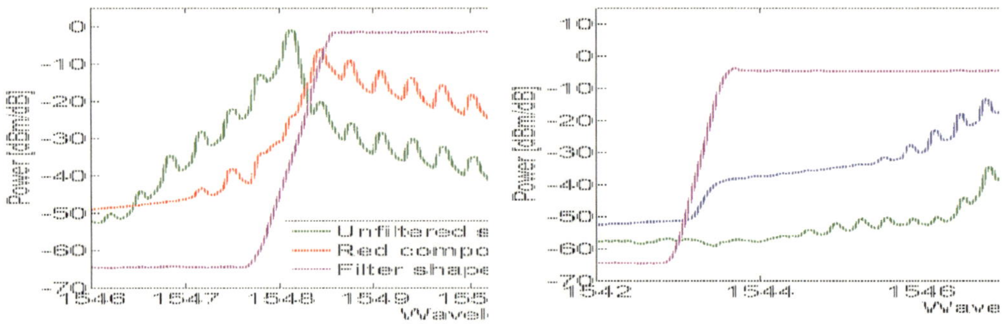

Fig. 7. Filtered and non-filtered spectra's at the SOA output

In the case of filtering out the red components, these distortions are filtered out, however for the case of blue component filtering the operating conditions had to be greatly altered (8 dB drop in pump power, and 30% drop in DC bias current for the SOA), as any distortions will be included in the broad filtered output signal.

The resulting eye patterns and Bit Error Rate (BER) vs. received power given in Fig. 8, indicate that these specific filter characteristics, especially the sharp roll-off and large band-width, greatly improve the performance of the scheme, compared with previous works. For red filtered WC there is a negligible negative penalty for BER worse than 10^{-7} but it is apparent that there is an error floor which brings the penalty for a BER of 10^{-9} to 0.5 dB. The error floor arising from the noise of the SOA is more dominant for the case of the red filtered WC since there is a power difference of 8dB between the blue and red 1st order side bands while the noise floor is the same. For the blue filtered results, a penalty of 0.7 dB is obtained and no error floor was observed.

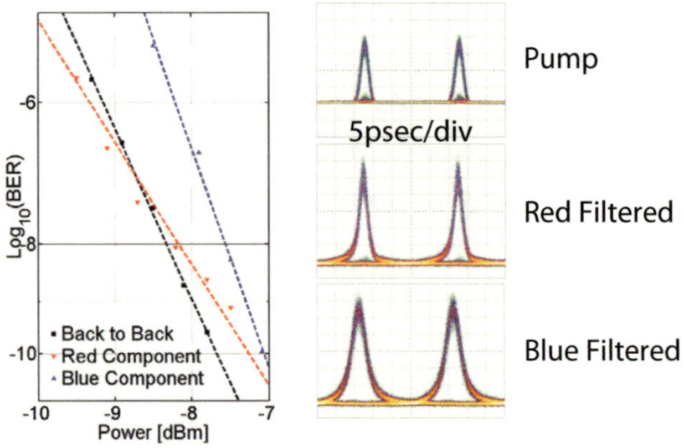

Fig. 8. BER (left) and eye patterns for B2B (top) and Red and Blue filtered (middle and bottom respectively) Wavelength converted signals

The eye patterns in Fig. 8 give an indication on the respective time domain performance for red and blue filtering. The filtering of the red components results in a much faster response with a FWHM of around 3 ps (only 1 ps more than for the original pulses, Fig. 8 top right). However for the case of filtering out the blue chirp components, which are strongly dependent on the slow recovery time of the SOA, the observed eye is much wider having a FWHM of around 4.5 ps and a pulse base duration of 12 ps.

80Gb/s wavelength conversion:

The pump signal entering the SOA is centered around 1560 nm and has a power of 0.7 dBm. The CW probe signal was at 1548.1 nm with a power of 6.7 dBm. The same SOA was used also for this experiment. At the output of the SOA a sharp flat top 6.15 nm wide Band Pass Filter (BPF) was place, centered on 1544.63 nm. The filter has a roll-off greater than 60 dB/nm and an insertion loss of 4.5 dB. After filtering, the 80 Gb/s signal is time demultiplexed to the 40 Gb/s original PRBS bit rate using Electro Absorption Modulator (EAM) gating, converted back to the electrical domain and tested for errors.
In Fig. 9, the inverted (before filter) and non-inverted spectra (taken directly after the BPF) are both shown. Notice the strong attenuation incurred by the CW signal (>35 dB) compared to the 9 dB (extra 4.5 dB due to detuning) attenuation of the 1st side band and no extra attenuation on higher order modulation side-bands. Also visible is the SOA noise floor at around -45 dBm, around the higher order side-bands. This noise together with the minimal impact on the 1st order side-band (-18 dBm) give an OSNR >25 dB, sufficiently good for the low penalty measured.

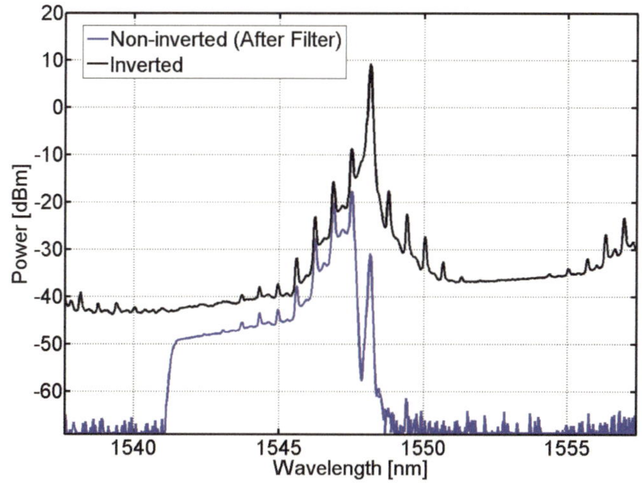

Fig. 9. Spectra of the converted signal at the output of the SOA before and after the filter

In Fig. 10 the BER for the two 40 Gb/s tributaries are shown (red lines) compared to their back to back counterparts (blue line). Also shown for comparison are the pump and probe eye patterns. The measured penalty is 0.5 dB and the eye is broadened from a 2 ps FWHM to about 4.5 ps, similar to what was measured for the experiment carried out at 40Gb/s. However the converted signal suffers from poorer OSNR leading to an observable change in BER slope.

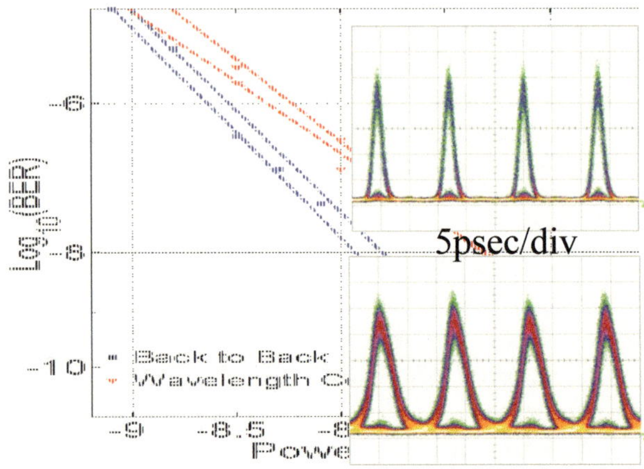

Fig. 10. BER (left) and eye patterns for B2B (blue, top right) and Wavelength converted signal (red, bottom right) respectively

3. Four Wave Mixing based wavelength convertors

The use of third order non-linearities for the purpose of all optical wavelength conversion has been demonstrated over the years in many different non-linear media. The efficiency and bandwidth of these phenomenon is governed by the third order non-linear susceptibility denoted by $\chi(3)$ and is dependent on the polarization, power, frequency detuning and dispersion of the non-linear medium used (Agrawal, 2002). In order to enhance FWM special phase matching and quasi phase matching techniques have been employed with exceptional bandwidth and efficiency demonstrated for devices based on periodically poled LiNbO3 devices (Yamawaku et al, 2003). Similarly, careful tailoring of single mode fiber dispersion, has also allowed for highly efficient FWM in highly non-linear fibers (HNLF) (Tanemura et al, 2004). However in the stride for small foot-print and low power options, a more useful solution is the employment of an SOA as a non-linear medium, as it offers integration potential and may contribute significant signal gain to offset the negative conversion efficiency.

Early studies of the nature of FWM in semiconductor traveling wave amplifiers has pointed out that the most dominant source of FWM in SOAs is the creation of gain and index gratings through the periodic modulation of the injected carriers in the device by the traveling pump and probe waves (Agrawal, 1987). Early demonstrations of wavelength conversion based on degenerate FWM in SOAs, date to the early 90', and were dedicated to the methodical characterization of the convertors in terms of conversion penalty and equivalent noise figure (Mecozzi et al, 1995; Summerfield & Tucker, 1996). In order to reduce the conversion penalty as well as lower the effective noise figure of the convertors, power levels of pump and probe signals was set so that the SOA was deeply saturated. However, high power levels, usually resulted in unwanted 2nd and 3rd order mixing products which enforced limitations on spectral spacing of pump and probe signals, especially so, for cases where multicasting conversion was demonstrated (Contestabile et al,

2004). Due to the relatively poor conversion efficiencies and high noise contribution of SOAs, the obtainable OSNR is quite limited. Thus, although FWM wavelength conversion does not suffer from time domain limitations, such as those present when performing conversion based on carrier dynamics, error free conversion for bit rates above 40Gb/s was never demonstrated. Furthermore, FWM is critically dependant on polarization alignment of pump and probe. This implies that for polarization multiplexed signals, an ever more popular bandwidth enhancement technique, FWM cannot be used in a simple manner (Contestabile et al, 2009).

In the following section recent results on FWM in SOA are detailed. These experiments focused on using a single SOA to obtain simultaneous conversion of two independent data channels. Various modulation formats and modulation speeds are explored, and a polarization insensitive set-up is also suggested.

Simultaneous FWM in SOAs (Gallep et al, 2010)

The key to the successful demonstration of simultaneous conversion of two independent data signals using FWM is proper power equalization of the input data signals as well as of the strong CW pump. Low optical power for the two input channels prevents the onset of deleterious FWM products which interfere with the converted products. However, higher input powers improve Optical Signal-to-Noise Ratio (OSNR) of the converted channels, which is essential for error-free operation. Above the optimal input power levels, used in the experiments described below, any increase in the modulated inputs does not enhance the performance but decreased the FWM efficiency due to power splitting into the non-degenerated and secondary FWM products. Similar considerations are also applied to the choice of CW-pump power: the pump must be strong enough to clamp the gain, minimizing any XGM that might be introduced by intensity modulated data inputs, and to reduce the ASE floor. On top of this optimization process, in this demonstration, it was important to obtain similar performance of the converter for single and dual operation, as it will allow for asynchronous operation. This constraint also led to a non optimal choice of probe power levels and in some cases single conversion performance could have been much better at the cost of degrading the performance of dual conversion. The choice of wavelengths (ITU channels) took into consideration the effect of unwanted conversion products between the pump and the data channels. In general the most suitable arrangement of data channels and CW pump was found to be such that the data channels are up-converted and that the spacing between them is twice as that of the spacing between the CW pump and the data channel closest to it (CW - ITU X, Data 1 - ITU X-1, Data 2 - ITU X-3). Implementation of down-conversion schemes (conversion to longer wavelengths) is not possible due inferior OSNR, 10dB lower than that achieved for up-conversion, as was pointed out already in Agrawal's seminal work of 1987 (Agrawal, 1987).

Mixed modulation formats 10 Gb/s (ASK+ PSK)

Fig. 11 presents the experimental setup used for the case of PSK and ASK simultaneous conversion at a bit rate of 10Gb/s in both channels.

The two laser sources at 1558.17 nm (-12 dBm) and 1556.55 nm (-17 dBm), ITU channels #24 and #26, were modulated with PSK and ASK respectivly at a rate of 10 Gb/s (NRZ PRBS 2^{31}-1 data sequence) and combined at the SOA input with a much stronger CW signal at 1555.75 nm (ch.#27). The polarization controllers (PC) after the lasers were carefully

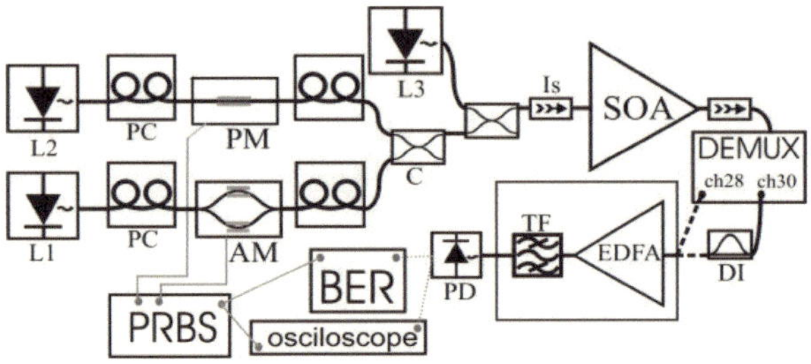

Fig. 11. Experimental setup for dual-channel (PSK + ASK) simultaneous lambda-conversion in a single SOA

adjusted to achieve the lowest insertion loss through the modulators, and the PCs just after them are used to align the polarization of the CW pump with the probe signals to maximize the FWM process. The SOA, ultra-nonlinear device with MQW structure (CIP), was biased at 500 mA, with a saturation output power of 15dBm and small signal gain >30 dB. At the output, the converted channels were filtered by an ITU-grid DEMUX (100 GHz spacing). To enable the bit-error rate (BER) versus received optical power measurements in similar conditions, the back-to-back and the converted signals were amplified by a low noise EDFA (10 dB gain, 4 dB noise figure) and filtered again (1.5 nm-window) to remove excessive ASE. The converted PSK signal was further processed by passing through a Delayed Interferometer (DI) to convert phase into amplitude modulation before detection. For the 10+10 Gb/s case the BER measurements were taken using a 10 Gb/s APD receiver.

The optical spectrum at the SOA's input and output as well as the eye diagrams and the BER vs. optical power at the receiver for the 10+10 Gb/s, ASK+PSK, are shown in Fig. 12. BER vs received optical power performance of a single converted channel is as good as the original data signal (back-to-back). Even in the presence of a 2nd converted channel the observed degradation is within the measurement error and in any case does not exceed 0.3dB.

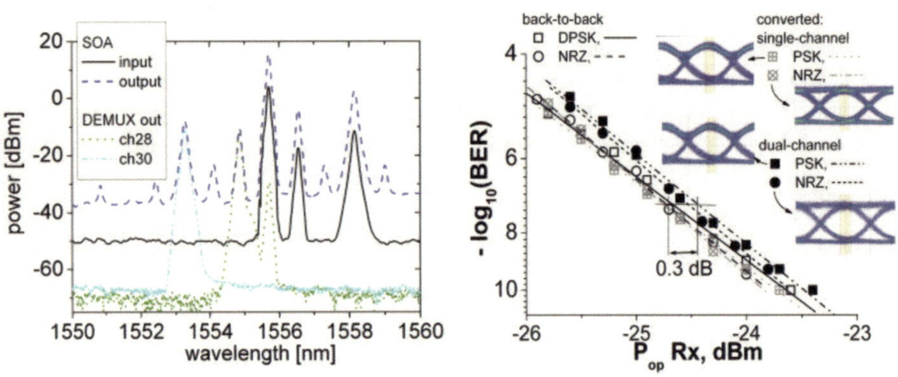

Fig. 12. Simultaneous lambda-conversion, 10+10 Gb/s (PSK+ASK): optical spectra for SOA input and output, DEMUX ch. #28 and #30 (left); eye diagrams and BER curves for single and dual conversion operation

The spectra at Fig.12 (left) illustrates the required spectral positioning of input data channels and CW pump as discussed above with the CW at ITU-grid channel #27, the ASK channel at channel #26 and the PSK channel at channel #24, avoiding interfering cross-channel products. The input PSK channel required more power (+5dB) than the ASK channel since the FWM efficiency drops the further the signal is detuned from the CW pump. In any case, this penalty-free performance is obtained for low input peak-powers (<-10 dBm) and a very modest - 2 dBm CW pump.

Mixed bit rates ASK (10+20Gb/s,10+40Gb/s)

The setup used for mixed bit rate ASK signals required several minor adaptations in comparison to the setup in Fig.11. An amplitude modulator (AM) after L2 replaced the phase modulator (PM) and a 40-Gb/s PIN photodetector replaced the APD receiver for all measured BER curves (also for the 10 Gb/s channels). In addition, the selected wavelength channels were slightly shifted in the ITU grid, but maintaining relative positioning: the CW pump at channel #28 (1554.94 nm) and the two modulated carriers at ch.#27 (1555.75 nm, L1) and ch.#25 (1557.36 nm, L2). This shift was required to better align the outputs to a 200 GHz DEMUX used to filter the converted channels out. Fig. 13 shows the measured BER vs. received power for NRZ converted channels at 10 and 20 Gb/s using a 2^{31}-1 bits long PRBS data sequence. Optimal input power levels for data carriers were found to be below -15 dBm and the CW pump was set at +7 dBm. Both positioning of the 10 and 20 Gb/s input data channels with respect to the CW pump were tested: close to (conversion from channel #27 to ch.#29) and apart (from ch.#25 to ch.#31). From Fig. 13, the 20 Gb/s channel presents error free operation, with 1 dB degradation of required optical power at the receiver for the same BER performance when being the closest (100 GHz) to the CW probe.

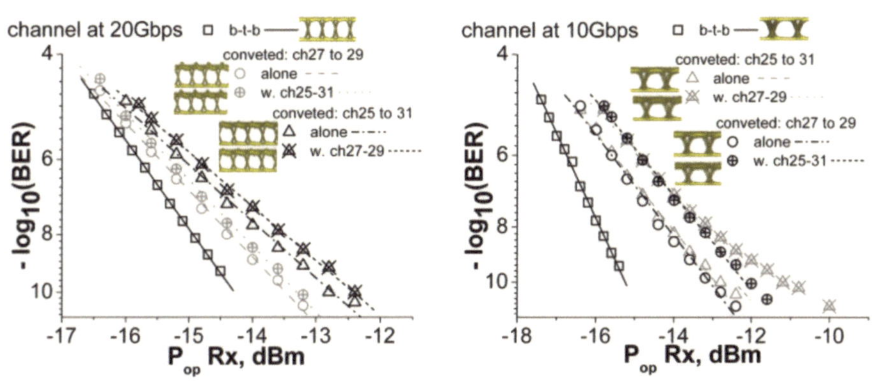

Fig. 13. Simultaneous lambda-conversion 20+10 Gb/s ASK: eye diagrams and BER curves of 20 (left) and 10 (right) Gb/s channels

When placed further away (300 GHz) the power penalty increases to 2 dB. A very small difference (0.1-0.3 dB) exists between single and dual-channel operation modes. For the 10 Gb/s channel, when placed closer to the CW pump, a power penalty of 2 dB was measured for single conversion and an extra 1.1 dB in dual-channel mode. When placed further away from the pump (ITU ch.#25), a power penalty of 2.2 dB penalty was observed and when a 2nd channel (ch.#27) was turned on simultaneously an error floor was observed around a BER ~ 10^{-12}; at a BER of 10^{-11} a 4-dB penalty was obtained. The detected noise floor is mainly due the noise from spurious FWM over the converted channel and the limited OSNR.

For the 40+10 Gb/s case (Fig.14), the converted 40 Gb/s channel shows an error floor above BER=10^{-12} regardless of the presence of a 2nd 10 Gb/s input channel. This noise floor is mostly the result of overshoots appearing at the higher ("1") bit-level and the limited OSNR at the SOA's output.

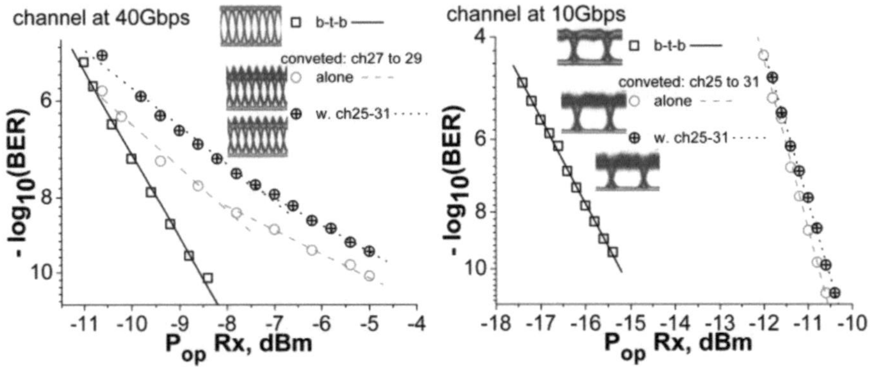

Fig. 14. Simultaneous lambda-conversion 40+10 Gb/s ASK: eye diagrams and BER curves of 40 (left) and 10 (right) Gb/s channels

A 4-dB total penalty was obtained at BER=10^{-11}, with an added 1dB penalty when the 2nd channel is turned on. The 10 Gb/s channel was measured to have a 4 dB penalty due mostly to noise over the high-level (see on inset in Fig. 14) with no difference between the single and the dual-channel operation.

For the case of simultaneous conversion of 40 and 10 Gb/s channels it was impossible to switch the respective positions of 40 and 10 Gb/s channels since the obtainable FWM efficiency and OSNR for the 40 Gb/s, when placed further away from the CW pump, could not deliver error-free operation.

Polarization insensitive simultaneous conversion (Gallep et al, 2010)

The experimental setup is presented in Fig.15(a). Each data carrier (lasers L1 and L2) passes in a polarization controller (PC) to optimize its modulator performance, with each channel modulated with PRBS 2^{31}-1 sequences and combined by a 100GHz WDM Multiplexer (MUX) with two CW probes (L3 and L4). These probes have equal power and are arranged in orthogonal polarization by passing through PCs and in a polarization beam-splitter (PBS). The combined signal is sent to the SOA, with optical isolators preventing multiple reflections. The PCs just after the modulators are used to change the relative polarization (mis)matching between the data channels and the CW carriers, and so compare the best and worst cases. The PC after the PBS is used to equalize the CW channels' gain in the SOA as well as their own degenerated FWM products' amplitudes.

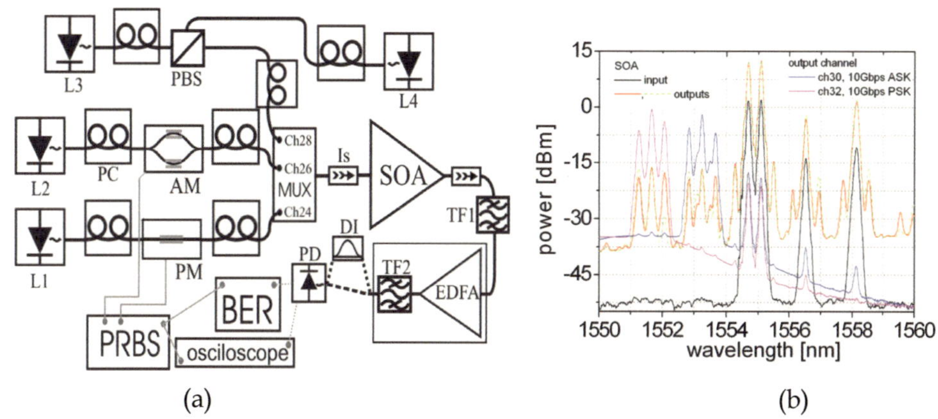

Fig. 15. Dual-channel polarization-robust AOWC: (a) the experimental setup and (b) optical spectra, for 10+10 Gb/s (ASK+PSK) operation.

For the 10+10Gb/s test the CW probes were within the ITU-grid channel number 28 (1554.94nm), and are 0.4nm apart (λL3=1554.74nm, λL4=1555.14nm); the data channels are located at channel #26 (L2, ASK) and #24 (L1, PSK). Channels #25 and #27 cannot be used since some FWM products due the interactions of the two CW probes with the input data channels are contained within their bandwidth. With this input spectra arrangement the output (converted) channels fits channels. #30 (ASK) and #32 (PSK), and are extracted by a tunable filter (TF1, 0.9nm wideband). The spectra are plotted in Fig.15(b). Once filtered the signals are further amplified by a low noise EDFA amplifier with a gain of 10dB and a noise figure of 4dB and another tunable 1.5nm wide filter (TF2) is used to remove excessive ASE before reaching the photo-diode. The detected signal is connected to a Bit-Error Rate (BER) tester to measure the performance and to an oscilloscope to obtain the eye-diagrams. The PSK channel also passes a properly tuned delay-interferometer (DI) to convert the data into ASK format.

Although the two detuned CW probes have orthogonal polarizations, some interaction between them still exists leading to FWM components on both sides of the probe signals (these signals are -30dB lower than the probes' power level). The degenerated FWM products due the interaction of each CW probe with each input channel and its replicas lead to a less then trivial spectral composition of the output channels spectra, each one with three adjacent carriers. The individual spectra contain the central (main) component, which is stable in power and two adjacent components who vary as the relative input optical polarization is changed. The small sub-peaks in the valleys in between the output ASK channels' 3 peaks (red and blue lines in Fig.15b) are due to ch.24 (PSK) FWM products, and so exclude the possibility of using the same wavelength scheme for ASK+ASK operation. The eye-diagrams and BER performance for the dual channel operation with 10+10Gb/s is shown in Fig.16, for the converted channels in the best and worst polarizations, alone or with the other carrier, as well as the back-to-back performance. The ASK output (Fig.16a) has maximum penalty of 1.5dB for the worst case, with polarization dependence between zero (single) and 0.9dB (dual channel). The PSK channel (Fig.16b) has maximum penalty of 2dB for the best case and 3dB for the worst, with minimum polarization dependence (respectively 0.3dB and 0.5dB), but presents an error-floor at 10^{-10}.

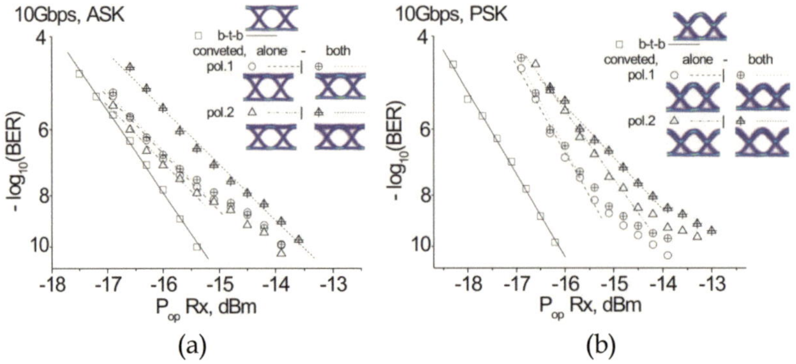

Fig. 16. BER curves and eye-diagrams (10Gbps) for dual channel conversion: (a) the ASK channel (b) the PSK channel.

This floor can be related to power fluctuation and time jitter after the PSK-to-ASK conversion in the 10GHz delayed-interferometer. With its 3 sub-carriers 50GHz apart, the output channel has reasonable part of its energy slightly detuned from the optimum point in the DI transfer function, and so some pattern dependence appears. The same setup (Fig.15a) was used to test the 20+10Gb/s, both channels in ASK. The same procedure was followed, but with the 20Gb/s input channel carrier (L2) located at ITU ch.#27, the 10Gb/s (L1) at ch.#24 and the CW probes in the ch.#29 band, and so the converted channels filtered out in the ch.#31 and ch.#34 band, with the extra channel spacing needed to avoid some 2nd order FWM that in the previous channel spacing overlapped with ch.#31's band. Fig.17 shows the eye diagrams and BER curves for the 20Gb/s (a) and 10Gb/s (b) channels and the optical spectra (c). The 20Gb/s and 10Gb/s channels have respectively maximum penalty of 2.8dB and 5.5dB for the worst polarization case, and polarization dependence respectively below 0.6dB and 0.2dB. The difference between single and dual-channel operation is larger (1.5dB) for the 10Gb/s channel in comparison with the 20Gb/s channel where it is bellow 1dB.

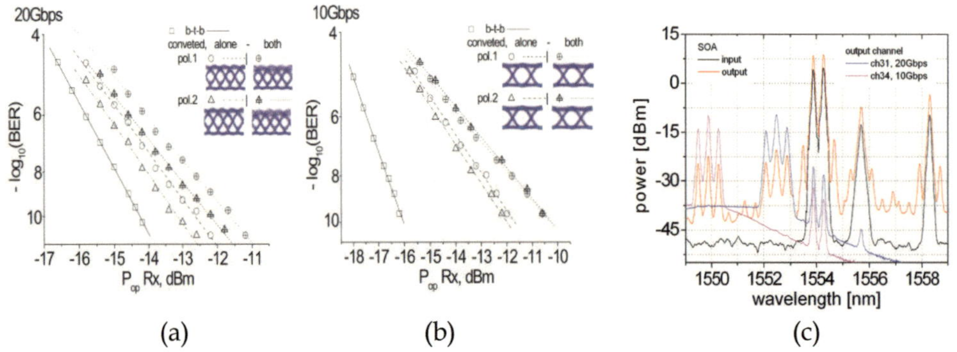

Fig. 17. BER curves and eye-diagrams for dual channel conversion (ASK+ASK): (a) the 20G channel, (b) the 10G channel; (c) optical spectra.

4. Quantom Dot SOAs

In the sections above, we have described in details how SOAs can be used for all optical wavelength conversion. One major limitation of SOAs which degrades the performance of most of the demonstrations is the relatively high noise floor of the amplifiers. This is usually the result of the narrow gain bandwidth of the semiconductor material making the SOA, which results in substantial Amplified Spontaneous Emission (ASE) noise. Quantum dot SOAs on the other hand operate in a fundamentally different manner when compared to bulk SOAs.

The predicted superiority of quantum-dot SOAs originates from the following physical properties of quantum dots. First, gain saturation occurs primarily due to spectral hole burning even for moderate peak power smaller than 20 dBm commonly used in optical communication systems ((a) Sugawara et al, 2001; (b) Sugawara et al, 2001; (c) Sugawara et al, 2001; (d) Sugawara et al, 2001; Sugawara et al, 2002). This is due to 'slow' carrier relaxation to the ground state of about 1–100 ps (Bhattacharya, 2000; Sugawara, 1999; Marcinkevicius & Leon, 2000). The response time of gain saturation is 100 fs–1 ps ((b) Sugawara et al, 2001; Sugawara et al, 2002), which is enough for a gigabit to sub-terabit optical transmission systems. Moreover, the pattern effect is negligible, owing to the compensation of the spectral holes by the carriers relaxing from the excited states including the wetting layer, i.e. the upper states work as carrier reservoirs ((a) Sugawara et al, 2001; (b) Sugawara et al, 2001; (c) Sugawara et al, 2001; (d) Sugawara et al, 2001; Sugawara et al, 2002). Second, spatial isolation of dots prevents the transfer of carriers among dots, leading to negligible cross talk between different wavelength channels under gain saturation, when the channels are separated by more than homogeneous broadening of the single-dot gain, which is about 10–20meV at 300° K (Sakamoto & Sugawara, 2000; Sugawara et al, 2000). Third, interaction of two different wavelength channels via spatially isolated and energetically non-resonant quantum dots within the same homogeneously broadened spectral hole, causes cross gain modulation and may be used for switching functions such as wavelength conversion (Sugawara et al, 2002).

These features provide a striking contrast to bulk or quantum-well SOAs. In conventional SOAs, gain saturation occurs primarily not through spectral hole burning but rather through a reduction in the total density of carriers even for optical power levels lower than 20 dBm. This is mainly due to ultrafast intra-band carrier to carrier scattering which takes place at time constants lower than 100 fs (Kuwatsuka et al, 1999). As a result, the response time of such amplifiers to sharp changes in carrier concentration, which may be caused by a strong optical pump signal for example, is dominated by carrier recombination lifetime which is of the order of 0.1–1 ns (Agrawal & Olsson, 1989), limiting the signal processing speed. Remarkable cross talk occurs between different wavelength channels because of the continuous energy states.

Demonstration of all optical wavelength conversion using QDSOA have, similar to those carried out using bulk and quantum well devices, been carried out exploiting both conversion based on carrier dynamics (XGM and XPM) as well as those based on parametric processes (FWM). For the case of FWM, it was shown that unlike bulk or quantum well semiconductor amplifiers, where conversion efficiency to longer wavelengths is generally much lower than that in the opposite direction, this property is drastically improved, and the asymmetry between conversion directions is eliminated. This is attributed to the reduction in linewidth enhancement factor due to the discreteness of the electron states in quantum dots (Akiyama et al, 2002). Due to the scarcity of QDSOA devices, and especially for QDSOA in the popular 1.55μm communications window, much more has been written in the form of numerical and analytical studies then actual experimental results, and in the experimental field most attention was given to pump and probe experiment, focusing on conversion efficiency and bandwidth, rather on BER performance and receiver sensitivity penalty.

For conversion based on XGM and XPM, the predicted picosecond recovery time scale has prompted a large research effort in this type of convertors. Below we detail some recently measured results of multi-casting achieved with QDSOAs at bit rates up to 40Gb/s, however recent work on this subject has also shown that using XGM and XPM good Q values for RZ eyes can be obtained up to a bit rate of 160Gb/s (Contestabile et al, 2009).

XGM+XPM based 1 to 4 Multicasting using QDSOA (Raz et al, 2008):

As explained above QDSOAs exhibit very fast recovery of gain, since ground states are filled within 0.1-1 psec. The enhanced blue chirped nature of XPM effects in QD-SOAs, when compared to bulk SOA, can be observed in Fig. 18 where the spectra's of wavelength converted inverted pulses from both a QD-SOA and a bulk SOA at 40-Gb/s RZ-PRBS are plotted. The output pulses resulting from XGM and XPM in the QD-SOAs (solid line) have a distinctly uneven spectral distribution of blue and red chirped components compared with a bulk SOA (dashed line), favoring the blue components, suggesting that the red shift is much shorter due to reduced recovery time for intra-dot processes (Akiyama et al, 2007).

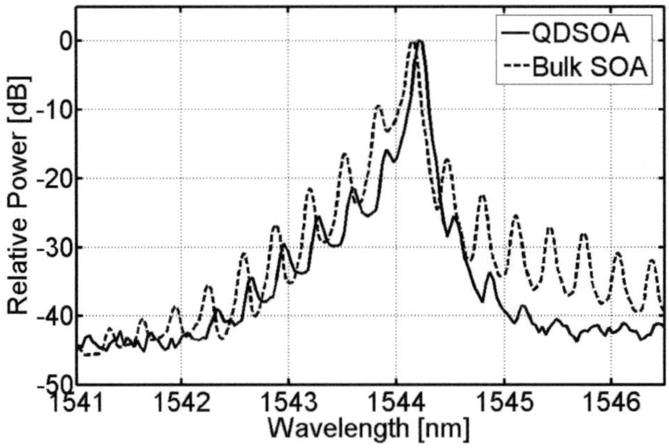

Fig. 18. The spectrum at output of a bulk SOA (dashed) compared to that of the QD-SOA (solid)

For this specific demonstration the QD-SOA we use was fabricated on an all active InP wafer. The QD material was grown on an n-type InP (100) substrate by metal-organic vapor-phase epitaxy. In the active region, five-fold stacked InAs QD layers separated by 40-nm-thick Q1.25-layer of InGaAsP were placed in the center of a 500-nm-thick lattice-matched Q1.25 waveguide core (Nötzel et al, 2006). The SOA was based on a deeply etched waveguide structure with a width of 1.6 µm, insuring single mode operation, and had a length of 2.2 mm. In order to avoid lasing, the optical waveguides were tilted by 7° and the chip-facets were anti-reflection coated. The QD-SOA was biased at 300 mA of current and cooled to 13°C, by a thermo-electric-cooler element aided by a water cooler, the device exhibited stable and repeatable performance. Main device performance merits for the operation current of 300mA include large 3dB bandwidth (>90 nm), high saturation output power (>13 dBm) and low chip noise figure (<7 dB) as well ultra-fast 10-to-90% recovery time (<10 ps at moderate bias currents). Fiber to fiber gain was approx. -10dB due to high coupling losses and field mismatch between the lensed fiber and the 2 micron wide deeply etched InP waveguides.

Experiment details and results:

The experimental set-up is shown in Fig. 19. The mode-locked fiber ring laser (MLFRL) emits pulses with a 1.3 ps full width at half maximum (FWHM) at 1560 nm. The channel separation for the CW probes was chosen to be 4.8 nm (≈600GHz) due to the large optical bandwidth of the short RZ pulse. The power of the modulated pump-beam in the waveguide was 7 dBm (assuming 6 dB insertion-losses at the input and output of the device) and the power for each CW signal was 3 dBm.

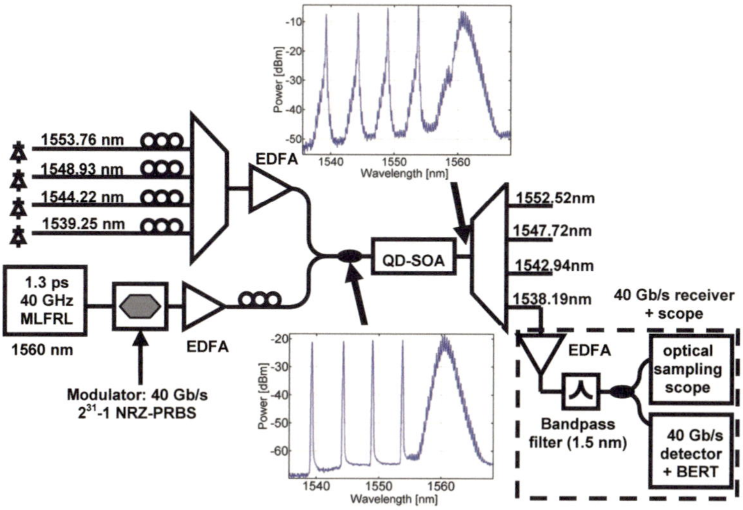

Fig. 19. Experimental set-up and optical spectra's at the QD-SOA in/output port.

Polarization controllers where used independently for each source to optimize polarization at the QD-SOA input. It is visible from the spectra of Fig. 19 that the OSNR at the QD-SOA output was larger than 42 dB. At the QD-SOA output, the channels are separated by a telecomm grade demultiplexer (DeMux). The central wavelengths of the CW signals are chosen to be +1.2 nm (≈150 GHz) detuned with respect to the central wavelengths of the DeMuX. The DeMuX had a 0.8 nm flat-top pass-band and >30 dB channel isolation. While the sharp optical filter was essential in obtaining a non-inverted output pulse, its limited pass-band resulted in a considerable pulse broadening from 1.3 to 7 ps FWHM (Fig. 20 bottom right). At the DeMuX output, the signal was further amplified and filtered to remove ASE-noise. The signals were then detected and tested for errors. BER curves for the pump signal (dashed) and the 4 converted signals (solid), as well as for a single channel under similar OSNR conditions (dash-dotted) were taken and are plotted in Fig. 20.

The measured penalty at 10-10 is in between 2 and 2.5 dB, and that for the single channel case is 2 dB. The best performing channel for the 1×4 wavelength conversion case, is that to the shortest wavelength (λ4=1539.25 nm), since it has only one adjacent signal. This channel's performance is also obtained for a 1×1 wavelength conversion (see Fig. 20 dash-dotted). The direct non-inverted error-free and low penalty, 1×4 multi-wavelength conversion, demonstrated is possible because the QD-SOA has high saturation power as

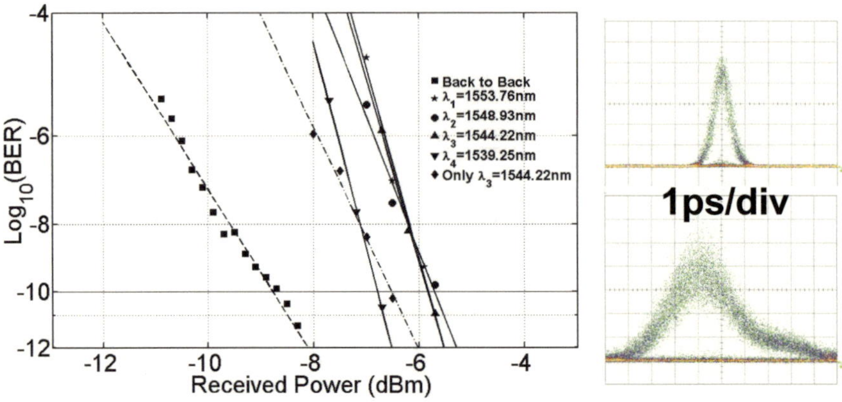

Fig. 20. Left: BER results for all the channels. Right: Eye patterns for the input and non-inverted WC output (top and bottom respectively)

well as low ASE noise. Thus when filtered out, the blue-chirp components of the output pulses have sufficient OSNR to ensure a small power penalty. The different slopes for the Back to Back and wavelength converted signals are due to the considerable broadening of the optical eye. The distorted eye is mainly due to the narrow pass bandwidth of the DeMuX and can be overcome by using a wider bandwidth filter.

5. Conclusions and future research

Semi-conductor optical amplifiers are fantastic devices for analog manipulation of light signals. Through diverse mechanisms, detailed in this chapter, it was shown that data signals at very high bit rates can be seamlessly copied from one optical carrier frequency to another. By manipulating the electrical carriers injected into the semi-conducting materials, data signals for arbitrary modulation formats have been shown to be converted through either the depletion of carriers (section 2 and 4) or the creation of index of refraction gratings (section 3). In most of the detailed demonstrations the resulting output signals are somewhat degraded, mostly due to reduced signal to noise ratio, resulting in a certain receiver sensitivity penalty. In section 4 we have also discussed how the use of novel materials, such as self assembled Quantum Dot amplifiers, can results in fast devices with improved signal to noise ratio and therefore support conversion of high speed signals.

From the overview given here it is clear that all optical wavelength conversion is a very useful application of photonic technologies. If one aims to accomplish equivalent translation of information between two optical frequencies, by moving through the electrical signal plain, the required circuit is far more complex circuit (especially so, for bit rates above 40Gb/s) and would consume far larger amounts of electrical power. Despite the obvious advantage of all optical signal processing via SOAs and the enormous research interest evident from the hundreds of scholarly publications written over the past 30 years, little if any of these methods and ideas were adopted for commercial applications. Only in the past two years there has been a slight change in the willingness to adopt such novel concepts with two high profile projects such as the MOTOR chips fabricated by UCSB (Nicholes et al, 2009) and the joint NTT & Alcatel Lucent all optical packet switch (Chiaroni et al, 2010).

In order to make SOA based all optical wavelength convertors into a main stream solution, several key developments need to take place. SOAs are essentially similar to semiconductor lasers which are being fabricated and mass manufactured for many years. The additional fiber pigtail on the second facet of the edge emitting structure, as well as the required anti-reflection coating, imply some added work and processing, but not much. Still the price of an SOA is around 2-3 orders of magnitude higher simply because the production volumes are very small. With this type of cost entry point, even the very simple SOA based wavelength convertor is more expensive then connecting, for example, two transceivers back-to-back to obtain wavelength conversion. So it is essential that prices of SOAs are sharply reduced. In addition, the noise level of SOAs remains a point of concern in case multiple hop optical networks are considered. The accumulated noise from several SOAs concatenated in an optical network, may lower the OSNR to intolerable levels. To improve signal quality, efficient optical signal re-generators need to be developed. Alternatively, further advancement of Quantum Dot amplifiers, with their low noise floor and broad gain spectrum, may prove to be extremely useful in promoting all optical wavelength conversion. On the networking side, new data transfer protocols may need to be invented which rely less on data buffering and more on exploitation of the frequency domain for the purpose of contention resolution. With these advancement there is good reason to believe that all optical wavelength conversion based on semi-conductor optical amplifiers can be am integral part of any future all optical network.

6. References

K. Inoue, "Crosstalk and its power penalty in multichannel transmission due to gain saturation in a semiconductor laser amplifier," J. Lightwave Technol., vol. 7, pp. 1118–1124, July 1989

J. Mark and J.Mork, "Sub-picosecond gain dynamics in InGaAsP optical amplifiers: Experiment and theory", Appl. Phys. Lett., 61 , 2281-2283 (1992)

CIP white paper on applications for QW SOAs, http://www.ciphotonics.com/PDFs_Jan08/SOA_Application_note_v3b.pdf

J. M. Wiesenfeld, B. Glance, J. S. Perino, A. H. Gnauck, "Wavelength Conversion at 10 Gb/s using a Semiconductor Optical Amplifier", IEEE Phot. Tech. Lett. Vol.5, No. 11,pp.1300-1303, 1993

Yasunori Miyazaki(1,2), Kazuhisa Takagi(1,2), Keisuke Matsumoto(1,2), Toshiharu Miyahara(1,2) Satoshi Nishikawa(1,2), Tatsuo Hatta(1,2), Toshitaka Aoyagi(1,2), and Kuniaki Motoshima(1,2), Polarization-Insensitive 40G-NRZ Wavelength Conversion Using SOA-MZI, Frontiers in optics 2007, FMI1]

H. Y. Yu, D. Mahgereftech, P. S. Cho, and J. Goldhar, "Optimization of the frequency response of a semiconductor optical amplifier wavelength converter using a fiber Bragg grating," J. Lightw. Technol., vol. 17, no. 2, pp. 308–315, Feb. 1999

Y. Dong, L. Lu, H. Wang, and S. Xie, "Improving performance using waveguide filter and optimal probe and signal powers for all-optical wavelength conversion," in Proc. Optical Fiber Communication (OFC), Baltimore, MD, Mar. 5–10, 2000, pp. 69–71, 2000

J. Leuthold, C. H. Joyner, B. Mikkelsen, G. Raybon, J. L. Pleumeekers, B. I. Miller, K. Dreyer, and C. A. Burrus, "100 Gb/s all-optical wavelength conversion with integrated SOA delayed-interference configuration," Electron. Lett., vol. 36, no. 13, pp. 1129–1130, Jun. 2000.

Y. Ueno, S. Nakamura, and K. Tajima, "Nonlinear phase shifts induced by semiconductor optical amplifiers with control pulses at repetition frequencies in the 40–160-GHz range for use in ultrahigh-speed all-optical signal processing," J. Opt. Soc. Amer. B, Opt. Phys., vol. 19, no. 11, pp. 2573–2589, Nov. 2002.

M. L. Nielsen and J. Mørk, "Increasing the modulation bandwidth of semiconductor-optical-amplifier-based switches by using optical filtering," J. Opt. Soc. Amer. B, Opt. Phys., vol. 21, no. 9, pp. 1606–1619, Sep. 2004.

S. Nakamura, Y. Ueno, and K. Tajima, "168-Gb/s all-optical wavelength conversion with a symmetric-Mach–Zehnder-type switch," IEEE Photon. Technol. Lett., vol. 13, no. 10, pp. 1091–1093, Oct. 2001.

Y. Liu, E. Tangdiongga, Z. Li, H. de Waardt, A. M. J. Koonen, G.D. Khoe and H.J.S. Dorren, Xuewen Shu and Ian Bennion, "Error-free 320 Gb/s SOA-based Wavelength Conversion using Optical Filtering", Proceedings of the OFC 2005, PDP28

J. Leuthold, D. M. Marom, S. Cabot, J. J. Jaques, R. Ryf, and C. R. Giles, "All-Optical Wavelength Conversion Using a Pulse Reformatting Optical Filter," Journal of Lightwave Technology, 22 , 186-192 (2004)

J. Mork and A. Mecozzi, "Theory of the ultrafast optical response of active semiconductor waveguides," Journal of Optical Society B, 13, 1803-1815 (1996)

M.L. Nielsen, J. Mørk, R. Suzuki, J. Sakaguchi, and Y. Ueno, "Experimental and theoretical investigation of the impact of ultra-fast carrier dynamics on high-speed SOA-based all-optical switches," Optics Express, 14, 331-347 (2006)

J. Mark and J.Mork, "Sub-picosecond gain dynamics in InGaAsP optical amplifiers: Experiment and theory", Appl. Phys. Lett., 61 , 2281-2283 (1992)

J. Mork and J. Mark, "Carrier heating in InGaAsP laser amplifiers due to two-photon absorption", Appl. Phys. Lett., 64 , 2206-2208 (1994)

O.Raz, J. Herrera, H.J.S. Dorren," Enhanced 40 and 80 Gb/s wavelength conversion using a rectangular shaped optical filter for both red and blue spectral slicing", Optics Express, Vol. 17, no. 3, pp. 1184-1193, 2009

Sheik-Bahae, D. C. Hutchings, D. J. Hagan, and E. W. Van Stryland, IEEE J. Quantum Electron. 27, 1296 (1991).

Y. Liu, E. Tangdiongga, Z. Li, S. Zhang, H. de Waardt, G. D. Khoe and H. J .S. Dorren, "80 Gbit/s Wavelength conversion using semiconductor optical amplifier and optical bandpass filter," Electronic Letters 41, 487-489 (2005)

Govind. P. Agrawal (2002), "*Fiber-optic Communication Systems*", John Wiley&Sons, 0-471-21571-6, USA

Jun Yamawaku et al, "Selective wavelength conversion using PPLN waveguide with two pump configuration", in conference on Lasers and Electro-optics 2003, CWB5

Takuo Tanemura et al, "Highly Efficient Arbitrary Wavelength Conversion Within Entire C-Band Based on Nondegenerate Fiber Four-Wave Mixing", IEEE Phot. Tech. Lett., Vol. 16, no. 2, pp.551-553, 2004

Govind P. Agrawal, "Four-wave mixing and phase conjugation in semiconductor laser Media", Opt. Lett., Vol.12, No.4, pp.260-262, 1987

A. Mecozzi, S. Scotti, A. D'Ottavi, E. Iannone, and P. Spano, "Four-Wave Mixing in Traveling-Wave Semiconductor Amplifiers", IEEE JQE, Vol. 31, no. 4, pp.689-699, 1995

M. A. Summerfield, R. S. Tucker, "Optimization of pump and signal powers for wavelength converters based on FWM in semiconductor optical amplifiers", IEEE Phot. Tech. Lett, Vol.8, no. 10, pp. 1316-1318, 1996

G. Contestabile, M. Presi, E. Ciaramella, "Multiple Wavelength Conversion for WDM Multicasting by FWM in an SOA", IEEE Phot. Tech. Lett, Vol. 16, no. 7, 2004

G. Contestabile et al, "Transparency of FWM in SOAs to Phase/Amplitude and Polarization", in Optical Fiber Communication Conference, paper OThM6, 2009

Cristiano M. Gallep, Harmen J.S. Dorren and Oded Raz Members IEEE," Four Wave Mixing Based Dual Wavelength Conversion in a Semiconductor Optical Amplifier", IEEE Phot. Tech. Lett, in press

C.M. Gallep, O. Raz, H.J.S. Dorren, "Polarization Independent Dual Wavelength Converter Based on FWM in a Single Semiconductor Optical Amplifier", Proceedings of the OFC 2010, OWP2, 2010

(a) M. Sugawara, N. Hatori, T. Akiyama, Y. Nakata, H. Ishikawa, Japan. J. Appl. Phys. 40 L488, 2001

(b) M. Sugawara, N. Hatori, T. Akiyama, Y. Nakata, Proc. Indium Phosphide and Related Materials '01 p 358, 2001

(c) M. Sugawara, N. Hatori, T. Akiyama, Y. Nakata, Proc. Indium Phosphide and Related Materials '01 p 471, 2001

(d) M. Sugawara, N. Hatori, T. Akiyama, Y. Nakata, H. Ishikawa, Proc. 4th Pacific Rim Conf. on Lasers and Electro-Optics p I-260, 2001

M. Sugawara, T. Akiyama, N. Hatori, Y. Nakata, Proc. Contemporary Photonics Technology 2002 p 139, 2002

A. Sakamoto, M. Sugawara, IEEE Photon. Technol.Lett. 12 pp. 107-109, 2000

M. Sugawara, K. Mukai, Y. Nakata, H. Ishikawa, Phys.Rev. B 61 pp.7595, 2000

P. Bhattacharya, IEEE JSTQE Vol. 6 pp. 426, 2000

A.V. Uskov, J. McInerney, Appl. Phys. Lett. Vol. 72 pp.58, 2000

M. Sugawara (ed) 1999 Semiconductors & Semimetals vol 60 (San Diego, CA: Academic) ch 5 p 209, 1999

S. Marcinkevicius, R. Leon, Appl. Phys. Lett. 76 2406, 2000

H. Kuwatsuka, T. Shimoyama, H. Ishikawa, IEEE JQE. Vol. 35 pp.1817, 1999

G.P. Agrawal, N.A. Olsson, IEEE JQE. Vol. 25 pp. 2297, 1989

T. Akiyama, H. Kuwatsuka, , N. Hatori, Y. Nakata, H. Ebe, M. Sugawara, "Symmetric Highly Efficient (0 dB) Wavelength Conversion Based on Four-Wave Mixing in Quantum Dot Optical Amplifiers", IEEE Phot. Tech. Lett, Vol. 14, no. 8, 2002

G. Contestabile, A. Maruta, S. Sekiguchi, K. Morito, M. Sugawara, K. Kitayama, "160 Gb/s cross gain modulation in quantum Dot SOA at 1550 nm", Proceedings of ECOC 2009, PDP. 1.4, 2009

O.Raz, J. Herrera, N. Calabretta, E. Tangdiongga, S. Anantathanasarn, R. Nötzel, H.J.S. Dorren, "Non-Inverted Multiple Wavelength Converter at 40GBs Using 1550nm Quantum Dot SOA", Electronic Letters, Vol. 44, No. 16, pp. 988-989, 2008

T. Akiyama et al, "Quantum-dot semiconductor optical amplifiers", Proceedings of the IEEE, 95, 757-1766 (2007)

R. Nötzel et al, "Self assembled InAs/InP quantum dots for telecom applications in the 1.55 µm wavelength range: Wavelength tuning, stacking, polarization control, and lasing (Review Paper)", Jpn. J. Appl. Phys., 45, 6544 (2006)

Steven C. Nicholes, Milan L. Mašanović, Biljana Jevremović, Erica Lively, Larry A. Coldren, and Daniel J. Blumenthal, "The World's First InP 8x8 Monolithic Tunable Optical Router (MOTOR) Operating at 40 Gbps Line Rate per Port", proc. of OFC 2009

Dominique Chiaroni, Géma Buforn Santamaria, Christian Simonneau, Sophie Etienne, Jean-Christophe Antona, Sébastien Bigo, Jesse Simsarian, "Packet OADMs for the next generation of ring networks ",Bell Labs Technical Journal Volume 14, Issue 4, pages 265-283, Winter 2010

5

The Application of Semiconductor Optical Amplifiers in All-Optical Wavelength Conversion and Radio Over Fiber Systems

Lin Chen, Jianjun Yu, Jia Lu, Hui Zhou and Fan Li
Hunan University,
China

1. Introduction

Wavelength conversion has been suggested as one of the key functions for wavelength-division-multiplexing (WDM) optical networks and photonic switch blocks. Several methods, such as a self-phase modulation (SPM), a cross-gain modulation (XGM), and cross-phase modulation (XPM), can be used to realize all optical wavelength conversion (AOWC) [1-29]. However, four-wave mixing (FWM) based on nonlinear media, such as optical fiber and semiconductor optical amplifier (SOA), is considered to be the most promising scheme because it is fully transparent to the signal bit rate and modulation format. AOWC in SOA has many advantages such as easily compatible and highly covert efficiency. AOWC based on FWM in SOA for regular signal such as ON/OFF keying (OOK) signal has already investigated maturely but not for OFDM signals.

This chapter discusses the performance for OFDM signal in AOWC based on FWM in an SOA. We found the result for OFDM signal is the same as that of OOK signal. Multiple frequency mm-wave generation is one of the key techniques in radio over fiber (ROF) system. Many methods can generate multiple frequency mm-wave such as using optical carrier suppression (OCS), suppression of odd-order sidebands, multi-cascaded external modulators and so on. Some references have proposed that multiple frequency mm-wave can be generated by using SOA based on FWM effect and discuss polarization insensitive in SOA. This chapter also introduces this method to generate mm-wave and discusses the polarization insensitive all-optical up-conversion for ROF system based on FWM in a SOA. We have proposed and experimentally investigated polarization insensitive all-optical up-conversion for ROF system based on FWM in a SOA. One method is that a parallel pump is generated based on odd-order optical sidebands and carrier suppression using an external intensity modulator and a cascaded optical filter. Therefore, the two pumps are always parallel and phase locked, which makes the system polarization insensitive. This scheme has some unique advantages such as polarization insensitive, high wavelength stability, and low-frequency bandwidth requirement for RF signal and optical components. The other method is where co-polarized pump light-waves are generated by OCS modulation to keep the same polarization direction and phase locking between two pumps. This scheme also has excellent advantages such as small size, high-gain, polarization insensitivity, and low-frequency bandwidth requirement for RF signal and optical components, and high wavelength stability. The results of above two mentioned experiments show that the scheme based on dual-pump FWM in a SOA is one of the most promising all-optical up-conversions for radio-over-fiber systems.

2. OFDM signal generation in our system description

In this section the basic functions of the generation in our system are described. The OFDM baseband signals are calculated with a Matlab program including mapping 2^{15}-1 PRBS into 256 QPSK-encoded subcarriers, among them, 200 subcarriers are used for data and 56 subcarriers are set to zero as guard intervals. The cyclic prefix in time domain is 1/8, which would be 32 samples every OFDM frame. Subsequently converting the OFDM symbols into the time domain by using IFFT and then adding 32 pilots signal in the notch. The guard interval length is 1/4 OFDM period. 10 training sequences are applied for each 150 OFDM-symbol frame in order to enable phase noise compensation. At the output the AWG low-pass filters (LPF) with 5GHz bandwidth are used to remove the high-spectral components. The digital waveforms are then downloaded to a Tektronix AWG 610 arbitrary waveform generator (AWG) to generate a 2.5Gb/s electrical OFDM signal waveform.

3. AOWC based on FWM in SOA for OFDM signal

AOWC has been regarded as one of the key techniques for wavelength-division-multiplexing (WDM) optical networks and photonic switch blocks and it can enhance the flexibility of WDM network management and interconnection [30-35]. Nowadays, there are some main techniques for wavelength conversion, which include XGM [33], XPM [34] and FWM [35-38].FWM is considered to be the most promising scheme because it is fully transparent to the signal bit rate and modulation format.

OFDM is as one of the key techniques for 4G (the Fourth Generation Mobile Communication System), immune to fiber dispersion and polarization mode dispersion in optical fiber communication [39-42]. AOWC based on FWM in SOA for regular signal, such as OOK signals, has already been investigated but not for OFDM signals.

We have theoretically analyzed and experimentally demonstrated three schemes for pumping, including single-pump, orthogonal-dual-pump and parallel-dual-pump based on the FWM effect for OFDM signal in SOA for wavelength conversion. Analysis result shows that: (1) the new converted wavelength signal carry the original signal, (2)single-pump scheme is sensitive to polarization, while orthogonal-dual-pump and parallel-dual-pump schemes are insensitive to polarization, (3)parallel-dual-pump scheme has the highest wavelength conversion efficiency, (4)Conversion efficiency of the converted signals are proportional to the amplitudes of the input signal and the pumps. In the single pump scheme, the conversion efficiency depends on the polarization angle between the pump and signal lightwave. In these dual-pump schemes, the conversion efficiency also depends on the frequency spacing between the pumps or between the signal and pump lightwave.

3.1 Theory and result

Figure 1 shows the configuration of all-optical wavelength conversion systems based on FWM for OFDM signal in a SOA. In the system, OFDM signal can be modulated on to a light wave generated from a distributed feedback laser diode(DFB-LD1) by an external intensity modulator (IM),two pumps are generated from DFB-LD2 and DFB-LD3, the modulated signal light wave and pump light waves are coupled and then amplified by EDFA before they are injected into the SOA for FWM process. After wavelength conversion and optical filtering by a circulator and a FBG, the new converted signal carried original signal can be obtained.

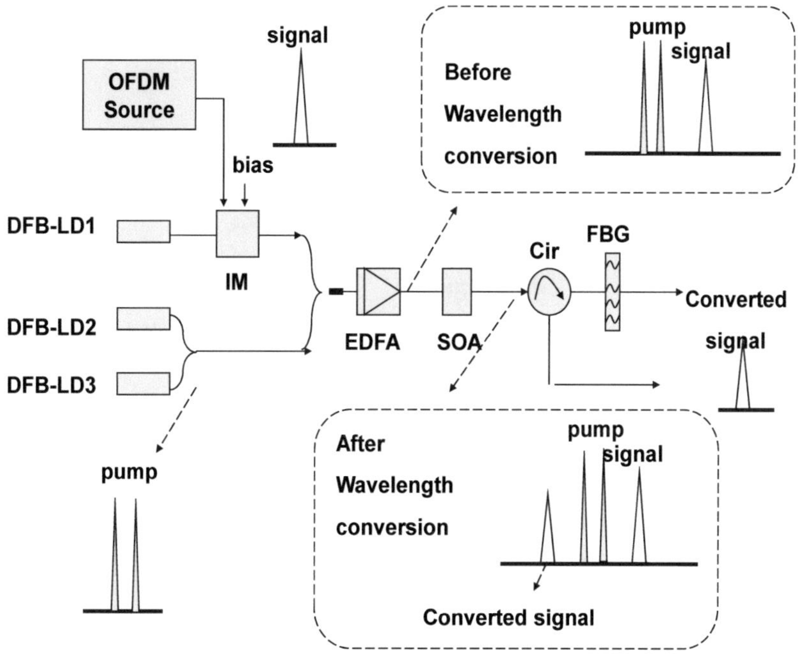

Fig. 1. Configuration of all-optical wavelength conversion systems based on FWM in a SOA. DFB-LD: Distributed feedback-laser diode. FBG: Fiber bragg grating. IM: Intensity modulator. SOA: Semicondoctor optical amplifier. Cir: Circulator.

Fig. 2 shows the principle of all-optical wavelength conversion systems based on FWM effect in an SOA. We build a coordinate system: for simplicity, the signal is assumed to be aligned with the X axis(horizontal orientation),Y axis(vertical orientation), pump1 is at some angle θ with respect to the X axis, and pump2 is at some angle ϕ with respect to X axis. After being amplified by an SOA, the optical field of pump light waves can be expressed as $E_i(\omega_i,\vec{r},t)=E_i(\omega_i,\vec{r})\exp j\left(k_i z-\omega_i t+\phi_i\right)$ (i=1,2). Here, k_i , ω_i = ϕ_i represent optical wave vector, angle frequency and phase, respectively. i=1,2 represent pump1 and pump2. The optical field of modulated signal light wave can be expressed as follows:

$$E_3(\omega_3,\vec{r},t)=A_3E_3(\omega_3,\vec{r})\exp j(k_3z-\omega_3t+\phi_3) \tag{1}$$

Here, A_3 represents the amplitude of the signal light wave. According to the principle of the four wave mixing effect, it can be envisaged as pairs of light waves to generate a beat, which modulate the input fields to generate upper or lower sidebands.

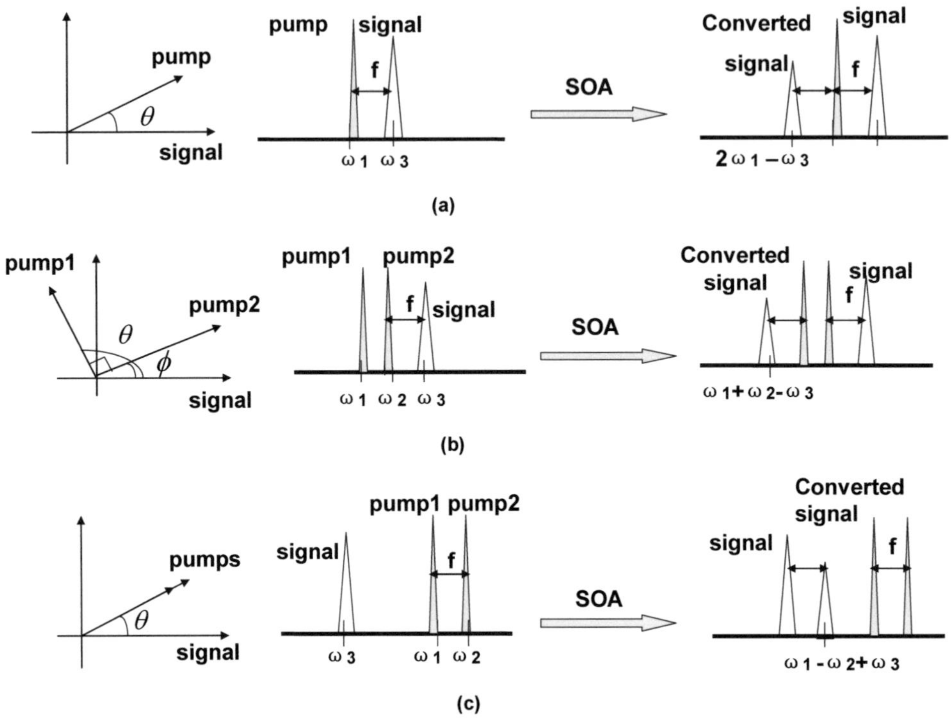

Fig. 2. Principle of all-optical wavelength conversion based on FWM effect. (a)single pump. (b)orthogonal pump. (c) parallel pump

3.1.1 Principle of single-pump configuration for wavelength conversion

In the single-pump configuration, a signal light wave and a pump light wave generate a beat $|\omega_1 - \omega_3|$, and its amplitude can be expressed as:

$$\alpha = r(\omega_1 - \omega_3)[(\vec{A}_1\vec{A}_3^*)\exp j(\omega_1 - \omega_3)t + (\vec{A}_1^*\vec{A}_3)\exp j(\omega_3 - \omega_1)t] \tag{2}$$

The beat $|\omega_1 - \omega_3|$ modulates ω_1 to produce upper and lower sidebands around ω_1 with frequency span of $|\omega_1 - \omega_3|$ and the optical field can be expressed as:

$$\begin{aligned}\vec{\mathrm{E}}_s &= \alpha\vec{E}_1(\omega_1) \\ &= r(\omega_1 - \omega_3)[(\vec{A}_1\vec{A}_3^*)\exp j(\omega_1 - \omega_3)t + (\vec{A}_1^*\vec{A}_3)\exp j(\omega_3 - \omega_1)t]\vec{E}_1(\omega_1) \\ &= r(\omega_1 - \omega_3)A_1A_3\cos\theta\vec{A}_1\left\{e^{j(2\omega_1-\omega_3)t+(2\phi_1-\phi_3)} + e^{j(\omega_3 t+\phi_3)}\right\}\end{aligned} \tag{3}$$

The beat $|\omega_1 - \omega_3|$ modulates ω_3 to produce upper and lower sidebands around ω_3 with a frequency span of $|\omega_1 - \omega_3|$ and the optical field can be expressed as:

$$\begin{aligned}\vec{\mathrm{E}}_s &= \alpha\vec{E}_3(\omega_3) \\ &= r(\omega_1 - \omega_3)[(\vec{A}_1\vec{A}_3^*)\exp j(\omega_1 - \omega_3)t + (\vec{A}_1^*\vec{A}_3)\exp j(\omega_3 - \omega_1)t]\vec{E}_3(\omega_3) \\ &= r(\omega_1 - \omega_3)A_1A_3\cos\theta\vec{A}_3\left\{e^{j(\omega_1 t+\phi_1)} + e^{j(2\omega_3-\omega_1)t+(2\phi_3-\phi_1)}\right\}\end{aligned} \tag{4}$$

What we are interested in is the optical frequency $2\omega_1 - \omega_3$, which is contributed by $|\omega_1 - \omega_3|$ modulating ω_1 and ω_3 .Then, the optical field of new generated frequency wavelength can be expressed as:

$$\vec{\mathrm{E}}_{2\omega_1-\omega_3} = (E_1.E_3^*)E_1 = A_1A_3r(\omega_1-\omega_3)\cos\theta\vec{A}_1e^{j[(2\omega_1-\omega_3)t+(2\varphi_1-\varphi_3)]} \quad (5)$$

Here, A_1 and A_3 represent the amplitudes of pump and newly converted signal light wave after four wave mixing effect, respectively. $r(\omega_1-\omega_3)$ is the conversion efficiency coefficient which is proportional to the frequency difference. On the basis of Eq. (5), we can derive the expression of optical power of the new signal as follows:

$$P_{2\omega_1-\omega_3} = A_1^2A_3^2r^2(\omega_1-\omega_3)\cos(\theta) \quad (6)$$

From Eq. (6) we can see that the output optical power is dependent on the frequency difference and the polarization angle between the pump and signal lightwave. The greater the frequency difference, the lower the conversion efficiency. When the polarization of the pump and the signal light are parallel, the output optical power takes maximum value. When the polarization of the pump and the signal light are orthogonal, the output optical power takes minimum value. From the above analysis, it appears that single-pump configuration is a polarization sensitive system.

3.1.2 Principle of orthogonal-pump configuration for wavelength conversion

In the orthogonal-pump configuration, three light waves with the frequencies of ω_1 , ω_2 and ω_3 generate three beats $|\omega_1-\omega_2|$, $|\omega_1-\omega_3|$ and $|\omega_2-\omega_3|$, each beat will modulate each input lightwave and generate two sidebands

The amplitude of beat $|\omega_1-\omega_3|$ can be expressed as:

$$\alpha = r(\omega_1-\omega_3)[(\vec{A}_1\vec{A}_3^*)\exp j(\omega_1-\omega_3)t + (\vec{A}_1^*\vec{A}_3)\exp j(\omega_3-\omega_1)t] \quad (7)$$

The beat $|\omega_1-\omega_3|$ modulates ω_2 to produce upper and lower sidebands around ω_2 with frequency span of $|\omega_1-\omega_3|$ and the optical field can be expressed as:

$$\begin{aligned}\vec{\mathrm{E}}_s &= \alpha\vec{E}_3(\omega_3)\\ &= r(\omega_1-\omega_3)[(\vec{A}_1\vec{A}_3^*)\exp j(\omega_1-\omega_3)t + (\vec{A}_1^*\vec{A}_3)\exp j(\omega_3-\omega_1)t]\vec{E}_2(\omega_2)\\ &= r(\omega_1-\omega_3)A_1A_3\cos\theta\vec{A}_2\left\{e^{j[(\omega_2+\omega_1-\omega_3)t+(\phi_2+\phi_1-\phi_3)]} + e^{j[(\omega_2+\omega_3-\omega_1)t+(\phi_2+\phi_3-\phi_1)]}\right\}\end{aligned} \quad (8)$$

The amplitude of the beat $|\omega_2-\omega_3|$ can be expressed as:

$$\alpha = r(\omega_2-\omega_3)[(\vec{A}_2\vec{A}_3^*)\exp j(\omega_2-\omega_3)t + (\vec{A}_2^*\vec{A}_3)\exp j(\omega_3-\omega_2)t] \quad (9)$$

The beat $|\omega_2-\omega_3|$ modulates ω_1 to produce upper and lower sidebands around ω_1 with the frequency span of $|\omega_2-\omega_3|$ and the optical field can be expressed as:

$$
\begin{aligned}
\vec{E}_s &= \alpha \vec{E}_1(\omega_1) \\
&= r(\omega_2-\omega_3)[(\vec{A}_2\vec{A}_3^*)\exp j(\omega_2-\omega_3)t + (\vec{A}_2^*\vec{A}_3)\exp j(\omega_3-\omega_2)t]\vec{E}_1(\omega_1) \\
&= r(\omega_2-\omega_3)A_2A_3\cos(\phi)\vec{A}_1\left\{e^{j[(\omega_1+\omega_2-\omega_3)t+(\phi_1+\phi_2-\phi_3)]} + e^{j[(\omega_1+\omega_3-\omega_2)t+(\phi_1+\phi_3-\phi_2)]}\right\}
\end{aligned} \tag{10}
$$

What we are interested in is optical frequency $\omega_1+\omega_2-\omega_3$, which is contributed by $|\omega_1-\omega_3|$ modulating ω_2 and $|\omega_2-\omega_3|$ modulating ω_1 .Thus, after a SOA the optical field of newly generated frequency wavelength can be expressed as:

$$
\begin{aligned}
\vec{E}_{\omega_1+\omega_2-\omega_3} &= (E_1.E_3^*)E_2 + (E_2.E_3^*)E_1 \\
&= [r(\omega_1-\omega_3)A_1A_3\cos\theta\vec{A}_2 + r(\omega_2-\omega_3)\cos(\phi)A_2A_3\vec{A}_1]e^{j[(\omega_1+\omega_2-\omega_3)t+(\phi_1+\phi_2-\phi_3)]}
\end{aligned} \tag{11}
$$

When $\theta-\phi=\frac{\pi}{2}$, it means that the signal and pump are orthogonally polarized, namely,

$$\cos\phi = \cos(\frac{\pi}{2}-\theta) = \sin\theta \tag{12}$$

Eq. (11) can be written as following:

$$
\begin{aligned}
\vec{E}_{\omega_1+\omega_2-\omega_3} &= (E_1.E_3^*)E_2 + (E_2.E_3^*)E_1 \\
&= [r(\omega_1-\omega_3)A_1A_3\cos\theta\vec{A}_2 + r(\omega_2-\omega_3)A_2A_3\sin(\theta)\vec{A}_1]e^{j[(\omega_1+\omega_2-\omega_3)t+(\phi_1+\phi_2-\phi_3)]}
\end{aligned} \tag{13}
$$

Here, A_1, A_2 and A_3 represent the amplitudes of pumps and newly converted signal light wave after four wave mixing effect, $r(\omega_1-\omega_3)$ and $r(\omega_2-\omega_3)$ represent the conversion efficiency coefficient , which is inversely proportional to the frequency difference. From Eq. (13) ， It can be seen that the output power of the optical frequency is:

$$P_{\omega_1+\omega_2-\omega_3} = A^2{}_3[r^2(\omega_1-\omega_3)A_1^2A_2^2\cos^2\theta + r^2(\omega_2-\omega_3)A_2^2A_1^2\sin^2\theta] \tag{14}$$

Because $|\omega_1-\omega_3| \approx |\omega_2-\omega_3|$, we obtain: $r(\omega_1-\omega_3) \approx r(\omega_2-\omega_3)$
Therefore, Eq. (13) can be simplified to

$$P_{\omega_1+\omega_2-\omega_3} = A^2{}_3r^2(\omega_1-\omega_3)A_1{}^2A_2{}^2(\cos^2\theta+\sin^2\theta) = A_1{}^2A_2{}^2A^2{}_3r^2(\omega_1-\omega_3) \tag{15}$$

It can be seen that output signal optical power is independent of θ, that is to say, the orthogonal-dual-pump configuration is a polarization insensitive system, and its optical power relies on $r(\omega_1-\omega_3)$ with the interval of pump and signal light wave frequency increasing, the optical power gradually decrease.

3.1.3 Principle of parallel-dual-pump configuration for wavelength conversion

In the parallel-dual-pump configuration, three light waves with frequency of ω_1, ω_2 and ω_3 generate three beats $|\omega_1-\omega_2|$, $|\omega_1-\omega_3|$ and $|\omega_2-\omega_3|$, each beat will modulate each input lightwave and generate two sidebands.

The amplitude of beat $|\omega_1 - \omega_2|$ can be expressed as:

$$\alpha = r(\omega_1 - \omega_2)[(\vec{A}_1\vec{A}_2^*)\exp j(\omega_1 - \omega_2)t + (\vec{A}_1^*\vec{A}_2)\exp j(\omega_2 - \omega_1)t] \quad (16)$$

The beat $|\omega_1 - \omega_2|$ modulates ω_3 to produce upper and lower sidebands around ω_3 with the frequency span of $|\omega_1 - \omega_2|$ and the optical field can be expressed as:

$$\begin{aligned} \vec{E}_s &= \alpha\vec{E}_3(\omega_3) \\ &= r(\omega_1 - \omega_2)[(\vec{A}_1\vec{A}_2^*)\exp j(\omega_1 - \omega_2)t + (\vec{A}_1^*\vec{A}_2)\exp j(\omega_2 - \omega_1)t]\vec{E}_3(\omega_3) \\ &= r(\omega_1 - \omega_2)A_1A_2\cos\theta\vec{A}_3\left\{e^{j[(\omega_3+\omega_1-\omega_2)t+(\phi_3+\phi_1-\phi_2)]} + e^{j[(\omega_3+\omega_2-\omega_1)t+(\phi_3+\phi_2-\phi_1)]}\right\} \end{aligned} \quad (17)$$

The amplitude of beat $|\omega_3 - \omega_2|$ can be expressed as:

$$\alpha = r(\omega_3 - \omega_2)[(\vec{A}_3\vec{A}_2^*)\exp j(\omega_3 - \omega_2)t + (\vec{A}_3^*\vec{A}_2)\exp j(\omega_2 - \omega_3)t] \quad (18)$$

The beat $|\omega_3 - \omega_2|$ modulates ω_1 to produce upper and lower sidebands around ω_1 with the frequency span of $|\omega_3 - \omega_2|$ and the optical field can be expressed as:

$$\begin{aligned} \vec{E}_s &= \alpha\vec{E}_1(\omega_1) \\ &= r(\omega_3 - \omega_2)[(\vec{A}_3\vec{A}_2^*)\exp j(\omega_3 - \omega_2)t + (\vec{A}_3^*\vec{A}_2)\exp j(\omega_2 - \omega_3)t]\vec{E}_1(\omega_1) \\ &= r(\omega_3 - \omega_2)A_3A_2\cos\theta\vec{A}_1\left\{e^{j[(\omega_1-\omega_2+\omega_3)t+(\phi_1-\phi_2+\phi_3)]} + e^{j[(\omega_1+\omega_2-\omega_3)t+(\phi_1+\phi_2-\phi_3)]}\right\} \end{aligned} \quad (19)$$

What we are interested in is the optical frequency $\omega_1 - \omega_2 + \omega_3$, which is contributed by the beat $|\omega_1 - \omega_2|$ modulateing ω_3 and beat $|\omega_3 - \omega_2|$ modulateing ω_1

$$\begin{aligned} \vec{E}_{\omega_1-\omega_2+\omega_3} &= (E_1.E_2^*)E_3 + (E_3.E_2^*)E_1 \\ &= [r(\omega_1 - \omega_2)A_1A_2\cos(\theta-\phi)\vec{A}_3 + r(\omega_3 - \omega_2)A_3A_2\cos(\phi)\vec{A}_1]e^{j[(\omega_1-\omega_2+\omega_3)t+(\varphi_1-\varphi_2+\varphi_3)]} \end{aligned} \quad (20)$$

Here, A_1, A_2 and A_3 represent the amplitudes of pumps and new converted signal light wave after the four wave mixing effect, $r(\omega_1 - \omega_2)$ and $r(\omega_3 - \omega_2)$ represent conversion efficiency coefficient , which is inversely proportional to the frequency difference.
When $\theta = \phi$, it means that signal and pump are parallel polarized and there is $r(\omega_1 - \omega_2) >> r(\omega_3 - \omega_2)$ because $(\omega_1 - \omega_2)$ is much smaller than $(\omega_3 - \omega_2)$. Therefore, Eq. (20) depends largely on the first term and the second term can be basically ignored. Therefore, the signal polarization has little effect one the output optical power and Eq. (19) reduces to

$$\vec{E}_{\omega_1-\omega_2+\omega_3} = [A_1A_2r(\omega_1 - \omega_2)\vec{A}_3 + A_3A_2r(\omega_3 - \omega_2)\vec{A}_1\cos(\varphi)]e^{j[(\omega_1-\omega_2+\omega_3)t+(\phi_1-\phi_2+\phi_3)\_]} \quad (21)$$

From Eq. (21), It can be seen that the power of the optical frequency is

$$P_{\omega_1-\omega_2+\omega_3} = A_1^2A_2^2A_3^2[r^2(\omega_1 - \omega_2) + r^2(\omega_3 - \omega_2)\cos^2(\phi)] \quad (22)$$

We can see that if the signal light polarization direction is parallel to the pump light polarization ($\phi = 0$), the output power takes a the maximum; whereas, if the signal light polarization direction is orthogonal to that of the pump ($\varphi = \frac{\pi}{2}$), the output power takes a minimum. Therefore, the converted signal power depends on the frequency interval between pumps, signal and pump and the polarization angle between them. However, we can conclude that parallel-dual-pump configuration is polarization insensitive system.

Through the above analysis above, output optical power of the structures of single-pump, orthogonal-dual-pump and parallel double-pump is:

$$\begin{aligned} P_{2\omega_1-\omega_3} &= A_1{}^2 A_3{}^2 r^2(\omega_1-\omega_3)\cos^2(\theta) \\ P_{\omega_1+\omega_2-\omega_3} &= A_1{}^2 A_2{}^2 A^2{}_3 r^2(\omega_1-\omega_3) \\ P_{\omega_1-\omega_2+\omega_3} &= A_1^2 A_2^2 A_3^2[r^2(\omega_1-\omega_2)+r^2(\omega_3-\omega_2)\cos^2(\phi)] \end{aligned} \tag{23}$$

At first, it seems from the above equations that the new wavelength converted signal carries the original signal.

Secondly, because the conversion efficiency coefficient is inversely proportional to the frequency interval, such a relationship $r(\omega_1-\omega_2) >> r(\omega_3-\omega_2)$ that the parallel pump has the highest wavelength conversion efficiency.

Finally, OFDM as one of the key techniques for 4G, is immune to fiber dispersion and polarization mode dispersion in optical fiber communication. We investigated AOWC based on FWM in a SOA for OFDM signal, which is of great significance. If we introduce OFDM signal into a AOWC, A_3 represents the amplitudes of OFDM signal light wave, it is a time-related functions, we can see from the above formula that the new converted wavelength signal carry the original OFDM signal. Therefore, the performance for OFDM signal in AOWC based on FWM in a SOA is the same as that of OOK signal.

3.2 Experimental setup

Fig. 3 shows the experimental configuration setup and results for an all-optical wavelength conversion based on the single pump FWM effect in a SOA. Two continuous lightwaves generated by the DFB-LD1 and DFB-LD2 at 1544.25nm and 1544.72nm, are used for the pump light and signal light. AWG produces 2.5Gb/s based on the orthogonal phase-shift keyed modulation OFDM signal and its electrical spectrum is shown in Fig. 3 (a). The CW light generated by DFB-LD1 at 1544.72nm signal light is modulated via a single-arm LN-MOD biased at 2.32V.The half-wave voltage ($v\pi$) of the LN-MOD is 7.8V, its 3dB bandwidth is greater than 8GHZ, and its extinction ratio is greater than 25dB.The 2.5 Gbit/s optical signals and the pump signal are combined by a optical coupler (OC) before an erbium-doped fiber amplifiers (EDFA) which is used to boost the power of the two signals. The optical spectra before and after SOA are shown in Fig. 3 (b) and (c), respectively. The optical power of the signal light, pump lights are 5.38dBm, 8.8dBm and 8.0dBm, respectively. As shown in Fig3(c), wavelength of the converted signal is 1543.78nm, optical signa-to-noise power ratio(OSNR) is 25dBm.The wavelength conversion efficiency is -15dB.A FBG with a 3dB bandwidth of 0.15nm and a TOF with a 0.5 nm bandwidth is used to filter out the converted signal. The converted OFDM signal is send to 10Gb/s optical receiver. The OFDM signal detected from optical receiver is sent to a real-time oscilloscope for data collection.

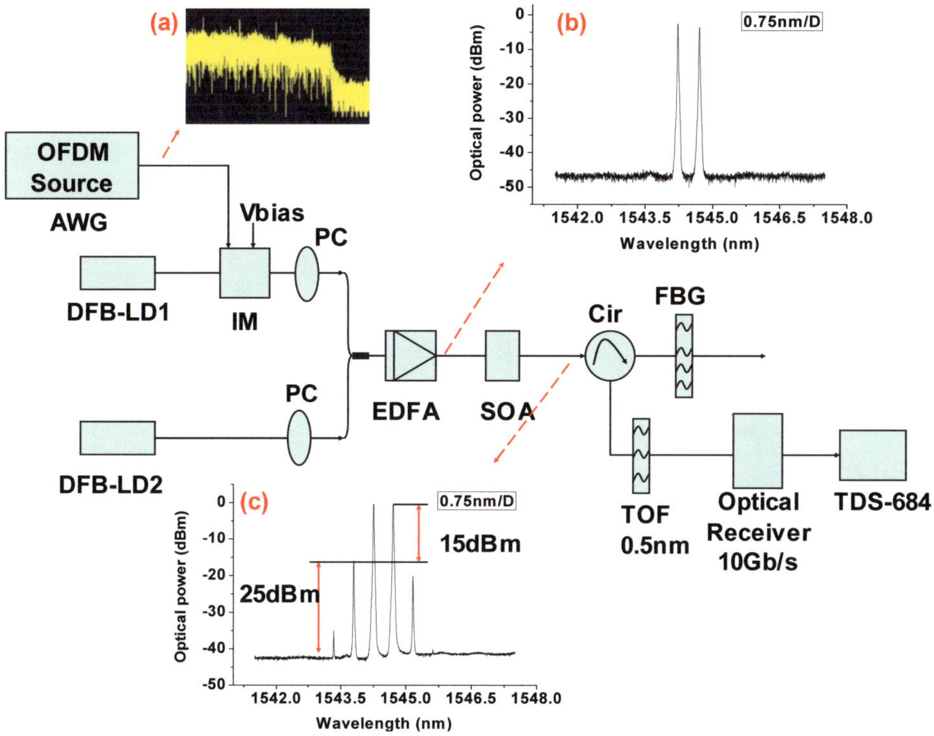

Fig. 3. Configuration of experimental setup and results for all-optical wavelength conversion based on single pump FWM effect in SOA.(a) Electrical spectra of the OFDM signal; (b)optical spectral of the combined signals before SOA; (c)optical spectra of signal after SOA. DFB-LD:distributed feedback laser diode; FBG:Fiber bragg grating, IM:Intensity modulator, SOA:Semicondoctor optical amplifier ,Cir:Circulator; TOF: Tunable optical filter

Fig. 4 shows the experimental configuration setup and results for the all-optical wavelength conversion based on the single pump FWM effect in a SOA. Two continuous lightwaves generated by the DFB-LD2 and DFB-LD3 at 1544.15nm and 1544.65nm, are used for the pump lights. AWG produces 2.5Gb/s based on the orthogonal phase-shift keyed modulation OFDM signal, and its electrical spectrum is shown in Fig4 (a). The CW light generated by DFB-LD1 at 1545.05nm is modulated via a single-arm LN-MOD biased at 1.62V.The half-wave voltage ($v\pi$) of the LN-MOD is 7.8V, its 3dB bandwidth is greater than 8GHz and its extinction ratio is greater than 25dB.The 2.5 Gbit/s optical signals and the pump signals are combined by a optical coupler (OC) before EDFA to boost the power of the two signals. The optical spectra before and after SOA are shown in Fig.4 (b) and (c), respectively. The optical power of the signal light and pump lights are 5.7dBm, 11.6dBm and 11.6dBm, respectively. As shown in Fig4(c), the wavelength of the converted signal is 1543.76nm, optical signal-to-noise power ratio(OSNR) is 25dBm.The wavelength conversion efficiency is -15dB.A FBG with a bandwidth of 0.15nm and a TOF with a 0.5 nm bandwidth is used to filter out the converted signal. The converted OFDM signal is sent to the 10Gb/s optical receiver. The OFDM signal detected from optical receiver is then sent to the real-time oscilloscope for data collection.

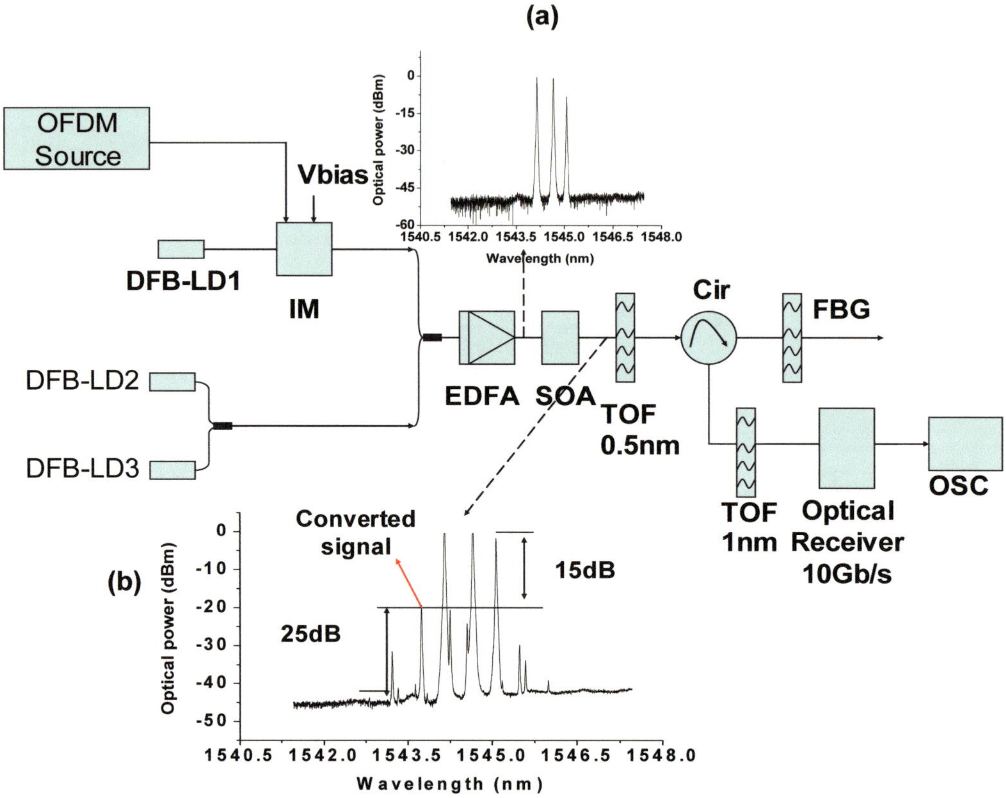

Fig. 4. Configuration of experimental setup and results for all-optical wavelength conversion based on orthogonal-dual-pump FWM effect in SOA.(a) Electrical spectra of the OFDM signal;(b)optical spectral of the combined signals before SOA;(c)optical spectra of signal after SOA. DFB-LD:distributed feedback laser diode; FBG:Fiber bragg grating, IM:Intensity modulator, SOA:Semicondoctor optical amplifier ,Cir:Circulator; TOF: Tunable optical filter. OSC: oscillator.

Fig. 5 shows the experimental configuration setup and results for the all-optical wavelength conversion based on the single pump FWM effect in a SOA. Two continuous lightwaves generated by the DFB-LD2 and DFB-LD3 are used for the pump lights. AWG produces 2.5Gb/s based on the orthogonal phase-shift keyed modulation OFDM signal, its electrical spectrum is shown in Fig.5 as inset (i). The CW light generated by DFB-LD1 at 1544.72nm signal light is modulated via a single-arm LN-MOD biased at 1.62V.The half-wave voltage ($v\pi$) of the LN-MOD is 7.8V, its 3dB bandwidth is greater than 8GHz, and its extinction ratio is greater than 25dB.The 2.5 Gbit/s optical signals and the pump signals are combined by an optical coupler (OC) before an EDFA to boost the power of the two signals. The optical spectra before and after a SOA are shown in Fig.5 (a) and (b), respectively. The optical power of the signal lightwave and pump lightwaves are 2.0dBm, 6.5dBm and 8.9dBm, respectively. As shown in Fig5(b), wavelength of the converted signal is 1543.78nm, optical signal-to-noise power ratio(OSNR) is 23dBm.The wavelength conversion efficiency is -17dB.A FBG with a 3dB bandwidth of 0.15nm and a TOF with 0.5 nm bandwidth is used to filter out the converted signal. The converted OFDM signal is sent to the 10Gb/s optical receiver. The OFDM signal detected from optical receiver is sent to real-time oscilloscope for data collection. The received electrical spectrum is shown in Fig.5 as inset (ii).

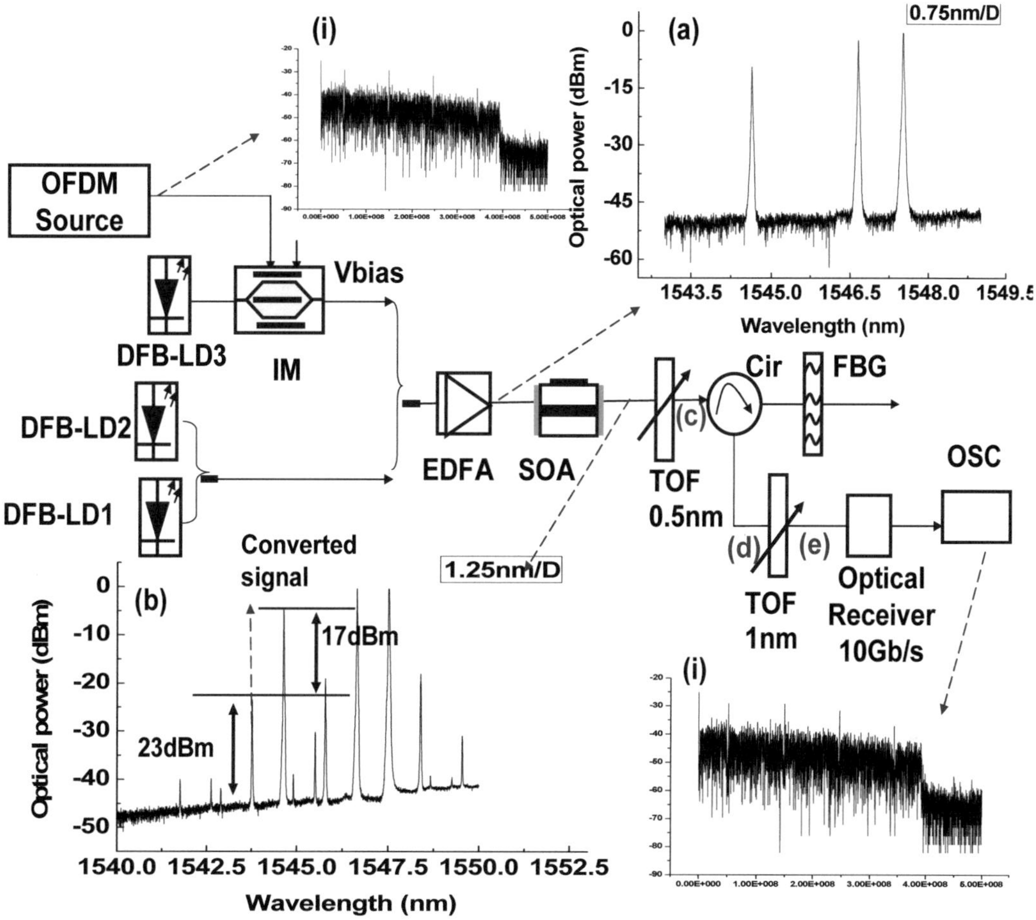

Fig. 5. Configuration of experimental setup and results for all-optical wavelength conversion based on parallel-dual-pump FWM effect in SOA.(i) Electrical spectra of the original OFDM signal; (ii) Electrical spectra of the converted OFDM signal; (a)optical spectral of the combined signals before SOA;(b)optical spectra of signal after SOA. DFB-LD:distributed feedback laser diode; FBG:Fiber bragg grating, IM:Intensity modulator ,SOA:Semicondoctor optical amplifier ,Cir:Circulator; TOF: Tunable optical filter. OSC: oscillator.

3.3 Experimental results

3.3.1 The comparison of Conversion efficiency

In the experiment, we measured the original signal and the pump optical power, the optical signal-to-noise ratio and the conversion efficiency of the three configurations as following table 1 show:

	Single-pump	Orthogonal-pump	parallel-pump
Original signal	5.38dBm	5.7dBm	2.0dBm
Pump1	8.8dBm	11.6dBm	6.5dBm
Pump2	8.0dBm	11.6dBm	8.9dBm
OSNR	25dB	25dB	23dB
Conversion efficiency	-15dB	-15dB	-17dB

Table 1. Comparison of three configurations

From the table we can see that the when three configurations in terms of optical signal-to-noise ratio and conversion efficiency are similar, the original signal light and pumped optical power of parallel-double-pump configuration are minimal, that means this scheme has the highest conversion efficiency. So, the experimental results are agreed well with the theoretical analysis.

3.3.2 The comparison of power penalty

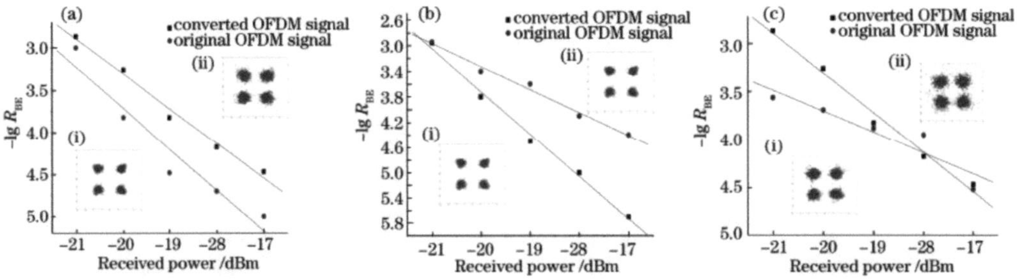

Fig. 6. The bit error rate (BER) curves and received constellations of three configuration

From the Fig. 6 we can see that the power penalty of parallel-dual-pump configuration is minimal compared to the other two configurations. After wavelength conversion, the converted signal is still OFDM signal, and the difference of received constellations between original and converted signal of parallel-dual-pump configuration is minimal, that is to say, this configuration has the smallest bit error rate (BER)

In conclusion, FWM based on SOA is considered to be the most promising scheme because it is fully transparent to the signal bit rate and modulation format, combined with OFDM signals, it can enhance the performance of optical networks, and is of significance for realizing all optical networks. On the other side, orthogonal-dual-pump and parallel-dual-pump schemes are polarization insensitive schemes. We can employ these schemes for all-optical up-conversion for ROF system.

4. The application of SOA in ROF system

The ROF converge two most important conventional communication technologies: radio frequency (RF) for wireless and optical fiber for wired transmission. It can afford huge bandwidth and communication flexibility, besides it can transmit wireless or wired signal to long distance region. So it becomes a very attractive technology in access network [43-48]. However, we still have to solve many problems in ROF system, such as simplify the base station and generation of high-frequency millimeter (mm)-wave [44, 47-48], In order to generate high-frequency mm-wave, we have tried many types of schemes [48-53]. In Ref. [50], we proposed a novel scheme to generate frequency-quadruple optical mm-wave radio-over-fiber system based on suppression odd-sideband by using one external modulator and cascaded Fiber Bragg Grating (FBG) filter.

Apart from this, there are many other suggestions to generate high-repetitive frequency millimeter (mm)-wave. Some researchers suggest we can use the nonlinear effects of some medium to generate mm-wave, such as XPM, SPM and FMW.FWM has the unique advantage of being transparent to the modulation format and the bit rate, which is of critical importance when handling analog or digital signals with speed of hundreds gigabit per second (difficult with XGM and XPM) [4, 13, 36, 54]. The medium which researchers are most interested in are HNLF and SOA [6].

In Ref. [55], H. Song etc. use the SOA-MZI (semiconductor optical amplifier Mach-Zehnder interferometer) to realize frequency up conversion to mm -wave, this just use the XPM effect in the SOA to generate the mm-wave. Many researchers preferred to use the FWM to generate mm-wave [6, 56-60].The reason is we can get cost-effective mm-wave by using FWM effect compared with other two nonlinear effects of high nonlinear medium. In 2006, J. Yao etc. proposed millimeter-wave frequency tripling based on FWM in a SOA [56]. In this article two signals are not phase and polarization locked , in order to keeping the phase of the two signal locked they used the optical phase-locked loop (OPLL), but they can not ensure the polarization of two signal and this directly leads to the low conversion efficiency . To improve the conversion efficiency, A. Wiberg used the OCS intensity modulation to generate two phase-locked wavelengths, and then he used these two wavelengths as two pumps to generate two new sidebands through the FWM effect in HNLF [61]. Through this scheme, a frequency six times of the electrical drive signal is obtained. It also improved the conversion efficiency. In this scheme A. Wiberg used HNLF instead of SOA, as we know the FWM in SOA has the following advantages against in HNLF:

a. In order to generate the two new sidebands with high power, the power of two pumps must be very high and the HNLF length should be long, which makes the system bulky and costly;
b. When pump power is very high, other nonlinear except FWM such as simulated Brillouin scattering (SBS), SPM and XPM may appear which will degrade the conversion efficiency.

To avoid difficult caused by using FWM effect of HNLF, J. Yao etc. suggested to use SOA instead of HNLF, and the pump are also two phase-locked generated by OCS intensity modulation [57].Then S. Xie etc. also proposal some scheme based on FWM effect of SOA to generate mm-wave [58-60].In Ref. [58], they use two cascaded optical modulators and FWM effect in SOA to generate a 12 times microwave source frequency with high spectral purity. First they generated frequency-quadruple optical mm-wave, then the optical lightwave is injected into SOA to get a 12 times microwave source frequency mm-wave. Since only one integrated MZM can also generate a frequency-quadruple optical mm-wave [62], So P.T. Shih etc. used only one MZM and SOA to get a 12 times microwave source frequency mm-wave [63]. What we have discussed above is just FWM effect of SOA when only two signals injected into SOA. Many researchers also investigated what will happen if they inject three signals (two of them are pump signals, another is probe signal) into SOA. They found FWM can also occur when certain conditions are met [64].

H. J. Kim accomplished all-optical up-conversion for ROF system through FWM in SOA because of its positive conversion efficiency and wide LO frequency bandwidth [64]. However, only double frequency mm-wave is generated and polarization sensitivity of this FWM system is not discussed in [64].Recently, polarization insensitive FWM in nonlinear optical fiber based on co-polarized pump scheme has been demonstrated in [6, 35], which is an effective way to increase the system stability. In Ref. [6, 35], two pumps are generated from different laser sources; therefore, the phase is not locked. Moreover, two polarization controllers (PC) are used to keep the two lightwaves to have the same polarization direction. We have investigated whether FWM is polarization sensitive in SOA based on co-polarized pump scheme, we proved FWM is polarization insensitive with parallel pump [52, 65-66], Configuration of experimental setup and results for all-optical wavelength conversion based on parallel-pump FWM effect in SOA shows in Fig.5.Three DFB generate pump and probe signals, two of them are used as pumps, another is probe. We must keep the pumps phase-locked and parallel. We try to change the polarization direction of the probe, the converted lightwaves are constant. We can see two converted signals in both sides of the probe signal, then we get any two of three lightwaves, we generate mm-wave after the optical to electrical conversion. And then we find the similar conclusion with orthogonal pumps [53].

In Ref. [57], we experimentally demonstrate all-optical up-conversion of radio-over-fiber signals based on a dual-pump four wave mixing in a SOA for the first time. The co-polarized pump light-waves are generated by OCS modulation to keep the same polarization direction and phase locked between two pumps. The proposed scheme to realize all-optical up-conversion based on FWM in a SOA is shown in Fig. 7. It is similar to the up-conversion scheme by nonlinear optical fiber [6]. The OCS signal is generated by an

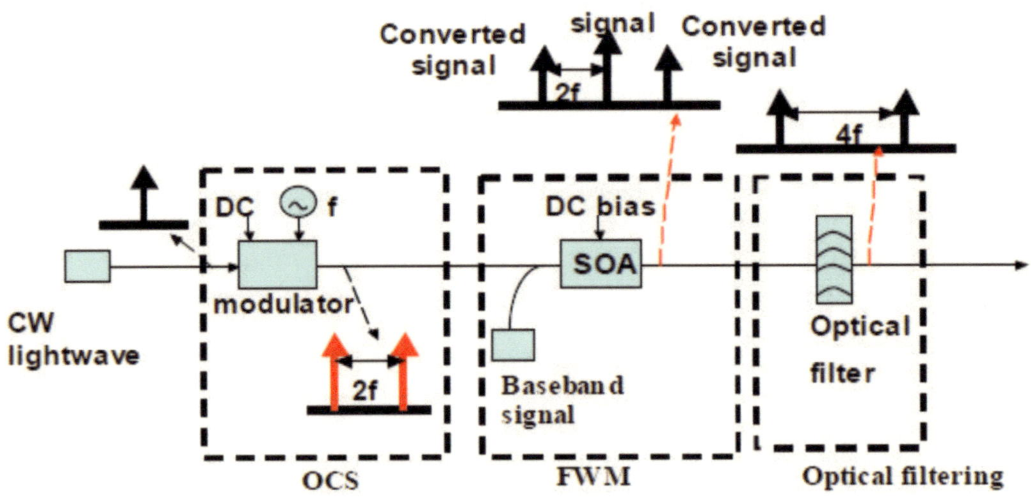

Fig. 7. The principle diagram of polarization-insensitive all-optical up-conversion based on FWM effect in a SOA. The repetitive frequency of the RF signal is f, and the IM's DC is biased at null point.

external intensity modulator (IM) biased at null point. The continuous wave (CW) lightwave generated by a DFB array is modulated via a single-arm IM driven by a RF sinusoidal wave signal with a repetitive frequency of f based on the OCS modulation scheme to generate two subcarriers with wavelength spacing of 2f. The generated two lightwaves, which will be used as pump signals, have the same polarization direction, optical power, and locked phase. The two converted signals with channel spacing of 4f can be obtained after FWM effect in SOA. The two converted signals have the same polarization direction and locked phase as well. When the pumps and original signal are removed by optical filters, the all-optical up-converted signals carried by 4f optical carrier are achieved.

Fig. 8 shows the experimental setup for single channel up conversion. In the central office (CO), the continuous lightwave generated by the DFB-LD0 at 1550nm is modulated by a single-arm LN-MOD biased at $v\pi$ and driven by an 10GHz LO to realize OCS. The repetitive frequency of the LO optical signal is 20GHz, and the carrier suppression ratio is larger than 20 dB. The high sidebands are removed by a 50/100 GHz IL, and the optical spectrum is shown in Fig. 8 as inset (i). The CW generated by DFB-LD1 at 1543.82 nm is modulated via another LN-MOD driven by 2.5-Gb/s pseudorandom binary sequence data with a length of $2^{31}-1$ to generate regular OOK non-return-to-zero (NRZ) optical signals. The 2.5 Gbit/s optical signals and the 20 GHz OCS pump signals are combined by a 3-dB OC before two individual EDFA are used to boost the power of the two signals respectively. The SOA is

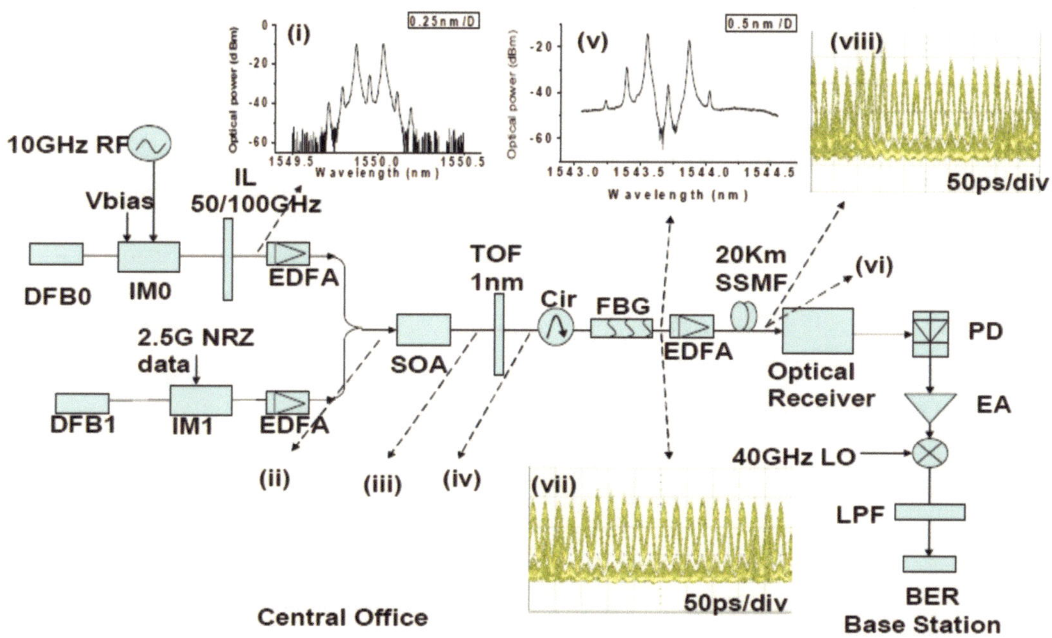

Fig. 8. Experimental setup and results for all-optical up-conversion base on four-wave mixing in SOA, Cir: optical circulator, TOF: tunable optical filter, EA: electrical amplifier. Inset (i): OCS signals after IL; (ii): the combined signals after OC; (iii): combined signals after SOA; (iv): converted DSB signals after 1nm TOF; (v): converted OCS signal after FBG; (vi): converted OCS signal after transmission; (vii): the mm-wavesignals before transmission; (viii): the mm-wave signals after transmission

injected by 200 mA, which gives 25 dB gain and 10 dBm saturation optical power. This SOA has polarization sensitivity smaller than 0.5 dB. The DSB signals with 40-GHz spaces between first order sidebands are created from 2.5-Gbit/s NRZ signals by the FWM effect in the SOA. A TOF with 1nm optical bandwidth is used to remove the pump signals while get the DSB signals. The carrier suppression signals with 40-GHz space are generated by using an optical circulator and a FBG. The FBG has a 3 dB reflection bandwidth of 0.2 nm. The optical spectra after FBG are shown in Fig. 8 as inset (v).

The OCS signals are detected by an optical receiver after transmission over 20-km SMF-28 before they are amplified by an EDFA. The eye diagrams of the 40-GHz optical mm-wave signals before and after 20-km SSMF-28 transmission can be seen in Fig. 8 insets (vii) and (viii), respectively. At the base station (BS), the optical signals are detected by an optical receiver. A TOF with 0.5 nm bandwidth is used to remove the ASE noise. After the optical receiver, the mm-wave signal with the down-link data is detected by an optical-electrical (O/E) converter with a 3-dB bandwidth of 40-GHz and amplified by a narrow-band electrical amplifier (EA), after these we get a frequency-quadruple of LO optical mm-wave. This scheme has excellent advantages such as small size, high-gain, polarization insensitivity, and low-frequency bandwidth requirement for RF signal and optical components, and high wavelength stability. 2.5 Gbit/s baseband signal has been successfully up-converted to 40 GHz carrier in this scheme. The experimental results show that the scheme based on dual-pump FWM in a SOA is one of the most promising all-optical up-conversions for ROF systems.

We then propose and experimentally investigate another polarization insensitive all-optical up-conversion scheme for ROF system based on FWM in a SOA [65]. In this scheme the parallel pump is generated based on optical odd-order sidebands and carrier suppression using an external intensity modulator and a cascaded optical filter. Therefore, the two pumps are always parallel and phase locked, which makes system polarization insensitive. After FWM in a SOA and optical filtering, similar to single sideband (SSB) 40GHz optical millimeter-wave is generated only using 10GHz RF as LO.As we know, SSB modulation is a good option to overcome fiber dispersion [67], we will improve mm-wave performance in this scheme.

Fig. 9 shows the principle of polarization-insensitive all-optical up-conversion for ROF systems based on parallel pump FWM in a SOA. In the central station, an IM and a cascaded optical filter are employed to generate quadruple frequency optical mm-wave, in which the odd-order sidebands and the optical carrier are suppressed. Obviously the generated two second-order sidebands have the parallel polarization direction and phase locked. Then the two pumps are combined with the signal lightwave by using an OC. The two converted new signals can be obtained after FWM process in the SOA. A TOF is used to suppress the pump signal. In this scheme, the optical signal similar to DSB signal is generated, which includes two converted signals and original signal after FWM in SOA. However, when one sideband is removed by an optical filter or optical interleaver(IL), the remaining signal is SSB-like signal, which includes one up-converted sideband and original signals. As we know that SSB signal can realize dispersion free long distance transmission. In the base station, the optical quadruple repetitive frequency mm-wave will be generated when they are detected by O/E converter after transmission.

Fig.10 shows the experimental setup for all-optical up-conversion in [65]. The lightwave generated from the DFB laser at 1543.8nm is modulated by the IM1 driven by a 10GHz sinusoidal wave. The IM1 is DC-biased at the top peak output power when the LO signal is

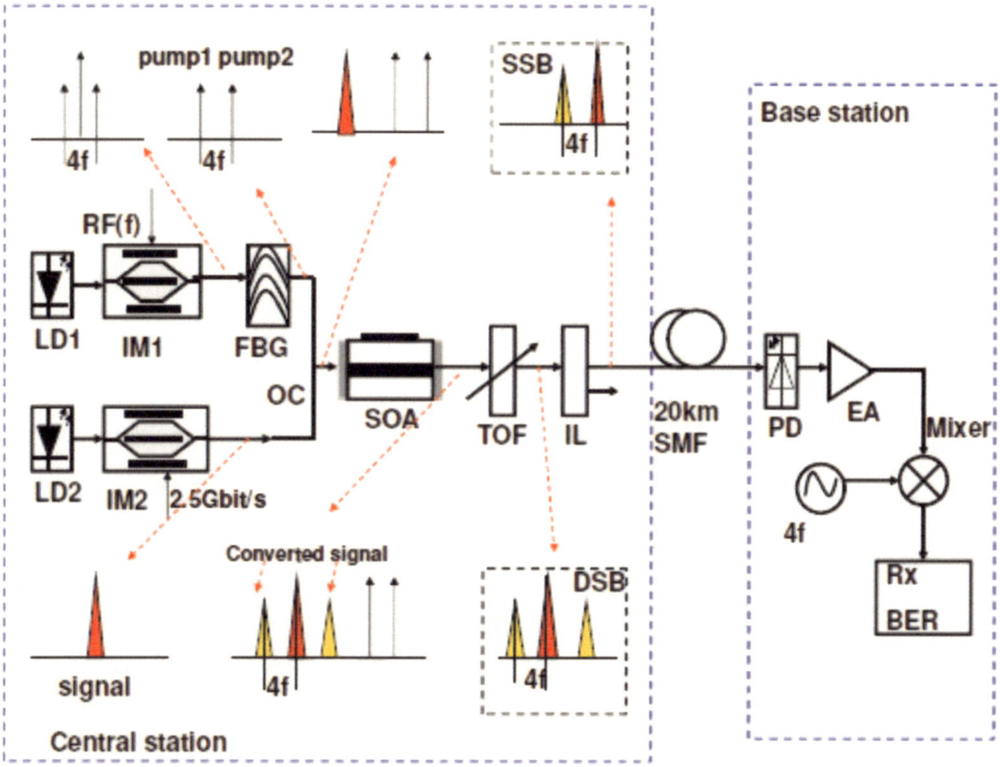

Fig. 9. The principle of polarization-insensitive all-optical up-conversion for ROF system based on parallel pump FWM in a SOA. FBG: fiber Bragg grating. OC: optical coupler. SOA: semiconductor optical amplifier. TOF: tunable optical filter. IL: interleaver. EA: electrical amplifier. IM: intensity modulator. SSB: single sideband. DSB: double sideband. PD: photodiode.BER: bit error ratio. RX: receive

removed. 10GHz RF microwave signal with a peak-to-peak voltage of 12V. The half-wave voltage of the IM is 6V; by this way, the odd-order modes are suppressed. The optical spectrum after IM1 is shown in Fig. 10 as inset (i). We can see that the first-order sidebands are suppressed and the frequency spacing between the second-order modes is equal to 40GHz. The carrier is removed by using a FBG. The output optical spectrum of FBG is shown in Fig. 10 as inset (ii). The two second-order sidebands are used as two parallel pumps. Because two pumps come from one laser, the pumps always have the same polarization direction and phase locked. The CW lightwave from another DFB laser at 1537.9nm is modulated by the second IM2 driven by 2.5Gbit/s electrical signal with a PRBS length of $2^{31}-1$ to generate regular NRZ optical signal.

The 2.5Gbit/s NRZ optical signals and two pump signals are combined by an OC before the EDFA. The optical spectra before and after the SOA are shown in Fig. 10 as inset (iii) and (iv), respectively. The SOA has 3-dB gain bandwidth of 68-nm, small signal fiber-to-fiber gain of 28-dB at 1552nm, polarization sensitivity smaller than 1dB, and noise figure of 6-dB at 1553nm. After the SOA, new up-converted signals are generated due to FWM, which is shown in Fig.10 as inset (iv). Then a tunable optical filter (TOF) with a bandwidth of 0.5nm is used to suppress the pump signals. The optical spectrum after the TOF is shown in Fig. 10

as inset (v). We can see that the converted and original signals are kept. The DSB signals with 80GHz frequency spacing between two converted sidebands are generated. In order to obtain the SSB signals, a 50/100 optical interleaver is used to remove one sideband. The optical spectrum after optical interleaver is shown in Fig. 10 as inset (vi). We can see that 2.5Gbit/s OOK signals are carried by the SSB-like signals with 40GHz frequency spacing between the converted signals and carrier, namely, the optical quadruple frequency mm-wave carried 2.5GHz signals is obtained. The power delivered to the fiber is 2dBm. After transmission over 20km SMF-28, the optical mm-wave is detected by O/E conversion via a photo-diode (PD) with a 3-dB bandwidth of 50 GHz. This scheme has some unique advantages such as polarization insensitive, high wavelength stability, and low-frequency bandwidth requirement for RF signal and optical components. 40GHz optical mm-wave SSB-like signal is generated by using 10GHz LO.

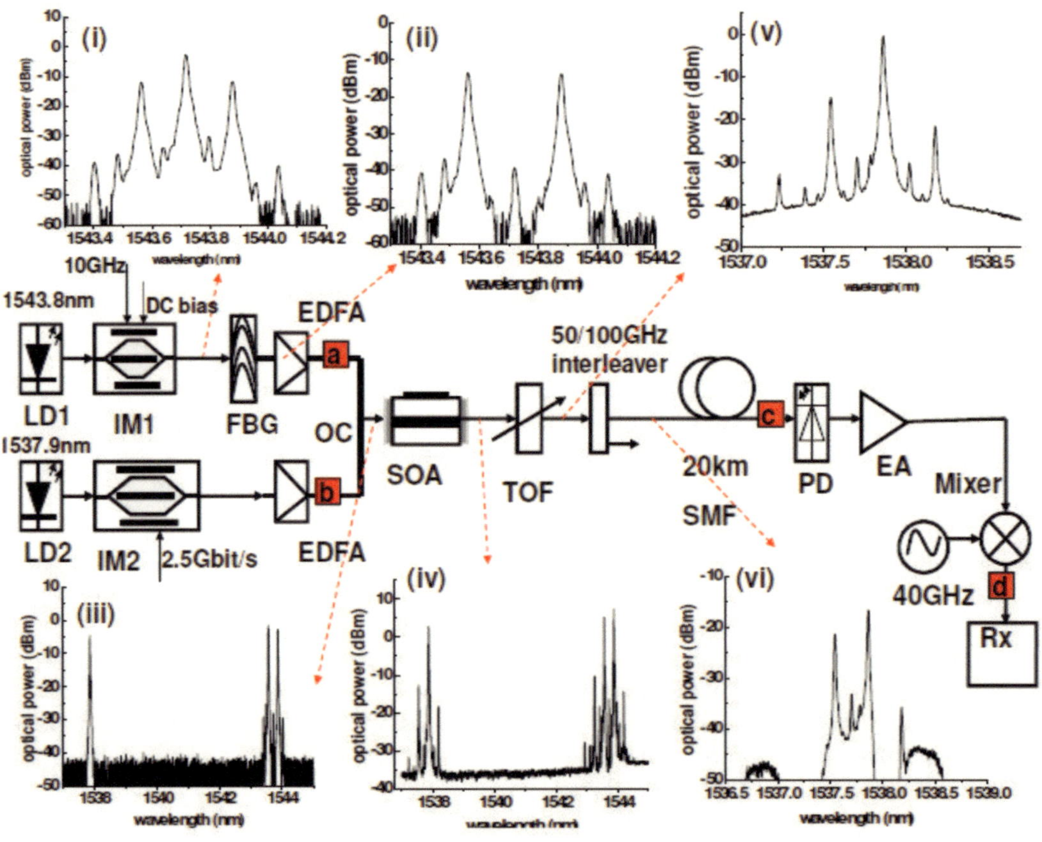

Fig. 10. Experimental setup and optical spectra for optical signal up-conversion. FBG: fiber Bragg grating. OC: optical coupler. SOA: semiconductor optical amplifier. TOF: tunable optical filter. EA: electrical amplifier. IM: intensity modulator. SSB: single sideband. DSB: double sideband. PD: photo-diode. BER: bit error ratio. EDFA: erbium-doped optical fiber amplifier.

In conclusion, we have obtained the following conclusions: (1) SOA can be used to generate high-repetitive mm-wave in ROF system; (2) we can use an IM to generate two pump instead of two independent DFB-LD, they may show good performance due to their phase-locked; (3) RoF system is polarization insensitive based on above mentioned method.

5. Conclusion

This chapter has theoretically and experimentally discussed the AOWC based on FWM effect in SOA for OFDM signals. The rules for OFDM signal is the same as that of regular OOK signal. We will investigate the FWM effect between the sub-carrier of OFDM signal which generate noise in the system in future. The application of SOA in RoF system has investigated. We experimentally proposed two polarization insensitive RoF systems. These schemes also have excellent advantages such as small size, high-gain, polarization insensitivity, and low-frequency bandwidth requirement for RF signal and optical components, and high wavelength stability.

6. Reference

[1] A. E. Kelly, D. D. Marcenac, and D. Nesset, "40 Gbit/s wavelength conversion over 24.6 nm using FWM in a semiconductor optical amplifier with an optimized MQW active region," *Electron. Lett.*, vol. 33, no. 25, pp. 2123-2124, Dec. 1997.

[2] W. Wang, H. N. Poulsen, L. Rau, H. F. Chou, J. E. Bowers, and D. J. Blumenthal, "Raman-enhanced regenerative ultrafast all-optical fiber XPM wavelength converter," *IEEE J. Lightw. Technol.*, vol. 23, no. 3, pp.1105-1115, Mar. 2005.

[3] J. Yu, X. Zheng, C. Peucheret, A. Clausen, H. Poulsen, and P. Jeppesen, "All-optical wavelength conversion of short pulses and NRZ signals based on a nonlinear optical loop mirror," *IEEE J. Lightw. Technol.*, vol. 18, no. 7, pp. 1007-1017, Jul. 2000.

[4] C. Porzi, A. Bogoni, L. Poti, and G. Contestabile, "Polarization and wavelength-independent time-division demultiplexing based on copolarized-pumps FWM in an SOA," *IEEE Photon. Technol. Lett.*, vol. 17, no. 3, pp. 633-635, Mar. 2005.

[5] J. Yu, Z. Jia, Y. K. Yeo, and G. K. Chang, "Spectrally non-inverting wavelength conversion based on FWM in HNL-DSF and its application in label switching optical network," in *Proceedings of. 25th ECOC*, pp. 32–35,2005.

[6] J. Ma, J. Yu, C. Yu, Z. Jia, X. Sang, Z. Zhou, T. Wang, and G. K. Chang, "Wavelength conversion based on four-wave mixing in high-nonlinear dispersion shifted fiber using a dual-pump configuration," *IEEE J. Lightw. Technol.*, vol. 24, no. 7, pp. 2851-2858, Jul. 2006.

[7] J. Hansryd, P. A. Andrekson, M. Westlund, J. Li and P. Hedekvist," Fiber-based optical parametric amplifiers and their applications," *JSTQE*, vol. 8, no. 3, pp. 506-520, May. 2002

[8] J. H. Lee, "All-optical signal processing devices based on holey fibers," *IEICE Trans. Electron.*, vol. E88-C, pp. 327–334, Mar. 2005.

[9] S. Radic and C. J. McKinstrie, "Optical Amplification and Signal Processing in Highly Nonlinear Optical Fiber," *IEICE Trans. Electron.*, vol. E88-C, pp. 859-869, May. 2005

[10] M. P. Fok, C. Shu and D. J. Blumenthal, "40-Gb/s polarization multiplexed RZ-DPSK signal wavelength conversion using a 32-cm bismuth-oxide nonlinear fiber." *CLEOE-IQEC*, 2007

[11] W. Astar, A. S. Lenihan, and G.M. Carter, "Polarization-insensitive wavelength conversion by FWM in a highly nonlinear PCF of polarization-scrambled 10-Gb/s RZ-OOK and RZ-DPSK signals," *IEEE Photon. Technol. Lett.*, vol. 19, no. 20, pp. 1676-1678, Oct. 2007.

[12] S. Watanabe, S. Takeda, G. Ishikawa, H. Ooi, J. G. Nielsen, and C. Sonne, "Simultaneous wavelength conversion and optical phase conjugation of 200 Gbit/s (5x40 Gbit/s) WDM signal using a highly nonlinear fiber four-wave mixing," *in the Proceedings of ECOE*, Paper TH3A1-4, pp. 1525–1529,1997.

[13] J. Yu, M.-F. Huang, Z. Jia, L. Chen, J.-G. Yu and G. K. Chang, "Polarization-nsensitive all-optical upconversion for seamless integration optical core/metro/access networks with ROF systems based on a dual-pump FWM scheme", *IEEE J. Lightw. Technol.*, vol. 27, no. 14, pp.2605 - 2611, Jul. 2009.

[14] R. M. Jopson, and R. E. Tench, "Polarization- independent phase conjugation of lightwave signals," *Electron. Lett.*, vol. 29, no. 25, pp. 2216-2217, Dec. 1993.

[15] K. Inoue, " Polarization independent wavelength conversion using fiber four-wave mixing with two orthogonal pump lights of different frequencies." *J. Lightwave Technol.*, vol. 12, no. 11, pp. 1916-1920, Nov. 1994.

[16] K. K. Y. Wong, M. Marhic, K. Uesaka, and L. Kazovsky, "Polarization-independent two-pump fiber-optical parametric amplifier." *IEEE Photon. Technol. Lett.*, vol. 14, no. 7, pp. 911-913, Jul. 2002.

[17] S. Radic, C.J. Mckinstrie, R. M. Jopson, J. C. Centanni, Q. Lin and G. P. Agrawal, " Record performance of parametric amplifier constructed with highly nonlinear fibre," *Electron. Lett.*, vol. 39, no.11, pp. 838-839, May. 2003.

[18] Y. Wang, C. Y. Yu, T. Luo, L. S. Yan, Z. Q. Pan, and A. E. Willner, "Tunable all-optical wavelength conversion and wavelength multicasting using orthogonal polarized fiber FWM," *IEEE J. Lightw. Technol.*, vol. 23, no. 10, pp. 3331-3338, Oct. 2005.

[19] J. Ma, J. Yu, C. Yu, and Z. Zhou, "Reducing polarization sensitivity for all-optical wavelength conversion of the optical packets based on FWM in HNL-DSF using co-polarized pump scheme," *Opt. Commun.*, vol. 260, no.2, pp. 522-527, Nov. 2006.

[20] T. Hasegawa, K. Inouc, and K.Oda, "Polarization independent frequency conversion by fiber four-wave mixing with a polarization diversity technique," *IEEE Photon. Technol. Lett.*, vol. 5, no.8, pp. 947-949, Aug. 1993.

[21] K. K. Y. Wong, M. Marhic, K. Uesaka, and L. Kazovsky," Polarization-independent one-pump fiber-optical parametric amplifier," *IEEE Photon. Technol. Lett.*, vol. 14, no. 11, pp. 1506-1508, Nov. 2002.

[22] K. Uesaka, K. K. Y. Wong, M. E. Marhic, and L. G. Kazovsky "Polarization-insensitive wavelength exchange in highly-nonlinear dispersion-shifted fiber," *in the Optical Fiber Communications (OFC)*, (Anaheim, CA, 2002), Paper ThY3,2002.

[23] T. Tanemura, K. Katoh, and K. Kikuchi, "Polarization-insensitive asymmetric four-wave mixing using circularly polarized pumps in a twisted fiber," *Opt. Express*, vol. 13, no. 19, pp. 7497-7505, Sep. 2005.

[24] K. Inoue, "Polarization effect on four-wave mixing efficiency in a single mode fiber," *IEEE J. Quantum Electron.*, vol. 28, no. 4, pp. 883-894, Apr. 1992.

[25] J. Yu, X. Zheng, F. Liu, A. Buxens, and P. Jeppesen, "Simultaneous realization wavelength conversion and signal regeneration using nonlinear optical loop mirror," *Opt. Commun.*, vol. 175, no. 1, pp. 173-177, Feb. 2000.

[26] C. J. McKinstrie, H. Kogelnik, R. M. Jopson, S. Radic, and A. V. Kanaev, "Four-wave mixing in fibers with random birefringence," *Opt. Express*, vol. 12, no. 10, pp. 2033-2055, May. 2004.

[27] *M. Jinno, "*All optical signal regularizing/regeneration using a nonlinear fiber Sagnac interferometer switch with signal-clock walk-off," *IEEE J. Lightw. Technol.*, vol. 12, no. 9, pp. 1649-1659, Sep. 1994 。

[28] J. Yu, P. Jeppesen, and S. N. Knudsen, "80 Gbit/s pulsewidth-maintained wavelength conversion based on a HNL DSF–NOLM including transmission over 80 km of conventional SMF," *Electron. Lett.*, vol. 37, no. 9, pp. 577-579, Jan. 2001.

[29] M. Jinno and M. Abe, "All-optical regenerator based on nonlinear fiber Sagnac interferometer," *Electron. Lett.*, vol. 28, no. 14, pp. 1350-1352, Jul. 1992.

[30] B. Mukhedee, "WDM optical Communication networks: progress and challengers, "*IEEE JSAC,* vol. 18, no. 10, pp: 1810-1824,Oct. 2000

[31] M. Kovacevic and A. Acampora, "Benefits of wavelength translation in all-optical clear-channel networks," *IEEE J Select Areas Commun,* , vol. 14, no. 5, pp. 868-880, Jun. 1996

[32] S. J. B. Yoo, "Wavelength conversion technologies for WDM network applications," *IEEE J. Lightw. Technol.*, vol. 14, no. 6, pp. 955-966, Jun. 1996

[33] K. Obermann, S. Kindt, D. Breuer, and K. Petermann, "Performance analysis of wavelength converters based on cross-gain modulationin semiconductor-optical amplifiers," *IEEE J. Lightw. Technol.* vol. 16, no. 1, pp. 78-85, Jan. 1998

[34] T. Durhuus, C. Joergensen , B. Mikkelsen , R. J. S. Pedersen, and K. E. Stubkjaer, "All optical wavelength conversion by SOA in a Mach-Zehnder configuration," *IEEE Photon Technol Lett,* vol. 6, no. 1, pp. 1044-1135, Jan.. 1994

[35] J. Yu, Z. Jia, Y. K. Yeo, and G. K. Chang, "Spectrally non-inverting wavelength conversion based on FWM in HNL-DSF and its application in label switching optical network, " *ECOC05*, pp. 32-35, 2009

[36] J. P. R. Lacey, M. A. Summerfield and S. J. Madden, "Tunability of polarization-insensitive wavelength converrs based on four-wave mixing in semiconductor optical amplifiers", *IEEE J. Lightw. Technol.,* vol.16, no. 12, pp. 2419-2427, Dec. 1998

[37] J. Lu, L. Chen, Z. Dong, Z. Cao, and S. C. Wen, "Polarization insensitive wavelength conversion based on orthogona pump four-wave mixing for polarization multiplexing signal in high-nonlinear fiber", *IEEE J. Lightw. Technol..*,vol.27, no. 34, pp. 5767-5774, Dec. 2009

[38] J. Lu, Z. Dong, L. Chen and J. Yu, "Polarization insensitive wavelength conversion based on four-wave mixing for polarization multiplexing signal in high-nonlinear fiber," *Opt. Commun.,* 282, pp.1274-1280, 2009

[39] L. James, A. Jean, "Orthogonal-frequency-division multiplexing for dispersion compensation of long-haul optical systems," *Opt Express,* vol. 14, no. 6, pp. 2079-2084, Mar. 2006

[40] H. C. Bao, W. Shieh, "Transmission of wavelength-division-multiplexing channels with coherent optical OFDM," *IEEE Photon Technol Lett,* vol. 19, no. 12, pp. 922-924, Apr. 2007

[41] S. L. Jansen, I. Morita, T. C. W Schenk, N. Takeda, and H. Tanaka, "Coherent Optical 25.8-Gb/s OFDM Transmission Over 4160-km SSMF," *IEEE J. Lightw. Technol.,* vol. 26, no. 1, pp. 6-15, Jan. 2008

[42] L. Chen, J. Lu, Z. Dong, and J. Yu, "A radio-over-fiber system with photonics generated OFDM signals and wavelength reuse for upstream data connection," *In Proceeding of ICAIT,* Shenzhen, China, 1-3A-1, Aug. 2008

[43] J. Yu, "Novel radio-over-fiber configurations with centralized lightwave for uplink connection," in Signals, Systems and Electronics, 2007. ISSSE '07. International Symposium on. 2007, pp. 615-616, Jun.30-Aug. 2, 2007.

[44] L. Chen, Y. Shao, X. Lei, H. Wen, and S. Wen, "A novel radio-over-fiber system with wavelength reuse for upstream data connection," *IEEE Photon. Technol. Lett.,*, vol.19, no. 6, pp. 387-389, Mar. 2007..

[45] G. K. Chang, J. Yu, Z. Jia, and J. G. Yu, "Novel optical-wireless access network architecture for providing broadband wireless and wired services.".OFC 2006. Paper OFM1, Mar.5-10, 2006.

[46] J. Yu, Z. Jia,L. Yi, Y. Su, G. K. Chang, T. Wang, "Optical millimeter-wave generation or up-conversion using external modulators," *IEEE Photon. Technol. Lett.,* vol.18, no. 1, pp. 265-267, Jan. 2006.

[47] L. Chen, H. Wong, and S. Wen, "A radio-over-fiber system with a novel scheme for millimeter-wave generation and wavelength reuse for up-link connection," *IEEE Photon. Technol. Lett.,* vol.18, no. 19, pp.2056-2058, Oct. 2006.

[48] J. Yu, Z. Jia, T. Wang, and G. K. Chang, "Centralized lightwave radio-over-fiber system with photonic frequency quadrupling for high-frequency millimeter-wave generation," *IEEE Photon. Technol. Lett.,* vol.19, no. 19, pp. 1499-1501, Oct. 2007.

[49] J. He, L. Chen, Z. Dong, S, Wen, and J. Yu, "Full-duplex radio-over-fiber system with photonics frequency quadruples for optical millimeter-wave generation," *Optical Fiber Technology,* vol. 15, no. 3, pp.290-295, Jun, 2009.

[50] L. Liu, Z. Dong, Y. Pi, J. Lu, L. Chen, J. Yu, and S. Wen, "Radio-over-fiber system with frequency quadruple for millimeter-wave generation by external modulator," *Chinese Journal of Lasers,* vol. 36, no. 1, pp. 148-153, 2009.

[51] J. Lu, Z. Dong, L. Chen, J. Yu, and S. Wen, "High-repetitive frequency millimeter-wave signal generation using multi-cascaded external modulators based on carrier suppression technique," *Opt. Commun.,* vol. 281, no. 19, pp. 4889-4892, Oct. 2008.

[52] J. Lu, Z. Dong, Z. Cao, L. Chen, J. Yu, "All-optical Wavelength Conversion Based on Parallel Dual-pump Four-wave Mixing in Semiconductor Optical Amplifier for OFDM Optical Signal," *Acta Photonica Sinica,* vol. 38, no. 11, pp. 2857-2863, Nov. 2009.

[53] Z. Cao, Z. Dong, J. Lu, L. Chen, J. Yu, "All-optical Orthogonal2pump wavelength conversion of optical OFDM signal," *Journal of Optoelectronics ·Laser*, vol. 20, no.5, pp.622-627, May., 2009.

[54] J. Yu, J. Gu, X. Liu, Z. Jia and G. K. Chang,"Seamless integration of WDM-PON and wideband radio-over-fiber for 8 × 2.5Gb/s all-optical up-conversion using Raman-assisted FWM,"in Optical Communication," ECOC 2005,vol. 1, pp.80-90, 2005

[55] H. J. Song, J. S. Lee, and J. I. Song, "All-optical harmonic frequency up-conversion for a WDM radio over fiber system," *in Microwave Symposium Digest, 2004 IEEE MTT-S International*. pp. 405-407, Jun.6-11, 2004.

[56] Q. Wang, H. Rideout, F. Zeng, J. Yao, "Millimeter-wave frequency tripling based on four-wave mixing in a semiconductor optical amplifier," *IEEE Photon. Technol. Lett.,*, vol. 18, no.23, pp. 2460-2462, Dec. 2006.

[57] Q. Wang, F. Zeng, H. Rideout, J. Yao, "Millimeter-wave generation based on four-wave mixing in an SOA," *in Microwave Photonics, 2006. MWP '06. International Topical Meeting on*. pp. 1-4, Oct. 2006.

[58] T. Wang, M. Chen, H. Chen, and J. Zhang, "Millimeter-wave signal generation using two cascaded optical modulators and FWM effect in semiconductor optical amplifier," *IEEE Photon. Technol. Lett.*, vol.19, no.16, pp. 1191-1193, Aug., 2007.

[59] T. Wang, M. Chen, H. Chen, and S. Xie, "Millimeter-Wave Signal Generation Using Four-Wave Mixing Effect in SOA," OFC/NFOEC 2007, Anaheim,CA, pp.1-3, March 25-29, 2007.

[60] T. Wang, M. Chen, H. Chen, and S. Xie, "Millimetre-wave signal generation using FWM effect in SOA," *Electronics Letters*, vol. 43, no. 1, pp. 36-38, Jan., 2007.

[61] A. Wiberg, P. Millan, M. Andres, and P. Hedekvist, "Microwave-photonic frequency multiplication utilizing optical four-wave mixing and fiber Bragg gratings," *IEEE J. Lightw. Technol.* vol. 24, no.1, pp.329-334, Jan. 2006.

[62] C. T. Lin, P.T. Shih, J. Chen, W. Xue, P.Peng and S. Chi, "Optical millimeter-wave signal generation using frequency quadrupling technique and no optical filtering," *IEEE Photon. Technol. Lett.*, vol. 20, no. 12, pp.1027-1029, Jun. 2008.

[63] P. T. Shih, J. Chen, C. T. Lin, W. Jiang, H. Huang, P. Peng and S. Chi, "Optical millimeter-Wave signal generation via frequency 12-tupling," *IEEE J. Lightw. Technol.*, vol.28, no.1, pp. 71-78, Jan. 2010.

[64] H. Kim, H. Song, J. Song, "All-optical frequency up-conversion technique using four-wave mixing in semiconductor optical amplifiers for radio-over-fiber applications," *in Microwave Symposium, 2007. IEEE/MTT-S International*. Honolulu, HI, Jun. 3-8, pp. 67-70, 2007.

[65] J. Lu, Z. Dong, Z. Cao, L. Chen, S. Wen, and J. Yu, "Polarization insensitive all-optical up-conversion for ROF systems based on parallel pump FWM in a SOA," *Opt. Express*, vol. 17. no.9, pp.6962-6967, Apr. 2009

[66] Z. Dong. Z. Cao, J. Lu, L. Chen, S. Wen, Z. Jia, and G. K. Chang, "All-optical up-conversion of millimeter-wave signals for ROF system using optical carrier suppression-based dual-pump FWM in an SOA," in Optical Fiber Communication - incudes post deadline papers, *OFC 2009*, San Diego, CA, pp. 1-3, Mar. 22-26, 2009.

[67] J. Yu, M. Huang, Z. Jia, T. Wang, G. K. Chang, "A Novel Scheme to Generate Single-Sideband Millimeter-Wave Signals by Using Low-Frequency Local Oscillator Signal," *IEEE Photon. Technol. Lett.* Vol. 20, no.7,pp. 478-480, Apr. 2008.

6

Impact of Pump-Probe Time Delay on the Four Wave Mixing Conversion Efficiency in Semiconductor Optical Amplifiers

Narottam Das[1], Hitoshi Kawaguchi[2] and Kamal Alameh[1,3]
[1]*Electron Science Research Institute, Edith Cowan University,*
[2]*Graduate School of Materials Science, Nara Institute of Science and Technology,*
[3]*Department of Nanobio Materials and Electronics,*
Gwangju Institute of Science and Technology,
[1]*Australia*
[2]*Japan*
[3]*Republic of Korea*

1. Introduction

Four-wave mixing (FWM) in semiconductor optical amplifiers (SOAs) have attracted much attention especially for applications involving fast wavelength conversion and optical demultiplexing (Mecozzi *et al.*, 1995; Mecozzi & Mϕrk, 1997; Das *et al.*, 2000). The optimisation of the time delay between the input pump and probe pulses is crucial for maximising the FWM conversion efficiency and reducing the timing jitter (Inoue & Kawaguchi, 1998b; Das *et al.*, 2005). The FWM conversion efficiency is mainly limited by the gain saturation of the SOA, which is strongly dependent on the pulse duration, repetition rate of the input pulses, and time delay between pump and probe pulses. Therefore, it is important to analyse theoretically and investigate experimentally the dependence of the FWM conversion efficiency on the time delay between the input pump and probe pulses in SOAs. The impact of the time-delay between sub-picosecond optical pump and probe pulses on the FWM conversion efficiency in SOAs was experimentally observed for the first time and reported by Inoue and Kawaguchi (Inoue & Kawaguchi, 1998b). A preliminary theoretical analysis for evaluating the FWM conversion efficiency of SOAs was reported by Das *et al.* (Das *et al.*, 2005).
The optimization of the time delay between the optical pump and probe pulses is very important in order to achieve a high FWM conversion efficiency and minimize the timing jitter (Inoue & Kawaguchi, 1998b; Das *et al.*, 2005). Shtaif *et al.*, (Shtaif & Eisenstein, 1995; Shtaif *et al.*, 1995) have investigated analytically and experimentally the dependence of the FWM conversion efficiency for short optical pulses in SOA. They measured the FWM conversion efficiency for short optical pulses in the order of 10 ps and suggested that the high FWM conversion efficiency can be obtained when short optical pulses are used. The FWM characteristics for subpicosecond time-delays between the input optical pulses in SOAs have been reported. However, no theoretical analysis (Inoue & Kawaguchi, 1998b) was reported. Therefore, it is very important to analyze theoretically and measure

experimentally the FWM conversion efficiency in SOAs for subpicosecond optical pulses and optimize the time-delay between the input pump and probe pulses.

In this Chapter, we present a comprehensive analysis based on the finite-difference beam propagation method (FD-BPM) for evaluating the FWM conversion efficiency in SOAs and discuss the experimental results reported on the optimisation of the time delay between pump and probe pulses to maximize the FWM conversion efficiency. This comprehensive analysis is crucial not only for attaining high FWM conversion efficiency but also for characterizing the timing jitter. Simulation results demonstrate that, for short optical pulses, the time delay between the pump and probe pulses at the input of the SOA can be optimized to maximize the FWM conversion efficiency. Excellent agreement between the simulation and experimental results is achieved, demonstrating a high FWM conversion efficiency in SOAs when the time delay between the input pump and probe pulses is optimized.

The FD-BPM can obtain the FWM characteristics for different propagation scenarios using a modified nonlinear Schrödinger equation (MNLSE) (Hong *et al.,* 1996 & Das *et al.,* 2000), simply by changing only the combination of input optical pulses. These scenarios are: (1) single pulse propagation (Das *et al.,* 2008), (2) two input pulses (Das *et al.,* 2000), (3) multiplexing of several input pulses (Das *et al.,* 2001), (4) two input pulses with phase-conjugation (Das *et al.,* 2001), and (5) two input pulses with optimum time-delay between them (Das *et al.,* 2007).

The analyses are based on the MNLSE considering the group velocity dispersion (GVD), self-phase modulation, and two-photon absorption, with the dependencies on the carrier depletion, carrier heating, spectral-hole burning and their dispersions, including the recovery times in SOAs (Hong *et al.,* 1996). The simulation results show that the FWM conversion efficiency increases with the optimum time-delay between the input pump and probe pulses and are in excellent agreement with the experiment results.

2. Theoretical model

In this section, we briefly discuss the nonlinear effects in SOAs, the formulation of modified nonlinear Schrödinger equation (MNLSE), and the finite-difference beam propagation method (FD-BPM) used in the simulation, the nonlinear propagation characteristics of solitary pulses, the FWM characteristics in SOAs, and the FWM characteristics for various time-delays between the input pump and pulses.

2.1 Nonlinear effects in SOAs

There are different types of "nonlinear effects" in SOAs. Among them, four types of important optical "nonlinear effects" are explained. These are (i) spectral hole-burning (SHB), (ii) carrier heating (CH), (iii) carrier depletion (CD), and (iv) two-photon absorption (TPA).

Figure 1 shows the time-development of the population density in the conduction band after excitation. The arrow (pump) shown in Fig. 1 is the excitation laser energy. Below 100 fs, the SHB effect is dominant. SHB occurs when a narrow-band strong pump beam excites the SOA, which has an inhomogeneous broadening. SHB arises due to the finite value of intraband carrier-carrier scattering time (~ 50 - 100 fs), which sets the time scale on which a quasi-equilibrium Fermi distribution is established among the carriers in a band. After about 1 ps, the SHB effect is relaxed and the CH effect becomes dominant. The process tends

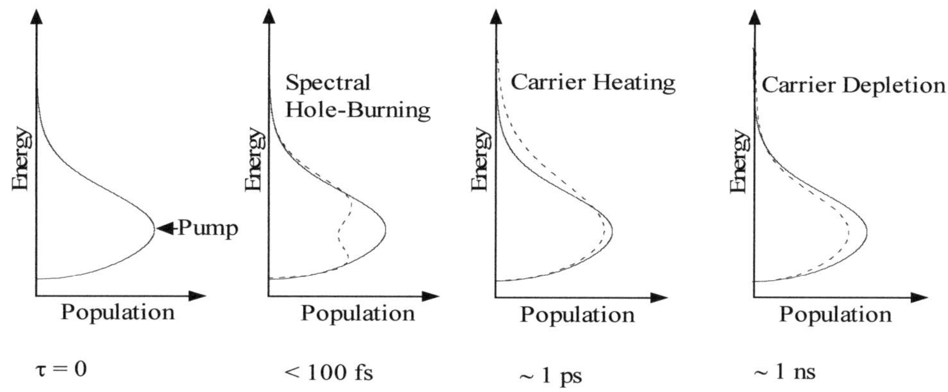

Fig. 1. Nonlinear effects in SOAs. The main effects are: (i) spectral hole-burning (SHB) with a life time of < 100 fs, (ii) carrier heating (CH) with a life time of ~ 1 ps, (iii) carrier depletion (CD) with a life time is ~ 1 ns, and (iv) two-photon absorption (TPA).

to increase the temperature of the carriers beyond the lattice's temperature. The main causes of heating the carriers are (1) the stimulated emission, since it involves the removal of "cold" carriers close to the band edge, and (2) the free-carrier absorption, which transfers carriers to high energies within the bands. The "hot"-carriers relax to the lattice temperature through the emission of optical phonons with a relaxation time of ~ 0.5 - 1 ps. The effect of carrier depletion (CD) remains for about 1 ns. The stimulated electron-hole recombination depletes the carriers, thus reducing the optical gain. The band to band relaxation also causes carrier depletion, with a relaxation time ~ 0.2 - 1 ns. For ultrashort optical pumping, the two-photon absorption (TPA) effect also becomes important. An atom makes a transition from its ground state to the excited state by the simultaneous absorption of two laser photons. All these mechanisms (effects) are taken into account in the simulation and the formulation of modified nonlinear Schrödinger equation (MNLSE).

2.2 Formulation of Modified Nonlinear Schrödinger Equation (MNLSE)

In this subsection, we will briefly explain the theoretical analysis of short optical pulses propagation in SOAs. Starting from Maxwell's equations (Agrawal, 1989; Yariv, 1991; Sauter, 1996), we reach the propagation equation of short optical pulses in SOAs which are governed by the wave equation (Agrawal & Olsson, 1989) in the frequency domain:

$$\nabla^2\overline{E}(x,y,z,\omega)+\frac{\varepsilon_r}{c^2}\omega^2\overline{E}(x,y,z,\omega)=0 \tag{1}$$

where, $\overline{E}(x,y,z,\omega)$ is the electromagnetic field of the pulse in the frequency domain, c is the velocity of light in vacuum and ε_r is the nonlinear dielectric constant which is dependent on the electric field in a complex form. By slowly varying the envelope approximation and integrating the transverse dimensions we arrive at the pulse propagation equation in SOAs (Agrawal & Olsson, 1989; Dienes *et al.*, 1996).

$$\frac{\partial V(\omega,z)}{\partial z}=-i\left\{\frac{\omega}{c}\left[1+\chi_m(\omega)+\Gamma\tilde{\chi}(\omega,N)\right]^{1/2}-\beta_0\right\}V(\omega,z) \tag{2}$$

where, $V(\omega,z)$ is the Fourier-transform of $V(t,z)$ representing pulse envelope, $\chi_m(\omega)$ is the background (mode and material) susceptibility, $\tilde{\chi}(\omega)$ is the complex susceptibility which represents the contribution of the active medium, N is the effective population density, β_0 is the propagation constant. The quantity Γ represents the overlap/ confinement factor of the transverse field distribution of the signal with the active region as defined in (Agrawal & Olsson, 1989).

Using mathematical manipulations (Sauter, 1996; Dienes *et al.*, 1996), including the real part of the instantaneous nonlinear Kerr effect as a single nonlinear index n_2 and by adding the two-photon absorption (TPA) term we obtain the MNLSE for the phenomenological model of semiconductor laser and amplifiers (Hong *et al.*, 1996). The following MNLSE (Hong *et al.*, 1996; Das *et al.*, 2000) is used for the simulation of FWM characteristics with optimum time-delays between pulses in SOAs:

$$\left[\frac{\partial}{\partial z}-\frac{i}{2}\beta_2\frac{\partial^2}{\partial\tau^2}+\frac{\gamma}{2}+\left(\frac{\gamma_{2p}}{2}+ib_2\right)|V(\tau,z)|^2\right]V(\tau,z)$$
$$=\left\{\frac{1}{2}g_N(\tau)\left[\frac{1}{f(\tau)}+i\alpha_N\right]+\frac{1}{2}\Delta g_T(\tau)(1+i\alpha_T)-i\frac{1}{2}\frac{\partial g(\tau,\omega)}{\partial\omega}\bigg|_{\omega_0}\frac{\partial}{\partial\tau}-\frac{1}{4}\frac{\partial^2 g(\tau,\omega)}{\partial\omega^2}\bigg|_{\omega_0}\frac{\partial^2}{\partial\tau^2}\right\}V(\tau,z) \tag{3}$$

We introduce the frame of local time τ (=t - z/v$_g$), which propagates with a group velocity v$_g$ at the center frequency of an optical pulse. A slowly varying envelope approximation is used in (3), where the temporal variation of the complex envelope function is very slow compared with the cycle of the optical field. In (3), $V(\tau,z)$ is the time domain complex envelope function of an optical pulse, $|V(\tau,z)|^2$ corresponding to the optical power, and β_2 is the GVD. γ is the linear loss, γ_{2p} is the two-photon absorption coefficient, b_2 (= $\omega_0 n_2/cA$) is the instantaneous self-phase modulation term due to the instantaneous nonlinear Kerr effect n_2, ω_0 (= $2\pi f_0$) is the center angular frequency of the pulse, c is the velocity of light in vacuum, A (= wd/Γ) is the effective area (d and w are the thickness and width of the active region, respectively, and Γ is the confinement factor) of the active region.

The saturation of the gain due to the carrier depletion is given by (Hong *et al.*, 1996)

$$g_N(\tau)=g_0\exp\left(-\frac{1}{W_s}\int_{-\infty}^{\tau}e^{-s/\tau_s}|V(s)|^2\,ds\right) \tag{4}$$

where, $g_N(\tau)$ is the saturated gain due to carrier depletion, g_0 is the linear gain, W_s is the saturation energy, τ_s is the carrier lifetime.

The spectral hole-burning (SHB) function $f(\tau)$ is given by (Hong *et al.*, 1996)

$$f(\tau)=1+\frac{1}{\tau_{shb}P_{shb}}\int_{-\infty}^{+\infty}u(s)e^{-s/\tau_{shb}}|V(\tau-s)|^2\,ds \tag{5}$$

where, $f(\tau)$ is the SHB function, P_{shb} is the spectral hole-burning saturation power, τ_{shb} is the spectral hole-burning relaxation time, and α_N and α_T are the linewidth enhancement factor associated with the gain changes due to the carrier depletion and carrier heating.

The resulting gain change due to the CH and TPA is given by (Hong *et al.*, 1996)

$$\Delta g_T(\tau) = -h_1 \int_{-\infty}^{+\infty} u(s) e^{-s/\tau_{ch}} (1 - e^{-s/\tau_{shb}}) |V(\tau - s)|^2 \, ds \\ - h_2 \int_{-\infty}^{+\infty} u(s) e^{-s/\tau_{ch}} (1 - e^{-s/\tau_{shb}}) |V(\tau - s)|^4 \, ds \tag{6}$$

where, $\Delta g_T(\tau)$ is the resulting gain change due to the CH and TPA, $u(s)$ is the unit step function, τ_{ch} is the carrier heating relaxation time, h_1 is the contribution of stimulated emission and free-carrier absorption to the carrier heating gain reduction and h_2 is the contribution of two-photon absorption.
The dynamically varying slope and curvature of the gain plays a shaping role for pulses in the sub-picosecond range. The first and second order differential net (saturated) gain terms are (Hong *et al.*, 1996),

$$\left.\frac{\partial g(\tau,\omega)}{\partial \omega}\right|_{\omega_0} = A_1 + B_1\left[g_0 - g(\tau,\omega_0)\right] \tag{7}$$

$$\left.\frac{\partial^2 g(\tau,\omega)}{\partial \omega^2}\right|_{\omega_0} = A_2 + B_2\left[g_0 - g(\tau,\omega_0)\right] \tag{8}$$

$$g(\tau,\omega_0) = g_N(\tau,\omega_0) / f(\tau) + \Delta g_T(\tau,\omega_0) \tag{9}$$

where, A_1 and A_2 are the slope and curvature of the linear gain at ω_0, respectively, while B_1 and B_2 are constants describing changes in A_1 and A_2 with saturation, as given in (7) and (8).
The gain spectrum of an SOA is approximated by the following second-order Taylor expansion in $\Delta\omega$:

$$g(\tau,\omega) = g(\tau,\omega_0) + \Delta\omega \left.\frac{\partial g(\tau,\omega)}{\partial \omega}\right|_{\omega_0} + \frac{(\Delta\omega)^2}{2} \left.\frac{\partial^2 g(\tau,\omega)}{\partial \omega^2}\right|_{\omega_0} \tag{10}$$

The coefficients $\left.\frac{\partial g(\tau,\omega)}{\partial \omega}\right|_{\omega_0}$ and $\left.\frac{\partial^2 g(\tau,\omega)}{\partial \omega^2}\right|_{\omega_0}$ are related to A_1, B_1, A_2 and B_2 by (7) and (8).
Here we assumed the same values of A_1, B_1, A_2 and B_2 as in (Hong *et al.*, 1996) for an AlGaAs/GaAs bulk SOA.
The time derivative terms in (3) have been replaced by the central-difference approximation in order to simulate this equation by the FD-BPM (Das *et al.*, 2000). In simulation, the parameter of bulk SOAs (AlGaAs/GaAs, double heterostructure) with a wavelength of 0.86 μm (Hong *et al.*, 1996) is used and the SOA length is 350 μm. The input pulse shape is sech^2 and is Fourier transform-limited.
The gain spectra of SOAs are important for obtaining the propagation and wave mixing (FWM and FWM with time-delay between the input pump and probe pulses) characteristics of short optical pulses. Figure 2 shows the gain spectra given by a second-order Taylor expansion about the pulse center frequency with derivatives of $g(\tau, \omega)$ by (7) and (8). The solid line shows the unsaturated gain spectrum having g_0 ($=g(\tau, \omega)$) of 92 cm^{-1} at ω_0. The dotted line is a saturated gain spectrum having $g_0/2$ (= 46 cm^{-1}) at ω_0, and the dashed-dotted line is a strongly saturated gain spectrum having $g(\tau, \omega)$ of 0 cm^{-1} at ω_0. The pump frequency ω_0 is set to near the gain peak, and the linear gain g_0 is 92 cm^{-1} at ω_0.

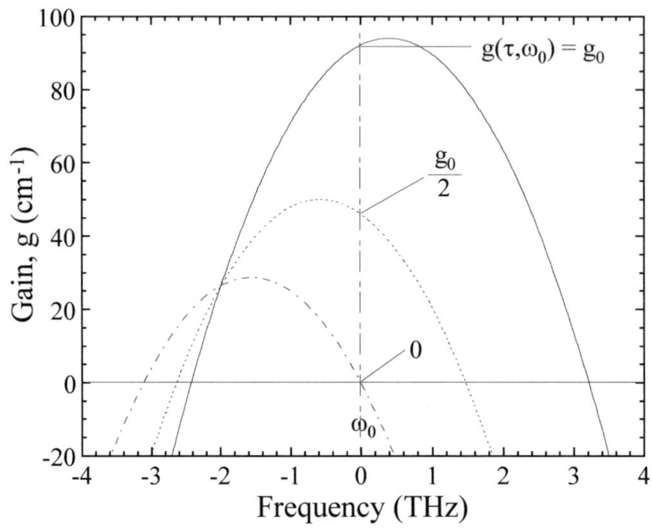

Fig. 2. The gain spectra given by the second-order Taylor expansion about the pulse center frequency, with the derivatives of $g(\tau,\omega)$ by (7) and (8). The upper curve is linear gain, g_0 = 92 cm^{-1}, middle curve is half-saturated gain, $g_0/2$ = 46 cm^{-1}, and lower curve is fully saturated gain, g_0 = 0.

The gain bandwidth is about the same as the measured value for an AlGaAs/GaAs bulk SOA (Seki *et al.*, 1981). If an InGaAsP/InP bulk SOA is used we can expect much wider gain bandwidth (Leuthold *et al.*, 2000). With a decrease in the carrier density, the gain decreases and the peak position is shifted to a lower frequency because of the band-filling effect.
This MNLSE was initially used by (Hong *et al.*, 1996) for the analysis of a "solitary pulse" propagation in an SOA. We used the same MNLSE for the simulation of FWM in an SOA using the FD-BPM. We have introduced a complex envelope function V (τ, 0) at the input side of the SOA for taking into account the two (pump and probe) pulses.

2.3 Finite-Difference Beam Propagation Method (FD-BPM)

To solve a boundary value problem by the finite-differences method, every derivative appearing in the equation, as well as in the boundary conditions, is replaced by the central differences approximation. Central differences are usually preferred because they lead to greater accuracy (Conte & Boor, 1980). In our simulations, we used the finite-differences (central differences) to solve the MNLSE for this research work.
Usually, the fast Fourier transformation beam propagation method (FFT-BPM) (Okamoto, 1992; Brigham, 1988) is used for the analysis of the optical pulse propagation in optical fibers by the successive iterations of the Fourier transformation and the inverse Fourier transformation. In the FFT-BPM, the linear propagation term (GVD term) and phase compensation terms (other than GVD, 1st and 2nd order gain spectrum terms) are separated in the nonlinear Schrödinger equation for the individual consideration of the time and frequency domain for the optical pulse propagation. However, in our model, equation (3) includes the dynamic gain change terms, i.e., the 1st and 2nd order gain spectrum terms which are the last two terms of the right-side in equation (3). Therefore, it is not possible to

separate equation (3) into the linear propagation term and phase compensation term and it is quite difficult to calculate equation (3) using the FFT-BPM. For this reason, we used the FD-BPM (Chung & Dagli, 1990; Conte & Boor, 1980). If we replace the time derivative terms of equation (3) by the below central-difference approximation, equation (11), and integrate equation (3) with the small propagation step Δz, we obtain the tridiagonal simultaneous matrix equation (12)

$$\frac{\partial}{\partial \tau} V_k = \frac{V_{k+1} - V_{k-1}}{2\Delta\tau}, \quad \frac{\partial^2}{\partial \tau^2} V_k = \frac{V_{k+1} - 2V_k + V_{k-1}}{\Delta\tau^2} \tag{11}$$

where, $V_k = V(\tau_k)$, $V_{k+1} = V(\tau_k + \Delta\tau)$, and $V_{k-1} = V(\tau_k - \Delta\tau)$

$$\begin{aligned} -a_k(z+\Delta z)\, V_{k-1}(z+\Delta z) + \{1 - b_k(z+\Delta z)\}\, V_k(z+\Delta z) - c_k(z+\Delta z)\, V_{k+1}(z+\Delta z) \\ = a_k(z)\, V_{k-1}(z) + \{1 + b_k(z)\}\, V_k(z) + c_k(z)\, V_{k+1}(z) \end{aligned} \tag{12}$$

where, $k = 1, 2, 3, \ldots\ldots\ldots\ldots, n$ and

$$a_k(z) = \frac{\Delta z}{2}\left[\frac{i\beta_2}{2\Delta\tau^2} + i\frac{1}{4\Delta\tau}\frac{\partial g(\tau,\omega,z)}{\partial\omega}\Big|_{\omega_0,\tau_k} - \frac{1}{4\Delta\tau^2}\frac{\partial^2 g(\tau,\omega,z)}{\partial\omega^2}\Big|_{\omega_0,\tau_k}\right] \tag{13}$$

$$\begin{aligned} b_k(z) = -\frac{\Delta z}{2}\Bigg[\frac{i\beta_2}{\Delta\tau^2} + \frac{\gamma}{2} + \left(\frac{\gamma_{2p}}{2} + ib_2\right)|V_k(z)|^2 - \frac{1}{2} g_N(\tau_k,\omega_0,z)(1+i\alpha_N) \\ - \frac{1}{2}\Delta g_T(\tau_k,\omega_0,z)(1+i\alpha_T) - \frac{1}{2\Delta\tau^2}\frac{\partial^2 g(\tau,\omega,z)}{\partial\omega^2}\Big|_{\omega_0,\tau_k}\Bigg] \end{aligned} \tag{14}$$

$$c_k(z) = \frac{\Delta z}{2}\left[\frac{i\beta_2}{2\Delta\tau^2} - i\frac{1}{4\Delta\tau}\frac{\partial g(\tau,\omega,z)}{\partial\omega}\Big|_{\omega_0,\tau_k} - \frac{1}{4\Delta\tau^2}\frac{\partial^2 g(\tau,\omega,z)}{\partial\omega^2}\Big|_{\omega_0,\tau_k}\right] \tag{15}$$

where, $\Delta\tau$ is the sampling time and n is the number of sampling. If we know $V_k(z)$, ($k = 1, 2, 3, \ldots\ldots\ldots\ldots, n$) at the position z, we can calculate $V_k(z+\Delta z)$ at the position of $z+\Delta z$ which is the propagation of a step Δz from position z, by using equation (12). It is not possible to directly calculate equation (12) because it is necessary to calculate the left-side terms $a_k(z+\Delta z)$, $b_k(z+\Delta z)$, and $c_k(z+\Delta z)$ of equation (12) from the unknown $V_k(z+\Delta z)$. Therefore, we initially defined $a_k(z+\Delta z) \equiv a_k(z)$, $b_k(z+\Delta z) \equiv b_k(z)$, and $c_k(z+\Delta z) \equiv c_k(z)$ and obtained $V_k^{(0)}(z+\Delta z)$, as the zeroth order approximation of $V_k(z+\Delta z)$ by using equation (12). We then substituted $V_k^{(0)}(z+\Delta z)$ in equation (12) and obtained $V_k^{(1)}(z+\Delta z)$ as the first order approximation of $V_k(z+\Delta z)$ and finally obtained the accurate simulation results by the iteration as used in (Brigham, 1988; Chung & Dagli, 1990).

Figure 3 shows the schematic diagram of the FD-BPM in time domain scale. Here, $\tau\,(= t - z / v_g)$ is the local time, which propagates with the group velocity v_g at the center frequency of an optical pulse and $\Delta\tau$ is the sampling time. z is the propagation direction and Δz is the propagation step. With this procedure, we used up to 3-rd time iteration for more accuracy of the simulations.

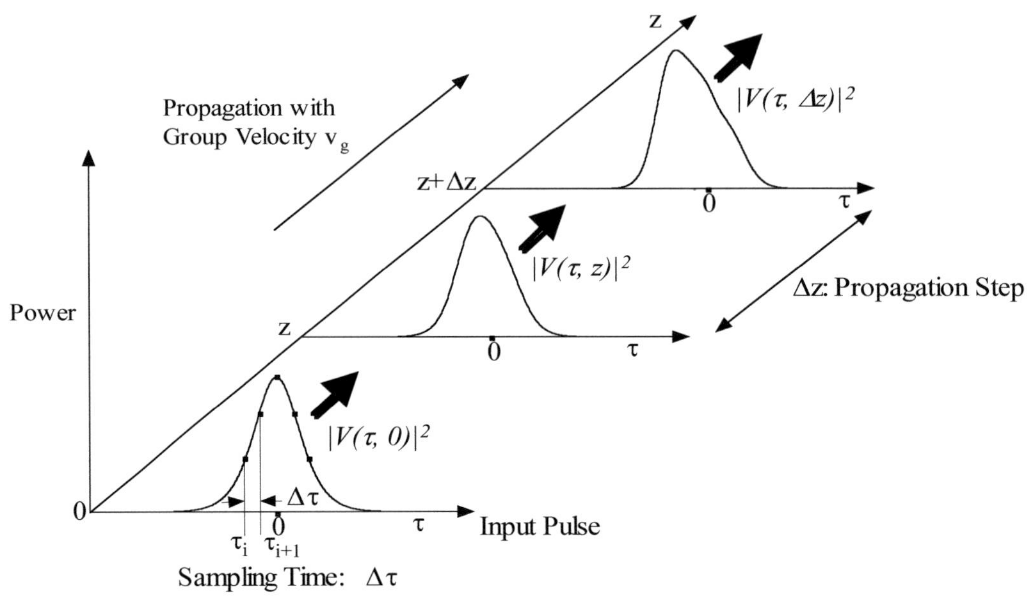

Fig. 3. Schematic diagram of FD-BPM in time domain. $\tau\,(=t-z/v_g)$ is the local time, which propagates with the group velocity v_g at the center frequency of an optical pulse and Δτ is the sampling time, and z is the propagation direction and Δz is the propagation step.

The FD-BPM (Conte & Boor, 1980; Chung & Dagli, 1990) is used for the simulation of several important charactreristics, namely, (1) single pulse propagation in SOAs (Das *et al.*, 2008; Razaghi *et al.*, 2009a & 2009b), (2) two input pulses propagating in SOAs (Das *et al.*, 2000; ; Connelly *et al.*, 2008), (3) multiplexing of several input pulses using FWM (Das *et al.*, 2001), (4) two input pulses with phase-conjugation propagating along SOAs (Das *et al.*, 2001), and (5) two propagating input pulses with time-delay between them being optimized (Das *et al.*, 2007).

2.4 Optical pulse propagation in SOAs

Optical pulse propagation in SOAs has drawn considerable attention due to its potential applications in optical communication systems, such as a wavelength converter based on FWM and switching. The advantages of using SOAs include the amplification of small (weak) optical pulses and the realization of high efficient FWM.

We analyzed the optical pulse propagation in SOAs using the FD-BPM in conjunction with the MNLSE, where several parameters are taken into account, namely, the group velocity dispersion, self-phase modulation (SPM), and two-photon absorption (TPA), as well as the dependencies on the carrier depletion, carrier heating (CH), spectral-hole burning (SHB) and their dispersions, including the recovery times in an SOA (Hong *et al.*, 1996). We also considered the gain spectrum (as shown in Fig. 1). The gain in an SOA was dynamically changed depending on values used for the carrier density and carrier temperature in the propagation equation (i.e., MNLSE).

Initially, (Hong *et al.*, 1996) used the MNLSE for the simulation of optical pulse propagation in an SOA by FFT-BPM (Okamoto, 1992; Brigham, 1988) but the dynamic gain terms were changing with time. The FD-BPM enables the simulation of optical pulse propagation taking into consideration the dynamic gain terms in SOAs (Das *et al.*, 2007; Razaghi *et al.*, 2009a & 2009b; Aghajanpour *et al.*, 2009). We used the modified MNLSE for optical pulse propagation in SOAs by the FD-BPM (Chung & Dagli, 1990; Conte & Boor, 1980). We used the FD-BPM for the simulation of FWM characteristics when input pump and probe pulses, which are delayed with respect to one another, propagate in SOAs.

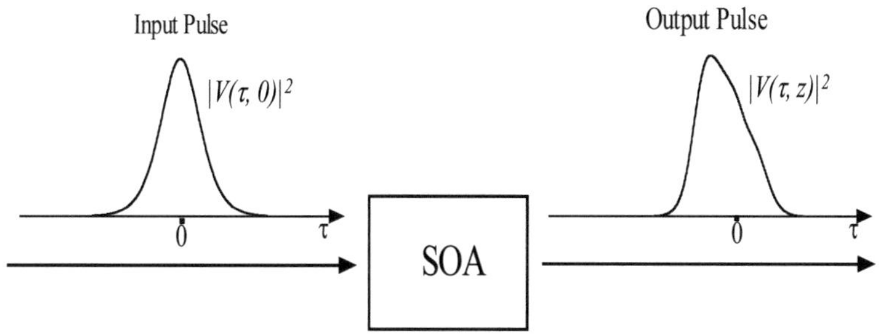

Fig. 4. Schematic diagram for the simulation of nonlinear propagation of single pulse's in SOA. Here, $|V(\tau,0)|^2$ and $|V(\tau,z)|^2$ are the intensity of input (z = 0) and output (after propagating a distance z) pulses of SOA.

Figure 4 illustrates the simulation model for nonlinear propagation characteristics of a single pulse in an SOA. An optical pulse is injected into the input side of the SOA (z = 0). Here, τ is the local time, $|V(\tau,0)|^2$ is the intensity (power) of input pulse at the input side of SOA (z = 0) and $|V(\tau,z)|^2$ is the intensity (power) of the output pulse at the output side of SOA after propagating a distance **z**. We also used this model to simulate FWM characteristics of SOAs for multi-pulse propagation.

Figure 5 shows the simulation results for single optical pulse propagation in an SOA. Figure 5(a) shows the temporal response of the propagated pulse for different output energy levels. In this simulation, the SOA length was 500 μm (other parameters are listed in Table 1) and the input pulse width was 1 ps. By increasing the input pulse energy, the output pulse energy increased until it saturated the gain of the SOA. The shift in pulse peak positions towards the leading edge (negative time) is mainly due to the gain saturation of the SOA (Kawaguchi *et al.*, 1999), because the gain experienced by the pulses is higher at the leading edge than at the trailing edge. Figure 5(b) shows the spectral characteristics of the propagating pulse at the output of the SOA for different output energy levels. These spectral characteristics were obtained by evaluating the fast Fourier transform (FFT) of the temporal pulse shapes shown in Figure 5(a). In Fig. 5(b) we also notice that by increasing the input pulse energy, the output pulse energy increases until the SOA is driven into saturation. The dips observed at the higher frequency side of the frequency spectra are due to the self-phase modulation characteristics of the SOA (Kawaguchi *et al.*, 1999). Also noticed that the output frequency spectra are red-shifted (the spectral peak positions are slightly shifted to the lower frequency side of the frequency spectra), and this is also attributed to the gain saturation of the SOA (Kawaguchi *et al.*, 1999). The simulation results are in excellent agreement with the experimental results reported by Kawaguchi *et al.* (Kawaguchi *et al.*, 1999).

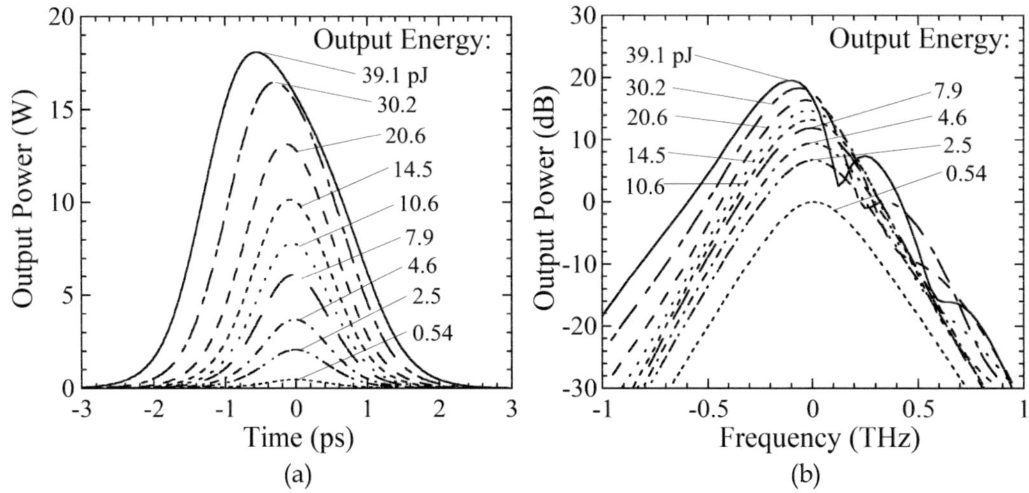

Fig. 5. (a) Temporal characteristics of the propagated pulses at the output of the SOA. Input pulse-width is 1 ps. By increasing the input pulse energy, the output energy increases until the SOA is driven into saturation. (b) Spectral characteristics of the output pulse for different output energy levels. The dips occurring at the higher frequency side are due to the self-phase modulation characteristics of the SOA.

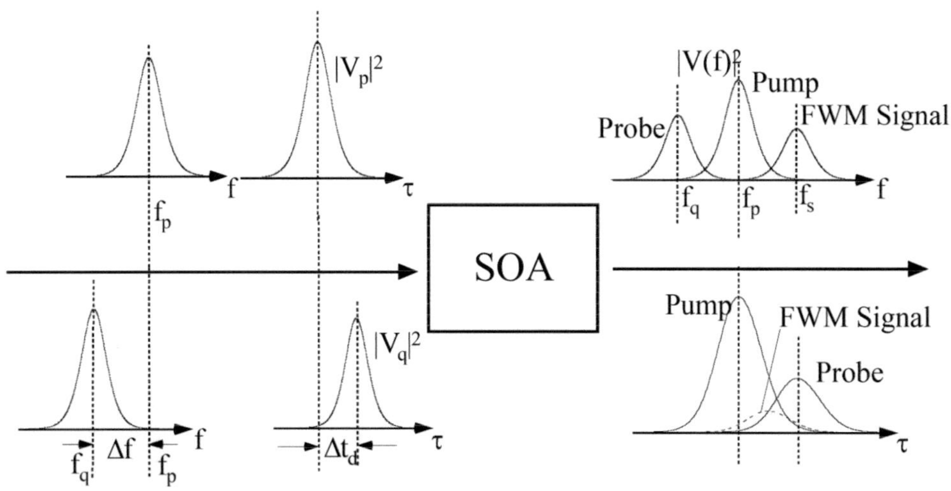

Fig. 6. Schematic diagram illustrating the input and output pump and probe pulses for the simulation of the FWM conversion efficiency in SOAs. The time delay between the input pump and probe pulses is Δt_d , and the detuning between the input pump and probe is $\Delta f = f_p - f_q$, f_p is the center frequency of the pump pulse and f_q is that of the probe pulse.

2.5 FWM characteristics with time-delays between input pulses in SOAs

When two optical pulses with different central frequencies f_p (pump) and f_q (probe) are injected into the SOA simultaneously, an FWM signal is generated at the output of the SOA at a frequency $2f_p - f_q$. Figure 6 shows the schematic diagram adopted for the simulation of FWM conversion efficiency in an SOA, illustrating the time delay $\Delta t_d = 0$ between the input pump and probe pulses, which are injected simultaneously into the SOA. For the analysis of the FWM conversion efficiency, the combined pump and probe pulse, V(τ), is given by

$$V(\tau) = V_p(\tau) + V_q(\tau \pm \Delta t_d)\exp(-i2\pi\Delta f \tau) \quad (16)$$

where, $V_p(\tau)$ and $V_q(\tau)$ are the complex envelope functions of the input pump and probe pulses respectively, $\tau\,(= t - z / v_g)$ is the local time that propagates with group velocity v_g at the center frequency of an optical pulse, Δf is detuning frequency and expressed as $\Delta f = f_p - f_q$, Δt_d is the time-delay between the input pump and probe pulses. The positive (plus) and negative (minus) signs in $V_q(\tau \pm \Delta t_d)$ correspond to the pump leading the probe or the probe leading the pump, respectively. Using the complex envelope function of equation (16), we solved the MNLSE and obtained the distribution of the probe and pump pulses as well as the output FWM signal pulse.
For the simulations, we used the parameters of a bulk SOA (AlGaAs/GaAs, double heterostructure) at a wavelength of 0.86 μm. The parameters are listed in Table 1 (Hong *et al.*, 1996). The length of the SOA was assumed to be 350 μm. All the results were obtained for a propagation step Δz of 5 μm. Note that, for any step size less than 5 μm the simulation results were almost identical (i.e., independent of the step size).

Name of the Parameters	Symbols	Values	Units
Length of SOA	L	350	μm
Effective area	A	5	μm^2
Center frequency of the pulse	f_0	349	THz
Linear gain	g_0	92	cm^{-1}
Group velocity dispersion	β_2	0.05	$ps^2\,cm^{-1}$
Saturation energy	W_s	80	pJ
Linewidth enhancement factor due to the carrier depletion	α_N	3.1	
Linewidth enhancement factor due to the CH	α_T	2.0	
The contribution of stimulated emission and FCA to the CH gain reduction	h_1	0.13	$cm^{-1}pJ^{-1}$
The contribution of TPA	h_2	126	$fs\,cm^{-1}pJ^{-2}$
Carrier lifetime	τ_s	200	ps
CH relaxation time	τ_{ch}	700	fs
SHB relaxation time	τ_{shb}	60	fs
SHB saturation power	P_{shb}	28.3	W
Linear loss	γ	11.5	cm^{-1}
Instantaneous nonlinear Kerr effect	n_2	-0.70	$cm^2\,TW^{-1}$
TPA coefficient	γ_{2p}	1.1	$cm^{-1}\,W^{-1}$
Parameters describing second-order Taylor expansion of the dynamically gain spectrum	A_1 B_1 A_2 B_2	0.15 -80 -60 0	$fs\,\mu m^{-1}$ fs $fs^2\,\mu m^{-1}$ fs^2

Table 1. Simulation parameters of a bulk SOA (AlGaAs/GaAs, double heterostructure) (Hong *et al.*, 1996; Das *et al.*, 2000).

3. Experimental setup

Figure 7 shows the experimental setup for the measurement of the FWM signal energy at the output of the SOA with time delays being introduced between input pump and probe pulses. This experimental setup is similar to the one that was used by Inoue & Kawaguchi (Inoue & Kawaguchi, 1998a). In this setup, we used an optical parametric oscillator (OPO) at 1.3 μm wavelength band as a light source. Here, an optical pulse train of 100 fs was generated at a repetition rate of 80 MHz by the OPO. The pump and probe pulses were obtained by filtering the output pulse of the OPO. The output was divided into a pump and a probe beam (pulse). Optical bandpass filters (4 nm) were inserted into the two beam passes to select the narrow wavelength component. After passing through the filters, the pulses were broadened to 550 fs width, which is close to the transform-limited secant hyperbolic shape. The time-delay between pump and probe pulses is given by the optical stage (as shown in Fig. 7 time delay stage) and it regulates the optical power. The two beams were combined and injected into the SOA. The beams were amplified by the SOA and then the FWM signals were generated at the output of the SOA. The FWM signal was selected from the SOA output using two cascaded narrow-band bandpass filters (3 and 4 nm) and detected by a photodiode. These 3 nm and 4 nm optical bandpass filters selected spectrally the FWM signal component from the output of the SOA. For the detection of FWM signal, we inserted a mechanical chopper into the probe beam path. The FWM signal was measured using the lock-in technique with 4 nm double-cavity bandpass filters. We adjusted the pump frequency to be 1.8 THz higher than that of the probe frequency, i.e., the pump-probe detuning was 1.8 THz.

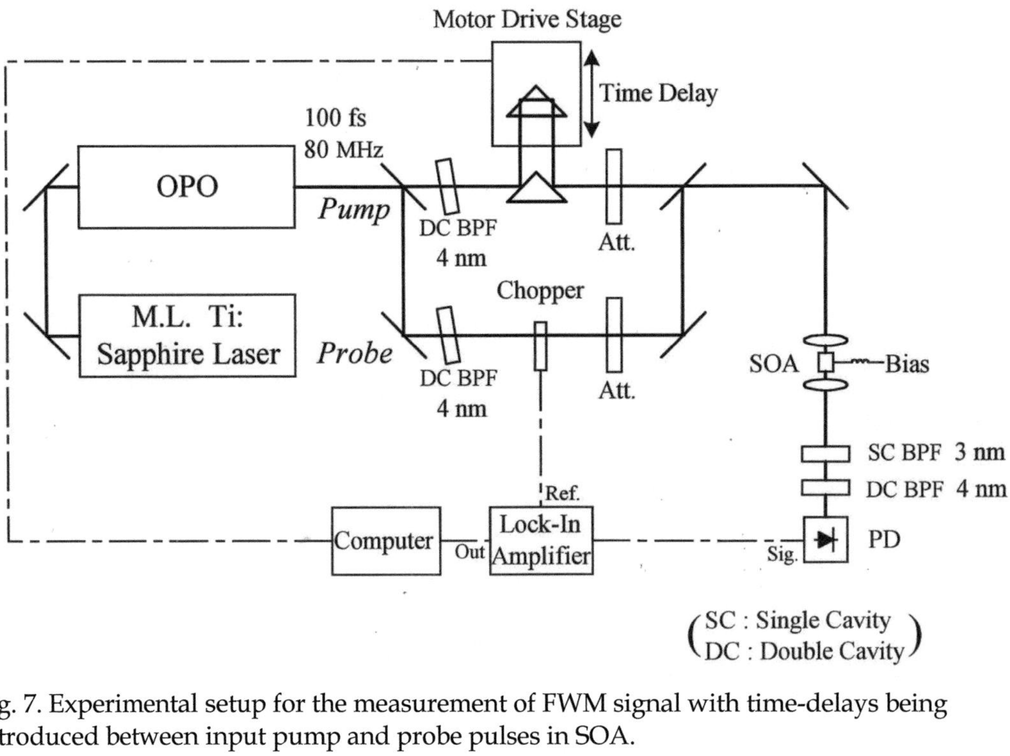

Fig. 7. Experimental setup for the measurement of FWM signal with time-delays being introduced between input pump and probe pulses in SOA.

4. Results and discussions

It was found that the optimum time-delay between the input pump and probe pulses shifts from the zero time-delay ($\Delta t_d = 0$) under the strong input pulse condition needed to achieve high FWM conversion efficiency in an SOA. These results are very important for the design of ultrafast optical systems that have high conversion efficiency and small timing jitter (Inoue & Kawaguchi, 1998b; Das *et al.*, 2005).

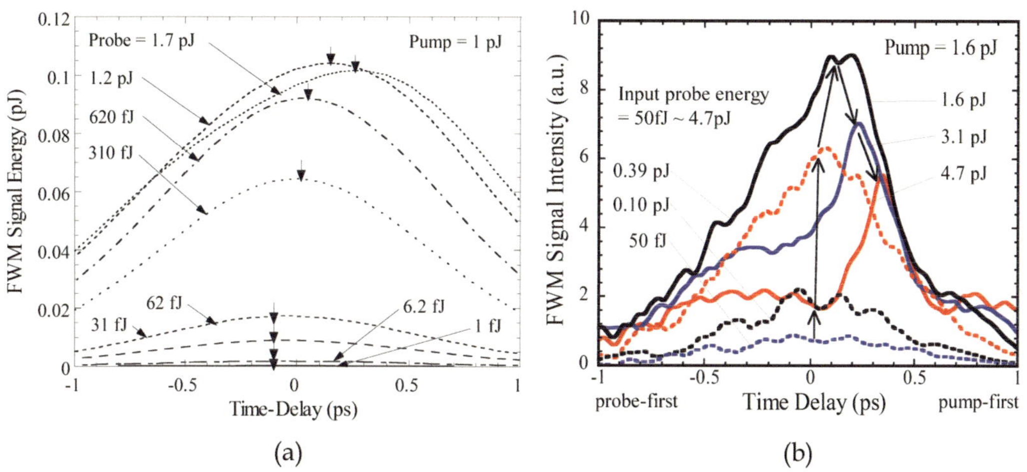

Fig. 8. (a) Simulation results: FWM signal energy (intensity) versus time-delay characteristics at the output of SOA. Input pulse-width is 1 ps and the pump-probe detuning is 3 THz. Input pump pulse energy was fixed at 1 pJ and input probe energies were varied from 1 fJ to 1.7 pJ. (b) Experimental results: FWM signal intensity versus time-delay. Input pulse-width is 550 fs and pump-probe detuning is 1.8 THz. Here, the input pump pulse energy was fixed at 1.6 pJ and input probe energies were varied from 50 fJ to 4.7 pJ.

Figure 8(a) shows the simulation results of the FWM signal energy (intensity) at the output of the SOA versus the time-delay between the input pump and probe pulses. Here, the plus time-delay refers to the pump pulse being injected before the probe pulse. The input pump energy was fixed at 1.0 pJ. With the increase of the input probe energy, the FWM signal intensity increased until the input probe energy of 1.2 pJ, which is comparable (nearly) to the input pump energy. For a higher input probe energy (1.7 pJ), the increase of the input probe energy decreased the FWM signal, and the peak position shifted towards the pump-first direction (Inoue & Kawaguchi, 1998b; Diez *et al.*, 1997) as illustrated by the arrows. This phenomenon is attributed to the optical nonlinear effects in the SOA, which limits the FWM conversion efficiency.

Figure 8(b) shows the experimental results of FWM signal intensity (energy) at the output of the SOA versus the time-delay between the input pump and probe pulses. For the measurements, a 1.3 μm OPO system was used as a light source (experimental set up is shown in Fig. 7). The measured optimum time-delay between input pump and probe pulses was measured using a multiple quantum well SOA of length 350 μm. The shapes of the input pump and probe pulses were sech² both had a pulse-width of 550 fs. The pump-probe detuning was set to 1.8 THz, and the input pump energy was fixed at 1.6 pJ. By increasing the input probe energy, the FWM signal intensity increased until the input probe energy became comparable to the input pump energy of 1.6 pJ. By further increasing the probe intensity, the FWM signal decreased and the peak position shifted to the pump-first direction as illustrated by the arrows. This demonstrated excellent agreement between the simulation and experimental results.

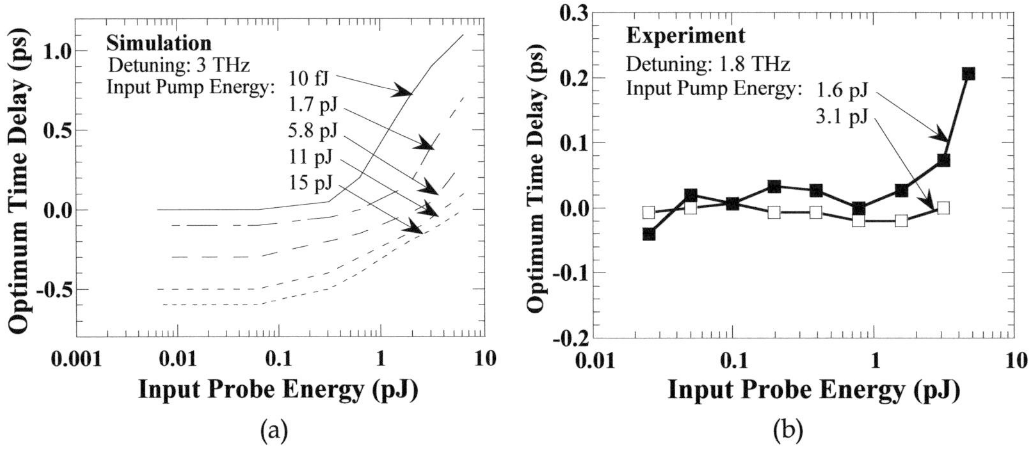

Fig. 9. (a) Simulation results: Optimum time delay versus input probe energy for different input pump energy levels. Input pulse-width = 1 ps; frequency detuning = 3 THz.
(b) Experimental results: Optimum time delay versus input probe energy for different input pump energy levels. Input pulse-width = 550 fs; frequency detuning = 1.8 THz.

Figure 9(a) shows the simulated optimum time delay between the pump and probe pulses versus the input probe energy, for an input pump and probe pulse-width of 1 ps, and a frequency detuning between pump and probe of 3 THz, and for different input pump energies. It is noticed that by increasing the input pump energy, the optimum time-delay reduces. However, increasing the input probe energy increases the optimum time delay. Figure 9(b) shows the measured optimum time delay versus the input probe energy for different pump energy levels. The input pulse-width was 550 fs, and the frequency detuning between pump and probe was 1.8 THz. The input pump energy levels were varied between 1.6 pJ and 3.1 pJ. Excellent agreement between the simulated and measured optimum-delay characteristics is observed for input probe energy levels above 0.1 pJ.

Figure 10(a) shows the simulated FWM conversion efficiency versus the input probe energy for different input pump energy levels. It is obvious that for a given input probe energy level, the FWM conversion efficiency increases when increasing the input pump energy level. On the other hand, for a given pump energy level, the FWM conversion efficiency decreases when the input probe energy is increased. Note that the dashed lines in Fig. 10(a) correspond to perfect pump-probe time overlap ($\Delta t_d = 0$), whereas solid lines correspond to optimum pump-probe time delays. Figure 10(b) shows the measured maximum FWM conversion efficiency (corresponding to optimum time delay between the pump and probe pulses) versus the input probe energy level for input pump energy levels 1.6 pJ and 3.1 pJ, respectively. It is noticed from Fig. 10(b) that for a low probe energy (below 1 pJ), the FWM conversion efficiency decreases with increasing the input pump energy, whereas, for a high input probe energy (above 1 pJ), the FWM conversion efficiency increases when the input pump energy increases. From Fig. 10(a) and Fig. 10(b), excellent agreement is seen between the simulated and measured results for the optimum FWM conversion efficiency.

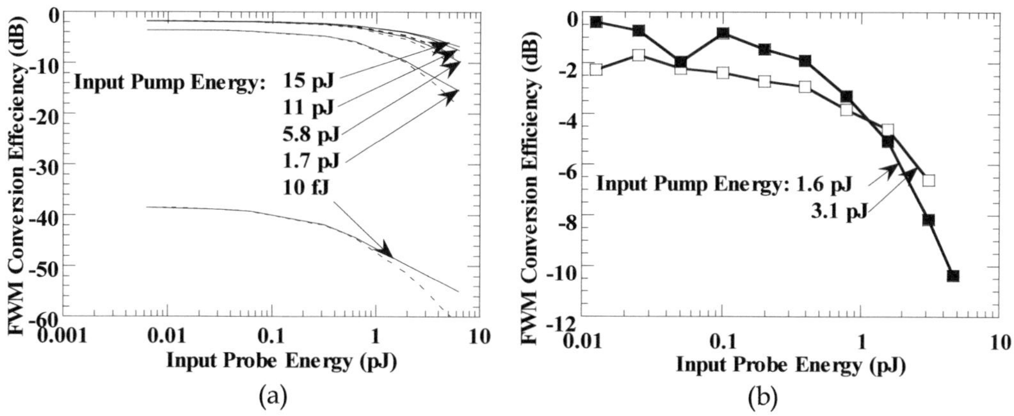

Fig. 10. (a) Simulation results: FWM conversion efficiency versus the input probe energy characteristics at the output of SOA. Dashed lines correspond to perfect pump-probe time overlap ($\Delta t_d = 0$), whereas solid lines correspond to optimum pump-probe time delays. (b) Experimental results: Optimum FWM conversion efficiency versus the input probe energy characteristics for different input pump energy levels.

5. Conclusion

We have presented an accurate analysis based on the FD-BPM, which optimizes the time delay between the input pump and probe pulses to maximise the FWM conversion efficiency in SOAs. We have shown that the gain saturation of the SOA degrades the FWM conversion efficiency. However, by optimizing the time delay between the pump and probe pulses, for a specific pulse duration and repetition rate, a high FWM conversion efficiency can be achieved. We have also simulated and experimentally measured the optimum time delay versus the input probe energy characteristics. Simulation and experimental results have confirmed that increasing the input probe energy increases the optimum time delay and that for a low probe energy, the FWM conversion efficiency decreases with increasing the input pump energy, whereas, for a high input probe energy, the FWM conversion efficiency increases when the input pump energy is increased.

6. Acknowledgments

The authors would like to thank Mr. Y. Ito and Mr. Y. Yamayoshi for their helpful contribution to this work. Authors acknowledge the support of the Department of Nano-bio Materials and Electronics, Gwangju Institute of Science and Technology, Republic of Korea.

7. References

Agrawal, G. P. (1989). *Nonlinear Fiber Optics.* Academic Press, Calif., ISBN 0-12-045142-5, San Diego.

Agrawal, G. P. & Olsson, N. A. (1989). Self-phase modulation and spectral broadening of optical pulses in semiconductor laser and amplifiers. *IEEE J. Quantum Electron.*, vol. 25, pp. 2297-2306, ISSN 0018–9197.

Aghajanpour, H.; Ahmadi, V. & Razaghi, M. (2009). Ultra-short optical pulse shaping using semiconductor optical amplifier. *Optics & Laser Technology,* vol. 41, pp. 654-658, ISSN 0030-3992.

Brigham, E. Oran (1988). *The Fast Fourier Transform and Its Applications*. Englewood Cliffs, N.J.: Prentice-Hall Inc. ISBN 0-13-307505-2.

Chung, Y. & Dagli, N. (1990). An Assessment of finite difference beam propagation method. *IEEE J. Quantum Electron.*, vol. 26, pp. 1335-39, ISSN 0018–9197.

Conte, S. D. & Boor, Carl de (1980). *Elementary Numerical Analysis: An Algorithmic Approach,* Third Edition, McGraw-Hill Book Company Co. ISBN, Singapore. ISBN 0070124477.

Connelly, M. J.; Barry, L. P., Kennedy, B. F. & Ried, D. A. (2008). Numerical analysis of four-wave mixing between picosecond mode-locked laser pulses in a tensile-strained bulk SOA. *Optical and Quantum Electronics,* vol. 40, pp. 411-418, ISSN 1572-817X.

Das, N. K.; Yamayoshi, Y. & Kawaguchi, H. (2000). Analysis of basic four-wave mixing characteristics in a semiconductor optical amplifier by beam propagation method. *IEEE J. Quantum Electron.* 36, 10, pp. 1184-1192, ISSN 0018–9197.

Das, N. K. & Karmakar, N. C. (2008). Nonlinear propagation and wave mixing characteristics of pulses in semiconductor optical amplifiers. *Microwave and Optical Technology Letters,* vol. 50, pp. 1223-1227, ISSN 0895-2477.

Das, N. K.; Karmakar, N. C., Yamayoshi, Y. & Kawaguchi, H. (2007). Four-wave mixing characteristics in SOAs with optimum time-delays between pump and probe pulses," *Microwave and Optical Technology Letters,* vol. 49, pp. 1182-1185, ISSN 0895-2477.

Das, N. K.; Yamayoshi, Y., Kawazoe, T. & Kawaguchi, H. (2001). Analysis of optical DEMUX characteristics based on four-wave mixing in semiconductor optical amplifiers. *IEEE /OSA J. Lightwave Technol.,* vol. 19, pp. 237-246, ISSN 0733-8724.

Das, N. K.; Kawazoe, T., Yamayoshi, Y. & Kawaguchi, H. (2001). Analysis of optical phase-conjugate characteristics of picosecond four-wave mixing signals in semiconductor optical amplifiers. *IEEE J. Quantum Electron.,* vol. 37, pp. 55-62, ISSN 0018–9197.

Das, N. K.; Karmakar, N. C., Yamayoshi, Y. & Kawaguchi, H. (2005). Four-wave mixing characteristics among short optical pulses in semiconductor optical amplifiers with optimum time-delays, *Proceedings of the 18th Annual Meeting of the IEEE Lasers and Electro-Optics Society 2005 (IEEE-LEOS2005),* pp. 127-128, ISBN 0-7803-9217-5, *Sydney, NSW, Australia*, October 2005, IEEE Press (USA).

Dienes, A.; Heritage, J. P., Jasti, C. & Hong, M. Y. (1996). Femtosecond optical pulse amplification in saturated media. *J. Opt. Soc. Am. B*, vol. 13, pp. 725-734, ISSN 0740-3224.

Diez, S.; Schmidt, C., Ludwig, R., Weber, H. G., Obermann, K., Kindt, S., Koltchanov, I. & Petermann, K. (1997). Four-wave mixing in semiconductor optical amplifiers for frequency conversion and fast optical switching. *IEEE J. Sel. Top. Quantum Electron.*, vol. 3, pp. 1131-1145, ISSN 1939-1404.

Inoue, J. & Kawaguchi, H. (1998a). Highly nondegenerate four-wave mixing among subpicosecond optical pulses in a semiconductor optical amplifier," *IEEE Photon. Technol. Lett.*, vol. 10, pp. 349–351, ISSN 1041-1135.

Inoue, J. & Kawaguchi, H. (1998b). Time-delay characteristics of four-wave mixing among subpicosecond optical pulses in a semiconductor optical amplifier. *IEEE Photon Technol. Lett.* 10, pp. 1566-1568, ISSN 1041-1135.

Hong, M. Y.; Chang, Y. H., Dienes, A., Heritage, J. P., Delfyett, P. J., Dijaili, Sol & Patterson, F. G. (1996). Femtosecond self- and cross-phase modulation in semiconductor laser amplifiers. *IEEE J. Sel Top Quant Electron.*, vol. 2, pp. 523-539, ISSN 1939-1404.

Kawaguchi, H.; Inoue J. & Kawazoe, T. (1999). Nonlinear propagation characteristics of subpicosecond optical pulses in semiconductor optical amplifiers. *Proceedings of the 9th European Conference on Integrated Optics and Technical Exhibition, ECIO'99*, pp. 455-458, ISBN, Italy, April 13-16, 1999, ECIO, Triono.

Leuthold, J.; Mayer, M., Eckner, J., Guekos, G., Melchior, H. and Zellweger, Ch. (2000). Material gain of bulk 1.55 μm InGaAsP/InP semiconductor optical amplifiers approximated by a polynomial model. *J. Appl. Phys.*, vol. 87, pp. 618-620, ISSN.

Mecozzi, A.; D'Ottavi, A., Iannone, E., & Spano, P. (1995). Four-wave mixing in travelling-wave semiconductor amplifiers. *IEEE J. Quantum Electron.*, vol. 31, pp. 689-699, ISSN 0018–9197.

Mecozzi, A. & Mϕrk, J. (1997). Saturation effects in nondegenerate four-wave mixing between short optical pulses in semiconductor laser amplifiers. *IEEE J. Sel Top Quantum Electron.*, vol. 3, pp. 1190-1207, ISSN 1939-1404.

Okamoto, K. (1992). *Theory of Optical Waveguides*, Corona Publishing Co., Tokyo; Ch. 7 (in Japanese). ISBN 4-339-00602-5.

Razaghi, M.; Ahmadi, A., Connelly, M. J. & Madanifar, K. A. (2009a). Numerical modelling of sub-picosecond counter propagating pulses in semiconductor optical amplifiers, *Proceedings of the 9th International Conference on Numerical Simulation of Optoelectronic Devices 2009 (NUSOD' 09)*, pp. 59-60, ISBN 978-1-4244-4180-8, GIST, Gwangju, South Korea, September 2009, IEEE Press (USA).

Razaghi, M.; Ahmadi, A., & Connelly, M. J. (2009b). Comprehensive finite-difference time-dependent beam propagation model of counter propagating picosecond pulses in a semiconductor optical amplifier. *IEEE /OSA J. Lightwave Technol.*, vol. 27, pp. 3162-3174, ISSN 0733-8724.

Sauter, E. G. (1996). *Nonlinear Optics.* John Wiley & Sons, Inc. New York. ISBN 0-471-14860-1.

Seki, K.; Kamiya, T. and Yanai, H. (1981). Effect of waveguiding properties on the axial mode competition in stripe-geometry semiconductor lasers. *IEEE J. Quantum Electron.*, vol. 17, pp. 706-713, ISSN 0018–9197.

Shtaif, M. & Eisenstein, G. (1995). Analytical solution of wave mixing between short optical pulses in semiconductor optical amplifier. *Appl Phys. Lett.* 66, pp. 1458-1460, ISSN 0003-6951.

Shtaif, M.; Nagar, R. & Eisenstein, G. (1995). Four-wave mixing among short optical pulses in semiconductor optical amplifiers. *IEEE Photon Technol. Lett.* 7, pp. 1001-1003. ISSN 1041-1135.

Yariv, A. (1991). *Optical Electronics*, 4th Edition, Saunders College Publishing, San Diego. ISBN 0-03-053239-6.

7

Pattern Effect Mitigation Technique using Optical Filters for Semiconductor-Optical-Amplifier based Wavelength Conversion

Jin Wang
Fraunhofer Institute for Telecommunications, Heinrich-Hertz-Institute
Germany

1. Introduction

The demand for bandwidth in telecommunication network has been increasing significantly in the last few years. It is to be expected that also in the next few years multimedia services will further increase the bandwidth requirement. To utilize the full bandwidth of the optical fibre, wavelength division multiplexing (WDM) and time-division multiplexing (TDM) techniques have been applied. In these two primary techniques, wavelength converters that translate optical signal of one wavelength into optical signals of another wavelength, see Fig. 1, have become key devices. In general, a wavelength converter has to be efficient, meaning that with low signal powers an error free converted signal can be obtained. Also, the wavelength converters have to be small, compact, and as simple (cheap) as possible.

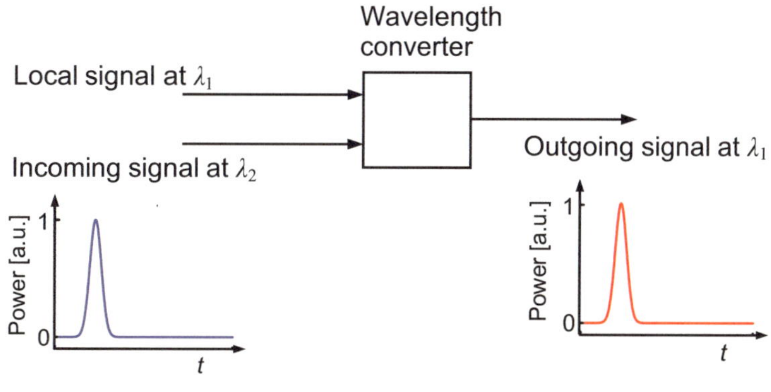

Fig. 1. A wavelength converter translates the incoming signal at λ_2 to the local signal at λ_1

Presently, large footprint optical-electrical-optical (o-e-o) translator units with large power consumption are used to perform wavelength conversion in optical cross-connects. Advantages of o-e-o methods are their inherent 3R (re-amplification, re-shaping, re-timing) regenerative capabilities and maturity. Conversely, the promise of all-optical wavelength conversion is the scalability to very high bit rates. All-optical wavelength converters (AOWC) can overcome wavelength blocking issues in next generation transparent networks and make possible reuse of the local wavelengths. All-optical wavelength converters can

also enable flexible routing and switching in the global and local networks, e.g. optical circuit-switching, optical burst-switching and optical packet-switching (Yoo, 1996).
Semiconductor optical amplifiers (SOAs) are closest to practical realization of all-optical wavelength converters (Yoo, 1996; Durhuus et al., 1996). SOA-based all-optical wavelength converters are compact, have low power consumption, and can be possibly operated at high speed. Indeed, a 320 Gbit/s SOA-based all-optical wavelength conversion has already been demonstrated (Liu et al., 2007). The advantages of using SOAs arise from the large number of stimulated emitted photons and free carriers, which are confined in a small active volume.
The SOA-based wavelength conversion works mostly via the variation of the gain and refractive index induced by an optical signal in the active region. The optical signal incident into the active region modifies the free carrier concentration. Thus, the optical gain and the refractive index within the active region are modulated. Other optical signals propagating simultaneously through the SOA also see these modulations of the gain and refractive index, being known as cross-gain modulation (XGM) and cross-phase modulation (XPM) (Yoo, 1996; Durhuus et al., 1996; Connelly, 2002). Thus, the information is transferred to another wavelength. In SOAs, a rich variety of dynamic processes drive the operation. These processes include the carrier dynamics between conduction band and valence band (interband dynamics) as well as inside of conduction band or valence band (intraband dynamics) (Connelly, 2002). They both affect the gain and the refractive index of the SOA, and thus the operation of the SOA-based wavelength converter. In addition, the fact that each of these effects has a specific lifetime leads to pattern dependent effects in the processed signals.
The pattern effect in the output signal out of an SOA is understood as follows. As subsequent incoming pulses are launched into a slow SOA, the carrier density is depleted continually. It recovers back to different levels and the amplifier gain also varies for different pulses, depending on the former bit pattern seen by the SOA. The unwanted pattern effect limits the implementation of the SOA-based wavelength converter at high speed.
The most practical approach to overcome pattern effects is to decrease the SOA recovery time by proper design (Zhang et al., 2006), optimum operation conditions (Girardin et al., 1998), an additional assisting light (Manning et al., 1994), and choice of new fast materials (Sugawara et al., 2002). Other approaches to mitigate the pattern effects are cascading several SOAs (Bischoff et al., 2004; Manning et al., 2006) or by using SOAs in a differential interferometer arrangement. Among them, the differential Mach-Zehnder interferometer (MZI) (Tajima, 1993), the differential Sagnac loop (Eiselt et al., 1995), the ultrafast nonlinear interferometer (UNI) (Hall & Rauschenbach, 1998) and the delay interferometer (DI) configurations (Leuthold et al., 2000), which exploit the XPM effect enable speeds beyond the limit due to the SOA carrier recovery times.
Recently, a new wavelength converter with an SOA followed by a single pulse reformatting optical filter (PROF) has been introduced (Leuthold et al., 2004b). In (Leuthold et al., 2004b), an experiment implementing a PROF based on MEMS technology demonstrated wavelength conversion at 40 Gbit/s, with record low input data signal powers of –8.5 and –17.5dBm for non-inverted and inverted operation. This is almost two orders of magnitudes less than typically reported for 40 Gbit/s wavelength conversions. The reason for the good conversion efficiency lies in the design of the filter. The PROF scheme exploits the fast chirp effects in the converted signal after the SOA and uses both the red- and blue-shifted spectral

components, while schemes with a single red- or blue-shifted filter (Leuthold et al., 2003; Nielsen & Mørk, 2006; Kumar et al., 2006; Liu et al., 2007) reject part of the spectrum. From an information technological point of view, rejection of spectral components with information should be avoided. Indeed, this scheme provides the best possible conversion efficiency for an SOA-based wavelength converter or regenerator.

The PROF scheme basically represents an optimum filter for the SOA response with the potential for highest speed operation. However, so far it is not clear, if these schemes with optical filters can successfully overcome pattern effects at highest speed.

In this chapter, we will show that the PROF scheme indeed and effectively mitigates SOA pattern effect. The pattern effect mitigation technique demonstrated here is based on the fact that the red chirp (decreasing frequency) and the blue chirp (increasing frequency) in the inverted signal behind an SOA have complementary pattern effects. If the two spectral components are superimposed by means of the PROF, then pattern effects can be successfully suppressed. An experimental implementation at 40 Gbit/s shows a signal quality factor improvement of 7.9 dB and 4.8 dB if compared to a blue- or red-shifted optical filter assisted wavelength converter scheme, respectively.

This chapter is organized as follows: In Section 2, an introduction of the SOA will be given. The basic SOA nonlinearities, which are used for wavelength conversion, will be reviewed. Also, the pattern effect in the SOA-based wavelength conversion will be discussed. In Section 3, the scheme of the SOA-based wavelength conversion assisted by an optical filter will be presented. The operation principle and experiment of this pattern effect mitigation technique are then explained and demonstrated.

2. Semiconductor-optical-amplifier based wavelength conversion

2.1 Semiconductor-optical-amplifier

Semiconductor-optical-amplifiers (SOAs) are amplifiers which use a semiconductor as the gain medium. These amplifiers have a similar structure to Fabry-Perot laser diodes but with anti-reflection elements at the end faces. SOAs are typically made from group III-V compound direct bandgap semiconductors such as GaAs/AlGaAs, InP/InGaAs, InP/InGaAsP and InP/InAlGaAs. Such amplifiers are often used in telecommunication systems in the form of fiber-pigtailed components, operating at signal wavelengths between 0.85 μm and 1.6 μm. SOAs are potentially less expensive than erbium doped fibre amplifier (EDFA) and can be integrated with semiconductor lasers, modulators, etc. However, the drawbacks of SOAs are challenging polarization dependences and a higher noise figure. Practically, the polarization dependence in the SOA can be reduced by an optimum structural design (Saitoh & Mukai, 1989).

A schematic diagram of a heterostructure SOA is given in Fig. 2. The active region, imparting gain to the input signal, is buried between the p- and n-doped layers, while the length of the active region is L. An external electrical current injects charge carriers into the active region and provides a gain to the optical input signal. An SOA typically has an amplifier gain of up to 30 dB.

Semiconductor amplifiers interact with the light, i.e. photons, in terms of electronic transitions. The transition between a high energy level W_2 in the conduction band (CB) and a lower energy level W_1 in the valence band (VB) can be radiative by emission of a photon with energy $hf = W_2 - W_1$ (h: Planck's constant, f: frequency), or non-radiative (such as thermal vibrations of the crystal lattice, Auger recombination). Three types of transitions,

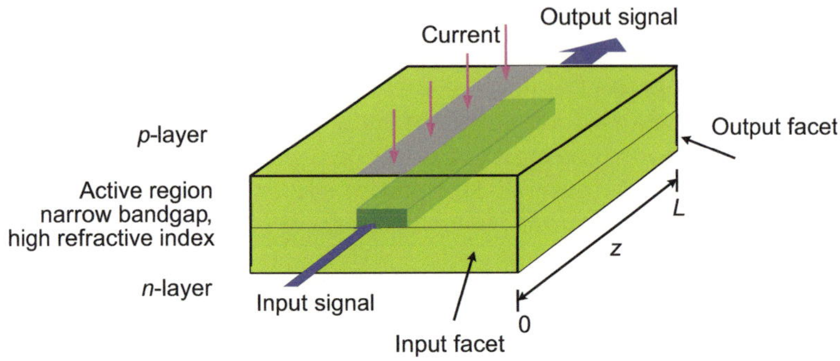

Fig. 2. Schematic diagram of an SOA

namely stimulated absorption, spontaneous emission and stimulated emission, are the basic mechanisms for all lasers and optical amplifiers. They are:

- Stimulated absorption: If an electromagnetic field exists, a photon with energy $hf = W_2 - W_1$ can be absorbed. Meanwhile, an electron can make an upwards transition from the lower energy level W_1 to the higher energy level W_2. The stimulated absorption rate depends on the electromagnetic energy density, and on the number of the electrons in the CB and the number of holes in the VB.
- Spontaneous emission: An electron in the CB can with a certain probability undergo a transition to the lower energy level W1 spontaneously, while emitting a photon with energy $hf = W_2 - W_1$, or loosing the transition energy through phonons or collisions. Obviously, the probability of the spontaneous emission is dependent on the number of the electron and hole pairs. These "spontaneously" emitted photons will be found with equal probability in any possible modes of the electromagnetic field. Thus a spontaneously emitted photon is regarded as a noise signal, since it represents a field with a random phase and a random direction.
- Stimulated emission: An incident photon can also induce with a certain probability a transition of the electron from the high energy level W_2 to the low energy level W_1. In this process a second photon is created. In contrast to spontaneous emission processes, this second photon is identical in all respects to the inducing photon (identical phase, frequency and direction). As does the stimulated absorption, the stimulated emission rate also depends on the incident electromagnetic energy density.

2.2 Basic SOA nonlinearities used for wavelength conversion

The SOA shows high nonlinearity with fast transient time. However, the high nonlinearity in the SOA is also accompanied with phase changes, which can distort the signals. This nonlinearity presents a most severe problem for optical communication applications, if the SOAs are used as "linear" optical amplifiers. Yet, high optical nonlinearity makes SOAs attractive for all-optical signal processing like all-optical switching and wavelength conversion. Indeed, the nonlinear gain and phase effects in an SOA, usually termed as cross-gain modulation (XGM) and cross-phase modulation (XPM), are widely used to perform wavelength conversion.

The XGM and XPM effects in an SOA are schematically shown in Fig. 3.

- XGM: As a strong control signal is launched into an SOA, it depletes the carrier in-side the SOA and saturates the gain of the SOA. If a weak probe signal (usually at a new wavelength) is also present in the SOA, the amplification of the probe signal is affected by the strong control signal, see the output power in Fig. 3. Thereby, the probe signal is modulated by the control signal. This effect is called cross-gain modulation (XGM).

- XPM: As the control signal causes a change of the carrier density inside the SOA, it also induces a change of the refractive index and a phase shift of the probe signal when passing through the SOA, see the output phase in Fig. 3. This effect is called cross-phase modulation (XPM). The XPM effect is usually measured in an interferometric configuration.

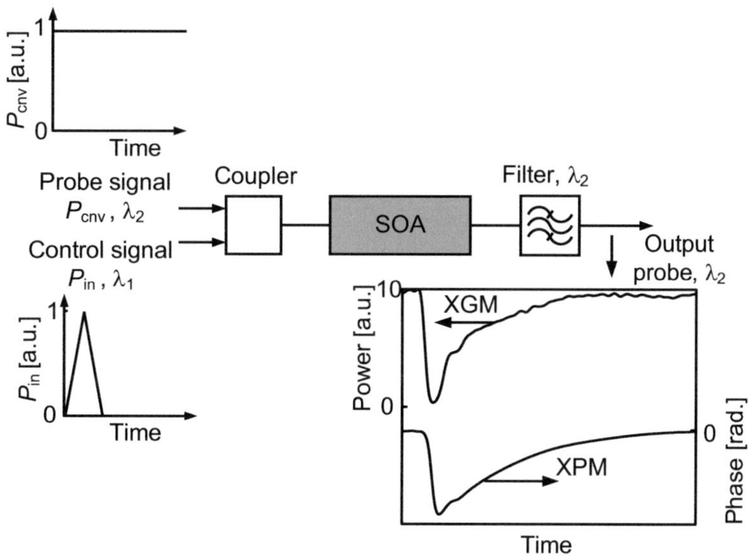

Fig. 3. XGM and XPM effects in an SOA. As an optical pulse of the input control signal at λ_1 is launched in the SOA, XGM and XPM effects are visualized by a cw probe signal at λ_2 at the output of the SOA, see left and right y-axes of the output signal.

In an SOA, the XGM and XPM are related phenomenologically by a linewidth enhancement factor α_H, also called Henry factor (Henry, 1982). The α_H-factor is defined as the ratio between the changes of the gain and the refractive index. In most publications (Henry, 1982; Agrawal & Olsson, 1989; Mecozzi & Mørk, 1997), the α_H-factor is assumed to be constant. This simplification indeed is valid only over a limited spectral range and for sufficiently small carrier density modulations in the SOA. Unfortunately, in most of the all-optical experiments, material properties change considerably and the α_H-factor varies significantly. In a new work (Wang et al., 2007), the temporal dynamics of the α_H-factor is investigated and it is shown that the α_H-factor changes during pump-probe experiments by more than an order of magnitude to the extent that even negative α_H-factors are observed.

After the control signal has died out, carriers recover. The carrier recovery process includes inter- and intraband carrier dynamics. The interband carrier dynamics limits the operation speed of the SOA. Especially, when subsequent control pulses are launched into a slow SOA (with a large carrier recovery time), the carrier density is progressively reduced. It recovers back to different levels and the amplifier gain also varies for different pulses, depending on the bit pattern seen by the SOA. This effect is usually called "pattern effect". The pattern effect is discussed with the help of the simulation and experiment in next section. The numerical model used in the simulation was given in (Wang et al., 2007), while the parameters were adapted for the current SOA and validated experimentally.

2.3 Pattern effect in SOA based wavelength conversion

The XGM experiment conditions have a great influence on the gain and phase dynamics. The gain and phase dynamics are practically characterized by a carrier recovery time. A large carrier recovery time of an SOA means that the SOA response is slow. In a XGM, a very slow SOA shows a pattern effect, which is a serious problem in high-speed all-optical communications.

As subsequent control pulses are launched into a slow SOA, the carrier density is depleted continually. It recovers back to different levels and the amplifier gain also varies for different pulses, depending on the former bit pattern seen by the SOA. In our XGM experiment and simulation, 160 Gbit/s control signals with different bit patterns were launched into an SOA with a length of 1.6 mm. The output powers and phase shifts are compared in Fig. 4(a) for a bit pattern "1111000000000000", and in Fig. 4(b) for a bit pattern "1011010000000000".

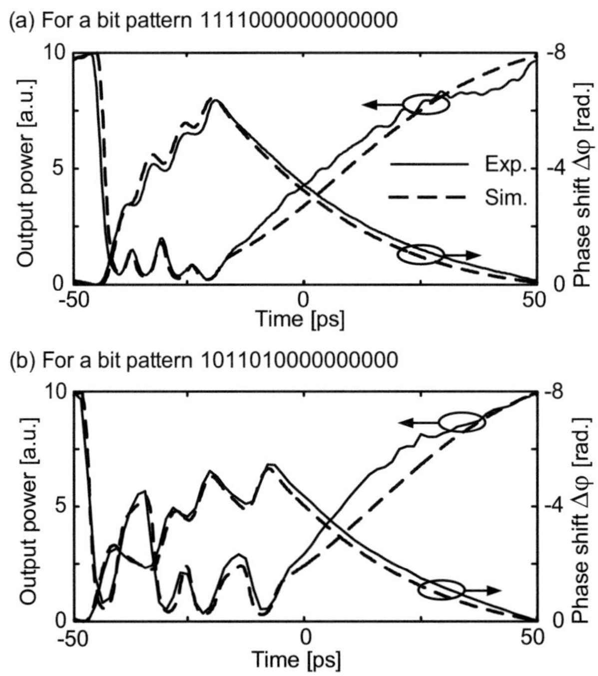

Fig. 4. Results of cross-gain modulation experiment with a bit pattern (a) 1111000000000000, and (b) 1011010000000000. Output power (left axis) and phase relaxation (right axis) are compared between experiment (solid lines) and simulation (dashed lines).

We see that the pattern effect will be a serious problem in high-speed signal processing. As shown in Fig. 4(a), the phase shift of subsequent bits increases but the relative phase shift between two subsequent bits is different. Sometimes, the pattern effect can be so strong that one bit "0" can not be recognized from bit "1"s, e.g. bit "0" in the bit pattern "1101" in Fig. 4(b).

Reducing pattern effect in an SOA can make SOA-based all-optical processing schemes working at high speed. Throughout this work, rather than decreasing the SOA recovery time by proper material design (Zhang et al., 2006) or choice of new fast materials (Sugawara et al., 2002), the main attention will be paid to mitigate pattern effect by optimum operation conditions and careful configuration design.

3. Pattern effect mitigation using an optical filter

3.1 SOA based wavelength conversion assisted by an optical filter

Recently, a new approach of a wavelength converter with an SOA followed by an optical filter was introduced and discussed, Fig. 5. For all-optical wavelength conversion, the input data signal P_{in} and the probe (cw) light P_{cnv} are launched into the SOA. In the SOA, the input data signal encodes the signal information by means of cross-gain and cross-phase modulations onto the cw signal. As a result, we obtain after the SOA an inverted signal P_{inv} at the wavelength of the cw light. The purpose of the subsequent passive optical filter with an appropriate amplitude and phase response is to reformat the signal P_{inv} into a new output-signal P_{cv} with a predetermined shape.

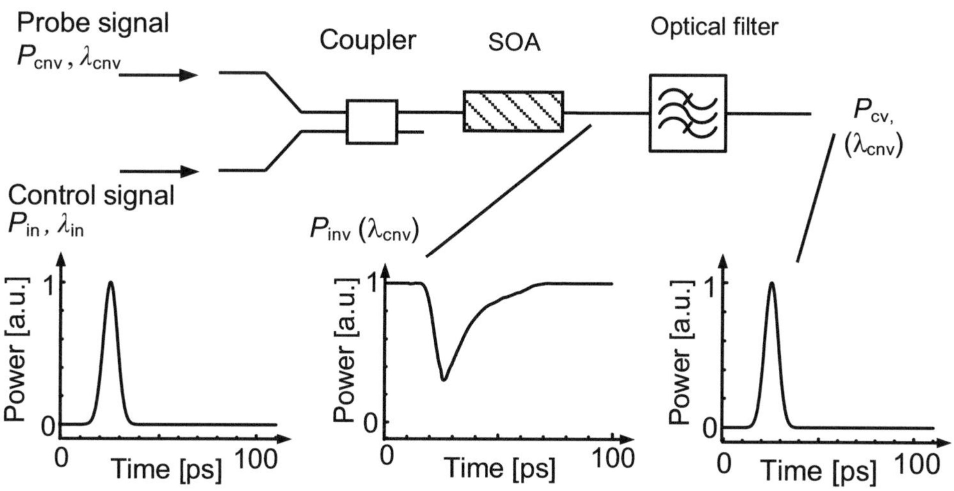

Fig. 5. All-optical wavelength converter based on an SOA followed by an optical filter.

Several filters have been proposed. These are the pulse-reformatting optical filter (PROF) (Leuthold et al., 2004a), the red-shift optical filter (RSOF) (Leuthold et al., 2003), the blue-shift optical filter (BSOF) (Leuthold et al., 2003; Nielsen et al., 2003; Liu et al., 2007), and a delay interferometer (DI) filter scheme (Leuthold et al., 2000; Leuthold et al., 2004b).
The PROF scheme can be understood as follows. As seen in Fig. 5 the probe signal P_{cnv} is cross-phase modulated by the control signal P_{in}, the P_{inv} signal after the SOA comprises a leading red-chirped (negative frequency variation with respect to time) component followed by a trailing blue-chirped (negative frequency variation with respect to time) component due to SOA carrier recovery. The P_{inv} signal also has a cw tone from the P_{cnv} signal. The PROF suppresses the cw tone and splits off the red and blue-chirped spectral components, which are subsequently recombined. This leads to a beating between the two signals that results in a strong and narrow converted pulse if the temporal delay between the two pulses

is adapted correctly. A schematic of the filter is depicted in Fig. 6(a) and the passband of the filter around the cw frequency f_{cw} is shown in the right part of Fig. 6(a).

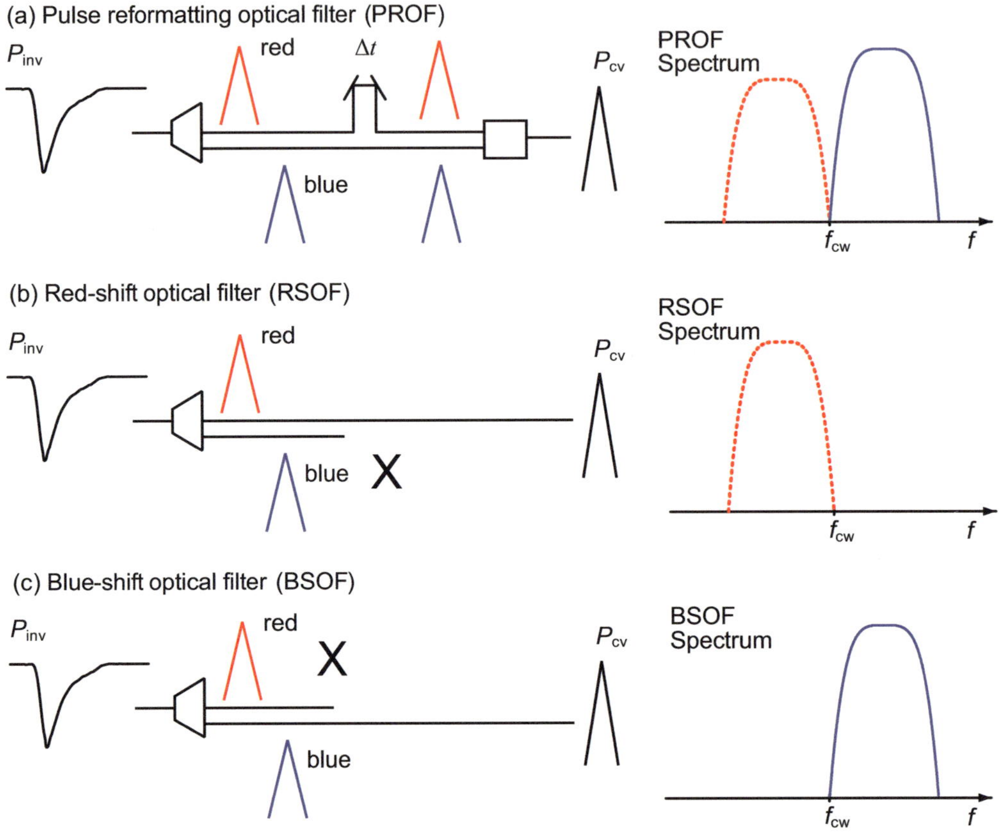

Fig. 6. Three different optical filter concepts for performing wavelength conversion. In the PROF scheme (a), the red- and blue-chirped spectral components are selected and combined with a proper temporal delay Δt between two paths. In the RSOF scheme (b), the red-chirped spectral component is selected and the blue-chirped spectral component is blocked (indicated with a symbol "X"). In the BSOF scheme (c), the blue-chirped spectral component is selected and the red-chirped spectral component is blocked.

The RSOF and BSOF schemes are similar to the PROF scheme, as shown in Fig. 6(b) and (c). Yet, either the red-chirped or the blue-chirped spectral components from the converted signals are selected by the respective filter, while the blue-chirped or the red-chirped spectral components are blocked as indicated by a symbol "X" in Fig. 6(b) and (c). With a proper choice of the BSOF's or RSOF's shape, the output signal P_{cv} shows a predetermined pulse shape.

The PROF, RSOF, and BSOF schemes exploit the fast chirp effects in the converted signal after the SOA, and are successful in performing high-speed wavelength conversion. However, so far it is not clear, if these schemes with optical filters can successfully overcome pattern effects at highest speed. In next section, we will investigate how to mitigate the pattern effect by using optical filters.

3.2 Pattern effect mitigation using a pulse reformatting optical filter

The scheme for the all-optical wavelength conversion with simultaneous pattern effect mitigation is shown in Fig. 7. The setup comprises an SOA followed by a pulse reformatting optical filter (PROF). The current PROF is built with discrete components, so that any parameters can be detuned and optimized individually. The PROF consists of a blue shifted optical filter (BSOF) and a red shifted optical filter (RSOF). An optical delay (OD) and a variable optical attenuator (VOA) were added after the RSOF. A conventional band-pass filter (BPF) has been added at the output of the PROF to curtail the spectrum of the converted signal to the ITU channel passband.

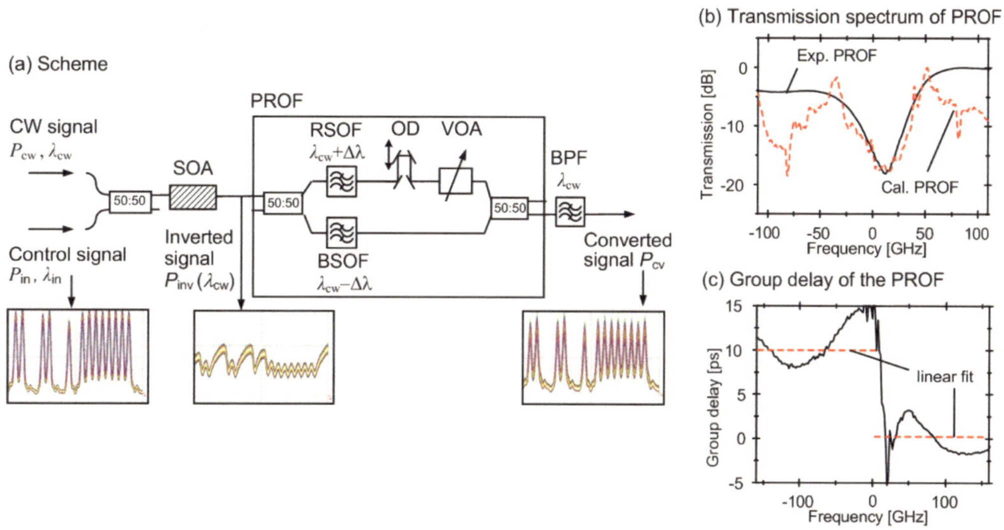

Fig. 7. (a) Wavelength conversion scheme with an SOA followed by a pulse reformatting optical filter (PROF), which transfers an inverted signal with strong pattern effect to non-inverted and pattern dependency removed signal. OD: optical delay, VOA: variable optical attenuator. (b) Transmission spectrum of PROF (dashed lines: calculated; solid line: as used in experiment). (c) Group delay of the PROF used in experiment. The dashed line is obtained from linear fitting.

The wavelength conversion with simultaneous pattern effect mitigation works as follows. A data signal modulates both the gain and the refractive index of the SOA thereby impressing the information in an inverted manner onto another continuous wave (cw) signal. Fig. 8(a) shows the pattern and power spectrum of the inverted cw signal behind the SOA. The leading edges of the converted light pulses are spectrally red-shifted (RS) whereas the trailing edges are blue-shifted (BS). In our scheme, we now split off the RS and the BS spectral components. The respective pattern and spectra are depicted in Fig. 8(b) and (c). The signal qualities of the respective signals are 13.0 dB and 7.9 dB. It can be noticed that the two signals have complimentary pattern effects. While the RS pattern shows an overshoot in the first bit in a sequence of '1's, the BS pattern undershoots particularly the first bit in a long sequence of '1's. When the two signals are combined - after introducing a proper delay and attenuation onto the RS signal - the two complimentary patterns compensate to each other, resulting in a non-inverted signal with a signal quality of as much as 17.8 dB. The pattern, eye diagram and the spectrum of the combined, wavelength converted signal are shown in Fig. 8(d).

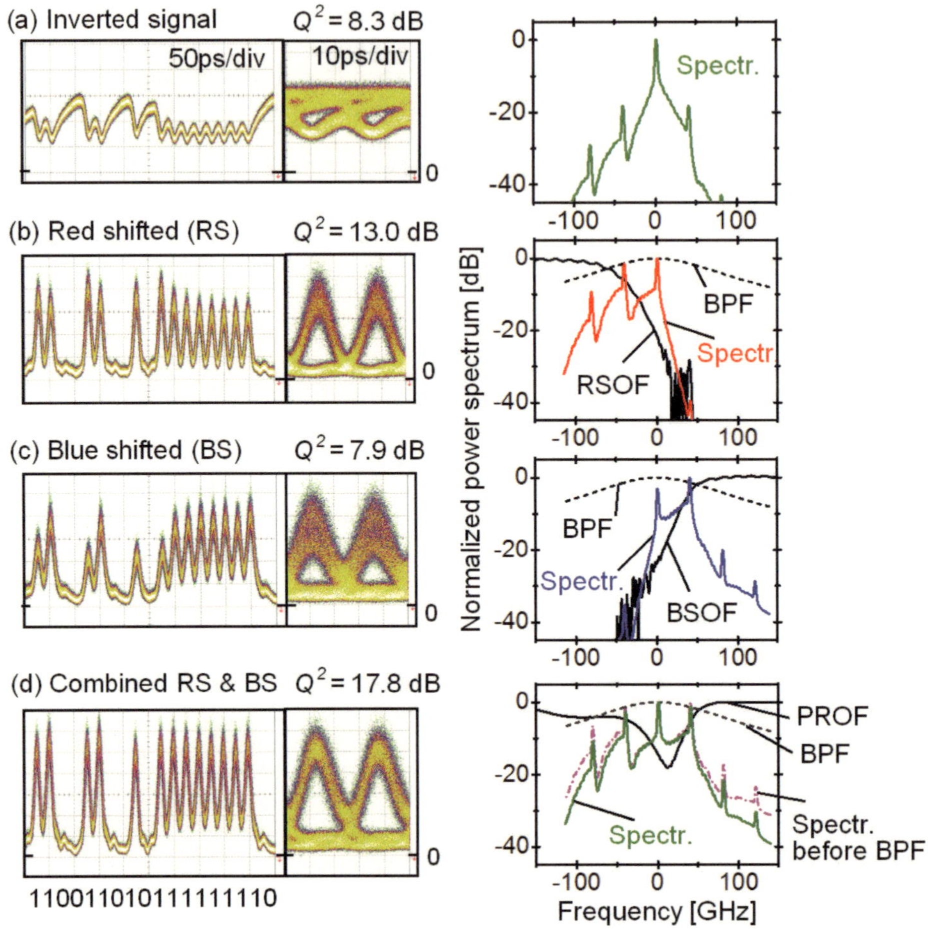

Fig. 8. Wavelength conversion experiment results, bit pattern (left column), eye diagram (middle column) and spectra with the filter shape (right column). Spectra are centred at the cw carrier frequency f_{cw}. (a) Inverted converted signal after the SOA, (c)-(d) RS, BS and combined signal.

In the experiment, a 33% RZ data signal at a bit rate of 40 Gbit/s with an average power P_{data} between –4 and 12.7 dBm and a carrier wavelength λ_{data} = 1530 nm was launched into the SOA. The pseudo-random data with a sequence length of 2^{11} -1 had a quality factor Q^2 = 19.8 dB. A cw signal with P_{cw} between –4 and 16 dBm and λ_{cw} = 1536.1 nm was coupled into the SOA. It became the wavelength converted signal. The bulk SOA of length L = 2.6 mm was biased at I = 750 mA. The non-saturated and saturated gain was 23 and 3 dB. The SOA recovery time (10% to 90%) was ~ 45 ps and polarization sensitivity was 0.5 dB. The plots in Fig. 8 have been recorded with a signal input power of 12.7 dBm and a cw power of 16 dBm. For this experiment we used two thin-film optical filters to realize the PROF. The parameters of the detuned filters are given in Table 1,

The optimum mode beating between red- and blue-chirped signals is obtained when two signals have similar amplitudes and the time delay between the red- and blue chirp vanishes (Leuthold et al., 2004). In our case we attenuated and delayed the red-chirped

Filter	Center λ [nm]	0.5 dB bandwidth [nm]	30 dB bandwidth [nm]	PMD [ps]
RSOF	1536.683	1.002	2.121	$\leq$ 0.10
BSOF	1535.352	1.035	2.148	$\leq$ 0.10

Table 1. Parameters of filters used. PMD: polarization mode dispersion.

signals by ~5 dB and 10 ps with respect to the blue-chirped signal. The transfer function of the optimized PROF has been measured. In Fig. 7(b) and (c), the transmission spectrum and group delay spectrum are depicted as solid lines. The ideal PROF filter transmission spectrum as derived from a calculation is plotted into the same figure (dashed lines). The PROF filter used in the experiment and the ideal filter are fairly similar. The Q^2-factors reported in Fig. 8 were recorded for the best combined signal. In our case, by varying the position of the filters, the Q^2-factor of the BS signal could be as good as 9.9 dB, while the Q^2-factor of the RS signal could not be improved beyond the 13 dB reported above.

The PROF scheme uses both the red- and blue-shifted spectral components, thus keeps all the information in the spectrum. This advantage leads a to a large input power dynamic range, in which a good signal quality is kept. In our experiments, we obtained a Q^2-factor of 15.8 dB with an input power (λ_{in} = 1530 nm) of as low as –4.3 dBm while a cw power (λ_{cw} = 1536.1 nm) is –4 dBm. At an input data power of 12.9 dBm, which is the available maximum in our experiment, and a cw power of 14.7 dBm, we also got a Q^2-factor of 18.2 dB. In fact, throughout this input power dynamic range of at least 17.2 dB with an adjustment of the cw power, we always got Q^2-factors above 15.6 dB, shown in Fig. 9. Note that the cw power adjustment for respective input data powers was stopped as long as we have a Q^2-factor above 15.6 dB. All of Q^2 values were measured with a bit-error rate test (BERT) and then interpolated from the inverse complementary error function. Yet, as it is not clear on what noise distribution we have after an all-optical wavelength conversion operation. The Q^2 values may differ based on the method used. However, the quantitative values given here are still indicative for the quality of the signal.

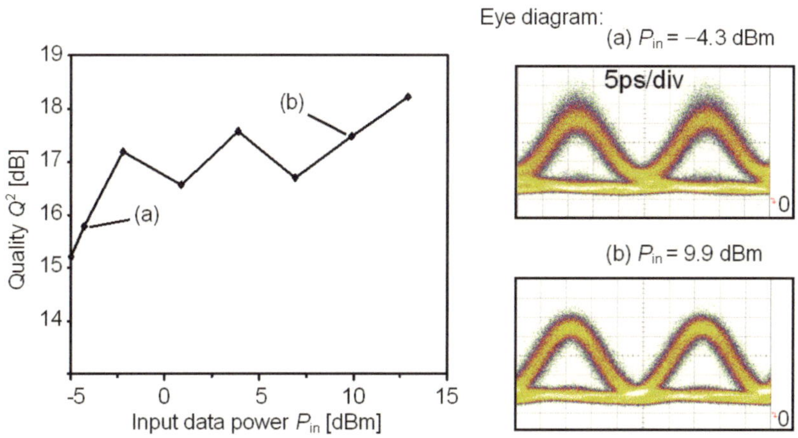

Fig. 9. Signal qualities of the converted signal for various input data powers without adaptation of filter parameters but with adaptation of the cw input power levels, while λ_{in} = 1530 nm and λ_{cw} = 1536.1 nm. Eye diagram for P_{in} = –4.3 dBm and 9.9 dBm are given in (a) and (b) respectively.

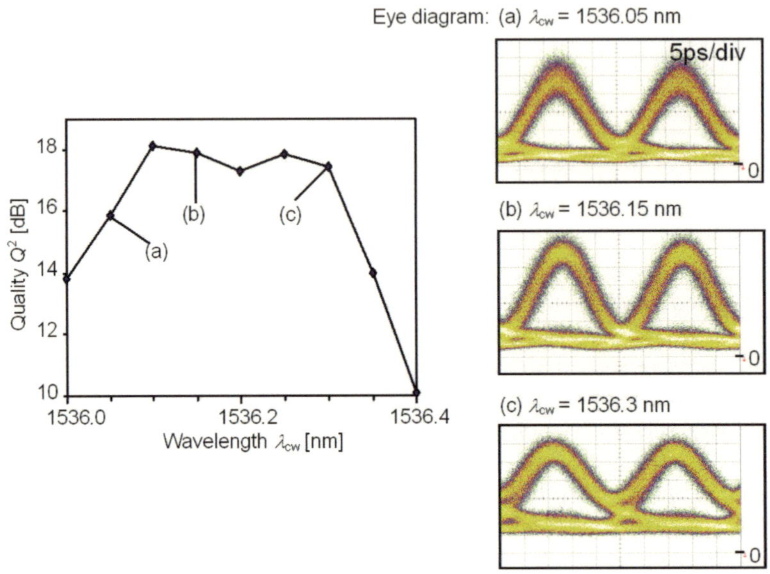

Fig. 10. Signal qualities of the converted signal for various cw wavelengths without adaptation of filter parameters, while the cw power P_{cw} = 14.7 dBm and the input power P_{in} = 12.9 dBm at λ_{in} = 1530 nm. Eye diagrams for λ_{cw} = 1536.05 dBm, 1536.15 dBm and 1536.3 nm are given in (a), (b) and (c) respectively.

Except a large input power dynamic range, the PROF scheme features a large tolerance of the operating wavelength. In the experiment, we kept the input power P_{in} = 12.9 dBm at λ_{in} = 1530 nm and the cw power P_{cw} = 14.7 dBm, while the operating wavelength of the cw signal λ_{cw} varied from 1536.0 nm to 1536.4 nm. The measured Q^2-factor is depicted in left part of Fig. 10. We see that a wavelength tolerance for a Q^2-factor above 15.6 dB is about 0.29 nm. Eye diagram for λ_{cw} = 1536.05 nm, 1536.15 nm and 1536.3 nm are given in Fig. 10(a), (b) and (c) respectively. All the eyes are clean and open.

3.3 Complementary pattern effects in the XPM-induced chirp

Now we investigate the complementary pattern effects in the XPM-induced red and blue chirp of the inverted signal. The complementary pattern effects have their origin in the carrier dynamics of the SOA. We therefore used an SOA model from (Wang et al., 2007) that encompasses both slow and fast carrier recovery effects. The parameters in (Wang et al., 2007) were adapted for the current SOA and validated experimentally. Two second-order Gaussian filters with a 3dB bandwidth of 120 GHz were used as the RSOF and the BSOF in the simulation to select the spectral components. Simulation results are given in Fig. 11. An exemplary input data signal with a bit pattern "0111", Fig. 11(a), is launched into the SOA. Fig. 11(b) shows the gain evolution with time. The gain decreases with the leading edge of each input pulse and recovers with the falling edge of each pulse. Further, with each sub-sequent bit the SOA gain gradually saturates. Due to Kramers-Kronig relation, a gain change is always accompanied by a phase-shift $\Delta\varphi_{NL}$, whose variation per time increment Δt in consequence determines the instantaneous frequency shift Δf (i.e. chirp, see formula of Fig. 11(c)). The leading edges of the input pulses induce a red chirp, while the trailing edges induce a blue chirp, Fig. 11(c). As we will see below, these two chirps have different pattern dependences.

We first study the pattern dependence of the signal after the RSOF. The first "1" bit in the pulse train induces strong carrier depletion and a large gain reduction Fig. 11(b). A strong carrier-depletion induces a large phase-shift and consequently a large red chirp on the leading edge, Fig. 11(c). For subsequent "1" pulses, the SOA does not fully recover. The gain for subsequent "1" pulses is smaller than for the first "1" bit. Therefore the carrier depletion decreases. As the gain reduction due to the carrier depletion decreases, the induced red chirp decreases as well, low part of Fig. 11(c). This decreasing red chirp however, by means of the RSOF filter transfer function (see 1,2,3 in Fig. 11(d)), and the increasing gain saturation create a bit pattern with decreasing power level, shown in Fig. 11(e) bottom.

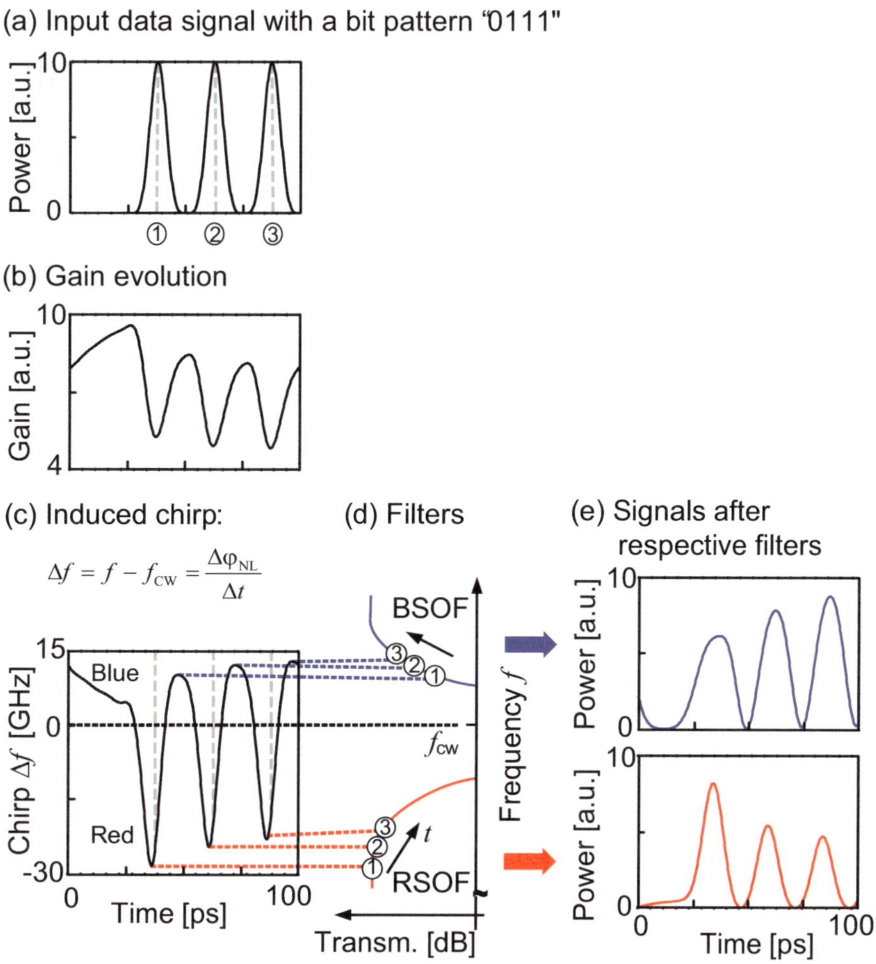

Fig. 11. Complementary pattern effects by frequency-amplitude conversion through differently centred filters. (a) Input data pulse train. (b) Simulated gain evolution in SOA and (c) induced frequency chirp in the inverted signal. (d) Schematic filter shape of the BSOF and the RSOF. (e) Pattern dependence in blue and red-spectral components after sending signal (b) with chirp (c) through filter plotted in (d). Dashed lines in (a), (b) and (c) indicate the time positions of the input three pulses.

The bit pattern of the signal behind the BSOF can be understood as follows. The gain recovers with the trailing edge of the first "1" pulse, creating a blue chirp. With subsequent bits launched into the SOA, the SOA is further saturated and the overall gain G decreases with following "1" bits. In turn the carrier depletion due to stimulated emission is now weaker while the current injection is unchanged. As a consequence, the carrier recovery rate is faster for subsequent pulses. This leads to a stronger blue-chirp as also observed in the simulation depicted in Fig. 11(c). By positioning the strongest blue-chirped signals closest to the centre of the BSOF passband, Fig. 11(d), transmission of the strongest chirped pulse is favoured. This leads to an increasing power level for the tails of longer patterns, shown in Fig. 11(e) top. The coherent superposition of the red- and blue-shifted signal with opposite patterns then leads to the pattern effect compensated output, Fig. 11(d). The scheme works well, as long as the frequency-amplitude conversion at the slope of the BSOF can overcompensate the gain saturation. Simulations show that results do not change for a 3 dB filter bandwidth of 80 to 220 GHz.

7. Conclusion

In this work, we investigate an SOA-based all-optical wavelength conversion scheme with an assisting optical filter after the SOA. The technique utilizes the complementary pattern effects of the XPM induced red and blue chirp in SOAs. By carefully selecting the optical filters and compensating the complementary pattern effects in the red and blue chirp to each other, a pattern effect mitigated converted signal is obtained. These advantages lead to a larger operation wavelength tolerance and a large input power dynamic range, in which a good signal quality is kept. This scheme is experimentally demonstrated having a ~0.3 nm operation wavelength range and over 17 dB input power dynamic range. This concept has been demonstrated at 40 Gbit/s, however, the general principle holds for higher speed.

8. References

Agrawal G. P. & Olsson N. A. (1989). Self-phase modulation and spectral broadening of optical pulses in semiconductor laser amplifiers. *IEEE J. Quantum Electron.*, Vol. 25, No. 11, p2297-2306, 1989.

Bischoff S.; Nielsen M. L. & Mørk J. (2004). Improving the all-optical response of SOAs using a modulated holding signal. *IEEE J. Lightw. Technol.*, Vol. 22, No. 5, 1303-1308, 2004.

Connelly M. J. (2002). *Semiconductor Optical Amplifiers*, Springer-Verlag, Boston, MA, 2002.

Durhuus, T.; Mikkelsen, B.; Joergensen, C.; Lykke Danielsen, S.; Stubkjaer, K.E. (1996). All optical wavelength conversion by semiconductor optical amplifiers. *IEEE J. Lightw. Technol.*, Vol. 14, No. 6, 942–954, 1996.

Eiselt, M.; Pieper, W.; Weber, H.G. (1995). SLALOM: Semiconductor laser amplifier in a loop mirror. *IEEE J. Lightw. Technol.*, Vol. 13, No. 10, 2099-2112, 1995.

Girardin, F. ; Guekos, G. ; Houbavlis, A. (1998). Gain recovery of bulk semiconductor optical amplifiers. *IEEE Photon. Technol. Lett.*, Vol. 10, No. 6, 784-786, 1998.

Hall K. L. & Rauschenbach K. A. (1998). 100-Gb/s bitwise logic. *Opt. Lett.*, Vol. 23, No. 8, 1271–1273, 1998.

Henry C. H. (1982). Theory of the linewidth of semiconductor lasers. *IEEE, J. Quantum Electron.*, Vol. QE-18, No. 2, 259-264, 1982.

Kumar, S. ; Zhang, B. ; Willner, A.E. (2006). Elimination of data pattern dependence in SOA-based differential-mode wavelength converters using optically-induced birefringence. *Proceedings of Optical Fiber Communication Conference*, 3 pp., Anaheim, California, USA, March 2006, IEEE Computer Society Press, Washington.

Leuthold, J. ; Joyner, C.H. ; Mikkelsen, B. ; Raybon, G. ; Pleumeekers, J.L. ; Miller, B.I. ; Dreyer, K. ; Burrus, C.A. (2000). 100-Gb/s all-optical wavelength conversion with integrated SOA delayed-interference configuration. *Electron. Lett.*, Vol. 36, No. 13, 1129-1130, 2000.

Leuthold, J. ; Ryf, R. ; Maywar, D.N. ; Cabot, S. ; Jaques, J. ; Patel, S.S. (2003). Nonblocking all-optical cross connect based on regenerative all-optical wavelength converter in a transparent network demonstration over 42 nodes and 16800 km. *J. Lightw. Technol.*, Vol. 21, No. 11, 2863-2870, 2003.

Leuthold, J. ; Marom, D.M. ; Cabot, S. ; Jaques, J.J. ; Ryf, R. ; Giles, C.R. (2004a) All-optical wavelength conversion using a pulse reformatting optical filter. *IEEE J. Lightw. Technol.*, Vol. 22, No. 1, 186-192, 2004.

Leuthold, J.; Moller, L. Jaques, J. Cabot, S. Zhang, L. Bernasconi, P. Cappuzzo, M. Gomez, L. Laskowski, E. Chen, E. Wong-Foy, A. Griffin, A. (2004b). 160 Gbit/s SOA all-optical wavelength converter and assessment of its regenerative properties. *Electron. Lett.*, Vol. 40, No. 9, 554-555, 2004.

Liu, Y.; Tangdiongga, E.; Li, Z.; de Waardt, H.; Koonen, A.M.J.; Khoe, G.D.; Shu X.; Bennion, I.; Dorren, H.J.S. (2007). Error-free 320-Gb/s all-optical wavelength conversion using a single semiconductor optical amplifier. *IEEE J. Lightw. Technol.*, Vol. 25, No. 1, 103-108, 2007.

Manning R. J. & Davies D. A. O. (1994). Three-wavelength device for all-optical signal processing. *Opt. Lett.*, Vol. 19, No. 12, 889-891, 1994.

Manning, R.J.; Yang, X.; Webb, R.P.; Giller, R.; Garcia Gunning, F.C.; Ellis, A.D. (2006). The 'turbo-switch' - a novel technique to increase the high-speed response of SOAs for wavelength conversion. *Proceedings of Optical Fiber Communication Conference*, 3 pp., Anaheim, California, USA, March 2006, IEEE Computer Society Press, Washington.

Nielsen M.; Lavigne B.; Dagens B. (2003). Polarity-preserving SOA-based wavelength conversion at 40 Gbit/s using bandpass filtering. *IEEE Electron. Lett.*, Vol. 39, No. 18, 1334-1335, 2003.

Nielsen M. & Mørk J. (2006). Bandwidth enhancement of SOA-based switches using optical filtering: theory and experimental verification. *Opt. Express*, Vol. 14, No. 3, 1260-1265, 2006.

Saitoh T. & Mukai T. (1989). Structural design for polarization-insensitive travelling-wave semiconductor laser amplifiers. *Opt. and Quantum Electron.*, Vol. 21, No. 1, s47-s58, 1989.

Sugawara M.; Akiyama T.; Hatori N.; Nakata Y.; Ebe H. & Ishikawa H. (2002). Quantum-dot semiconductor optical amplifiers for high-bit-rate signal processing up to 160 Gb/s and a new scheme of 3R regenerators. *Meas. Sci. Technol.*, Vol. 13, No. 11, 1683-1691, 2002.

Tajima K. (1993). All-optical switch with switch-off time unrestricted by carrier lifetime. *Jpn. J. Appl. Phys.*, Vol. 32, No. 12A, L1746-L1749, 1993.

Wang J.; Maitra, A.; Poulton, C.G.; Freude, W.; Leuthold, J. (2007). Temporal dynamics of the alpha factor in semiconductor optical amplifiers. *IEEE J. Lightw. Technol.*, Vol. 25, No. 3, 891-900, 2007.

Yoo, S. J. B. (1996). Wavelength conversion technologies for WDM network applications. *IEEE J. Lightw. Technol.*, Vol. 14, No. 6, 955-966, 1996.

Zhang, L.; Kang, I.; Bhardwaj, A.; Sauer, N.; Cabot, S.; Jaques, J.; Neilson, D.T. (2006). Reduced recovery time semiconductor optical amplifier using p-type-doped multiple quantum wells. *IEEE Photon. Technol. Lett.*, Vol. 18, No. 22, 2323-2325, 2006.

Part 3

Semiconductor Optical Amplifiers: Other Processing Functions

8

Chromatic Dispersion Monitoring Method Based on Semiconductor Optical Amplifier Spectral Shift Effect in 40 Gb/s Optical Communication Systems

Ming Chen
Research Institute of Optoelectronic Technology and School of Information and Communication, Guilin University of Electronic Technology
P. R. China

1. Introduction

The optical signals degrade as travel down the optical link due to the optical fiber properties such as chromatic dispersion, polarization mode dispersion, polarization dependent loss, polarization dependent gain and various fiber nonlinear effects, which are considered as limitations in the high-speed, lang-haul fiber communication systems. Where chromatic dispersion causes different wavelengths to travel at different group velocities in single-mode transmission fiber and it has become a major source of transmission degradation due to the continuing increase of the bit rate and distance in high-speed long-distance optical communication systems (Kaminow et al., 2008). In the long-haul optical fiber communication systems or optical fiber communication networks, the accumulated chromatic dispersion is managed by creating a dispersion "map", in which the designer of a transmission optical fiber link alternates elements that produce positive and the negative chromatic dispersion. In this dispersion "map", the dispersion has some nonzero value at each point along the optical fiber link, the degradations, from nonlinear effects such as four-wave-mixing (FWM) and cross phase modulation (XPM), are effectively eliminated, but the total accumulated dispersion is near to zero at the end of the optical fiber link (Kaminow & Li, 2002). It seems that there need not other dispersion compensation techniques any more. Unfortunately, chromatic dispersion changes with dynamic optical fiber network reconfiguration and variation with environmental conditions such as temperature in practice (Agrawal, 2002). This dynamical action causes dynamical residual chromatic dispersion in a dynamical fiber link. In addition, signal tolerance to accumulated chromatic dispersion diminishes as the square of the bit rate. Therefore, 40Gbit/s signals are 16 times more sensitive to chromatic dispersion than 10Gbit/s signals. The signal tolerance to chromatic dispersion is restricted about 50ps/nm in single channel speed 40Gbit/s fiber communication systems, so there requires more carefully chromatic dispersion management. The residual chromatic dispersion of a dynamical fiber link can easily extend the tolerance in those high speed fiber communication systems, they need more precise and dynamical monitoring and compensation methods (Pan, 2003).

Many novel and effective dispersion compensation methods have been proposed (Kaminow & Li, 2002) (Pan, 2003), and they are included, dispersion compensating fibers, linear-chirped fiber Bragg gratings etc., for fixed dispersion compensation. The dispersion compensating fibers have negative dispersion, which can compress the extended signal pulses due to the positive dispersion of the single-mode transmission fibers. These dispersion compensating fibers can be made by conventional optical fiber fabrication technics and can also be made by photonic crystals — a novel new optoelectronic technology (Sukhoivanov et al, 2009), called microstructured fibers or photonic crystal fibers with many periodic-arrayed air-holes and one or many defects in the fiber cross section (Bjarklev et al, 2003). Due to the complicated cross section structure, many excellent optical properties can be obtained by careful selecting the photonic crystal fiber structure parameters, such as photonic crystal fibers with large negative dispersion can be achieved. (Chen et al., 2010) Optical fiber Bragg gratings with linear-chirp have emerged as powerful tools for chromatic dispersion compensation because of their potential for low loss, small footprint, and low optical nonlinearities (Sukhoivanov et al, 2009). Fiber Bragg gratings are section of single-mode fiber in which the refractive index of the core is modulated in a periodic fashion as a function of the spatial coordinate along the length of the optical single-mode fiber (Kashyap et al, 2009). In chirped fiber gratings, the Bragg matching condition for different positions along the grating length, thus the different wavelength is reflected at different position. As a result, the extending signal pulses can be compressed by careful tailing the chirp profile of the fiber gratings (Kaminow & Li, 2002). Fiber Bragg gratings, achieved by sampled fabricating techniques, can be used for chromatic dispersion compensation in multichannel optical communication systems, such as wavelength division multiplexing (WDM) fiber communication system (Ibsen et al., 1998). And those with nonlinear chirped profiles can be used to achieve dynamical compensation in optical fiber communication systems with variable and unpredictable residual chromatic dispersion. When using a nonlinear chirp profile, the chromatic dispersion can be tuned by simply stretching the grating or heating the grating with heat-conduction coatings because of the sensitiveness of the reflection spectrum and group delay with grating structure, stress and temperature (Sun et al., 2006). Based on a thermally tunable nonlinear chirped fiber grating, we have achieved a tunable chromatic dispersion compensation system for optical fiber communication system with single channel speed 40Gbit/s and carrier suppressed return to zero (CSRZ) modulation format (Chen et al., 2007). The fiber grating is covered with uniform thin metal electric-conducting film which can add voltage to heat up. when the fiber grating is added voltage and has current in the electric-conducting film, the fiber grating is heated and then the temperature is changed, the reflection spectrum and group delay is also changed. We can control the group delay at certain wavelength by control the voltage added in the film-covered optical fiber grating. In our system, the measured chromatic dispersion can be varied from -60ps/nm to -260ps/nm for wavelength 1553.40nm.

The schematic common chromatic dispersion compensation system is shown in Fig.1. It is mainly consisted by a chromatic dispersion compensation module, a chromatic dispersion monitoring module and an optical fiber coupler, as shown in Fig.1. The optical signal with accumulated residual chromatic dispersion from optical fiber link is firstly sent into the chromatic dispersion compensation module. After compensating, the optical signal is sent into the optical fiber link again, and some power of the compensated optical signal is separated from the optical fiber link by the optical fiber coupler after the chromatic dispersion compensation module and is sent into the chromatic dispersion monitoring module, which includes an optical receiver, some electrical signal processing modules and relevant computer algorithms. This chromatic dispersion monitoring module can generates an electrical control signal for the chromatic dispersion compensation module according to one or certain parameters of the input optical signal. The one or certain parameters are relating to the accumulated residual chromatic dispersion of the fiber communication system links

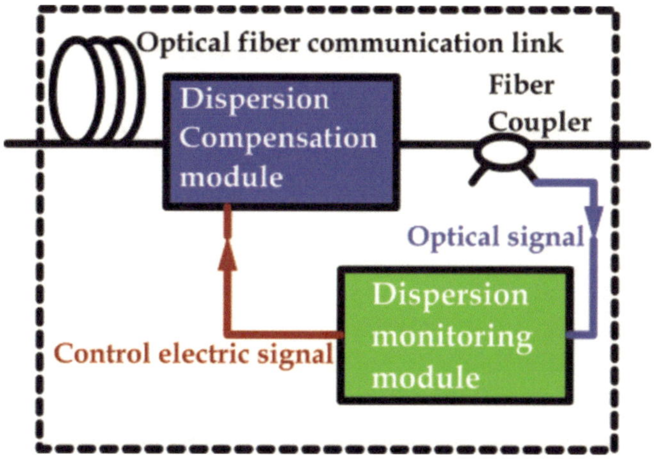

Fig. 1. Dispersion compensation system. The blue solid line denotes optical signal withdraw from optical fiber link after dispersion compensation and the red that denotes electric control signal come from dispersion monitoring module.

(Kaminow & Li, 2002) (Pan, 2003) (Hong, 2002).

Dynamical chromatic dispersion management has become a critical issue for high-bit-rate transmission systems, especially for systems with speeds beyond 10Gbit/s, and reconfigurable optical networks, because the accumulated chromatic dispersion can easily go beyond the optical communication systems' tolerance. Chromatic dispersion management, in high speed optical communication systems, is very difficult and needs effective high speed response chromatic dispersion monitoring methods, as shown in Fig.1. The range and precision of the monitoring methods decide the range and precision of the chromatic dispersion compensation systems (Seyed Mohammad Reza Motaghian Nezam, 2004). Previous works on chromatic dispersion monitoring have resulted in the development of numerous approaches, such as detecting the intensity modulating from phase modulation (Ji et al., 2004), modulating the frequency of the transmitted data signal and monitoring the clock deviation (Pan et al., 2001), inserting in-band subcarriers in the transmitter and monitoring their radio frequency tones (Ning et al., 2006) (Luo et al., 2006), adding an amplitude modulated double sideband subcarrier to the signal and measuring the phase delay between two subcarrier tones (Wang et al., 2006), extracting the clock component and measuring its radio frequency (RF) power (Inui et al., 2002), extracting two single sideband sideband components of the data signal and detecting their phase difference (Hirano et al., 2002), employing nonlinear optical detection (Wielandy et al., 2004) (Li et al., 2004), measuring the chromatic dispersion induced distortion using a peak detector (Ihara et al., 1999), and so on. In practice, both chromatic dispersion and polarization mode dispersion are all influent the performance of the high-speed optical fiber communication systems, and effective simultaneous monitoring methods for chromatic chromatic dispersion and polarization mode dispersion are necessary. We developed a novel and effective method to monitor chromatic dispersion and polarization mode dispersion simultaneously using two polarization-modulation pilot tones with different frequencies (Chen et al., 2007). It has been demonstrated that radio frequency (RF) output power increase with group velocity delay (GVD) and differential group delay (DGD) and the power radio of the two pilot tones increases with GVD and decreases with DGD, thus chromatic dispersion and polarization

mode dispersion can be distinguished and monitored simultaneously. This is an effective monitoring method in high-speed optical fiber communication systems.
In this chapter, we demonstrate an other novel in-line dynamical monitoring methods for chromatic dispersion based on the spectral shift effect of a semiconductor optical amplifier (Chen et al., 2007). This spectral shift effect is result of the self phase modulation effect in the semiconductor amplifier. Due to large nonlinearites of semiconductor optical amplifiers, the spectral shift effect is enhanced, and this effect is impacted by the residual chromatic dispersion of the optical fiber link, which is optical signal transmitted. Using an optical filter — a fiber grating, we can obtain the variational power of the spectrum of the optical signals, and then we can achieve the dynamical chromatic dispersion monitoring in line for a high speed optical fiber communication system.

2. Monitoring principle based on semiconductor optical amplifiers

In the past two decades, optical communication has changed the we communicate. It is a revolution that has fundamentally transformed the core of telecommunications, its basic science, its enabling technology, and its industry. The optical networking technology represents a revolution inside the optical optical communications revolution and it allows the letter to continue its exponential growth. Optical networking represents the next advance in optical communication technology. Semiconductor optical amplifier is a kind of key devices for all-optical networks (Dutta & Wang, 2006). The advances in research and many technological innovations have led to superior designs of semiconductor optical amplifiers. Semiconductor optical amplifiers are suitable for integration and can be used as signal amplification and functional devices, such as optical demultiplexing, wavelength conversion, and optical logic elements make them attractive for all-optical network and optical time division multiplexed systems (Kaminow & Li, 2002) (Kaminow et al, 2008).
The theory of pulse propagation in semiconductors is well known (Shimada et al, 1994). The semiconductor optical amplifiers are treated as a two-level system. When the carrier's intra-band relaxation time τ_{in} in the conduction band is induced, solving the problem become complex. Fortunately, the intra-band relaxation time τ_{in} is generally about 0.1 ps in semiconductor devices. It is supposed that the pulse width of input optical signals $\tau_p \geq 1.0ps$, solving this problem will become very simple. It is to said that the condition $\tau_p \gg \tau_{in}$ is always satisfied. In our research, this condition is easily satisfied. At the same time, given that the semiconductor optical amplifier cavity is very short, and the dispersion of the waveguide in the semiconductor optical amplifier can be neglected, and the we can obtained the equations that described transmission actions of the input pulses in semiconductor optical amplifiers as follows:

$$\frac{\partial P(z,\tau)}{\partial t} = (g(z,\tau) - \alpha_{int}) \cdot P(z,\tau), \tag{1}$$

$$\frac{\partial \phi(z,\tau)}{\partial z} = -\frac{1}{2}\alpha_{LEF} \cdot g(z,\tau), \tag{2}$$

$$\frac{\partial g(z,\tau)}{\partial \tau} = \frac{g_0 - g(z,\tau)}{\tau_c} - \frac{g(z,\tau) \cdot P(z,\tau)}{E_{sat}}, \tag{3}$$

where $P(z,\tau)$ and $\phi(z,\tau)$ denote instantaneous power and phase respectively, $g(z,\tau)$ is the saturation gain parameter, α_{int} is the loss coefficient of the semiconductor optical amplifier cavity, g_0 denotes the small signal gain, α_{LEF} is the line-width enhancement factor, $\tau = t - z/v_g$, v_g is the group velocity of the light, and E_{sat} is the saturation of the semiconductor optical amplifier. Equation (2) describes the self-phase modulation (SPM) of the semiconductor optical amplifier. Our chromatic dispersion monitoring method is based on this nonlinear effect.

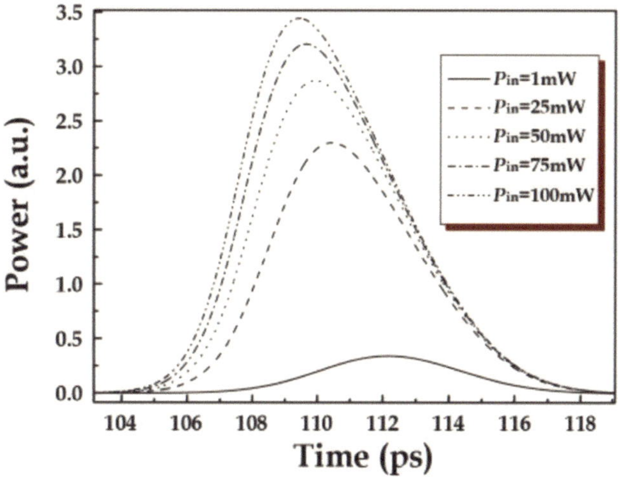

(a) Waveforms of the amplified Gauss pulses with different peak power.

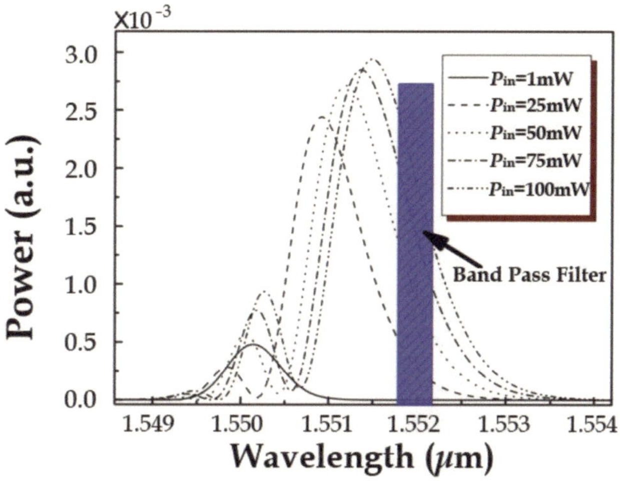

(b) Spectra of the amplified Gauss pulses with different peak power.

Fig. 2. Waveforms and corresponding spectra of the amplified Gauss pulses with different peak power after amplified by the semiconductor optical amplifier.

Without loss of generality, we show the principle of the chromatic dispersion monitoring methods using Gauss profile pulses due to their simplicity, although the optical pulses with the carrier suppressed return to zero modulation format cannot be approximated by Gauss profile pluses. Firstly, we study the influence on the shape and spectrum of input signal pulses in a semiconductor optical amplifier in theory. In our study, the small signal gain g_0 is 30dB, and the spontaneous carrier lifetime τ_c is 140ps. The line-width enhancement factor α_{LEF} is decided by the peaks of input signal pulses, and its typical values for semiconductor lasers and semiconductor optical amplifiers are in the range between 3 to 8 (Shimada et al, 1994). Let $\alpha_{LEF} = 5$ in our research. The input Gauss profile pulses can be written as:

$$A_{in}(\tau) = \sqrt{P_{in}} exp\Big(- \frac{1+iC}{2} \Big(\frac{\tau}{\tau_0} \Big)^{2m} \Big), \tag{4}$$

where P_{in} and C denote peak power and chirp parameter of the input signal pulses, respectively, m is the pulse amplitude. In order to further simplify this problem, we suppose the chirp paraments of the input signal pulses to be $C = 1$ and Gauss function of the order $m = 1$. In our theoretical system, the wavelength of the carrier light wave is 1550nm, the pulse width equals 0.2 bit period, and the single-channel speed of the optical communication system is 40Gbit/s. The numerical simulating software is Optisystem 6.0 from OptiWave® Inc. of Canada.

The waveform shapes and spectra curves of the transmitted pulses with different peak powers, after amplified by the semiconductor optical amplifier, are shown in Fig. 2. Figure 2(a) shows the waveform shapes and Figure 2(b) shows corresponding spectra curves of the amplified transmitted optical signal pulses with Gauss profile. In Figure 2(b), the blue shadowed part indicates the filter band of the band-pass optical filter, which can select corresponding frequencies' power to be detected in our chromatic dispersion monitoring method, which will be demonstrated in detail in the following parts of this chapter. From this figure, we can conclude that the amplified Gauss pulses lose their symmetry with increasing of the input light pulse peak power, the leading edge is more sharper compared with the trailing edge. This is because the leading edge experiences more larger gain than that of the trailing edge. The spectrum of the amplified optical signals develops a structure with multi-peaks, with the dominant spectral peak shifting to the long wavelength side (red shift) as the input pulse peak power increases. The physical mechanism behind the spectral shift and distortion is the self phase modulation, which occurs as a result of index nonlinearities induced by gain saturation effect.

Our dynamical chromatic dispersion monitoring method is based on the spectral effect resulted from self phase modulation effect of the semiconductor optical amplifier, as mentioned previously. As an optical signal pulses transmitting in an optical fiber link with chromatic dispersion, the peak power of the optical signal pulses are influenced by the chromatic dispersion. Because the self phase modulation is related to the peak power of the input pulses, the peak power decides the spectral shift effect. As shown in Figure 2(b), we can use a band-pass optical filter to obtain the corresponding frequencies' power and use it to accomplish online dynamical chromatic dispersion monitoring, because the power depends on chromatical dispersion of optical communication fiber links sensitively.

3. Experimental System

Figure 3 shows our dynamical chromatic dispersion monitoring system. The optical carrier comes from the continuous wave (CW) laser with center wavelength 1553.40nm and is sent into a 40Gbit/s pseudo random binary sequence (PRBS) system with suppressed return to zero modulation format is shown in Figure 4. As shown in this figure, the optical carrier frequency — 1553.40nm, is wholly suppressed, the frequency difference of the two first-order harmonic wave peaks is 40GHz, and the frequency difference is also 40GHz between the high-order (order>1) harmonic waves and the neighboring lower-order harmonic waves.

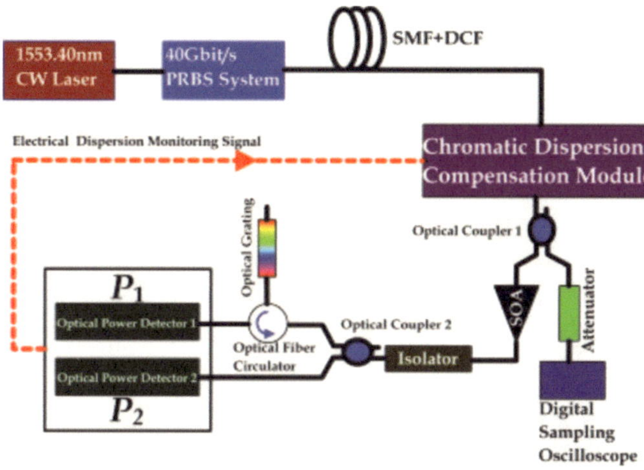

Fig. 3. Experimental system of dispersion monitoring based on semiconductor optical dispersion spectral shift effect.

Output optical signals from the pseudo random binary sequence system are transmitted into an optical fiber link that consists of some single-mode fibers with positive chromatic dispersion and some conventional dispersion compensation fibers with negative chromatic dispersion. In order to simulate the dynamical residual chromatic dispersion of a dynamical fiber links, we can obtain different chromatic dispersion values for the experiment by changing the length of single-mode fibers and that of the dispersion compensation fibers. The optical signals are then transmitted in to a dynamical chromatic dispersion compensation module, which can compensate the remnant chromatic dispersion using the monitoring signal from our proposed chromatic dispersion monitoring method. This dispersion compensation module is based on a thermally tunable optical fiber grating (Sun et al., 2006) (Chen et al., 2007). Output from compensation system, the optical signal stream is sent to an optical fiber coupler and is split into two signal streams with different optical power, the optical power ratio of the two signal streams is 20:80. One signal stream with large optical power is received by a digital sampling, oscilloscope after an attenuator. The other signal stream with small optical power is more further split into other two signal streams with the same optical power after going through a semiconductor optical amplifier and an isolator by a 3dB optical fiber coupler. one is received by an optical detector at an obtained optical power P_2 ; another is received by other optical detector at an obtained optical power P_1 after an optical fiber circulator and an optical fiber grating that can reflect parts of the spectrum denoted by part I, part II and part III, as shown in Figure 4. The semiconductor optical amplifier is a product of the Center for Integrated Photonics® (CIP) Ltd. of the United Kingdom. The product type is SOA-NL-OEC-1500.

If we only used the optical power P_1 to monitor the chromatic dispersion of the optical fiber communication links, it is influenced easily by the optical power fluctuation in the optical fiber communication system links. In order to avoid this influence, in our dynamical chromatic dispersion monitoring method, we use the radio (P_1/P_2) of the obtained optical power P_1 to the obtained optical power P_2 to monitor the remnant chromatic dispersion of a high speed optical fiber communication system, because the optical power radio is independent of the optical power variety in the optical fiber communication system links.

As mentioned previously, similar to the optical spectrum of Gauss profile pulses shown in Figure 2(b), the peak of the amplified output optical spectrum will shift toward the more longer wavelength side as the peak power of input pulses increases, as shown in Figure 5. The amplified output optical spectrum symmetry is lost. The optical power of the long wavelength side is higher than that of the short wavelength side.

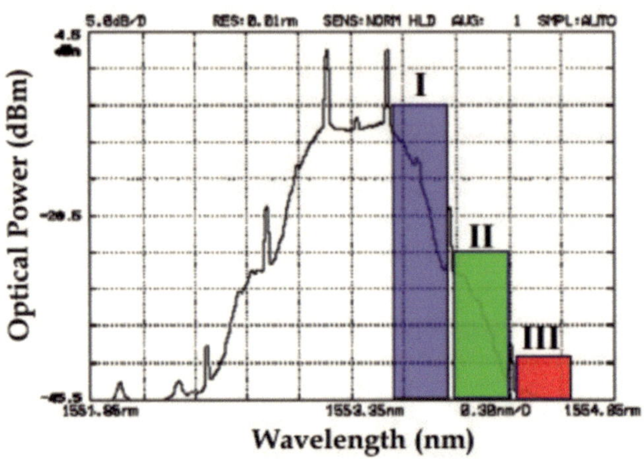

Fig. 4. Back to back spectrum of optical signals in high speed optical communication system with 40Gbit/s single-channel speed.

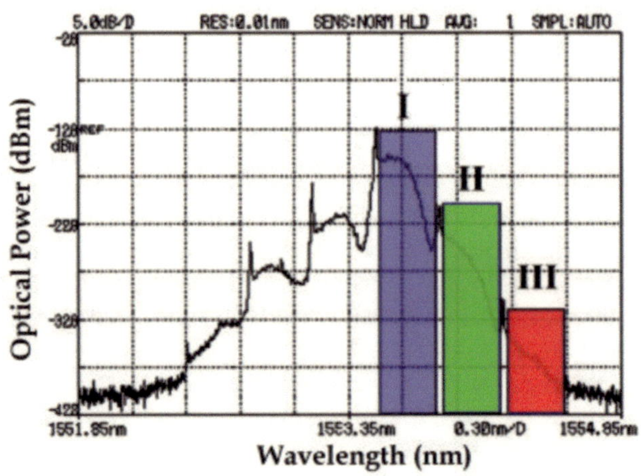

Fig. 5. Spectrum of output optical signals after amplified by the semiconductor optical amplifier in high speed optical communication system with 40Gbit/s single-channel speed.

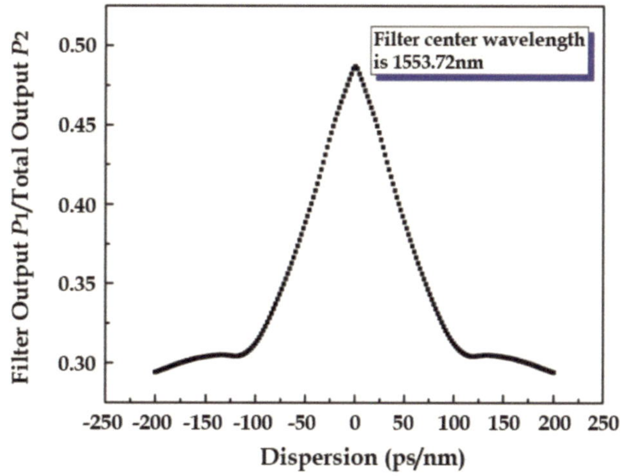

(a) Monitoring curve using optical fiber grating filter with center wavelength 1553.72nm

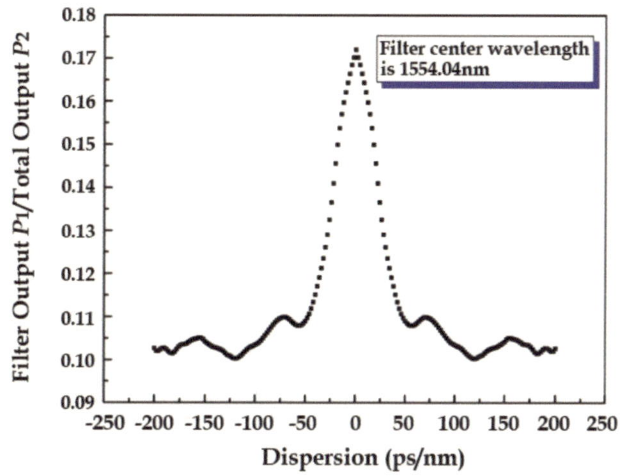

(b) Monitoring curve using optical fiber grating filter with center wavelength 1554.04nm

Fig. 6. Dispersion monitoring curves using optical fiber grating filter with center wavelengthes 1553.72nm and 1554.04nm, respectively.

To obtain an optimal chromatic dispersion monitoring signal, one needs to filter part of the output amplified spectrum to detect the optical power of spectral shift components. However, because the distribution of the spectral shift resulting from self phase modulation effect spans a wide frequency range, it needs an optimal scheme of the filter that can output the power of spectral shift components for chromatic dispersion monitoring. As shown in Figure 4, the power of each separate harmonic wave peak is higher than the shift frequencies' power due to the frequency shift effect. Thus, the separate harmonic wave peaks should be excluded from the filter pass-band. The short wavelength spectrum side is not suited for chromatic dispersion monitoring due to its multi-peaks structure, as shown previously. We divide the spectrum of the long wavelength side into three parts, (part I, part II and part III masked by three colored shadows, as shown in Figure 4 and 5) with frequency ranges 20-60 GHz, 60-100 GHz and 100-140GHz offset from center frequency (the wavelength is 1553.40 nm) of the optical spectrum, respectively. The spectral range part III beyond the wavelengths of part II is ignored due to low optical power.

It will be proved that the more narrow the band filter used, the more chromatic dispersion monitoring precision can be achieved in our method, but the output power will be too low to detect and can fail more easily due to the noise of the photoelectric diodes and optical amplifiers. In our method, the 3dB reflective band of the optical grating is 20 GHz; thus, we can obtain enough optical power to monitor chromatic dispersion and exclude the harmonic wave peaks from the pass band of the filter by careful choosing the filter center wavelength.

In order to obtain preferable monitoring conditions, we use two optical fiber gratings filters with center wavelengths of 1553.72nm and 1554.04nm respectively for our analysis and discussion. The 3dB reflective bands of the two optical fiber grating filters are all 20GHz, i.e. their reflective bands are all 0.16nm. The reflective band of the optical fiber grating filter with center wavelength 1553.72nm is stood in part I and other is located in part II, as shown in Figure 4 and Figure 5.

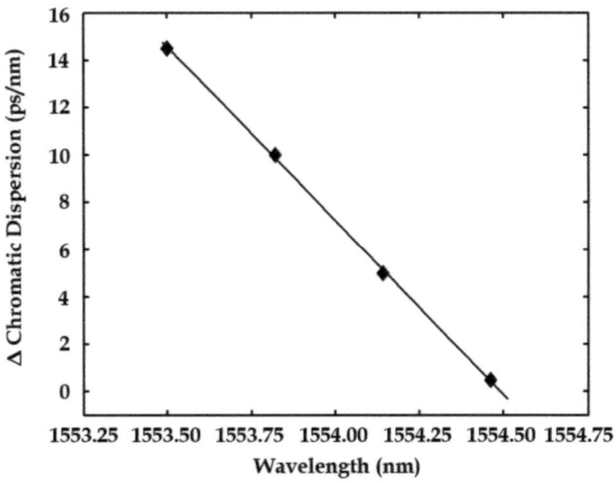

Fig. 7. Dependence of the chromatic dispersion monitoring precision on the filter center wavelength without the influence of the power of the signal peaks.

Figure 6 shows the chromatic dispersion curves in high speed optical fiber communication system with a single-channel speed of 40Gbit/s and suppressed return to zero (CSRZ) modulation format using the two optical fiber grating filters, which have mentioned above. Using the optical fiber grating filter with center wavelength 1553.72nm, the chromatic dispersion monitoring range is ±120ps/nm and the monitoring precision is about 10ps/nm, as shown in Figure 6(a). However, using the optical fiber grating filter with center wavelength 1554.04nm, the chromatic dispersion monitoring range is ±60ps/nm and the monitoring precision is higher than 5ps/nm, as shown in Figure 6(b). It can conclude that we can achieve more smaller chromatic dispersion monitoring range and more higher monitoring precision if we used an optical fiber grating filter with center wavelength located in part III of the suppressed return to zero modulation format signals' spectrum curve, as shown in Figure 4 and 5.

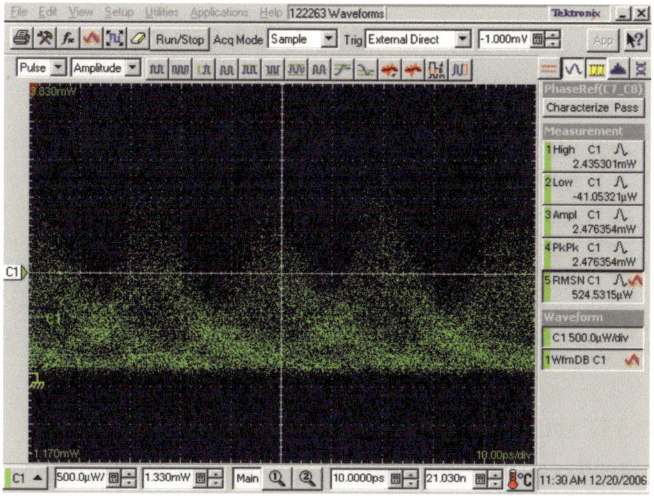

(a) Signal eye diagram of before chromatic dispersion compensation.

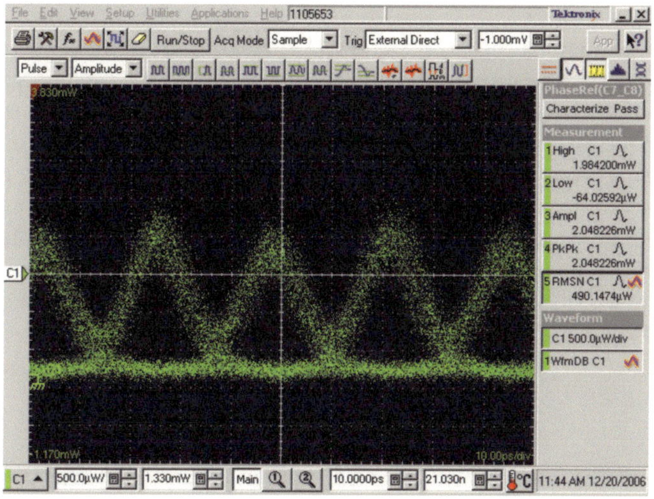

(b) Signal eye diagram of afert chromatic dispersion compensation.

Fig. 8. Signal eye diagrams tested by Tektronix® TDS8200 digital sampling oscilloscope of before and after chromatic dispersion compensation.

Figure 7 shows the dependence of the chromatic dispersion monitoring precision on the optical fiber grating filter center wavelength without the influence of the power peaks of the signals, because these peaks are excluded out of the pass band of the optical fiber grating filter by careful choosing the filter center wavelength. We concluded that the longer the center wavelength of the optical fiber grating filter used, the more chromatic dispersion monitoring precision can be achieved. In practice, we must choose an optimal optical fiber grating filter to obtain the optimal monitoring range and optimal monitoring precision for the dynamical chromatic dispersion in high-speed optical fiber communication systems. For a high speed optical communication system with a single-channel speed of 40 Gbit/s and suppressed return to zero modulation format, in the chromatic disoersion monitoring system, the best filter with a center wavelength of 1554.04 nm can be selected.

The eye diagrams of the optical fiber communication system with remnant chromatic dispersion of 60 ps/nm, before and after chromatic dispersion compensation, are shown in Figure 8. These eye diagrams were obtained by a Tektronix® TDS8200 digital sampling oscilloscope. Figure 8(a) is an eye diagram before chromatic dispersion compensation, and Figure 8(b) is a corresponding eye diagram after chromatic dispersion compensation. It can be concluded that our dynamical dispersion monitoring method, based on semiconductor optical amplifier spectral shift effect, is preferable for a high speed optical fiber communication system with a single-channel speed of 40 Gbit/s and suppressed return to zero modulation format.

4. Conclusion and Discussion

We demonstrated a dynamical chromatic dispersion monitoring method for high speed optical fiber communication systems. This method is based on the spectral shift resulting from self phase modulation of semiconductor optical amplifier. The more longer the wavelength components used for chromatic dispersion monitoring, the more monitoring precision of this method can be achieved, but the monitoring range becomes small simultaneously. Thus, in practice we must carefully consider the chromatic dispersion monitoring range and monitoring precision at the same time. This can be achieved by choosing an optimal optical fiber grating filter. For a high speed optical fiber communication system with a single channel speed of 40Gbit/s and suppressed return to zero format modulation, we use the optical fiber grating, with center wavelength and band width of 1554.04 nm and 20GHz respectively, as the optical filter. The chromatic dispersion monitoring range is $\pm$ 60 ps/nm and the chromatic dispersion monitoring precision is higher than 5 ps/nm in our method. Therefore, this technique is promising for use in remnant chromatic dispersion online monitoring in 40Gbit/s optical communication systems. In addition, it can be used for other high speed optical fiber communication systems by minimized modification.

5. Acknowledgments

The author thanks the Foundation of Guangxi Key Laboratory of Information and Communication and the foundation from the National Key Laboratory of Electromagnetic Environment of P. R. China for their supports. Most of the research work of this Chapter demonstrated is finished in Department of Electronic Engineering of Tsinghua University of P. R. China, during the postdoctoral stage of Prof. Ming Chen. The author thanks the Prof. S. Z. Xie, the tutor of the author's postdoctoral stage, for his lots of kindly supports, and because of the lucky opportunity from Prof. Xie, the author can live in the very beautiful Tsinghua Yuan two years. The author would like to thank Prof. M. H. Chen, Dr. H. W. Chen, Dr. Y.J. Zhang and all of the members in the Prof. Xie's research group for their meaningful discussion and suggestion. Ming Chen's e-mail addresses are m_chen@126.com or mchen@guet.edu.cn.

6. References

Kaminow, I. P.; Li, T., Willner, A. E. (2008). *Optical Fiber Telecommunictions V B: Systems and Networks*, Elsevier Inc., ISDN:978-0-12-374172-1.

Kaminow, I. P.; Li, T. (2002). *Optical Fiber Telecommunictions IV B: Systems and Impairments*, Academic Press, ISDN:0-12-395173-9.

Agrawal, G. P. (2008). *Fiber-Optic Communication Systems, 3rd Ed.*, John Wiley & Sons, Inc., ISDN:0-471-22114-7.

Pan, Z. Q. (2003). Overcoming fiber dispersion effects in high-speed reconfigure wavelength division multiplexing optical communication systems and networks. Dissertation for PhD, University of Southern California, Los Angeles, California, USA.

Bjarklev, A.; Broeng, J.; Bjarklev, A. S. (2003). *Photonic Crystal Fibres*, Kluwer Academic Publishers, ISDN:1-4020-7610-X.

Sukhoivanov, I. A.; Guryev, I. V. (2009). *Photonic Crystals: Physics and Practical Modeling*, Springer-Verlag, ISDN:978-3-642-02645-4.

Kashyap, R. (2009). *Fiber Bragg Gratings, 2nd Ed.*, Acdemic Press, ISDN:0-12-372579-8.

Ibsen, M.; Durkin, M. K.; Cole, M. J.; Laming, R. I. (1998). Sinc-sampled fiber Bragg gratings for indential multiple wavelength operation. *IEEE Photonics Technology Letters*, Vol.10, No.6, (June 1998) 842-845.

Sun, J.; Dai, Y. T.; Chen, X. F.; Zhang, Y. J.; Xie, S. Z. (2006). Thermally tunable dispersion compensator in 40Gb/s system using FBG fabricated with linearly chirped phase mask. *Optics Express*, Vol.14, No.1, (January 2006) 44-49.

Chen, M.; He, L.; Yang, S.; Zhang, Y.; Chen, H.; Xie, S. (2007). Chromatic sispersion and PMD monitoring and compensation techniques studies in optical communication systems with single channel speed 40Gbit/s and CSRZ format. *Optics Express*, Vol.15, No.12, (June 2007) 7667-7676.

Chen, M.; Yang, Q.; Li, T.S.; Chen, M.S.; He, N. (2010). New high negative dispersion photonic crystal fiber. *Optik*, Vol.121, No.10, (June 2010) 867-871.

Hong, I. W. (2002). Dispersion compensation in fiber optic communication systems. Thesis for the Degree of Master of Science, San Jose State University, San Jose, California, USA.

Seyed Mohammad Reza Motaghian Nezam. (2004). Chromatic and polarization mode dispersion monitoring for equalization in optical fiber communication systems. Dissertation for PhD, University of Southern California, Los Angeles, California, USA.

Ji, H. C.; Park, k. J.; Lee, J. H; Chung, H. S.; Son, E. S.; Han, K. H; Jun, S. B.; Chung, Y. C. (2004). Optical performance monitoring techniques based on pilot tones for WDM networks applications. *Journal of Optical Networking*, Vol.3, No.7, (July 2004) 510-533.

Pan, Z.; Yu, Q.; Xie, Y.; Havstad, S. A.; Willner, A. E.; Starodubov, D. S.; Feinberg, J. (2001). Chromatic dispersion monitoring and automated compensation for NRZ and RZ data using clock regeneration and fading without adding signaling. *Conference of 2001 Optical Fiber Communication*, Vol.3, Wh5-1.

Ning, G.; Shum, P.; Aditya, S.; Liu, N.; Gong, Y. D. (2006). On-line simultaneous monitoring of polarization and chromatic dispersion. *Applied Optics*, Vol.45, No.12, (December 2006) 2781-2785.

Luo, T.; Yu, C.; Pan, Z.; Wang, Y.; Arieli, Y.; Willner, A. E. (2005). Diepersive effects monitoring for RZ data by adding a frequecy-shifted carrier along the orthogonal polarization state. *IEEE Journal of Lightwave Technology*, Vol.23, No.10, (October 2005) 3295-3301.

Wang, Y.; Pan, Z; Sahin, A.; Yan, L; Yu, C.; Willner, A. (2006). In-line chromatic dispersion monitoring using optically-added phase-modulated in-band tones for 10Gg/s system. *Tech. Dig. Optical Fiber Communications (OFC 2003)*, 404-406.

Luo, T.; Yu, C.; Pan, Z.; Wang, Y.; Arieli, Y.; Willner, A. E. (2005). Diepersive effects monitoring for RZ data by adding a frequecy-shifted carrier along the orthogonal polarization state. *IEEE Journal of Lightwave Technology*, Vol.23, No.10, (October 2005) 3295-3301.

Hirano, A.; Kuwahara, Miyamoto, Y. (2002). A novel dispersion compensation scheme based on phase comparison between to SSB signals generated from a spectrally CS-RZ signal. *Tech. Dig. Optical Fiber Communications (OFC 2002)*, 196-197.

Wielandy, S.; Fishteyn, M.; Zhu, B. Y. (2004). Optical performance monitoring using nonlinear detection. *IEEE Journal of Lightwave Technology*, Vol.22, No.3, (March 2004) 784-793.

Li, S. P.; Kuksenkov, D. V. (2004). A novel dispersion monitoring technique based on four-wave mixing in optical fiber. *IEEE Photonics Technology Letters*, Vol.16, No.3, (March 2004) 942-944.

Ihara, T.; Oikawa, Y. (1999). Detection of, and compensation for, waveform change due to chromatic dispersion. U.S. Patent No.5,999,289, assigned to Fujitsu, Ltd.(Dec 7, 1999).

Chen, M; Zhang, Y. J.; He, L.; Si, Z; Yang, S.; Sun, J.; Zhang, Y.; Chen, H.; Xie, S. (2007). Simultaneous monitoring methods for chromatic dispersion and polarization mode dispersion based on polarization modulation. *Journal of Optics A: Pure Applied Optics*, Vol.9, No.4, (April 2007) 320-324.

Chen, M.; He, L.; Dai, Y.; Yang, S.; Chen, H.; Xie, S. (2007). Chromatic dispersion monitoring method based on semiconductor optical amplifier spectral shift in 40Gbit/s optical communication systems. *Optical Engineering*, Vol.46, No.11, (November 2007) 115008-1-115008-6.

Dutta, N. K.; Wang, Q. (2006). *Semiconductor Optical Amplifiers*, World Scientific, ISDN:981-256-397-0.

Kaminow, I. P.; Li, T. (2002). *Optical Fiber Telecommunictions IV A: Components*, Academic Press, ISDN:0-12-395172-0.

Kaminow, I. P.; Li, T., Willner, A. E. (2008). *Optical Fiber Telecommunictions V A: Components and Subsystems*, Elsevier Inc., ISDN:978-0-12-374171-4.

Shimada, S.; Ishio, H. (1994). *Optical Amplifiers and their Applications*, John Wiley & Sons., ISDN:978-0-47-194005-0.

9

Slow and Fast Light in Semiconductor Optical Amplifiers for Microwave Photonics Applications

Perrine Berger[1], Jérôme Bourderionnet[2], Daniel Dolfi[3], Fabien Bretenaker[4] and Mehdi Alouini[5]
[1,2,3]*Thales Research and Technology*
[4]*Laboratoire Aimé Cotton, CNRS - Université Paris Sud*
[5]*Institut de Physique de Rennes*
France

1. Introduction

The generation of continuously tunable optical delays and tunable phase shifts is a key element in microwave photonics. Among the targeted applications, one can quote the filtering of microwave signals, the synchronization of optoelectronics oscillators, and the control of optically fed phased array antennas. With these applications in view, large efforts are currently done in order to develop delay lines and phase shifters based on slow and fast light effects. To date, one of the most mature approaches for integration in real field systems is that based on Coherent Population Oscillations (CPO) in Semiconductor Optical Amplifiers (SOAs). This approach offers compactness, continuous tunability of the delay or phase shift through injected current control, and possible high-level parallelism.

Slow and fast light capability of semiconductor devices has been first studied in the past decade while bearing in mind the delays, the phase shifts and the bandwidth they can offer. Consequently, in a first section, we present the recent advances in architectures based on slow and fast light in SOAs for microwave photonics applications. After a brief introduction about microwave photonics, we present the physical interpretation of the different architectures proposed in the literature. We point out the underlying physics, common to these architectures, and evidence the advantages and drawbacks of each of them. However, within the scope of integration in a realistic radar system, it is also required to study the impact of these slow and fast light architectures on the performances of the microwave photonics link. In particular, the RF transfer function, the generation of spurious signals by harmonic and intermodulation products, and the intensity noise, have to be studied in order to compute the Spurious Free Dynamic Range (SFDR), a key characteristic in microwave photonics. Consequently, in a second section, we present the tools to simulate and understand the RF transfer function, the generation of spurious signals through harmonic distortion and intermodulation products, and the intensity noise at the output of a SOA.

In a third section, we use the models presented in the previous part in order to investigate the dynamic range of a microwave photonics link including an architecture based on slow and fast light in SOAs. We focus on the architecture using a SOA followed by an optical filter. The models are experimentally validated and the influence on the microwave photonics link is discussed.

2. Background and context

2.1 Slow and fast light for microwave photonics

This section presents microwave photonics and explains why slow and fast light can be useful for this applied research field.

2.1.1 Microwave photonics link including a slow and fast light device

Microwave photonics realizes processing of microwave signals ($\Omega/2\pi \simeq 0.1$ - 35 GHz) in the optical domain using photonic devices. Indeed optics offer some advantages compared to electronics for the addressing and processing of microwave signals: the most sticking asset is the low loss transport along an optical fiber (0.2 dB/km) compared to a coaxial cable (1000 dB/km !). As illustrated in Fig. 1, the basic architecture of a microwave photonics link is composed of a laser, which creates the optical carrier ($\lambda_0 \simeq 1.5$ μm). The optical carrier is modulated by the microwave signal either directly, or through an external Mach Zehnder modulator. Optical devices (represented on Fig. 1 by "slow and fast light device") process the modulated carrier. At the end of the link, a fast photodiode retrieves the processed microwave signal.

The aim of our study is to introduce a slow and fast light device in a microwave photonics link, as it is represented in Fig. 1. Let us consider a monochromatic microwave signal (whose angular frequency is Ω) and a linear modulator, the optical field E after the modulator is then composed of the optical carrier $E_0(z)e^{-i\omega_0 t}$ and two sidebands $E_1(z)e^{-i(\omega_0+\Omega)t} + E_{-1}(z)e^{i(\omega_0-\Omega)t}$.

After propagation through the slow and fast light device, the microwave signal retrieved by the photodiode, at the angular frequency Ω, is:

$$\begin{aligned} M_1^{OUT} &= \sum_{\omega_p-\omega_q=\Omega} E_p E_q^*, \qquad (1)\\ &= E_1 E_0^* e^{i(k_1-k_0)L} + E_{-1}^* E_0 e^{i(k_0-k_{-1})L}, \text{ when both sidebands are detected,}\\ &= E_1 E_0^* e^{i(k_1-k_0)L} \text{ or } E_{-1}^* E_0 e^{i(k_0-k_{-1})L}, \text{ when only one sideband is detected,} \end{aligned}$$

where L is the length of the dispersive medium (described by $k_i = k(\omega_i)$).

To characterize a microwave photonics link, we study the microwave transfer function through the complex parameter $S_{21} = \frac{M_1^{OUT}}{M_1^{IN}}$, whose magnitude and phase can be expressed as:

$$|S_{21}| = G^2, \qquad (2)$$

$$arg(S_{21}) = \Delta k L, \qquad (3)$$

where G is the optical gain of the link, and ΔkL the phase shift introduced by the dispersive medium (for example $\Delta k = k(\omega_0 + \Omega) - k(\omega_0)$ when only the sideband E_1 is detected). Consequently, by governing the dispersion and in particular the optical group velocity $v_g = \frac{d\omega}{dk}$ of the slow and fast light medium, it is possible to induce a controlled phase shift or delay on the retrieved microwave signal, which is an important function in microwave photonics, as it is illustrated in the following sections.

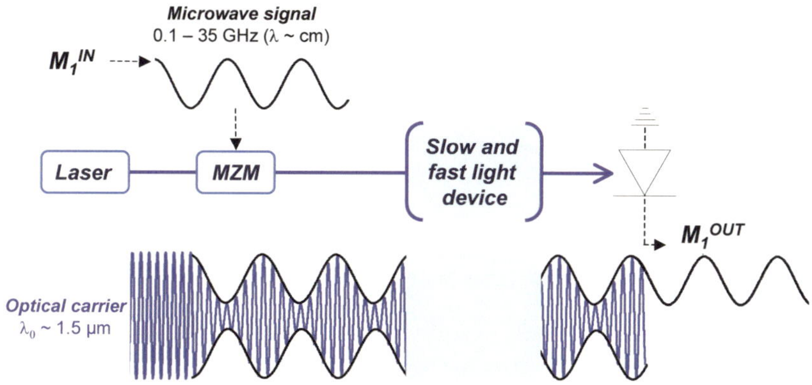

Fig. 1. Microwave photonics link including a slow and fast light device. MZM=Mach Zehnder Modulator.

2.1.2 Tunable true time delay line for microwave photonics

If a tunable true time delay is introduced on a microwave signal at the angular frequency Ω, the phase shift introduced by the slow and fast light medium must be:

$$\Delta kL = \tau\Omega \text{ with } \tau \text{ tunable.} \tag{4}$$

This definition is illustrated on Fig. 2.

The generation of tunable true time delays over a large instantaneous RF bandwidth Δf_{RF} is a key function in the processing of microwave signals, where the instantaneous bandwidth Δf_{RF} will reach 10% to 30% of the operating frequency $f_{operating}$. To illustrate this point, let us focus on the addressing of active antennas. Active antennas offer a better precision of the radiation direction, a better gain, and lower secondary lobes than classic antennas.They are composed of a matrix of radiating elements whose input signals can be controlled in magnitude and phase. The radiating pattern and the main radiating direction are controlled through the phase shifter of each radiating element. The phases applied to the different emitters define the phase plane in which all the RF signals constructively interfere. The main radiating direction is then perpendicular to this plan. Angling the beam is then equivalent to tuning the phase plane, which is controlled by the RF phase shifters. However, the delays introduced by these phase shifters are frequency dependent: consequently, a change in frequency introduces an uncontrolled beam squint. Tunable true time delay lines is an answer to these problems.

The generation of true time delays, tunable up to the nanosecond (or larger), on an instantaneous bandwidth reaching the GHz range is necessary. Optical solutions have already been proposed, by geometrically modifying the optical path. However these techniques does not match the real field requirements because of the weight, the cost and the limited number of achievable delays. Slow and fast light media could offer a complementary solution, by offering a fast, continuous, and precise control of delays.

2.1.3 Tunable phase shifter for microwave photonics

A microwave photonics including a slow and fast light medium can also be used as an optically tunable RF phase shifter. The phase shift introduced by the slow and fast light medium must be:

$$\Delta kL = \phi \text{ covering } [0, 2\pi] \tag{5}$$

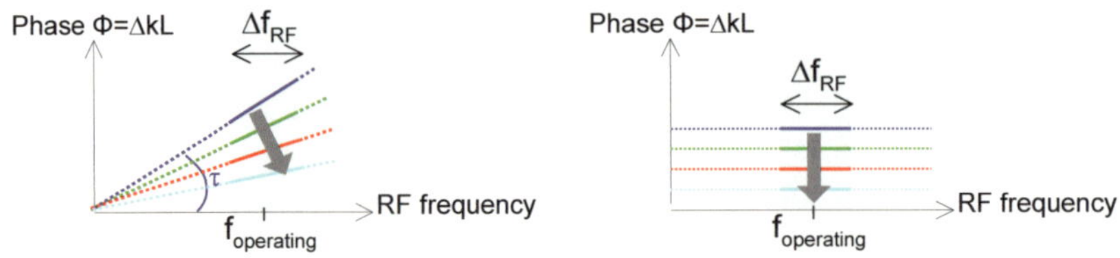

(a) Tunable true time delay (b) RF Phase shifter

Fig. 2. (a) Illustration of tunable true time delay. (b) Illustration of RF phase shifter.

This definition is illustrated on Fig. 2.
This component exists in the microwave world (contrary to the true time delay lines). However, more and more RF functions are developed by photonic means. Consequently, tunable RF phase shifter in the optical domain is required in order to avoid useless Optical/Electronical and Electronical/Optical conversions. Among the targeted applications, one can quote the filtering of microwave signals: on each arm a phase shifter over an instantaneous bandwidth Δf_{RF} is required; or optoelectronics oscillators, either for synchronization or tunability: in this case, only a phase shifter at the operating frequency is required.
To date, one of the most mature approaches of slow and fast light medium for integration in real field systems is that based on Coherent Population Oscillations (CPO) in semiconductor optical amplifiers (SOAs). This approach offers compactness, continuous tunability of the delay or phase shift through injected current control, and possible high-level parallelism. We present in the following paragraphs the basic concepts of this technique.

2.2 Coherent Population Oscillations (CPO) in Semiconductor Optical Amplifier (SOA)

This section introduces the basic concepts and the main equations involved in Coherent Population Oscillations (CPO) in a Semiconductor Optical Amplifier (SOA), and explains how this phenomenon induces slow and fast light. We describe the SOA behavior by a phenomenological model initially described for semiconductor lasers by Agrawal & Dutta (1993), and which was initially used to describe CPO in SOA (Agrawal, 1988; Mørk et al., 2005). It gives a good insight in the involved phenomena. However, this model, initially developed for lasers, must be cautiously used in the case of SOAs: its limitations are discussed in section 4, where a more rigorous model is developed. From this phenomenological model, we derive the equations of propagation and analyze the dispersion properties and in particular the changes in group velocity induced by CPO.

2.2.1 Phenomenological model of SOA

This phenomenological model of SOA, well described by Agrawal & Dutta (1993) is based on experimental observations of the behavior of semiconductor lasers.
The main assumption of the model consists in considering that the variations of the material gain g and optical index n, caused by a small variation of the carrier density ΔN, are proportional to ΔN. The material gain g and the optical index n can thus be expressed as:

$$g = \bar{g} + \Delta g, \tag{6}$$

$$n = \bar{n} + \Delta n, \tag{7}$$

with $\bar{N}$ and $\bar{g} = g(\bar{N})$ the static carrier density and gain, $\bar{n} = n(\bar{(N)})$, the static optical index, and Δg and Δn the variations of the material gain and the optical index caused by a small change ΔN of the carrier density. We assume then:

$$\boxed{\Delta g \propto \Delta N \text{ and } \Delta n \propto \Delta N.} \tag{8}$$

We introduce the differential gain a:

$$\Delta g = a\Delta N, \tag{9}$$

and the linewidth enhancement factor α introduced by Henry (1982) to model the index gain coupling in semiconductor material: $\alpha = -2k_0 \frac{\Delta n}{\Gamma \Delta g}$. Then the variation Δn can be written as:

$$\Delta n = -\frac{\alpha}{2k_0}\Gamma \Delta g, \tag{10}$$

with $k_0 = \frac{\omega}{c}$, and Γ the confinement factor. We complete this phenomenological model with a rate equation, which incorporates all the mechanisms by which the carriers are generated or lost in the active region:

$$\boxed{\frac{dN}{dt} = \frac{I}{qV} - \frac{N}{\tau_s} - \frac{g|E|^2}{\hbar\omega_0},} \tag{11}$$

where I is the injected current, $|E|^2$ is the optical intensity inside the SOA, τ_s is the carrier lifetime, V is the volume of the active region, q is the elementary charge, and ω_0 is the angular frequency of the optical carrier E_0.

2.2.2 Coherent population oscillations

CPOs are induced by an optical carrier which is modulated in intensity at the angular frequency Ω: $|E|^2 = M_0 + M_1 e^{-i\Omega t} + c.c.$. Note that the injected current can also be modulated (see section 3.3).

If the optical carrier power is large enough, it implies a gain saturation. Indeed, the rate equation (11) shows that oscillations of the carriers (CPO) are induced : $N = \bar{N} + \Delta N e^{-i\Omega t} + c.c.$. At the first order, we can assume a linear variation of the gain and optical index (Eq. 8):

$$g = \bar{g} + \Delta g e^{-i\Omega t} + c.c., \tag{12}$$

$$n = \bar{n} + \Delta n e^{-i\Omega t} + c.c. \tag{13}$$

Finally the Eqs. 9 and 11 lead to the expression of the gain variation Δg (and then Δn thanks to Eq. 10):

$$\Delta g = \frac{A}{1 + M_0/U_s - i\Omega\tau_s}, \tag{14}$$

where M_0 is the DC component of the optical intensity, U_s the saturation intensity, and for sake of clarity, A is considered here as a constant (for example $A = -\bar{g}M_1/U_s$ when only the optical intensity is modulated, see section 4).

The optical index is then time and frequency dependent due to the CPO, which implies slow and fast propagation of the light, as we explain in the following paragraphs.

2.2.3 Equations of propagation

The optical field $\mathcal{E}(z,t)$ verifies the following wave equation:

$$\frac{\partial^2}{\partial^2 z}\mathcal{E}(z,t) = \frac{1}{c^2}\frac{\partial^2}{\partial^2 t}\epsilon_{sc}\mathcal{E}(z,t), \tag{15}$$

where ϵ_{sc} is the relative permittivity.

We expand the optical field as $\mathcal{E}(z,t) = \Re\left\{\sum_p E_p(z)e^{i(\tilde{\beta}_p z-\omega_p t)}\right\}$, where $\tilde{\beta}_p$ is the complex propagation constant. E_0 accounts for the optical carrier, and $E_{\pm 1}$ the modulation sidebands at $\omega_{\pm 1} = \omega_0 \pm \Omega$. The complex propagation constant can be expressed as:

$$\tilde{\beta} = k_0\tilde{\mu} = k_0\sqrt{\epsilon_{sc}}, \tag{16}$$

with $k_0 = \frac{\omega}{c}$ and $\tilde{\mu}$ the complex optical index, which can be written as:

$$\tilde{\mu} = n + i\frac{-\Gamma g + \gamma}{2k_0}, \tag{17}$$

where n is the real refractive optical index, and γ holds for internal losses. From Eqs. 12, 13 and 17, the complex optical index $\tilde{\mu}$ can be expanded as: $\tilde{\mu} = \tilde{\mu}_0 + \Delta\tilde{\mu}e^{-i\Omega t} + c.c.$, with:

$$\begin{aligned}\tilde{\mu}_0 &= \bar{n} + \frac{i}{2k_0}\left(-\Gamma\bar{g} + \gamma\right),\\ \Delta\tilde{\mu} &= -\frac{\alpha + i}{2k_0}\Gamma\Delta g.\end{aligned} \tag{18}$$

Lastly, we derive the equations of propagation from Eqs. 15, 18 and $\epsilon_{sc} = \tilde{\mu}^2$:

$$\begin{aligned}\frac{dE_0}{dz} &= \frac{1}{2}\left(\Gamma\bar{g} - \gamma\right)E_0,\\ \frac{dE_1}{dz} &= \frac{1}{2}\left(\Gamma\bar{g} - \gamma\right)E_1 + \frac{1 - i\alpha}{2}\Gamma\Delta g E_0,\\ \frac{dE_{-1}}{dz} &= \frac{1}{2}\left(\Gamma\bar{g} - \gamma\right)E_{-1} + \frac{1 - i\alpha}{2}\Gamma\Delta g^* E_0.\end{aligned} \tag{19}$$

2.2.4 Slow and fast light induced by CPO

We have shown in the previous paragraph that the modulation of the optical intensity leads to CPO, which induces a frequency dependence of the complex optical index. We illustrate here how it induces slow and fast light.

We define the real, n_r, and imaginary, n_{im}, parts of the index: $\tilde{\mu} = n_r + in_{im}$. The Eqs. 14 and 18 lead to the following expressions of the variations of the real and imaginary parts of the optical index, induced by CPO:

$$\Delta n_r = \Re\{\Delta\tilde{\mu}\} = -\frac{cA}{2w}\frac{\alpha(1 + M_0/U_s) - \Omega\tau_s}{(1 + M_0/U_s)^2 + (\Omega\tau_s)^2}, \tag{20}$$

$$\Delta n_{im} = \Im\{\Delta\tilde{\mu}\} = -\frac{cA}{2w}\frac{(1 + M_0/U_s) + \alpha\Omega\tau_s}{(1 + M_0/U_s)^2 + (\Omega\tau_s)^2}, \tag{21}$$

where Ω is the angular frequency of the CPO, w is the angular frequency of the considered optical field component, and $w = w_0 + \Omega$ with w_0 the angular frequency of the optical carrier.

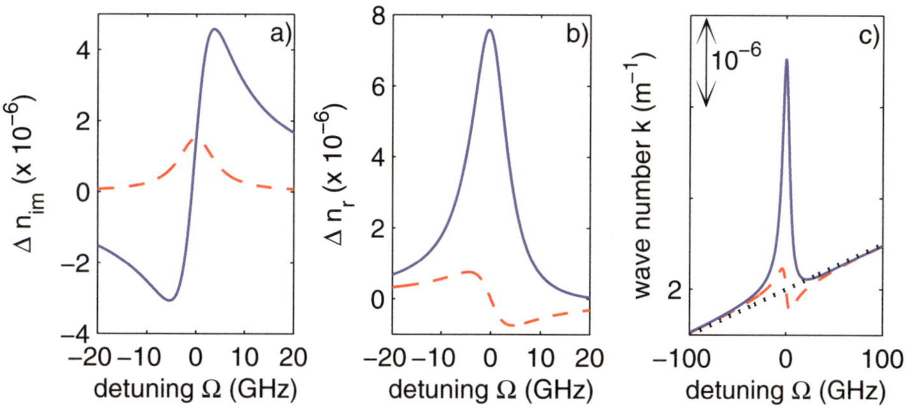

Fig. 3. Variations of (a) the imaginary part Δn_{im} of the complex optical index (proportional to the absorption), (b) the real part Δn_r of the complex optical index, and (c) the wave number k. In blue, for semiconductor material (where $\alpha \neq 0$), and in dashed red line, for an equivalent 2-level atomic resonance (where $\alpha = 0$). Parameters: $\alpha = 5$ (blue line) or $\alpha = 0$ (red dashed line), $M_0/U_s = 1$, $A = -4000m^{-1}$, $\tau_s = 450ps$, $n_{sc} = 3$, $\lambda_0 = 1.5\mu m$.

The variations of the imaginary Δn_{im} and real Δn_r parts of the optical index are displayed on Fig.3(a) and (b). Δn_{im} and Δn_r are related by the Kramers-Kronig relationship. We illustrate that for semiconductor material, due to the coupling index-gain (modeled by the factor α), the variation of the imaginary Δn_{im} optical index, proportional to the variation of the absorption, is asymmetric: it has been first observed by Bogatov et al. (1975). The variation of the optical index with respect to the frequency is then very different from the case of 2-level atomic resonance (represented in dashed red line in Fig. 3).

The frequency dependence of the optical index leads to a strong dispersion in the vicinity of the frequency of the optical carrier: as illustrated on Fig.3(c), the real wave number $k(\omega) = \frac{n(\omega)\omega}{c}$ differs then from the "normal" refraction $\frac{n_{sc}\omega}{c}$ (represented by the black dotted straight line). This is associated with a variation of the group velocity of the light, which can be defined as $v_g = \frac{d\omega}{dk}$. Consequently, CPO create here ($A < 0$) "fast light" for positive detuning $\Omega > 0$ ($\frac{dk}{d\omega} < 0$), and "slow light" for negative detuning $\Omega < 0$ ($\frac{dk}{d\omega} > 0$). In comparison, a 2-level atomic resonance create essentially fast light at low detuning: $|\Omega| <$ a few GHz. As the amplitude of CPO is controlled by either the input optical power (M_0) or the injected current (though the parameter A), the group velocity can be tuned and controlled, which opens the possibility to conceive optical delay lines and optical phase shifters for microwave signals, as we will explain it in the following part.

The arrival time τ (to a detector for example) of a signal propagating through this slow and fast light medium has to be carefully handled. Indeed it is equal to the group delay simply defined as $\tau_g = L\frac{dk}{d\omega}|_{\bar{\omega}}$, with $\bar{\omega}$ the central optical frequency of the signal, only if the modulation before the photodiode is Single-SideBand (SSB), and if the bandwidth of the signal is smaller than the CPO bandwidth (< 1 GHz). We will highlight this last point in the following part. For further details on the definition of the arrival time for optical pulses, you can refer to (Peatross et al., 2000).

3. Advances in architectures based on slow and fast light in SOAs for microwave photonics applications

In the previous part, we saw the underlying physics of the slow and fast light generated by Coherent Population Oscillations (CPO) in SOAs. We present here the different architectures based on this phenomenon, which are proposed for microwave photonics applications. In the first section, we present an architecture of an optical delay line. In the second section, we show that by changing the modulation format before the photodiode, the latter architecture becomes an optical phase shifter for microwave signals. Lastly, we present an alternative set-up to realize an optical RF phase shifter, by using forced CPO.

3.1 SOA-based optical delay line

3.1.1 Set-up, experimental results and equations

This architecture is the first proposed in literature (Mørk et al., 2005; Pesala et al., 2006). The set-up is quite simple: the slow and fast light device (represented in Fig. 1) included in the microwave link is a single SOA. The corresponding experimental results are presented on Fig. 4. The microwave gain $|S_{21}|$ present a high pass filter behavior, usually observed in SOAs (Boula-Picard et al., 2005), and is associated to an interesting phase shift, which presents a linear variation at low frequency, with a slope tunable through the injected current or the optical input power: a key characteristic to set up a tunable delay line (see Fig. 2).
These results are not in adequation with the expected behavior of a semiconductor material in which the coupling index-gain is significant ($\alpha \neq 0$) (displayed in Fig. 3), but it is similar to the material where $\alpha = 0$: indeed a negative delay (which is then an advance, associated to a so called "fast light") is detected at low frequency ($<$ GHz). However it can be easily explained by the double-sideband modulation format before the detector. Indeed, as both the modulation sidebands are detected, the RF retrieved signal at the angular frequency Ω is $M_1 = E_1 E_0^* + E_{-1} E_0^*$, and Eq. 19 leads to:

$$\frac{dM_1}{dz} = [-\gamma + \Gamma\bar{g}]\, M_1 + \Gamma \Delta g M_0, \tag{22}$$

with Δg deduced from the rate equation (11):

$$\Delta g = -\frac{\bar{g} M_1 / U_s}{1 + M_0/U_s - i\Omega\tau_s}, \tag{23}$$

with $U_s = \frac{\hbar\omega}{\Gamma a \tau_s}$. We notice then that the contribution of the coupling gain-index is canceled out when both the sidebands are detected, which explains that the gain and phase shift are similar to a material where $\alpha = 0$.

3.1.2 Analysis of the different contributions. Physical interpretation

We integrate Eq. 22 over a small slice dz, whose length is noted L, to derive an analytical expression of the RF transfer function $S_{21} = \frac{M_1(L)}{M_1(0)}$:

$$\frac{M_1(L)}{M_1(0)} = \mathcal{A} + \frac{1}{2}\mathcal{G}^{opt}, \tag{24}$$

with $\mathcal{A} = 1 - \gamma L + \Gamma \bar{g} L$ corresponding to the optical amplification (average RF gain) and $\mathcal{G}^{opt}$ the contribution due to CPO, which are induced by the modulation of the optical intensity:

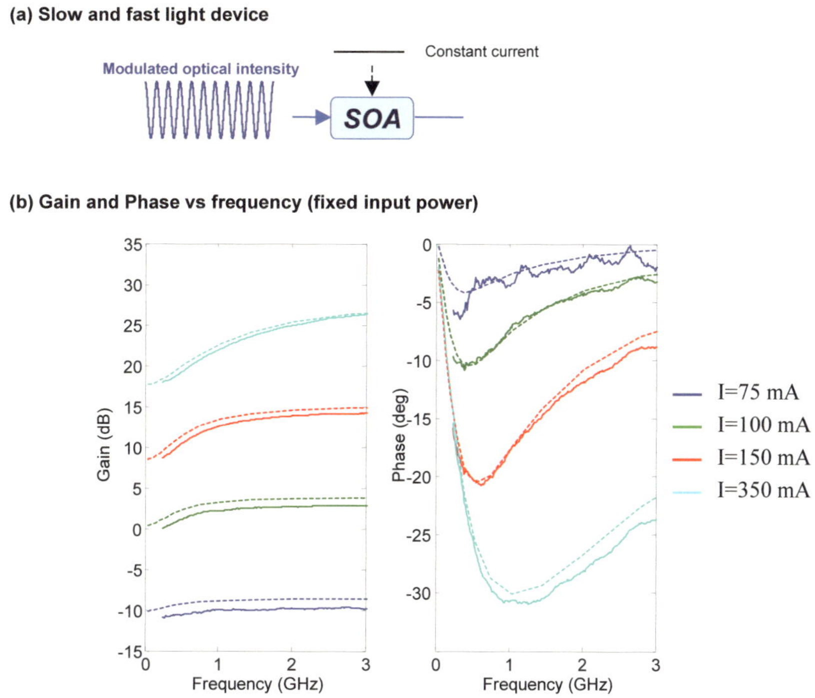

Fig. 4. SOA-based optical delay line. (a) Slow and fast light device. (b) Experimental results: gain and phase shift of the RF signal, at a fixed input optical power (1 mW), for different currents. Results extracted from (Berger, Alouini, Bourderionnet, Bretenaker & Dolfi, 2010).

$$\mathcal{G}^{opt} = \Gamma \Delta g L M_0(0)/M_1(0) = \frac{-2\Gamma g0LM_0/U_s}{1 + M_0/U_s - i\Omega\tau_s}. \qquad (25)$$

We represent the module and the argument of the amplification $\mathcal{A}$ and the CPO contribution $\mathcal{G}^{opt}$ on Fig. 5. The optical amplification $\mathcal{A}$ is in phase with the incident signal, and constant with respect to the RF frequency. At low frequency $\Omega\tau_s < 1$, $\mathcal{G}^{opt}$ is in antiphase with the incident modulated signal. Indeed, the carrier density is modulated by saturation due to the modulated optical intensity: a larger number of carriers will decay near the maximum intensity than near the minimum. Consequently the carrier density is modulated in antiphase with respect to the incident signal, generating a gain also in antiphase. At high frequency $\Omega\tau_s > 1$, the carriers can no longer follow the optical modulation, and the efficiency of the gain modulation decreases. This explains the low pass filter behavior of $\mathcal{G}^{opt}$ displayed on Fig. 5. Moreover, in Fig. 5(a), we note that the amplification $\mathcal{A}$ is always dominant compared to $\mathcal{G}^{opt}$: consequently at low frequency $\Omega\tau_s < 1$, the modulated gain $\mathcal{G}^{opt}$ in antiphase with respect to the incident signal decreases the total gain. This explains the dip at low frequency on the module of the RF transfer function, which is associated with an interesting phase shift, opening the possibility to conceive a tunable delay line.

3.1.3 Evaluation of the performances for a tunable delay line

We assume the plane-wave approximation: $M_1(z) = \tilde{M}_1 e^{\tilde{\beta}z}$. Even if it is not truly accurate, it is helpful to understand the essential physics and to evaluate the order of magnitude of

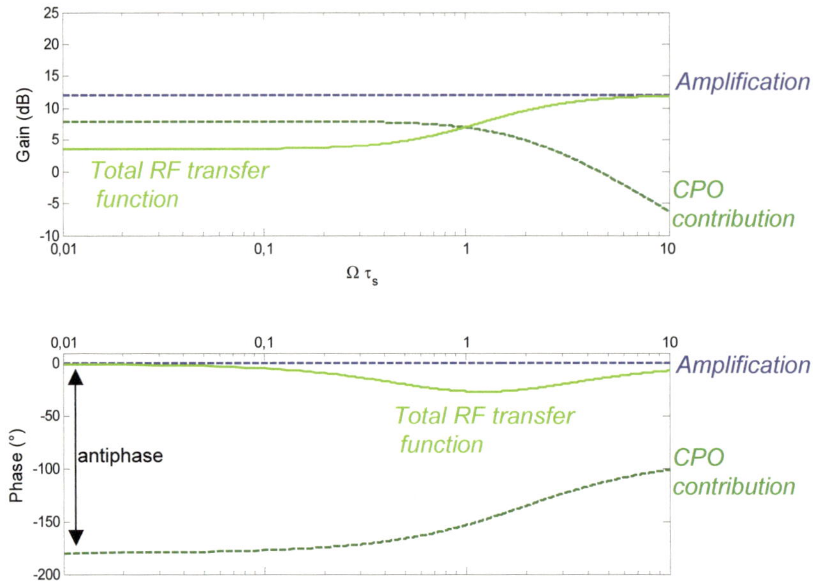

Fig. 5. Contributions and resulting total transfer function.

achievable delay and bandwidth. From Eq. 22, we can deduce the complex propagation constant $\tilde{\beta}$:

$$\tilde{\beta}L = -i\left(-\gamma L + \Gamma\bar{g}L - \frac{\Gamma\bar{g}LM_0/U_s}{1 + M_0/U_s - i\Omega\tau_s}\right). \tag{26}$$

After a small slice dz, if we assume that the gain compensates losses, the output signal is then: $G_{RF}M_1(0)e^{-i\Omega(t-\tau)}$, with:

$$\tau = -\frac{\Gamma\bar{g}LM_0/U_s}{(1 + M_0/U_s)^2 + (\Omega\tau_s)^2}\tau_s \tag{27}$$

$$\simeq -\frac{\Gamma\bar{g}LM_0/U_s}{(1 + M_0/U_s)^2}\tau_s \text{ at low frequency } \Omega\tau_s < 1 + M_0/U_s. \tag{28}$$

Consequently, at low frequency, CPO introduce a true time advance τ (negative delay, independent on the frequency Ω): Eq. 28 gives the limit of the achievable delay and bandwidth.

It is possible to make a rough evaluation of the maximal achievable bandwidth-delay product, by assuming the saturation parameters (U_s, τ_s, $\Gamma\bar{g}L$) constant with the current and the input optical power :

$$(\Delta f_{RF} * \tau)_{max} < \frac{1}{2\pi}\frac{M_0/U_s}{1 + M_0/U_s}(\Gamma\bar{g}L)_{max}. \tag{29}$$

The higher the input optical power M_0 is, the higher the bandwidth-delay product is. A SOA with a high gain is suitable. However the maximal advance (negative delay) achievable for a given current is: $\tau_M(I) = -\frac{1}{4}\Gamma\bar{g}L\tau_s$ for a strong input optical power $M_0/U_s \sim 1$. The delays are tunable from 0 to $\tau_M(I_{max})(< 0)$, as the gain $\bar{g}L$ governed by the injected current I. The delay is null at the transparency, and can be positive when the SOA is in the absorption regime (but it is associated with high losses on the RF signal). In order to refine the performances, and take into account saturation effects, we use the mean saturation

parameters (U_s, τ_s, $\Gamma\bar{g}L$) along the SOA, calculated from the model presented in section 4, for a given current and input optical power. For example, a commercial SOA (InP/InGaAsP Quantum Well Booster Amplifier from COVEGA) has a maximal tunable advance of $\approx$ 516ps ($\approx$ 120ps) over an instantaneous bandwidth $<$ 590MHz ($<$ 410MHz) for an input optical power of 20mW (1mW).
The major axes of research are now to find the best material to increase these performances, and try to find an architecture which enables to translate these characteristics at any operating frequency.

3.2 CPO enhanced by index-grating coupling

3.2.1 Set-up, experimental results and equations

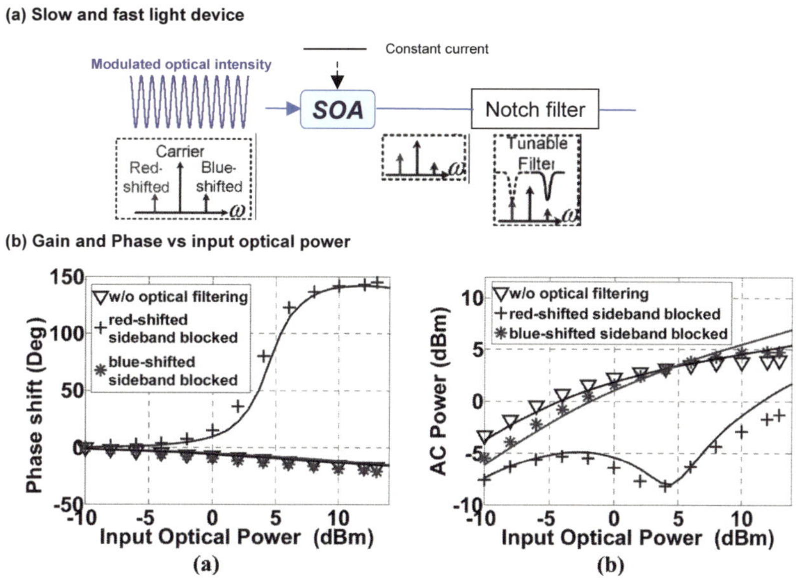

Fig. 6. CPO enhanced by index-grating coupling. (a) Slow and fast light device. (b) Experimental results: gain and phase shift of the RF signal, at a fixed current (1 mW), with respect to different input optical power. Graph arranged from (Xue et al., 2008).

As it is explained in the previous paragraph, when both the modulation sidebands are detected by the photodiode, the contribution of the index-gain coupling is canceled out. In order to benefit from the enhancement of the gain and index gratings by the index-gain coupling, Xue et al. (2008) analyze an architecture including an optical notch filter before the photodiode in order to select one sideband (see Fig. 6(a)). They showed that an enhanced phase shift up to $+150°$ is detected only when the red shifted sideband is blocked, compared to $-20°$ for other configurations (see Fig. 6(b)).
Consequently the RF retrieved signal at the Ω is either $M_r = E_1 E_0^*$ when the red-shifted sideband is blocked or $M_b = E_{-1} E_0^*$ when the blue shifted sideband is blocked. Eq. 19 leads to:

$$\frac{dM_r}{dz} = [-\gamma + \Gamma\bar{g}]\, M_r + \frac{1+i\alpha}{2}\Gamma\Delta g M_0, \quad (30)$$

$$\frac{dM_b}{dz} = [-\gamma + \Gamma\bar{g}]\, M_b + \frac{1-i\alpha}{2}\Gamma\Delta g M_0, \quad (31)$$

with Δg defined by Eq. 23.

3.2.2 Analysis of the different contributions. Physical interpretation

We integrate Eq. 30 and Eq. 31 over a small slice dz, whose length is noted L, to derive an analytical expression of the RF transfer function $S_{21} = \frac{M_i(L)}{M_i(0)}$:

$$\frac{M_r(L)}{M_r(0)} = \mathcal{A} + \frac{1}{2}\mathcal{G}^{opt} + \frac{i\alpha}{2}\mathcal{G}^{opt}, \tag{32}$$

$$\frac{M_b(L)}{M_b(0)} = \mathcal{A} + \frac{1}{2}\mathcal{G}^{opt} - \frac{i\alpha}{2}\mathcal{G}^{opt}, \tag{33}$$

with $\mathcal{A} = 1 - \gamma L + \Gamma \bar{g} L$ and $\frac{1}{2}\mathcal{G}^{opt}$ expressed as in the previous paragraph (we assume $M_r(0) = M_b(0)$).

We represent the module and the argument of the amplification $\mathcal{A}$ and the dominant contribution due to CPO: $\pm\frac{i\alpha}{2}\mathcal{G}^{opt}$ on Fig. 7(a). Contrary to the previous architecture, the contribution enhanced by the gain-index coupling $\pm\frac{i\alpha}{2}\mathcal{G}^{opt}$, is dominant over the amplification $\mathcal{A}$ at low frequency. At high frequency, the amplification overcomes. Consequently, a transition between the contribution enhanced by the gain-index coupling $\pm\frac{i\alpha}{2}\mathcal{G}^{opt}$ and the amplification $\mathcal{A}$ is realized at a frequency designed as f_t. At the transition, the two configurations, when either the red-shifted or blue-shifted sideband is blocked, are not similar: the phase of the enhanced contribution goes from $+90°$ ($-90°$) to $+180°$ ($0°$), when the red (blue)-shifted sideband is blocked. Consequently, if the transition frequency is enough high, when the blue-shifted sideband is blocked, the contribution enhanced by the gain-index coupling $-\frac{i\alpha}{2}\mathcal{G}^{opt}$ is almost in phase with the amplification $\mathcal{A}$: the resulting phase of the RF transfer function is continuous at the transition (see Fig. 7(b)). However, when the red-shifted sideband is blocked, the contribution enhanced by the gain-index coupling $+\frac{i\alpha}{2}\mathcal{G}^{opt}$ is almost in antiphase with the amplification $\mathcal{A}$: at the transition, the resulting phase of the RF transfer function presents almost a π phase jump, associated with a dip in the magnitude of the RF transfer function. Moreover, the frequency of the transition f_t is tunable with either the injected current or the input optical power: consequently at a fixed frequency f, the phase of the RF transfer function jump from the phase $arg(+\frac{i\alpha}{2}\mathcal{G}^{opt}) \sim \pi$ $(f < f_t)$ to $arg(\mathcal{A}) = 0$ $(f > f_t)$, when the input optical power or the injected current is changed. This interpretation explains well the experimental results.

3.2.3 Evaluation of the performances for a RF phase shifter

We have shown in the previous paragraph that the transition is frequency tunable with the injected current or the optical input power. Consequently, for a fixed frequency, it is possible to realize a RF phase shifter. This technique has been applied to realize a tunable opto-electronic oscillator (OEO) (Shumakher et al., 2009a) and a tunable microwave filter (Xue, Sales, Mork & Capmany, 2009). Due to the involved mechanisms, it is clear that without taken into account propagation effects, the maximal achievable tunable phase shift is π per component. A 2π phase shift has been realized with cascaded SOAs by Xue, Sales, Capmany & Mørk (2009). Owing to the high value of the enhancement linewidth factor, the transition frequency f_t, and then the operating frequency, can be increased up to $\sim$ 40GHz.

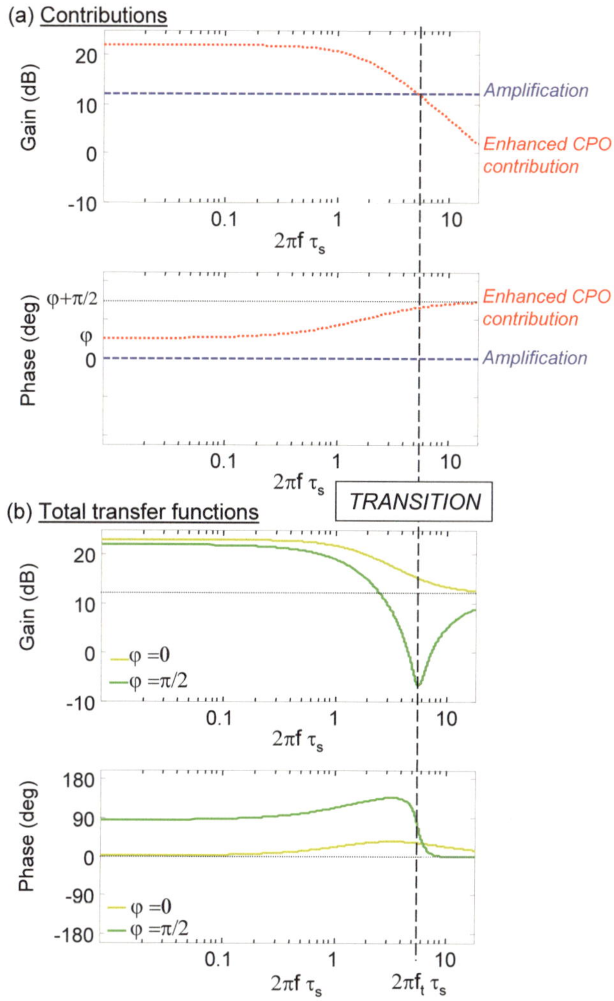

Fig. 7. (a) Contributions to the RF phase change: the red curves correspond both to either the CPO contribution enhanced by index-grating coupling $\frac{1}{2}\mathcal{G}^{opt}$ or the CPO contribution enhanced by forced CPO $\frac{1}{2}\mathcal{G}^{elec}$. When the red or blue shifted sideband is blocked: $\varphi = \pm\pi/2$; in the case of forced CPO architecture, $\varphi = \phi$. ϕ is the phase difference between the electronic and optical modulations. (b) Total transfer functions obtained from (a), when $\varphi = 0$ or $\pi/2$.

3.3 Forced CPO in SOA

We showed in the previous paragraph that the achievable phase shift experienced by an optically carried RF signal passing through the SOA has been recently increased up to π by optically filtering out the red-shifted modulation sideband before detection (Xue et al., 2008). However, this method involves the use of a very sharp optical filter and is consequently efficient mainly for relatively high frequencies (typically above 5 GHz). Furthermore, the insertion of the optical filter leads to a significant enhancement of noise (Shumakher et al., 2009b). Besides, four-wave mixing in SOA have been enhanced by modulating the current of the SOA (Capmany et al., 2002), but the resulting phase shift has never been studied before the recent theoretical study by Anton et al. (2009), which points out the enhancement of phase shift by forced CPO. The first experimental observation and a simple physical interpretation of forced CPO is presented in (Berger, Bourderionnet, de Valicourt, Brenot, Dolfi, Bretenaker & Alouini, 2010).

3.3.1 Set-up, experimental results and equations

The slow and fast light device is here a SOA whose current is modulated at the same RF frequency than the input optical power. The input microwave signal modulates then both the injected current of the SOA and the input optical power. An RF attenuator enables to control the modulation depth of the injected current (Fig. 8(a)). The phase difference between the two modulating signals is maintained to 90°. The current of the SOA is set to a fixed current (in the illustration of Fig. 8: 80 mA). The gain and the phase shift introduced by the SOA are then measured for different modulation depths of the injected current, by introducing an attenuation η on the modulation. The results are shown in Fig. 8(b). Two different regimes can be identified with respect to the frequency. At high frequencies, the response (gain and phase) of the modulated SOA is similar to the response of the non-modulated SOA (usual CPO behavior, represented by the black dashed curve). On the contrary, at low RF frequencies, forced CPO occur, and the phase tends to a value between 90° and 180° at low frequency. This brings us to define a transition frequency, f_t, below which the phase of the signal can reach 180° and above which the phase is close to 0°. These results are then very similar to those obtained by filtering out one of the sideband before the photodiode. Indeed we show in the following paragraph than the physical interpretation is almost the same.

Both the sidebands are detected by the photoreceiver, the equation of propagation is then the same as Eq. 22. However, CPO are not only induced by the modulated optical intensity $M_0 + M_1 e^{i\Omega t} + c.c.$, they are also induced by the modulated current $I = I_0 + \eta I_1 e^{i\Omega t + \phi} + c.c.$, with ϕ the phase shift with respect to the modulation of M_0. Consequently, Δg is expressed as:

$$\Delta g = \frac{I_1 e^{i\phi}/I_s - \bar{g} M_1/U_s}{1 + M_0/U_s - i\Omega\tau_s}, \tag{34}$$

with $U_s = \frac{\hbar\omega}{\Gamma a \tau_s}$ and $I_s = \frac{qV}{\Gamma a \tau_s}$.

3.3.2 Analysis of the different contributions. Physical interpretation

We integrate Eq. 22 by taking into account Eq. 34 over a small slice dz, whose length is noted L, to derive an analytical expression of the RF transfer function $S_{21} = \frac{M_1(L)}{M_1(0)}$:

$$\frac{M_1(L)}{M_1(0)} = \mathcal{A} + \frac{1}{2}\mathcal{G}^{opt} + \frac{1}{2}\mathcal{G}^{elec}, \tag{35}$$

with $\mathcal{A}$ and $\mathcal{G}^{opt}$ expressed as in the previous paragraphs and $\mathcal{G}^{elec}$ represents the contribution of forced CPO, generated by the modulated current:

$$\mathcal{G}^{elec} = \frac{1}{M_1(0)} \frac{I_1 e^{i\phi} M_0/I_s}{1 + M_0/U_s - i\Omega\tau_s}. \tag{36}$$

The module and the argument of the amplification term $\mathcal{A}$ and the dominant contribution due to CPO: $\frac{1}{2}\mathcal{G}^{elec}$ present the same behavior than in the architecture when one of the sideband is blocked (Fig. 7). The phase of the enhanced contribution goes from ϕ to $\phi + 90°$. Consequently,

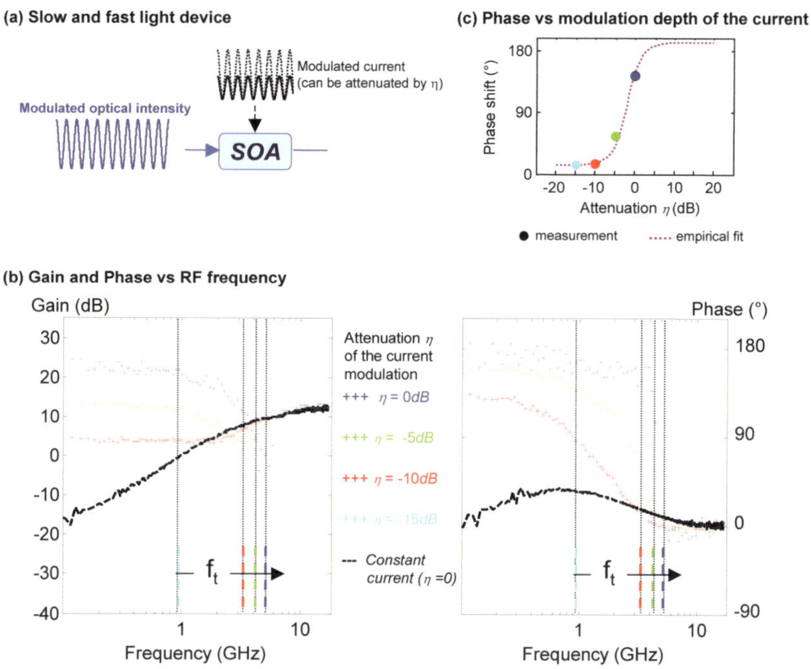

Fig. 8. Forced CPO architecture. (a) Slow and fast light device. (b) Experimental results: gain and phase shift of the RF signal, with respect to the RF frequency, at a fixed current (80 mA), and input optical power (1 mW), for various attenuation of the modulation of the current. (c) RF phase shifter at 3.2 GHz. Graph extracted from (Berger, Bourderionnet, de Valicourt, Brenot, Dolfi, Bretenaker & Alouini, 2010).

if the phase difference ϕ is set to $+90°$ (like in the experiment (Berger, Bourderionnet, de Valicourt, Brenot, Dolfi, Bretenaker & Alouini, 2010)), the configuration is similar to the architecture when the red-shifted sideband is blocked: at the transition between the dominant contribution due to CPO and the amplification, these latter are almost in antiphase, leading to an almost π phase shift and a dip in the gain. The frequency of the transition is tunable with the modulation depth of the current. In Fig. 8(b), we tag the transition frequency f_t for each attenuation.

3.3.3 Evaluation of the performances for a RF phase shifter

This behavior can be easily exploited to design an adjustable phase shifter, as shown in Fig. 8(c). In this example, the phase is controllable from 15° to 144° for a fixed frequency $f = 3.2$ GHz. These performances are similar to those achieved using sideband optical filtering before detection (Xue et al., 2008). In this architecture of forced CPO, a π phase shift could be achieved by increasing the current modulation depth. This could also increase the maximum operation frequency of the shifter. Indeed simulations show that a larger ratio between the modulation depth of the current and the optical modulation index would lead to a phase shifter whose operating frequency is close to 10 GHz. The main drawback of this architecture is that it involves the use of balanced path for the optical and electronic modulation, and a balanced hybrid $0° - 90°$ power divider, in order to conceive an optically tunable RF phase shifter.

4. Generalized CPO model (signal, distortion, noise) based on macroscopic parameters

In the previous part, we presented the main different architectures published in literature: we explained the essential physics of each of them through a physical interpretation and explained how it can be used as a tunable delay line or a tunable phase shifter. We also roughly evaluate the performances of these architectures as a phase shifter or delay line. Within the scope of integration in a realistic radar system, it is also required to study the impact of these slow and fast light architectures on the performances of the microwave photonics link. In particular, the RF transfer function, the generation of spurious signals by harmonic and intermodulation products, and the intensity noise, have to be studied in order to compute the Spurious Free Dynamic Range (SFDR), a key characteristic in microwave photonics. This part describes the predictive models to compute all of these figure of merits.

4.1 Model including dynamic saturation

We derived the equation of propagation in section 2.2 from a phenomenological model of the SOA explained in details by Agrawal & Dutta (1993). This phenomenological model relies on the assumption that both the variations of the material gain Δg and optical index Δn, caused by a small variation of the carrier density ΔN, are proportional to ΔN. This assumption is based on experimental observations in semiconductor lasers, and is justified in lasers above the threshold: the carrier density N changes little above threshold, and the linear variation is a reasonable approximation for small changes in N. Consequently, as far as SOAs are concerned, even if this phenomenological model explains well the essentials physics, as we show in the previous sections, it is not enough for an accurate and predictive model: indeed, this approximation does not give account of strong saturation conditions, with high gain and carrier density variations, which typically occur in quantum wells structures with strong carrier confinement, and can't describe in a predictive way the RF transfer response of the SOA for any operating conditions (injected current, input optical power).

We propose here to describe the model of Berger, Alouini, Bourderionnet, Bretenaker & Dolfi (2010) which takes into account the carrier density variation along the propagation axis and its influence on the differential gain a and the carrier lifetime τ_s. We chose to detail it because it relies on key ideas which makes the predictability and the accuracy of this model, but also of the following models about the distortion and the noise.

Our central hypothesis is that the differential gain a and the carrier lifetime τ_s can be determined as functions of the DC component of the optical intensity M_0 solely, allowing these dependencies to be determined from gain measurements.

Let us first suppose that we fulfill the small signal condition. In this case, the stimulated emission is negligible compared to the spontaneous emission, leading to the unsaturated steady state solution of the rate equation Eq. 11:

$$\frac{I}{q\,L\,S_{act}} = \frac{\bar{N}}{\tau_s}, \tag{37}$$

where L is the length of the SOA, S_{act} is the area of the active section of the SOA. Moreover, we also suppose in this case that the carrier density $\bar{N}$ is constant along the SOA. These hypothesis are justified in moderate length SOAs because the amplified spontaneous emission does not saturate the gain.

Under these conditions, a measurement of the small signal modal gain Γg_0 versus I will be equivalent, owing to Eq. 44, to a determination of the modal gain $\Gamma \bar{g}$ versus $\bar{N}/\tau_s$. Here, Γ is the ratio S_{act}/S of the active to modal gain areas in the SOA.

A last relationship between $\frac{\bar{N}}{\tau_s}$ and M_0 is then required to determine the modal gain $\Gamma \bar{g}$ as a function of M_0. It is obtained by substituting $\Gamma g(\frac{\bar{N}}{\tau_s})$ in the saturated steady state solution of the carriers rate equation Eq. 11:

$$\frac{I}{q\,L\,S_{act}} - \frac{\bar{N}}{\tau_s} - \frac{\Gamma \bar{g}(\frac{\bar{N}}{\tau_s})}{\hbar\omega}\frac{M_0}{\Gamma} = 0, \tag{38}$$

where the injected current I is now fixed by the operating conditions.

Added to the previous relationship between $\Gamma \bar{g}$ and $\frac{\bar{N}}{\tau_s}$, the Eq. 45 gives another expression of $\Gamma \bar{g}$ as a function of $\frac{\bar{N}}{\tau_s}$, $\frac{M_0}{\Gamma}$ and I. Consequently, $\Gamma \bar{g}$ and $\frac{\bar{N}}{\tau_s}$ can be known with respect to the local intensity $\frac{M_0(z)}{\Gamma}$ and the injected current I.

To solve Eqs. 19, we need to express $\bar{N}$ as a function of $\frac{M_0(z)}{\Gamma}$ and I. This is equivalent to express $\bar{N}$ with respect to $\frac{\bar{N}}{\tau_s}$ since $\frac{\bar{N}}{\tau_s}$ is known as a function of $\frac{M_0(z)}{\Gamma}$ and I. Consequently, we model our SOA using the well-known equation:

$$\frac{\bar{N}}{\tau_s} = A\bar{N} + B\bar{N}^2 + C\bar{N}^3, \tag{39}$$

where A, B, and C, which are respectively the non-radiative, spontaneous and Auger recombination coefficients, are the only parameters that will have to be fitted from the experimental results.

Using Eq. 39 and the fact that we have proved that $\bar{N}/\tau_s$ and $\Gamma \bar{g}$ can be considered as function of $\frac{M_0(z)}{\Gamma}$ and I only, we see that $\bar{N}$, $\Gamma a = \Gamma \frac{\partial \bar{g}}{\partial \bar{N}}$, and $\frac{U_s}{\Gamma} = \frac{\hbar\omega}{\Gamma a \tau_s}$ can also be considered as functions of $\frac{M_0(z)}{\Gamma}$ and I. This permits to replace Eqs. 19 by the following system:

$$\frac{dM_0}{dz} = \left(\Gamma \bar{g}(\frac{M_0(z)}{\Gamma}, I) - \gamma\right) M_0, \tag{40}$$

$$\frac{dE_1}{dz} = \frac{1}{2}\left(\Gamma \bar{g}(\frac{M_0(z)}{\Gamma}, I) - \gamma\right) E_1 + \frac{1 - i\alpha}{2}\Gamma \Delta g(\frac{M_0(z)}{\Gamma}, I) E_0, \tag{41}$$

$$\frac{dE_{-1}^*}{dz} = \frac{1}{2}\left(\bar{g}(\frac{M_0(z)}{\Gamma}, I) - \gamma\right) E_{-1}^* + \frac{1 + i\alpha}{2}\Gamma \Delta g(\frac{M_0(z)}{\Gamma}, I) E_0^*, \tag{42}$$

with:

$$\Delta g(\frac{M_0(z)}{\Gamma}, I) = \frac{M_1/U_s(\frac{M_0(z)}{\Gamma}, I)}{1 + \Gamma M_0/U_s(\frac{M_0(z)}{\Gamma}, I) - i\Omega\tau_s(\frac{M_0(z)}{\Gamma}, I)} \tag{43}$$

Eqs. 40, 41 and 42 are then numerically solved: Eq. 40 gives $\frac{M_0(z)}{\Gamma}$, with the initial condition $\frac{M_0(0)}{\Gamma} = \sqrt{\gamma_i}\frac{P_{in}}{S_{act}}$, where P_{in} is the optical input power. $\frac{M_0(z)}{\Gamma}$ can be then introduced into Eqs. 41, 42. It is then possible to simulate the optical fields E_1, E_{-1}, or the RF signal M_1 (which is equal either to $E_0^* E_1 + E_0 E_{-1}^*$, or to $E_0^* E_1$, or to $E_0 E_{-1}^*$, depending on the modulation format before the photodiode).

It is important to note that the recombination coefficients A, B and C are the only fitting parameters of this model. Once obtained from experimental data, they are fixed for any other experimental conditions. Moreover, the only geometrical parameters that are required are the length L of the SOA and the active area cross section S_{act}. The derivation of a predictive model, independent of the experimental conditions (current and input optical power) is then possible, provided that the simple measurements of the total losses and the small signal gain versus the current are conducted. The above model lies in the fact that first, the spatial variations of the saturation parameters are taken into account, and second, their values with respect to the local optical power are deduced from a simple measurement. These keys ideas lead to a very convenient model of the microwave complex transfer function of the SOA, and then of the slow light properties of the component. It can be easily used to characterize commercial components whose design details are usually unknown. We illustrate the accuracy and the robustness of the model in the part 5. Lastly, it is worth mentioning that in order to compute the complex transfer function of an architecture including a SOA and a filter, the complex transfer function of the filter has to be then applied to the output field compounds E_k computed by the previous model (Dúill et al., 2010a).

4.2 Distortion model

The model we present in this part is a generalization of the former one. It enables to take into account higher order coherent population oscillations due to large signal modulation, or the non-linearities at the input of the SOA (from the Mach-Zehnder that modulates the optical beam for example), and can be used to compute the harmonic generation and the intermodulation products. The detailed model is presented in (Berger, Bourderionnet, Alouini, Bretenaker & Dolfi, 2009).

4.2.1 Harmonic generation

In order to find the level of the generated harmonics, we first consider that the input optical field is modulated at the RF frequency Ω. $|E|^2$, g and N are hence all time-periodic functions with a fundamental frequency of Ω. They can therefore be written into Fourier harmonic decompositions:

$$|E(z,t)|^2 = \sum_{k=-\infty}^{+\infty} M_k(z)e^{-ik\Omega t}, \tag{44}$$

$$N(z,t) = \bar{N}(z) + \sum_{\substack{k=-\infty \\ k\neq 0}}^{+\infty} N_k(z)e^{-ik\Omega t}, \tag{45}$$

$$g(z,t) = \bar{g}(z) + a(z) \sum_{\substack{k=-\infty \\ k\neq 0}}^{+\infty} N_k(z)e^{-ik\Omega t} \tag{46}$$

where $\bar{N}(z)$ and $\bar{g}(z)$ respectively denote the DC components of the carrier density and of the optical gain. $a(z)$ is the SOA differential gain, defined as $a(z) = \partial\bar{g}/\partial\bar{N}$. Defining g_k as the oscillating component of the gain at frequency $k\Omega$, and considering only a finite number K of harmonics, the carrier rate equation (Eq. 11) can be written as:

$$\hbar\omega\left(\frac{I}{qV} - \frac{\bar{N}}{\tau_s}\right) = \alpha_0\bar{g} + \sum_{\substack{p+q=0 \\ p,q\in[-K,K] \\ p\neq 0}} g_p M_q, \tag{47}$$

$$0 = \alpha_i g_i + \sum_{\substack{p+k=i \\ p,q\in[-K,K] \\ p\neq i}} g_p M_q, \quad \text{for } i \neq 0 \text{ and } i \in [-K, K] \tag{48}$$

where $\alpha_k = U_s(1 + M_0/I_s - ik\Omega\tau_s)$, and $\alpha_0 = M_0$ is the DC optical intensity. U_s denotes the local saturation intensity and is defined as $U_s = \hbar\omega/a\tau_s$. It is worth mentioning that α_k is obtained at the first order of equation (Eq. 11), when mixing terms are not considered. It is important to note that in the following, $\bar{N}$, $\bar{g}$, a, τ_s, U_s, and consequently the α_k's are all actually functions of z. Their variations along the propagation axis is then taken into account, unlike most of the reported models in which effective parameters are used (Agrawal, 1988; Mørk et al., 2005; Su & Chuang, 2006).
In order to preserve the predictability of the model, $\bar{g}$, U_s and τ_s has to be obtained as in the small signal case. However, in the case of a large modulation index, an iterative procedure has to be used: in a first step, we substitute $\bar{N}/\tau_s$, U_s and τ_s in (47) by their small signal values $\bar{N}/\tau_s^{(0)}$, $U_s^{(0)}$ and $\tau_s^{(0)}$. The gain components $\bar{g}$ and g_k can be then extracted from Eqs. 47 and 48. Similarly to the small signal case, using equations (39) and (47), we obtain $\bar{N}/\tau_s^{(1)}$, $U_s^{(1)}$ and $\tau_s^{(1)}$ as functions of I, A, B, C and $M_k(z)$. This procedure is repeated until convergence of $\bar{N}/\tau_s^{(n)}$, $U_s^{(n)}$ and $\tau_s^{(n)}$, which typically occurs after a few tens of iterations.
The propagation equation (Eq 19) can now be expressed as:

$$\frac{dE_k}{dz} = \frac{1}{2}(\bar{g} - \gamma_i) E_k + \frac{1 - i\alpha}{2} \sum_{\substack{p+q=k, \\ -K<p,q<K}} \Gamma g_p E_q, \tag{49}$$

From these equations it is straightforward to deduce the equation for the component M_k of the optical intensity, either if the modulation is single-sideband or double-sideband. For numerical simulations, it is very useful to express the Eqs. 47, 48 and 49 in a matrix formulation. The expressions can be found in (Berger, Bourderionnet, Alouini, Bretenaker & Dolfi, 2009).
In the case of a real microwave photonics link, the harmonics at the input of the SOA, created by the modulator, has to be taken into account. By using the reported model, the third harmonic photodetected power, can be evaluated with:

$$H_3 = 2R\eta_{ph}^2 |M_{3,out} \times S|^2 \tag{50}$$

where R and η_{ph} are respectively the photodiode resistive load (usually 50Ω) and efficiency (assumed to be equal to 1). S denotes the SOA modal area.

4.2.2 Intermodulation distortion

Intermodulation distortion (IMD) calculation is slightly different from what has been discussed in the above section. Indeed, the number of mixing terms that must be taken into account is significantly higher. For radar applications a typical situation where the IMD plays a crucial role is that of a radar emitting at a RF frequency Ω_1, and facing a jammer emitting at Ω_2, close to Ω_1. Both Ω_1 and Ω_2 are collected by the antenna and transferred to the optical carrier through a single electro-optic modulator. The point is then to determine the nonlinear frequency mixing due to the CPO inside the SOA. In particular, the mixing products at frequencies $\Omega_2 - \Omega_1$ (or $\Omega_1 - \Omega_2$) and $2\Omega_2 - \Omega_1$ (or $2\Omega_1 - \Omega_2$) — respectively called second

(IMD$_2$) and third (IMD$_3$) order intermodulation distortions — have to be evaluated at the output of the SOA.
The main difference with harmonic calculation is that the optical intensity, and hence the SOA carrier density N, and the SOA gain g are no longer time-periodic functions of period Ω, but also of period $\delta\Omega = \Omega_2 - \Omega_1$.

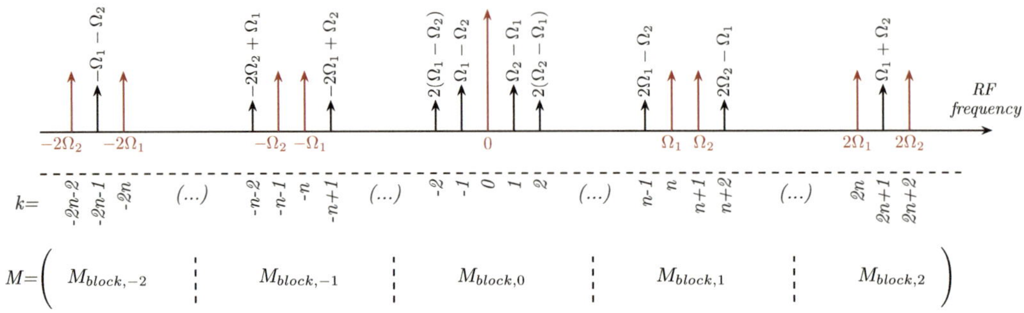

Fig. 9. Set of significant spectral components of $|E|^2$, N and g, and associated index k in their Fourier decompositions. n is defined such as $\Omega_1 = n\delta\Omega$. Graph extracted from (Berger, Bourderionnet, Alouini, Bretenaker & Dolfi, 2009).

We consider a typical radar frequency Ω_1 of 10GHz, and a frequency spacing $\delta\Omega$ of 10MHz. Here, for intermodulation distortion calculation, we assume that only the spectral components at $\Omega_{1,2}$, $2\Omega_{1,2}$, and all their first order mixing products significantly contribute to the generation of IMD$_2$ and IMD$_3$, as illustrated in figure 9. The M_k's and the g_k's are then reduced in 19 elements vectors which can be gathered into blocks, the jth block containing the mixing products with frequencies close to $j \times \Omega_1$. The Eqs. 47, 48 and 49 can be then be written as matrices in block, and the full procedure described in the previous can be applied in the same iterative way to determine the g_k's, U_s and τ_s, and to finally numerically solve the equation (49). Detailed matrices are presented in (Berger, Bourderionnet, Alouini, Bretenaker & Dolfi, 2009). Similarly to equation (50), the photodetected RF power at $2\Omega_2 - \Omega_1$ is then calculated through:

$$\mathrm{IMD}_3 = 2R\eta_{ph}^2 |M^{out}_{2\Omega_2-\Omega_1} \times S|^2 . \tag{51}$$

We explained in this section how to adapt the predictive small-signal model including dynamic saturation, in order to compute the harmonics and the intermodulation products, while keeping the accuracy and predictability of the model. It is worth noticing that in a general way, the propagation of the Fourier compounds of an optically carried microwave signal into the SOA can be seen as resulting from an amplification process and a generation process by frequency mixing through CPO. We will see in part 5 how these two effects, which are in antiphase, can be advantageously used to linearize a microwave photonics link.
In order to compute the dynamic range of a microwave photonics link, the only missing characteristic is the intensity noise.

4.3 Intensity noise

The additional intensity noise can be extracted from the model of the RF transfer function described in section 4.1. The principle is detailed in (Berger, Alouini, Bourderionnet, Bretenaker & Dolfi, 2009b). Indeed, when the noise is described in the semi-classical beating

theory, the fields contributing to the intensity noise are the optical carrier and the spontaneous emission. We define the input spontaneous emission power density as the quantum noise source at the input of SOA, which can be extracted from a measurement of the optical noise factor. The input intensity is then composed of:
(1) a spontaneous-spontaneous beat-note which is only responsive to the optical gain.
(2) a carrier-spontaneous beat-note, which can be considered as an optical carrier and a sum of double-sideband modulation components at the frequency Ω (Olsson, 1989). However, the right-shifted and the blue-shifted sidebands at Ω are incoherent. Consequently, the double sidebands at Ω has to be taken into account as two independent single-sideband modulations. Their respective contributions to the output intensity noise can be then computed from the model of the RF transfer function described in section 4.1. All the contributions are finally incoherently summed.
The relative intensity noise and the noise spectral density can be then easily modeled from the RF transfer function described in section 4.1. It is interesting to observe that first this model leads to an accurate description of the output intensity noise (Berger, Alouini, Bourderionnet, Bretenaker & Dolfi, 2009b). Secondly, we can show that the relative intensity noise after a SOA (without optical filter) is proportional to the RF transfer function, leading to an almost constant carrier-to-noise ratio with respect to the RF frequency (Berger, Alouini, Bourderionnet, Bretenaker & Dolfi, 2009a): the dip in the gain associated to tunable delays, does not degrade the carrier-to-noise ratio. However, it is not anymore valid when an optical filter is added before the photodiode (Duill et al., 2010b; Lloret et al., 2010), due to the incoherent sum of the different noise contributions.

5. Dynamic range of slow and fast light based SOA link, used as a phase shifter

We focus here on the study of a single stage phase shifter consisting of a SOA followed by an optical notch filter (ONF), which attenuates the red shifted modulation sideband (see section 3.2). In order to be integrated in a real radar system, the influence of such an architecture on the microwave photonics link dynamic range has to be studied. The large phase shift obtained by red sideband filtering is however accompanied by a significant amplitude reduction of the RF signal at the phase jump. An important issue in evaluating the merits of the filtering approach is its effect on the linearity of the link. Indeed, similarly to the fundamental signal whose characteristics evolve with the degree of filtering, it is expected that attenuating the red part of the spectrum should affect the nonlinear behavior of the CPO based phase shifter. The nonlinearity we consider here is the third order intermodulation product (IMD3). This nonlinearity accounts for the nonlinear mixing between neighboring frequencies f_1 and f_2 of the RF spectrum, and refers to the detected RF power at frequencies $2f_2 - f_1$ and $2f_1 - f_2$. Since these two frequencies are close to f_1 and f_2, this quantity is of particular importance in radar and analog transmission applications, where IMD3 is the dominant detrimental effect for MWP links (Ackerman, 1994).
To this aim, the predictions of the model presented in the previous part are compared with experimental results (RF complex transfer function, intermodulation products IMD3). Then we use our predictive model to find out the guidelines to optimize a microwave photonics link including a SOA based phase shifter.

5.1 Experimental confirmation of the model predictions

The experimental set-up for IMD3 measurement is depicted on Fig. 10. The RF tones are generated by two RF synthesizers at $f_1 = 10\,\text{GHz}$ and $f_2 = 10.01\,\text{GHz}$. The two RF signals are

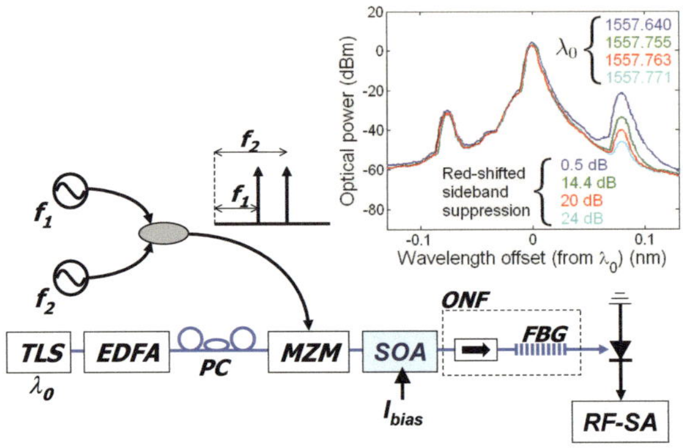

Fig. 10. Experimental setup for IMD3 measurement. EDFA, Erbium-Doped Fiber Amplifier; PC, Polarization Controller. The red-shifted sideband attenuation is varied from 0.5 dB to 24 dB (inset). Extracted from (Berger, Bourderionnet, Bretenaker, Dolfi, Dúill, Eisenstein & Alouini, 2010).

combined to drive a zero-chirp Mach-Zehnder modulator (MZM). A single frequency tunable laser source (TLS) feeds the MZM. The optically carried signal is coupled to the phase shifting element that comprises a commercial SOA, an optical notch filter made of an isolator, and a fiber Bragg grating (FBG). The 3 dB spectral bandwidth of the FBG is 0.2nm. The SOA gain and the output saturation power are respectively 21.3 dB and 15 dBm at 500 mA bias current. The optical power at the SOA input was set to 10 dBm which ensures operation in strong saturation conditions, which is favorable for phase shifting applications (Shumakher et al., 2009b). The various degrees of red-shifted optical sideband filtering are obtained by tuning the optical carrier wavelength with respect to the central wavelength of the notch filter (see Fig. 10, inset). Finally, to get rid of all photodiode nonlinearities, an optical attenuator was placed before the detector. The detected RF signal is then sent to a RF spectrum analyzer which records the output RF power at the four frequencies f_1, f_2, $2f_2 - f_1$ and $2f_1 - f_2$ for IMD3 evaluation. The RF phase shift at 10 GHz is also measured using a vector network analyzer (VNA). For our measurements, we considered four different wavelengths, that is, four degrees of optical filtering (see inset of Fig. 10).

The plots presented in Fig. 11 correspond respectively from left to right to a red-shifted sideband optical attenuation of 0.5, 14.4, 20 and 24 dB. The model predictions are in very good agreement with the experimental results, for the RF phase, the RF output power and the output IMD3 power.

The most important result of Fig. 11 is that theory and experiment agree remarkably well to evidence the presence of a dip in power-versus-current characteristics of both the RF signal and the IDM3. Although the two dips do not occur at the same bias current (respectively 200 mA and 120 mA for fundamental RF signal and IMD3), they behave similarly with respect to the degree of filtering. Both dips decrease as the red-shifted sideband filtering diminishes, and completely disappear in the absence of filtering. Indeed, as we explained in section, the filtering reveals the index-gain coupling. Near the dip, the output signal comes equally from the amplification of the input signal (contribution in phase), and from CPO enhanced by index-coupling, (contribution almost in antiphase at 10 GHz) (see Fig. 7).

However, the dip in the IMD3 doesn't lead directly to an improvement of the dynamic range as it doesn't appear for the same operating conditions than the phase shift. A trade-off has to be found between the available phase shift and the dynamic range.

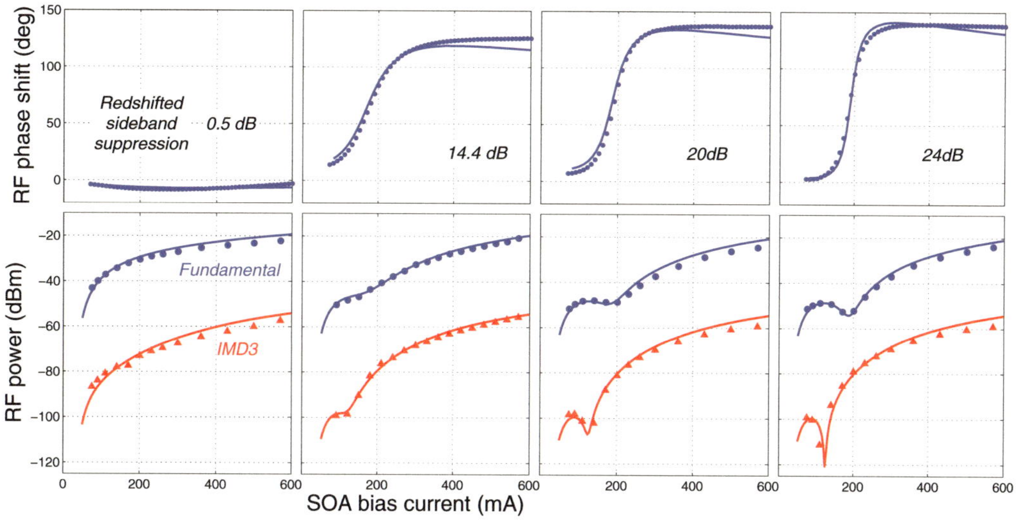

Fig. 11. Top: RF phase shift at 10 GHz versus SOA bias current; Bottom: RF power at fundamental frequency f_1 (in blue), and at $2f_2 - f_1$, (IMD3, in red). From left to right, red-shifted sideband attenuation increases from 0.5 dB to 24 dB. Symbols represent experimental measurements, and solid lines show theoretical calculations. Extracted from (Berger, Bourderionnet, Bretenaker, Dolfi, Dúill, Eisenstein & Alouini, 2010).

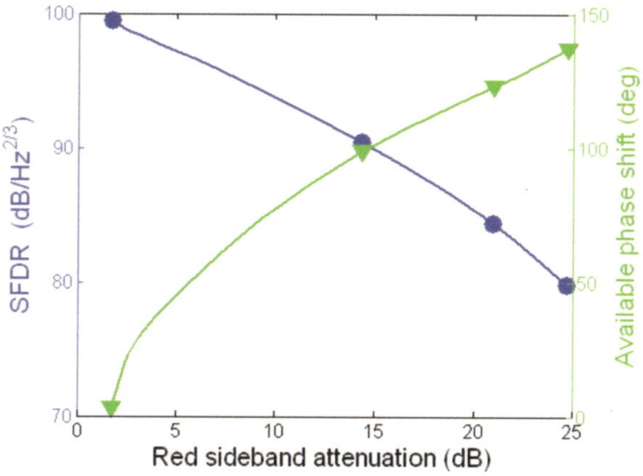

Fig. 12. In blue: Spurious Free Dynamic Range (SFDR); in green: available phase shift. Both are represented with respect to the red sideband attenuation. The model prediction is represented by a line, the dots are the experimental points.

5.2 Influence of the optical filtering on the performance of the phase shifter

To this aim, we compute the Spurious-Free Dynamic Range (SFDR), which is the key figure of the dynamic range in microwave photonics (Ackerman, 1994). It is defined as the RF power range where the intermodulation products IMD3 are below the noise floor. We represent in Fig. 12 the SFDR and the available phase shift with respect to the red sideband attenuation. It appears that the best trade-off between the dynamic range and the available phase shift corresponds to the minimum strength of filtering which enables to reveal the index-gain coupling. With this non-optimized link, we reach a SFDR of $90dB/Hz^{2/3}$ for an available phase shift of 100 degrees.

5.3 Linearized amplification at high frequency

In a more general context, a SOA can be used to reduces the non-linearities of a microwave photonics link. Indeed, the input linearities (from the modulator for example) can be reduced by the nonlinearities generated by the gain in antiphase created by the CPO. It has already been demonstrated using a single SOA (without optical filter) at low frequency (2 GHz) (Jeon et al., 2002)). However with a single SOA, the gain in antiphase due to CPO is created only at low frequency (below a few GHz), as it is illustrated on Fig. 5. However, when the SOA is followed by an optical filter attenuating the red-shifted sideband, the gain in antiphase is created at high frequency, as it is illustrated on Fig. 7. This architecture enables then a linearization of the microwave photonics link well beyond the inverse of the carrier lifetime. Indeed we have experimentally demonstrated that a dip in the IMD3 occurs at 10 GHz (Fig. 11). However the instantaneous bandwidth is still limited to the GHz range.

6. Conclusion

We have reviewed the different set-ups proposed in literature, and we have given the physical interpretation of each architecture, aiming at helping the reader to understand the underlying physical mechanisms.

Moreover, we have shown that a robust and predictive model can be derived in order to simulate and understand the RF transfer function, the generation of spurious signals through harmonic distortion and intermodulation products, and the intensity noise at the output of a SOA. This model takes into account the dynamic saturation along the propagation in the SOA, which can be fully characterized by a simple measurement, and only relies on material fitting parameters, independent of the optical intensity and the injected current. In these conditions, the model is found to be predictive and can be used to simulate commercial SOAs as well. Moreover, we have presented a generalization of the previous model, which permits to describe harmonic generation and intermodulation distortions in SOAs. This model uses a rigorous expression of the gain harmonics. Lastly, we showed the possibility to use this generalized model of the RF transfer function to describe the intensity noise at the output of the SOA.

This useful tool enables to optimize a microwave photonics link including a SOA, by finding the best operating conditions according to the application. To illustrate this point, the model is used to find out the guidelines for improving the MWP link dynamic range using a SOA followed by an optical filter, in two cases: first, for phase shifting applications, we have shown that the best trade-off between the dynamic range and the available phase shift corresponds to the minimum strength of filtering which enables to reveal the index-gain coupling. Second, we have experimentally demonstrated and have theoretically explained how an architecture composed of a SOA followed by an optical filter can reduce the non-linearities of the modulator, at high frequency, namely beyond the inverse of the carrier lifetime.

7. References

Ackerman, E. (1994). *Photonic Aspects of modern radar*, Artech House.

Agrawal, G. P. (1988). Population pulsations and nondegenerate four-wave mixing in semiconductor lasers and amplifiers, *J. Opt. Soc. Am. B* 5(1): 147–159.

Agrawal, G. P. & Dutta, N. K. (1993). *Semiconductor Lasers*, Kluwer Academic, Boston.

Anton, M. A., Carreno, F., Calderon, O. G., Melle, S. & Arrieta-Yanez, F. (2009). Phase-controlled slow and fast light in current-modulated SOA, *Journal of Physics B: Atomic, Molecular and Optical Physics* 42(9): 095403 (8pp).

Berger, P., Alouini, M., Bourderionnet, J., Bretenaker, F. & Dolfi, D. (2009a). Influence of slow light effect in semiconductor amplifiers on the dynamic range of microwave-photonics links, *Slow and Fast Light*, Optical Society of America, p. SMB6.

Berger, P., Alouini, M., Bourderionnet, J., Bretenaker, F. & Dolfi, D. (2009b). Slow light using semiconductor optical amplifiers: Model and noise characteristics, *C. R. Physique* 10: 991–999.

Berger, P., Alouini, M., Bourderionnet, J., Bretenaker, F. & Dolfi, D. (2010). Dynamic saturation in semiconductoroptical amplifiers: accurate model, roleof carrier density, and slow light, *Opt. Express* 18(2): 685–693.

Berger, P., Bourderionnet, J., Alouini, M., Bretenaker, F. & Dolfi, D. (2009). Theoretical study of the spurious-free dynamic range of a tunable delay line based on slow light in soa, *Opt. Express* 17(22): 20584–20597.

Berger, P., Bourderionnet, J., Bretenaker, F., Dolfi, D., Dúill, S. O., Eisenstein, G. & Alouini, M. (2010). Intermodulation distortion in microwave phase shifters based on slow and fast light propagation in SOA, *Opt. Lett.* 35(16): 2762–2764.

Berger, P., Bourderionnet, J., de Valicourt, G., Brenot, R., Dolfi, D., Bretenaker, F. & Alouini, M. (2010). Experimental demonstration of enhanced slow and fast light by forced coherent population oscillations in a semiconductor optical amplifier, *Opt. Lett.* 35: 2457.

Bogatov, A. P., Eliseev, P. G. & Sverdlov, B. N. (1975). Anomalous interaction of spectral modes in a semiconductor laser, *Quantum Electronics, IEEE Journal of* 11: 510.

Boula-Picard, R., Alouini, M., Lopez, J., Vodjdani, N. & Simon, J.-C. (2005). Impact of the gain saturation dynamics in semiconductor optical amplifiers on the characteristics of an analog optical link,

Capmany, J., Sales, S., Pastor, D. & Ortega, B. (2002). Optical mixing of microwave signals in a nonlinear semiconductor laser amplifier modulator, *Opt. Express* 10(3): 183–189.

Dúill, S. O., Shumakher, E. & Eisenstein, G. (2010a). The role of optical filtering in microwave phase shifting, *Opt. Lett.* 35(13): 2278–2280.

Duill, S., Shumakher, E. & Eisenstein, G. (2010b). Noise properties of microwave phase shifters based on SOA, *Lightwave Technology, Journal of* 28(5): 791 –797.

Henry, C. (1982). Theory of the linewidth of semiconductor lasers, *Quantum Electronics, IEEE Journal of* 18(2): 259 – 264.

Jeon, D.-H., Jung, H.-D. & Han, S.-K. (2002). Mitigation of dispersion-induced effects using soa in analog optical transmission, *Photonics Technology Letters, IEEE* 14(8): 1166 – 1168.

Lloret, J., Ramos, F., Sancho, J., Gasulla, I., Sales, S. & Capmany, J. (2010). Noise spectrum characterization of slow light soa-based microwave photonic phase shifters, *Photonics Technology Letters, IEEE* 22(13): 1005 –1007.

Mørk, J., Kjær, R., van der Poel, M. & Yvind, K. (2005). Slow light in a semiconductor waveguide at gigahertz frequencies, *Opt. Express* 13(20): 8136–8145.

Olsson, N. (1989). Lightwave systems with optical amplifiers, *Lightwave Technology, Journal of* 7(7): 1071–1082.

Peatross, J., Glasgow, S. A. & Ware, M. (2000). Average energy flow of optical pulses in dispersive media, *Phys. Rev. Lett.* 84(11): 2370–2373.

Pesala, B., Chen, Z., Uskov, A. V. & Chang-Hasnain, C. (2006). Experimental demonstration of slow and superluminal light in semiconductor optical amplifiers, *Opt. Express* 14(26): 12968–12975.

Shumakher, E., Duill, S. & Eisenstein, G. (2009a). Optoelectronic oscillator tunable by an soa based slow light element, *Lightwave Technology, Journal of* 27(18): 4063–4068.

Shumakher, E., Dúill, S. O. & Eisenstein, G. (2009b). Signal-to-noise ratio of a semiconductor optical-amplifier-based optical phase shifter, *Opt. Lett.* 34(13): 1940–1942.

Su, H. & Chuang, S. L. (2006). Room temperature slow and fast light in quantum-dot semiconductor optical amplifiers, *Applied Physics Letters* 88(6): 061102.

Xue, W., Chen, Y., Öhman, F., Sales, S. & Mørk, J. (2008). Enhancing light slow-down in semiconductor optical amplifiers by optical filtering, *Opt. Lett.* 33(10): 1084–1086.

Xue, W., Sales, S., Capmany, J. & Mørk, J. (2009). Experimental demonstration of 360otunable rf phase shift using slow and fast light effects, *Slow and Fast Light*, Optical Society of America, p. SMB6.

Xue, W., Sales, S., Mork, J. & Capmany, J. (2009). Widely tunable microwave photonic notch filter based on slow and fast light effects, *Photonics Technology Letters, IEEE* 21(3): 167–169.

10

Photonic Integrated Semiconductor Optical Amplifier Switch Circuits

R. Stabile and K.A. Williams
Eindhoven University of Technology
The Netherlands

1. Introduction

The acceptance of pervasive digital media has placed society in the Exabyte era (10^{15} Bytes). However the data centres and switching technologies at the heart of the Internet have led to an industry with CO_2 emissions comparable to aviation (Congress 2007). Electronics now struggles with bandwidth and power. Electronic processor speeds had historically followed Gordon Moore's exponential law (Roadmap 2005), but have recently limited at a few thousand Megahertz. Chips now get too hot to operate efficiently at higher speed and thus performance gains are achieved by running increasing numbers of moderate speed circuits in parallel. A bottleneck is now emerging in the interconnection network. As interconnection is increasingly performed in the optical domain, it is increasingly attractive to introduce photonic switching technology. While there is still considerable debate with regard to the precise role for photonics (Huang et al., 2003; Grubb et al., 2006; Tucker, 2008; Miller 2010), new power-efficient, cost-effective and broadband approaches are actively pursued.
Supercomputers and data centers already deploy photonics to simplify and manage interconnection and are set to benefit from progress in parallel optical interconnects (Adamiecki et al., 2005; Buckman et al., 2004; Lemoff et al., 2004; Patel et al., 2003; Lemoff et al., 2005; Shares et al., 2006; Dangel et al., 2008). However, it is much more efficient to route the data over reconfigurable wiring, than to overprovision the optical wiring. Wavelength domain routing has been seen by many as the means to add such reconfigurability. Fast tuneable lasers (Gripp et al., 2003) and tuneable wavelength converters (Nicholes et al., 2010) have made significant progress, although bandwidth and connectivity remain restrictive so far. All-optical techniques have been considered to make the required step-change in processing speeds. Nonlinearities accessible with high optical powers and high electrical currents in semiconductor optical amplifiers (SOAs) create mixing products which can copy broadband information photonically (Stubkjaer, 2000; Ellis et al., 1995; Spiekman et al., 2000). When used with a suitable filter, these effects can be exploited to create photonic switches and even logic. However, the required combination of high power lasers, high current SOAs and tight tolerance filters is a very difficult one to integrate and scale. Hybrid electronic and photonic switching approaches (Chiaroni et al., 2010) are increasingly studied to perform broadband signal processing functions in the simplist and most power-efficient manner while managing deep memory and high computation functions electronically. This can still reduce network delay and remove power-consuming optical-electronic-optical conversions (Masetti et

al., 2003; Chiaroni et al., 2004). The SOA gate has provided the underlying switch element for the many of these demonstrators, leading to a new class of bufferless photonic switch which assumes (Shacham et al., 2005; Lin et al., 2005; Glick et al., 2005) or implements (Hemenway et al., 2004) buffering at the edge of the photonic network. Such approaches become more acceptable in short-reach computer networking where each connection already offers considerable buffering (McAuley, 2003). Formidable challenges still remain in terms of bandwidth, cost, connectivity, and energy footprint, but photonic integration is now striving to deliver in many of these areas (Grubb et al, 2006; Maxwell, 2006; Nagarajan & Smit, 2007).

This chapter addresses the engineering of SOA gates for high-connectivity integrated photonic switching circuits. Section 2 reviews the characteristics of the SOA gates themselves, considering signal integrity, bandwidth and energy efficiency. Section 3 gives a quantitative insight into the performance of SOA gates in meshed networks, addressing noise, distortion and crosstalk. Section 4 reviews the scalability of single stage integrated switches before considering recent progress in monolithic multi-stage interconnection networks in Section 5. Section 6 provides an outlook.

2. SOA gates

SOA gates exhibit a multi-Terahertz bandwidth which may be switched from a high-gain state to a high-loss state within a nanosecond using low-voltage electronics. The electronic structure is that of a diode, typically with a low sub-Volt turn on voltage and series resistance of a few Ohms. Photonic switching circuits using SOAs have therefore been relatively straight forward to implement in the laboratory. The required electrical power for the SOA gate is largely independent of the optical signal, thus breaking the link between rising energy consumption and rising line-rate which plagues electronics. SOA gates and the underlying III-V technologies also bring the ability to integrate broadband controllable gain elements with the broadest range of photonic components. A wide range of optical switch concepts based on SOAs have already been proposed to facilitate nanosecond timescale path reconfiguration (Renaud et al., 1996; Williams, 2007) performing favourably with the even broader range of high speed photonic techniques (Williams et al., 2005). Now we review the state of the art for the SOA gate technology itself, highlighting system level metrics in terms of signal integrity, bandwidth and power efficiency.

2.1 Signal integrity

The broadband optical signal into an amplifying SOA gate potentially accrues noise and distortion in amplitude and phase. Noise degrades signal integrity for very low optical input powers, while distortion can limit very high input power operation. The useful intermediate operating range, commonly described as the input power dynamic range (Wolfson, 1999), is therefore maximised through the reduction of the noise figure and increase in the distortion threshold. The signal degradation is generally characterised in terms of the additional signal power penalty required to maintain received signal integrity. Figure 1 quantifies power penalty degradation in terms of noise at low optical input powers and distortion at high optical input powers for the case of a two input two output 2x2 SOA switch fabric (Williams, 2006).

Noise originates primarily from the amplified spontaneous emission inherent in the on-state SOA gate. The treatment for optical systems has been most comprehensively treated for fiber amplifier circuits (Desurvire, 1994). The interactions of signals, shot noise, amplified

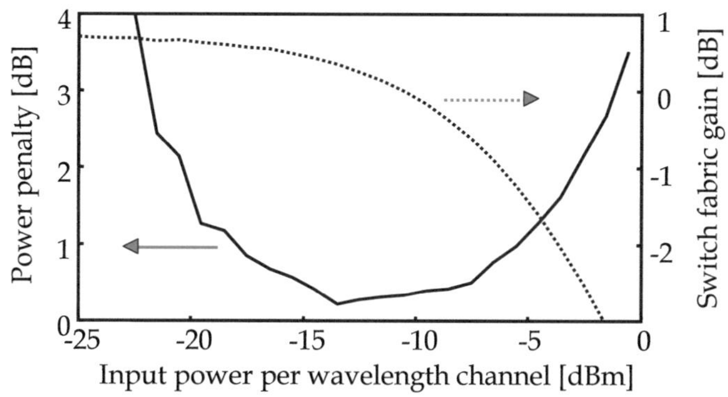

Fig. 1. Simulated input power dynamic range for a 2x2 SOA switch fabric (Williams, 2006)

spontaneous emission noise and the respective beat terms can require careful filtering and bandwidth management to ensure optimum performance. The alignment of optical signals with respect to the gain spectrum also impacts performance through the degree of population inversion. Noise may be managed through the minimisation of loss and the reduced requirement for high current amplifiers (Lord & Stallard, 1989). State of the art noise figures for fiber-coupled SOAs are of the order 6-8dB (Borghesani et al., 2003), depending on whether the structure is optimised for low-power input signals (pre-amplifiers) or power booster amplifiers (post-amplifiers). These values are higher than for fiber amplifiers, due to the losses in fiber to chip coupling and imperfect population inversion. The design focus has therefore been on reducing losses (Morito et al., 2005).

Distortion in the saturation regime results from the charge carrier depletion from the incoming data signal. When optical data signals are amplitude-modulated (on-off keyed), the signal can deplete charge carriers and therefore reduce gain on the timescale of the spontaneous lifetime. This leads to the time dependent patterning and therefore nonlinear distortions on the optical output signal waveform. This can be alleviated by changing the data format: Proposals range from wavelength keying (Ho et al., 1996; Kim & Chandrasekhar, 2000), wavelength domain power averaging (Mikkelsen et al., 2000; Shao et al., 1994), and wavelength coding (Roberts et al., 2005) for on-off keyed modulation. Increasingly popular constant power envelope formats (Wei et al., 2004; Cho et al., 2004; Ciaramella et al., 2008, Winzer, 2009) are also more resilient. Distortion is less evident for very low data rates where bit periods exceed the nanosecond time-scale spontaneous lifetime, and also for very high data rates where the longest sequence of bits are shorter than the spontaneous lifetime. Indeed, the optical transfer function can be considered as a notch filter and this mode of operation has already been exploited for noise suppression (Sato & Toba, 2001).

Pseudo random bit sequences are routinely used to assess data transmission. The longer 2^{31} patterns have been particularly important for point to point telecommunications links to stress-test all elements for the broadest bandwidth. The longest sequence of ones in a 2^{31} pattern remains at the same level for over 3ns for a 10Gbit/s sequence, and is thus sensitive to patterning (Burmeister & Bowers, 2006). However line rates of 100Gbit/s and above would lead to maximum length sequences shorter than the spontaneous lifetime. For higher line rates still, sophisticated optical multiplexing schemes are devised, and the concept of the pattern length becomes less meaningful: Wavelength multiplexing measurements commonly decorrelate replicas of the same signals (Lin et al., 2007), while optically multiplexed signals use calibrated interleavers available only for the shortest 2^7 pattern sequences (Albores et al., 2009). Packet switched test-beds impose more fundamental constraints: a 2^{31} sequence contains over two billion bits, far exceeding any likely data packet length. Codes for receiver power balancing and packet checking also limit the effective pattern lengths, and therefore shorter sequences are commonly used.

Techniques to increase the distortion threshold are readily understood through a manipulation of the steady state charge carrier rate equation. Equation 1 approximates the rate of change of charge carriers (left) in terms of the injected current, stimulated amplification, and spontaneous emission (right). The steady state condition is defined when the derivative tends to zero ($dN/dt \rightarrow 0$).

$$dN/dt = I/eV - \Gamma dg/dn(N-N_0)P - N/\tau_s \rightarrow 0 \tag{1}$$

The terms in Equation 1 correspond to the injected current I into active volume V. N represents the charge carrier density, Γ is the optical overlap integral describing the proportion of amplified light which overlaps with the active layer. dg/dn is the differential gain and N_0 is the transparency carrier density. τ_s is the charge carrier lifetime. By defining a gain term $G = dg/dn\ (N-N_0)$ it is possible to substitute out the unknown carrier density variable N in Equation 1 and derive an expression for gain saturation by rearranging equation (1):

$$G\,(1 + \Gamma\tau_s\, dg/dn\ P) = g\,(\tau_s I/eV - N_0) \tag{2}$$

In the linear limit, the photon density P tends to zero, and the right hand side variables may be approximated by one linear gain term $G_{linear} = g\ (\tau_s I/eV - N_0)$. A general expression for gain G may thus be defined in terms of a linear gain G_{linear}, photon density P and a photon density saturation term such that $G = G_{linear}/(1+P/P_{saturation})$. Saturation is now simply defined in terms of optical overlap integral Γ, carrier lifetime τ_s and differential gain dg/dn (Equation 3) and it turns out that each of these parameters can be exploited to reduce distortion.

$$P_{saturation} = (\Gamma\tau_s\, dg/dn)^{-1} \tag{3}$$

The optical overlap integral is defined by the waveguide design which has been chosen to confine the carriers and the optical mode. While bulk active regions offer the highest confinement, quantum wells (in reducing numbers) allow for an increase in distortion threshold with output saturation powers of order +15dBm and higher being reported (Borghesani et al., 2003; Morito et al., 2003). Quantum dot epitaxies allow even further reductions in optical overlap for the highest reported saturation powers (Akiyama et al., 2005). Tapered waveguide techniques additionally offer improved optical power handling (Donnelly et al., 1996; Dorgeuille et al., 1996). Optimising optical overlap does however have implications for current consumption, electro-optic efficiency and signal extinction in the off-state.

The carrier lifetime can be speeded up using an additional optical pump (Yoshino & Inoue, 1996; Pleumeekers et al., 2002; Yu & Jeppesen, 2001; Dupertuis et al., 2000). A natural evolution of this, gain clamping (Tiemeijer & Groeneveld, 1995; Bachman et al., 1996; Soulage et al., 1994), has also been extensively studied as a means to increase the distortion threshold. Here the amplification occurs within a lasing cavity and so an out-of-band oscillation defines the carrier density N at the threshold gain condition through fast stimulated emission. Gain clamping can increase the distortion threshold by several decibels (Wolfson, 1999; Williams et al., 2002) and can even be extended to allow variable gain (Davies et al., 2002).

The differential gain term in equation 2 describes how the change in complex dielectric constant amplifies the optical signal. This parameter may be engineered through epitaxial design. The associated differential refractive index modulation, commonly approximated by a line-width broadening coefficient, can also be exploited to suppress distortion. Fast chirped components may be precision filtered from slower chirped components in the output signal to enhance the effective bandwidth (Inoue, 1997; Manning et al., 2007). While the approach does remove energy from the optical signal, it also enables some of the most impressive line rates in all-optical switching (Liu et al., 2007).

2.2 Bandwidth

SOA gates may be characterised by a number of time-constants and bandwidths. The Gigahertz speed at which the circuit may be electronically reconfigured is determined primarily by the spontaneous recombination lifetime and any speed-up technique employed (section 2.1). While this time constant has an impact on the durations of packets and guard-bands in a packet-type network, this does not directly impact the signalling speed, where the multi-Terahertz optical gain bandwidth of the SOA becomes important. These limits are now discussed in the context of state of the art.

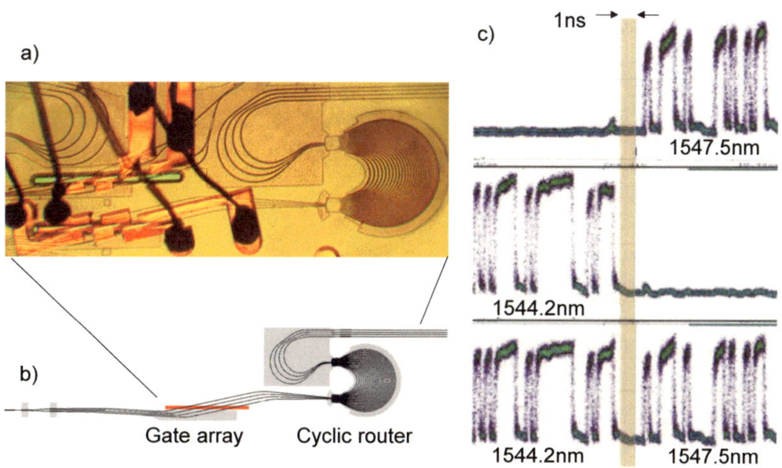

Fig. 2. Dynamic routing with nanosecond switching windows for a SOA cyclic router (Rohit et al., 2010):
a) The microscope photograph for the SOA gate array and arrayed waveguide cyclic router
b) The waveguide arrangement fot the single input, multiple output circuit
c) Time traces showing the selecting and routing of wavelength channels

The electronic switching time from high gain to high loss is limited primarily by the spontaneous recombination lifetime with reports routinely in the nanosecond range (Dorgeuille et al., 1998; Kikichi et al., 2003; Albores-Mejia et al., 2010; Rohit et al., 2010; Burmeister & Bowers, 2006), enabling comparable nanosecond duration dark guard bands between data packets. Figure 2 shows how such fast switching speeds can be exploited in the routing of data in a SOA-gated router. Schemes for label based routing have been reported using comparable approaches (Lee et al., 2005; Shacham et al., 2005).
Real time current control has been considered as a means to ensure optimum operating characteristics of the individual SOA gates. Techniques range from the monitoring of the narrow-band tone (Ellis et al., 1988) and broad-band data (Wonfor et al., 2001) on the SOA

electrodes themselves through to customised monitor diodes (Tiemeijer et al., 1997) and integrated power monitoring (Newkirk et al., 1992; Lee et al., 2005). Hierarchical approaches have also been proposed to enable the management of photonic parameters independently of the digital switch state (White et al., 2007). The possibility to react to thermal transients within the circuit, and even enable self calibration is increasingly important as circuit complexity evolves. This abstraction of the physical layer becomes increasingly important as network level functions such as self-configuration are considered (Lin et al., 2005).

Signalling line-rates of up to 40Gbit/s have been demonstrated using SOAs in a transmission environment (Brisson et al., 2002), and also for integrated switch elements (Burmeister & Bowers, 2006). To extend beyond 40Gb/s requires optical multiplexing. Here SOAs have been demonstrated for in-line amplification for multiwavelength transmission (Reid et al., 1998; Jennen et al., 1998; Sun et al., 1999). The early experiments operated the SOAs within the saturation regime, but later demonstrations in the linear regime with reduced crosstalk enable hundreds of Gbit/s WDM transmission (Spiekman et al., 2000).

Optically transparent networking becomes feasible once the circuit elements become polarisation insensitive. Polarisation properties are engineered through the design of the waveguide dimensions and the radiative transitions in the active media. The latter are tailored using epitaxially defined strain. A broad range of reports have demonstrated polarisation independent operation for both bulk (Emery et al., 1997; Dreyer et al., 2002; Morito et al., 2000; Kakitsuka et al., 2000; Morito et al., 2003; Morito et al., 2005) and quantum wells SOAs (Godefroy et al., 1995; Kelly et al., 1997; Ougazzadeu, 1995; Tiemeijer et al., 1996).

2.3 Energy

The energy efficiency for an interconnection network is commonly quantified in terms of energy requirement per bit and includes the full end-to-end digital power usage. This concise metric allows for a cross-comparison with electronic switching fabrics, and assists with the road-mapping for CMOS technology. Figure 3 shows schematic arrangement for two example photonic interconnection networks with electronic and photonic switching. Photonic links remove transmission losses from the comparison, allowing a focus on the switch technologies themselves. At the time of writing, state of the art vertical cavity laser array transceivers with multimode fibers enabled energy efficiencies of a few picoJoules per bits, and distributed feedback lasers on silicon are being developed for reduced power consumption single mode fiber transceivers. Transceiver technologies dominate the interconnect power budget and a prime motivator for optical switch research has now become the replacement of large numbers of power consuming transceivers with a smaller, data agnostic switch circuit, to remove power draining OEO conversions and excess packaging.

Photonic integration reduces optical losses by minimising the number of on-off-chip connections. This additionally improves noise performance and reduce operating gain for the SOA gates. This is important as it is the current used for amplification, non-radiative and spontaneous recombination which ultimately determines energy consumption. If the non-radiative currents become too high, and Joule heating in the resistive p-layers of the SOA gates becomes significant, this can lead to a spiralling reduction in available gain, and the need for significant heat extraction. Spot-size conversion (Morito et al., 2003) is increasingly implemented to remove the losses between the SOA chip and the off-chip network elements, such as the fiber patch-cords.

Cooler-free operation is now mandatory for data communications transceivers, but remains unthinkable in many high performance telecommunications links. Integrated circuits

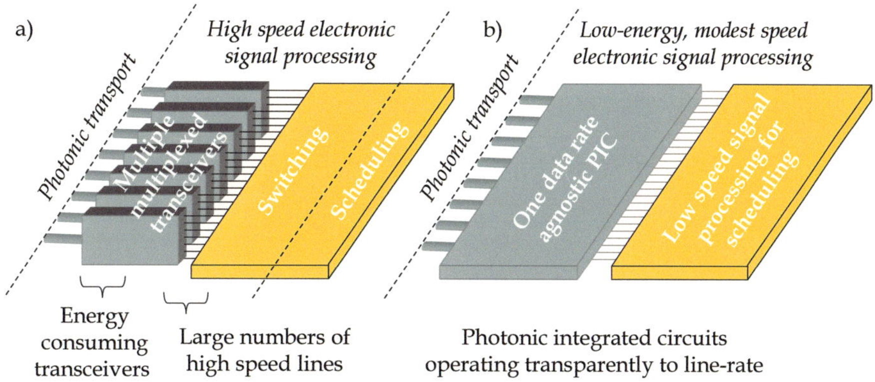

Fig. 3. Schematic diagrams highlighting the motivation for hybrid photonic switch matrices with electronic switch (left) and photonic switch (right)

exploiting semiconductor optical amplifiers are however well suited to uncooled operation due to the broad spectral bandwidth. Initial reports have been promising. Uncooled operation for a quantum dot SOA has been demonstrated for a wide temperature range up to 70 °C (Aw et al., 2008), providing 19dB of optical gain at high temperatures with negligible 0.1dB system penalty at 10Gb/s. Aluminium containing quaternaries, used for the highest performance uncooled 10Gb/s data communications lasers, have also been used for SOAs. These epitaxial designs allow for enhanced electronic confinement and therefore excellent electronic injection efficiency at high temperature. SOAs have also been operated at 45°C such that the packaged SOA module may operate with sub-Watt operating power over the temperature range 0-75°C (Tanaka et al., 2010).

3. Networks

High-connectivity, multi-port electronic switches exploit multi-stage interconnection networks (Dally & Townes, 2004; Kabacinski, 2005) and photonic networks are also set to benefit from such approaches. Figure 4 shows an example of a switch network proposed to allow the scaling of a SOA broadcast select architecture with four outputs per stage using the hybrid Clos/broadcast-and-select architecture (White et al., 2009).

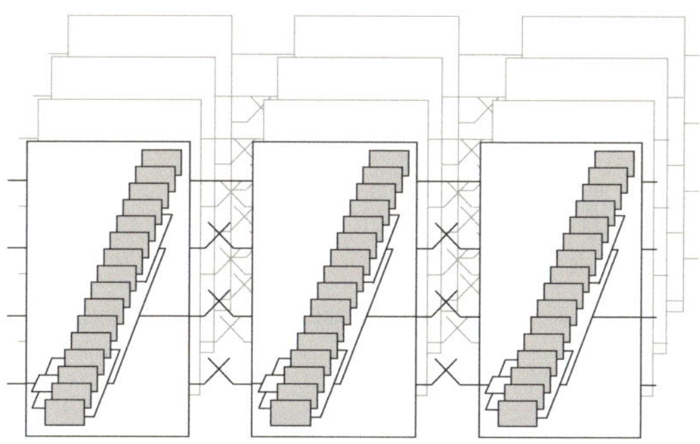

Fig. 4. An example multi-stage switch architecture showing parallel scaling and serial interconnection of SOA gates

A 4x4 broadcast and select switch using SOA gates is placed within each of the twelve switch cells. These are interconnected to each other in three stages to create the larger 16x16 network. Both serial and parallel interconnection of SOA gates is required for the multistage interconnection networks. The interactions between SOA elements in such an architecture is now considered, firstly in terms of signal evolution through the cascaded network, and secondly in terms of crosstalk from incompletely extinguished signals from interferer paths.

3.1 Cascaded networks

The concatenation of multiple SOAs in amplified transmission and switching networks can lead to aggregated noise and distortion. The build-up of noise between stages can be minimised through reduction in gain and loss (Lord & Stallard, 1989). Reflections at the inputs and outputs of the SOA gates were particularly problematic in the early literature (Mukai et al., 1982; Grosskopf et al., 1988; Lord & Stallard, 1989), but can now be minimised through integration (Barbarin et al., 2005) and facet treatments (Buus et al., 1991). The residual distortion of signals (Section 2.1) can additionally build up with increasing numbers of SOA gates, leading to a reduction in the input power dynamic range, and ultimately the power penalty itself.

The largest cascaded networks of SOAs have been studied using recirculating loops, where a signal is switched into and out of a loop with an amplifier and a loss element. The signal circulates for predetermined numbers of iterations - often this is varied as part of the study - and is then assessed for signal degradation. Up to forty cascades have been feasible while maintaining an eye pattern opening - good discrimination between logical levels - for 10Gb/s data sequences (Onishchukov et al., 1998). Studies have also considered transmission over individual fiber spools and field installed fiber spans. Figure 5 summarises many of the leading reports into signal degradation with increasing number of SOAs. Data points are included for a pioneering research teams including those at Philips (Kuindersma et al., 1996; Smets et al., 1997; Jennen et al., 1998) and Bell Labs (Olsson, 1989; Ryu et al., 1989). The evidence suggests that power penalty can be modest for reasonably low levels of cascaded amplifiers, with a steady degradation in penalty as cascade numbers approach ten or more SOAs even when circuits are operated with high levels of gain. It is

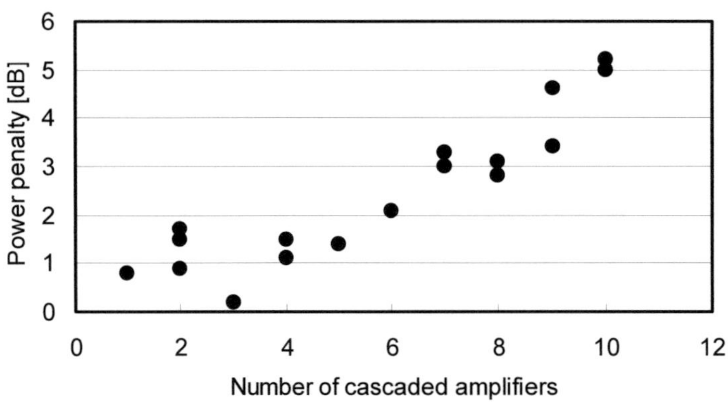

Fig. 5. Power penalty in transmission experiments for cascaded semiconductor optical amplifiers

worth noting that much of this data predates the innovative low distortion amplifier designs developed over the last decade. Operating parameters can nonetheless become increasingly stringent with important implications for control systems (Section 2.2).

3.2 Crosstalk

The aggregation of stray signals from disparate locations in a switch network leads to crosstalk. Contributions may be separated into coherent leakage, incoherent leakage, and cross gain modulation within co-propagating wavelength multiplexes.

Coherent crosstalk was identified as a particularly troublesome source of signal degradation for large-mesh, optically-transparent, telecommunications networks. Channels unintentionally combined with either remnants of themselves or other identical wavelengths lead to interferometric beating (Legg et al., 1994). Coherent crosstalk with long timescale fluctuations compromises threshold setting in receivers. The resulting beat noise incurs large power penalties and bit error floors (Gillner et al., 1999). If the path length differences are minimised to less than one bit period and the wavelengths are stable, as might be anticipated in a monolithic multistage network, phase difference becomes invariant and less problematic (Dods et al., 1997). Coherent crosstalk can incur an overhead of order 10dB on the crosstalk requirement (Goldstein et al., 1994; Goldstein & Eskildsen, 1995; Eskildsen & Hansen, 1997) and this has led some to suggest a -40dB extinction ratio requirement for telecommunication networks using an optical switch technology (Larsen and Gustavsson, 1997). Figure 6 summarises representative quantifying the role of crosstalk on signal degradation. Coherent crosstalk is identified with open symbols, while the closed symbols represent incoherent crosstalk measurements and calculations. The calculations performed by Buckman are also included for the cases of Gaussian and numerically determined distributions for incoherent crosstalk characteristics. It is evident from figure 6 that the level of crosstalk which may be accomodated is significantly higher for incoherent forms of crosstalk (Goldstein et al., 1994; Buckman et al., 1997; Yang & Yao, 1996; Jeong & Goodman, 1996; Albores-Mejia et al., 2009).

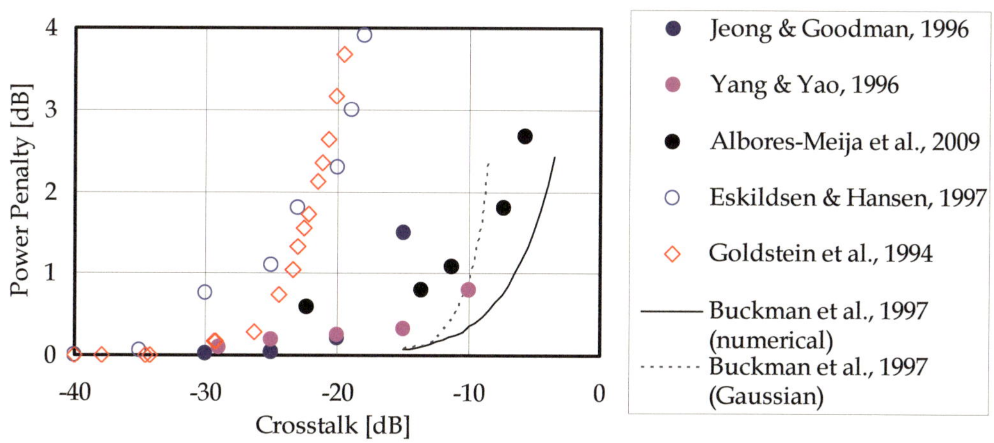

Fig. 6. Crosstalk incurred penalty in SOA networks

Switch extinction ratio is related to crosstalk at the circuit level. A worst case approximation for crosstalk build up in a given path is simply the sum of signal leakage contributions in each switch in the path (Saxtoft & Chidgey, 1993). Cumulated crosstalk ratio may be described as the product of the number of stages between an input and output N_{stages}, the number of interferer inputs at each stage with radix N_{radix}, and the extinction ratio of the switch element $X_{extinction}$:

$$\Sigma X_{crosstalk} = N_{stages} \cdot (N_{radix} - 1) \cdot X_{extinction} \tag{4}$$

While the approach can be a useful guide for low channel counts, this can lead to overestimated power penalty at high channel counts (Buckman et al., 1997) due to statistical averaging (see for example Section 2.1). Nonetheless extinction ratios achieved for SOA gates are commonly reported in the 40dB range (Larsen & Gustavsson, 1997; Varazza et al., 2004, Tanaka et al., 2009; Albores-Mejia et al., 2010; Stabile et al., 2010).

Inter-wavelength crosstalk has been studied across architectures. Many early switch architectures assumed one wavelength per switch element in multiwavelength fabrics, and this called for a multi-domain description of spatially- and spectrally-originating crosstalk (Gillner et al., 1999; Zhou et al., 1994; Zhou et al., 1996). Recent requirements for massive data capacities have led recent work to focus on multi-wavelength routing where inter-wavelength crosstalk can occur through cross gain modulation (Oberg & Olsson, 1988; Inoue, 1989; Summerfield & Tucker, 1999).

4. Multi-port switches

Creating multi-port switches from SOA gates requires additional interconnecting passive circuit elements. As the techniques and technologies for creating integrated power splitters, low-loss wiring, low-radius bends, corner mirrors and waveguide crossings have evolved, the levels of integration have allowed connectivity to increase from two to four and eight output ports.

4.1 Two port switch elements

The broadest range of switching and routing concepts have been demonstrated for the simplest two input two output multiport switches. The SOA gate based switches can be classified as interoferometric or as broadcast and select. The former should allow near complete coupling of optical power into the desired path, enabling the removal of unnecessary and undesirable energy loss. The latter allows a broader range of network functionality, including broadcast and multicast.

Interferometric schemes include the exploitation and frustration of multimode interference in matrices of concatenated 1x2 MMI switches (Fish et al., 1998), vertical directional couplers (Varazza et al., 2004) and gated arrayed waveguide grating based switches (Soganci et al., 2010). The first two approaches lend themselves well to cross-grid architectures and have been demonstrated at 4x4 connectivity. The incorporation of SOA gates with an interferometer also offers enhanced extinction ratio. The switched arrayed waveguide grating approach is also scalable, although only as a 1xN architecture.

Broadcast and select architectures have been more widely studied as they are intrinsically suited to conventional laser based processing methods and epitaxies. The SOA gates are able to overcome losses associated with the splitter network, allowing zero fiber-to-fiber insertion loss at modest currents. Selective area epitaxy has allowed the separate optimisation of active and passive circuit components required for insertion-loss-free operation (Sasaki et al., 1998; Hamamoto & Komatsu, 1995). The splitting and combining functions have been implemented using Y-couplers (e.g. Lindgren et al., 1990), multimode interference couplers (e.g. Albores-Mejia et al., 2009) or arrangements of total internal reflecting mirrors (e.g. Himeno et al., 1988; Gini et al., 1992; Burton et al., 1993; Sherlock et al., 1994; Williams et al., 2005). Chip footprints of below 1mm^2 have been acheived in this manner. Figure 7 shows the example of the mirrors created in an all active switch design interconnecting eight SOA gates in a cross-grid array. The input and output guides include a linearly tapered mode expander, which terminates at one of four splitters. The splitters comprised 45° totally internal reflecting mirror which partially intersect the guided mode. Part of the light is routed into the perpendicular guide and the remaining part is routed to the through path.

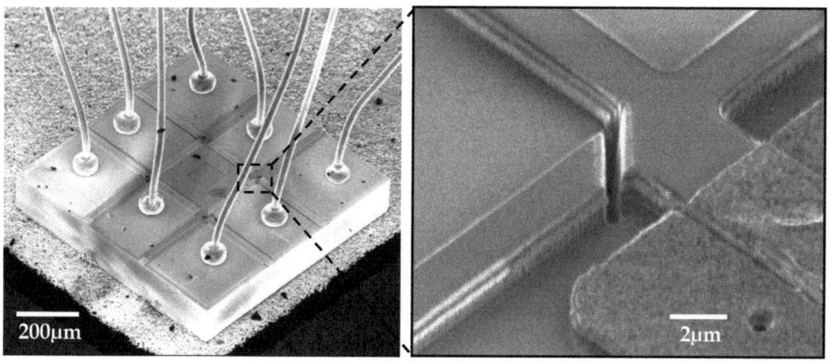

Fig. 7. Two port integrated switch circuit (left) within a footprint of under 1mm^2 using (right) ultracompact total internal reflecting mirrors (Williams at al., 2005)

Microbends offer a route to even further size reductions, while addressing a tolerance to fabrication variability (Stabile & Williams, 2010). Whispering gallery mode operation is predicted to give order of magnitude relaxation in required tolerances with respect to single mode microbends. Polarization conversion can also be maintained below 1% with appropriately designed structures.

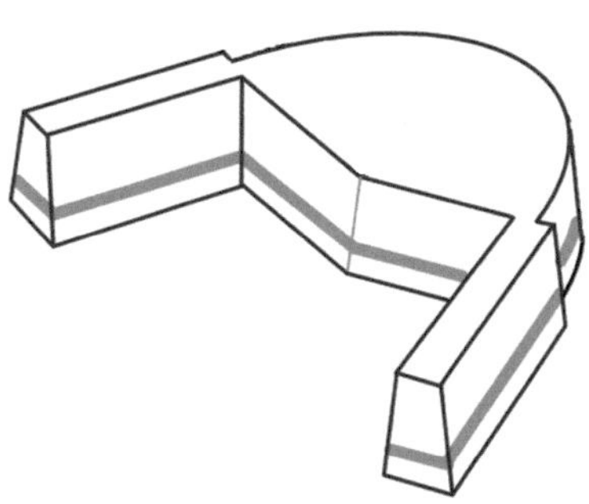

Fig. 8. Schematic diagram for a fabrication tolerant whispering gallery mode bend for high density switch circuits (Stabile & Williams, 2010)

Quantum dot epitaxies have also been considered to exploit anticipated advantages for broadband amplification, low distortion and low noise (Akiyama et al., 2005). The first monolithic 2x2 switch demonstration has been performed for the 1300nm spectral window (Liu et al., 2007) showing negligible power penalty of <0.1dB for 10Gb/s data routing. The first demonstrations in the 1550nm window followed, showing excellent power penalties of order 0.2dB for 10Gbit/s data routing (Albores-Mejia et al., 2009). Multiple monolithically integrated 2x2 circuits have also been demonstrated with 0.4-0.6dB penalty showing only a weak signal degradation as quantum dot circuit elements are incorporated in larger switch fabrics (Albores-Mejia et al., 2008).

4.2 Four port switch elements

Single stage four port switches have been implemented for a number of broadcast and select configurations (Gustavsson et al., 1992; Bachmann et al., 1996; Larsen & Gustavsson, 1997; van Berlo et al., 1995; Sasaki et al., 1998). Electrode counts of between sixteen and twenty-four result, depending on whether additional on-chip amplification is required to overcome circuit losses. This can add considerable complexity to circuit layout and is a potential limit to single stage scaling. The first transmission experiments were reported for 50 km distances at 2.488 Gbit/s, with less than 1 dB power penalty (Gustavsson et al., 1992) with an input power dynamic range of over 10dB. Wavelength division multiplexed transmission was also demonstrated with four 622 Mb/s wavelength channels spaced equally from 1548-1560nm (Almstrom et al., 1996). Field trials at 2.5 Gbit/s were performed with three switch circuits in a 160 km fiber-optic link. The majority of studies have been restricted to modest data capacities between one input port and one output port (Gustavsson et al., 1992; Gustavsson et al., 1993; Djordjevic et al., 2004).

Multi-port dynamic routing has recently been demonstrated for a 4x4 switch using a round-robin scheduler and nanosecond-speed control electronics (Stabile et al., 2010). Figure 8 shows the monolithic photonic circuit on the left, and the output signals on the right. The SOA gates are sequentially biased to enable the routing of the inputs to the outputs. The right hand figures show the time traces recorded for each of the outputs, showing data packets from each available input. Rotating priority (round-robin) path arbitration allows the simplest control algorithm with only one input clock signal, abstracting the photonic complexity from the logic control plane.

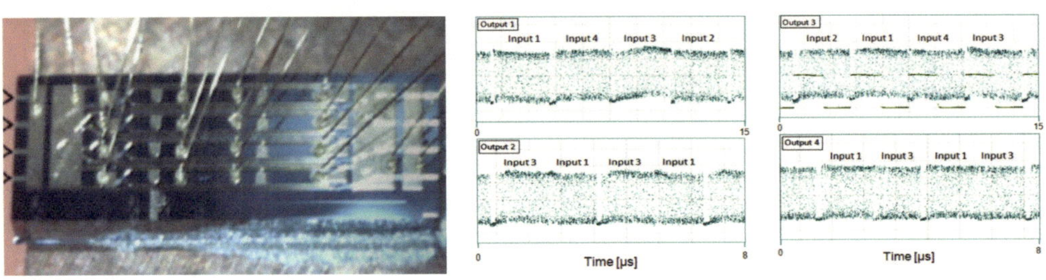

Fig. 8. Four port integrated switch circuit within 4mm^2 showing dynamic multi-path routing (Stabile et al., 2010)

Multi-path routing has also been assessed for wavelength multiplexed inputs to three ports in a discretely populated switching fabric. Field programmable gate arrays enabled the

synchronisation of switching and diagnostics. A power penalty in the range of 0.3–0.6 dB was observed due to multi-path crosstalk and a further power penalty in the range of 0.4–1.2 dB was incurred through dynamic routing (Lin et al., 2007).
Connection scaling studies have allowed insight into the available power margins for SOA switch fabrics operating at high line wavelength division multiplexed line-rates. The potential for single-stage 8×8 switches at a data capacity of 10×10 Gbit/s is predicted with a 1.6dB power margin, identifying a potential route to Tbit/s switch performance in a single-stage low-complexity switch fabric (Lin et al., 2006).

4.3 Eight port switch elements

Scaling to even higher levels of connectivity have been constrained by existing waveguide crossing and waveguide bend techniques, and this is most clearly evidenced by the dearth of single stage 8x8 switches. Researchers realising high connectivity single stage switches have therefore focussed efforts on 1x8 monolithic connectivity.
Array integration has been explored as the first step towards large scale monolithic integration (Dorgeuille et al., 1998; Suzuki et al., 2001; Sahri et al., 2001; Kikuchi et al., 2003; Tanaka et al., 2010). The packaged array of 32 gain clamped SOA gates (Sahri et al., 2001) has enabled the most extensive system level assessments in telecommunications test-beds (Dittmann et al., 2003). Implementation of arrays of eight gates have also led to the early demonstrations of 8×8 optical switching matrices based on SOA gate arrays with 1.28Tbit/s (8×16×10Gb/s) aggregate throughput (Dorgeuille et al., 2000). These approaches rely on fiber splitter networks.
Quantum dot all-active epitaxial designs (Wang et al., 2009) have been implemented using multi-electrode amplifiers to create the separate SOA gates. The input channel is split to the eight output gates by means of three stages of on-chip 1x2 MMI couplers. The use of low splitting ratios is expected to allow more reproducible optical output power balancing. The excellent measured power penalties allow the cascading of two stages which should enable 1x64 functionality.
Active-passive regrown wafers (Tanaka et al., 2009) have also been used to create compact monolithic 1x8 switches. The thin tensile-strained MQW active layers used for the SOA gates allow for an optimisation of output saturation power, noise, and polarization insensitivity. A compact circuit footprint is facilitated by using a high density chip to fiber coupling and through the use of a field flattened splitter to create a uniform split ratio 1x8 in a highly compact 250 μm structure. This approach exhibits an on-state gain of 14.3 dB which is largely wavelength and temperature insensitive. A path to path gain deviation of order 3.0 dB is also achieved. Extinction ratios of order -70dB were reported with an extensive input power dynamic range of 20.5 dB for 10-Gbit/s signals. The high levels of gain overcome the additional off-chip splitter losses which are incurred when combining eight such circuits to construct an 8x8 switching fabric (Kinoshita, 2009).

5. Multi-stage interconnection networks

A broad range of multi-stage networks have been studied for photonic networks (Beneš, 1962; Wu & Feng, 1980; Spanke & Beneš, 1987; Hluchyj & Karol, 1991; Shacham & Bergman, 2007). The constraints imposed in SOA gate based networks lead to a preference for smaller numbers of stages (Williams et al., 2008; White et al., 2009). Simulations are presented to

provide insight into the scalability of multi-stage photonic networks. Then examples of multistage networks are given for 2x2 and 4x4 building blocks, highlighting the state of the art for connectivity, the numbers of integrated stages and line-rate.
Numerical simulations for the physical layer have been performed using travelling wave amplifier modelling which inherently accounts for noise and distortion and allows for wavelength multiplexed system simulation (Williams et al., 2008). Connectivity limits for Tbit/s photonic switch fabrics are studied by scaling the number of splitters in a three stage switch fabric: An intermediate loss between each SOA gate accounts for the radix of the switch element. A 3.5dB loss describes each 1x2 splitter or coupler element in the circuit. Figure 9 summarises the dimensioning simulations by presenting input power dynamic range as a function of the number of splitters per stage. Power penalty contours are given for 1dB and 2dB power penalties to show tolerated inter-stage losses and therefore connectivity.

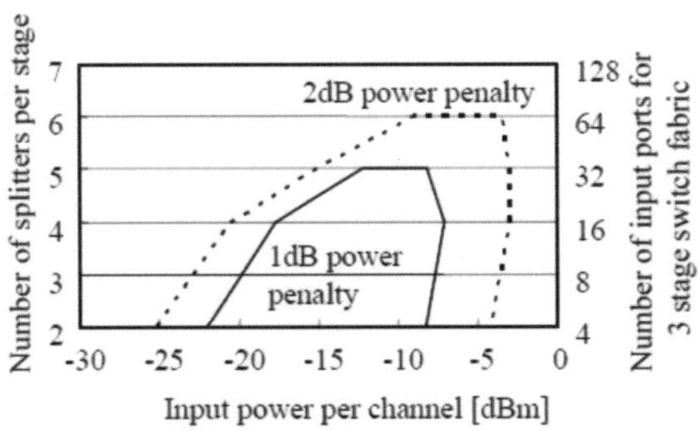

Fig. 9. Simulated power penalty in increasing connectivity SOA gate switching networks Optical data rate at 10λx10Gbit/s using on-off keyed data format (Williams et al., 2008)

Input power dynamic range for 10λx10Gb/s wavelength multiplexed data is seen to reduce both with the number of switch stages and the optical loss between each switch stage. The dynamic range specified for a 1dB power penalty over three stages is observed to exceed 10dB for the four splitter architectures, which is equivalent to a three stage 16×16 switch. For the case of six splitters, a 5dB dynamic range for 2dB power penalty is indicative of viable performance for a 64×64 interconnect based on 8×8 switch stages. Large test-beds exploiting multiple stages of discrete SOA gates have supported these findings. Wavelength multiplexed routing in a 12×12 switch exploiting three stages of concatenated 1×2 SOA-switches enables Terabit class interconnection (Liboiron-Ladouceur et al., 2006). Two stages of SOA gates are implemented in a 64×64 wavelength routed architecture proposed for supercomputers (Luijten & Grzybowski, 2009).
Connectivity for integrated photonic circuits has recently been increased to record levels through the use of the Clos-Broadcast/Select architecture highlighted in Figure 4. Three stages of four 4x4 switch building blocks were integrated within the same circuit (Wang et al., 2009) to demonstrate the first 16×16 port count optical switches using an all-active AlGaInAs quantum well epitaxy. Paths in the circuit have enabled 10Gbit/s routing with 2dB circuit gain and a power penalty of 2.5dB. The electrical power consumption of the all-active chip is estimated to be 12W for a fully operational circuit, which corresponds to a modest power density of 0.3W/mm^2. The power consumption could be approximately halved by replacing the current active shuffle networks with their passive equivalents.

Capacity has also recently been increased to record 320Gb/s line-rates per path for a multi-stage photonic interconnection network (Albores-Mejia et al., 2010). This represents both the leading edge in the number of monolithically integrated switching stages and the highest reported line rates through a switching fabric. Bit error rate studies show only modest levels of signal degradation. The circuit is presented in Figure 10. The N-stage planar architecture includes up to four serially interconnected crossbar switch elements in one path, and is representative of a broader class of 2x2 based multistage interconnection networks. The step change in line rate is believed to be attributable to the use of the active-passive epitaxial regrowth, which allows the separate optimisation of gates and routing circuits.

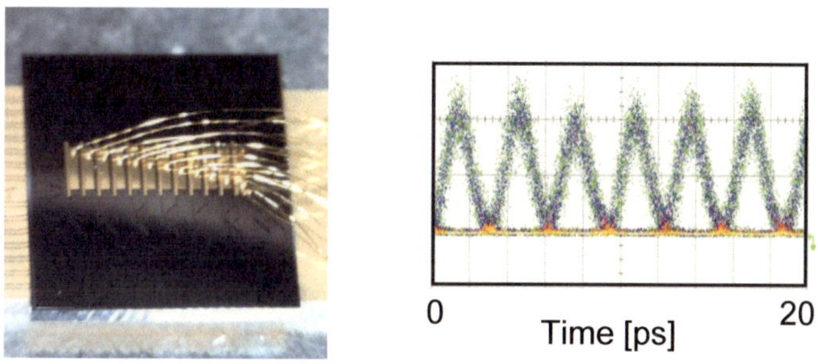

Fig. 10. Photograph of a four port multistage interconnection network, and right, the eye diagrams after four stages of integrated crossbars for 320Gb/s (Albores-Mejia et al., 2010)

6. Conclusion

Integrated photonics is poised to become a key technology where the highest signalling speeds are required. The numbers of integrated optoelectronic components which can be integrated on a chip can rise significantly, and with this, the sophistication of circuit functions can be expected to grow. The critical parameters required for high capacity, high connectivity switching circuits have now been demonstrated, and the challenge is to devise architectures that are able to simultaneously match performance with energy efficiency and integration. A symbiotic relationship between massive bandwidth photonic circuits and intelligent electronic control circuits could well evolve to create a generation of ultrahigh speed signal processors.

7. References

Adamiecki, A., M. Duelk and J.H. Sinsky, "25 Gbit/s electrical duobinary transmission over FR-4 backplanes", Electronics Letters, 41, 14, 826-827, (2005)

Akiyama, T., M. Ekawa, M. Sugawara, K. Kawaguchi, H. Sudo, A. Kuramata, H. Ebe and Y. Arakawa, "An ultrawide-band semiconductor optical amplifier having an extremely high penalty-free output power of 23 dBm achieved with quantum dots", Photonic Technology Letters, 17, 8, 1614-1616, (2005)

Albores-Mejia, A., K.A. Williams, T. de Vries, E. Smalbrigge, Y.S. Oei, M.K. Smit, S. Anantathanasarn, R. Notzel, "Scalable quantum dot optical Switch matrix in the 1.55 um wavelength range", Proceedings Photonics in Switching, Paper D-06-4 (2008)

Albores-Mejia, A., K.A. Williams, T. de Vries, E. Smalbrugge, Y.S. Oei, M.K. Smit and R. Notzel, "Integrated 2 x 2 quantum dot optical crossbar switch in 1.55 mm wavelength range", Electronics Letters, 45, 6, 313-314, (2009)

Albores-Mejia, A., K.A. Williams, F. Gomez-Agis, S. Zhang, H.J.S. Dorren, X.J.M. Leijtens, T de Vries, Y.S. Oei, M.J.R. Heck, L.M. Augustin, R. Notzel, D.J. Robbins, M.K. Smit, "160Gb/s serial line rates in a monolithic optoelectronic multistage interconnection network", proceedings 17th Annual IEEE Symposium on High-Performance Interconnects, New York, (2009)

Albores-Mejia, A., K.A. Williams, T. de Vries, E. Smalbrugge, Y.S. Oei, M.K. Smit, and R. Notzel, "Low power penalty monolithically-cascaded 1550nm-wavelength quantum-dot crossbar switches", proceedings Optical Fiber Conference, (2009)

Albores-Mejia, A.F. Gomez-Agis, H.J.S. Dorren, X.J.M. Leijtens, M.K. Smit, D.J. Robbins and K.A. Williams, "320Gb/s data routing in a monolithic multistage semiconductor optical amplifier switching circuits", proceedings European Conference on Optical Communications, Invited paper, We.7.E.1, (2010)

Albores-Mejia, A., F. Gomez-Agis, S. Zhang, H.J.S. Dorren, X.J.M. Leijtens, T. de Vries, Y.S. Oei, M.J.R. Heck, R. Notzel, D.J. Robbins, M.K. Smit and K.A. Williams, "Monolithic multistage optoelectronic switch circuit routing 160Gb/s line-rate data," Journal of Lightwave Technology, to appear, (2010)

Almstrom, E., C.P. Larsen, L. Gillner, W.H. van Berlo, M. Gustavsson and E. Berglind, "Experimental and analytical evaluation of packaged 4x4 InGaAsP/InP semiconductor optical amplifier gate switch matrices for optical networks", Journal of Lightwave Technology, 14, 6, 996-1004, (1996)

Aw E.T., H. Wang, M.G. Thompson, A. Wonfor, R.V. Penty, I.H. White, A.R. Kovsh, "Uncooled 2x2 quantum dot semiconductor optical amplifier based switch", Proceedings Conference on Lasers and Electro-Optics, (2008)

Bachmann, M., P. Doussiere, J.Y. Emery, R. N'Go, F. Pommereau, L. Goldstein, G. Soulage and A. Jourdan, "Polarisation-insensitive clamped-gain SOA with integrated spot-size converter and DBR gratings for WDM applications at 1.55um wavelength", Electronics Letters, 32, 22, 2076-2078, (1996)

Barbarin, Y., E.A.J.M. Bente, C. Margquet, E.J.S. Leclere, J.J.M. Binsma and M.K. Smit, "Measurement of reflectivity of butt-joint active-passive interfaces in integrated extended cavity lasers", Photonics Technology Letters, 17, 11, 2265-2267, (2005)

Beneš, V. "On rearrangeable three stage connecting networks," Bell Systems Technology Journal XLI, 1481–1492 (1962).

Borghesani, A., N. Fensom, A. Scott, G. Crow, L. Johnston, J. King, L. Rivers, S. Cole, S. Perrin, S. Scrase, G. Bonfrate, A. Ellis, I. Lealman, G. Crouzel, L.H.K. Chun, A. Lupu, E. Mahe, P. Maigne, "High saturation output (>16.5dBm) and low noise figure (<6dB) semiconductor optical amplifier for C band operation", proceedings Optical Fiber Communications Conference, paper 534-536, (2003)

Brisson, C., S. Chandrasekhar, G. Raybon and K.F. Drew, "Experimental investigation of SOAs for linear amplification in 40Gb/s transmission systems", proceedings Optical Fiber Communications Conference, WV6, (2002)

Buckman L.A., L.P. Chen and K.Y. Lau, "Crosstalk penalty in all-optical distributed switching networks", Photonics Technology Letters, 9, 2, 250-252, (1997)

Buckman-Windover, L.A., J.N. Simon, S.A. Rosenau, K.S. Giboney, G.M. Flower, L.W. Mirkarimi, A. Grot, B. Law, C.-K. Lin, A. Tandon, R.W. Gruhlke, H. Xia, G. Rankin,

M.R.T. Tan, and D.W. Dolfi, “Parallel-optical interconnects >100 Gb/s”, Journal of Lightwave Technology, 22, 9, 2055, (2004)

Burmeister, E.F. and J.E. Bowers, "Integrated gate matrix switch for optical packet buffering", Photonics Technology Letters, 18, 1, 103-105, (2006)

Burton J.D., P.J. Fiddyment, M.J. Robertson and P. Sully, "Monolithic InGaAsP-InP laser amplifier gate switch matrix", Journal of Quantum Electronics, 29, 6, 2023-2027, (1993)

Buus, J., M.C. Farries and D.J. Robbins, "Reflectivity of coated and tilted semiconductor facets", Journal of Quantum Electronics, 27, 6, 1837, (1991)

Chiaroni, D., F. Masetti, D. Verchère, A. Jourdan, G. Luyts , B. Pawels, "A comparison of electronic and optical packet/burst switching fabrics", proceedings Optical Fiber Communication Conference, Los Angeles, California, WF1, (2004)

Chiaroni, D., R. Urata, J. Gripp, J.E. Simsarian, G. Austin, S. Etienne, T. Segawa, Y. Pointurier, C. Simonneau,, Y. Suzaki, T. Nakahara, M. Thottan, A. Adamiecki, D. Neilson, J.C. Antona, S. Bigo, R. Takahashi, V. Radoaca, "Demonstration of the interconnection of two optical packet rings with a hybrid optoelectronic packet router", proceedings European Conference on Optical Communications, PD3.5, (2010)

Cho P., Y. Achiam, G. Levy-Yurista, M. Margalit, Y. Gross and J. Khurgin, "Investigation of SOA nonlinearities on the amplification of DWDM channels with spectral efficiency up to 2.5b/s/Hz", Photonics Technology Letters, 16, 3, 918-920 (2004)

Ciaramella, E., A. D'Ericco and V. Donzella, "Using semiconductor optical amplifiers with constant envelope WDM signals", J. Quantum Electronics, 44, 5, 403-409, (2008)

Congress report on server and data center energy efficiency: Public law 109-431 U.S. Environmental Protection Agency ENERGY STAR Program, August 2, (2007)

Cunningham, D. and W.G. Lane, Gigabit ethernet networking, Macmillan (1999)

Dally W.J. and B. Townes, Principles and practices of interconnection networks, Morgan Kaufmann (2004)

Dangel, R., C. Berger, R. Beyeler, L. Dellmann, M. Gmür, R. Hamelin, F. Horst, T. Lamprecht, T. Morf, S. Oggioni, M. Spreafico, and B.J. Offrein, " Polymer-waveguide-based board-level optical interconnect technology for datacom applications", Transactions on Advanced Packaging, 31, 4, 759-757, (2008)

Davies, A.R., K.A. Williams, R. V. Penty, I. H. White, M. Glick and D. McAuley, "Variable-gain operation of a gain-clamped sampled grating diode laser amplifier", proceedings on Conference on Lasers and Electro Optics, (2002)

Desurvire, E., Erbium doped fiber amplifiers, Wiley & Sons, (1994)

Dittmann, L., C. Develder, D. Chiaroni, F. Neri, F. Callegati, W. Koerber, A. Stavdas, M. Renaud, A. Rafel, J. Solé-Pareta, W. Cerroni, N. Leligou, L. Dembeck, B. Mortensen, M. Pickavet, N. Le Sauze, M. Mahony, B. Berde, and G. Eilenberger, "The European IST project DAVID: A viable approach toward optical packet switching", Journal on Selected Areas in Communications, 21, 7, 1026-1040 (2003)

Djordjevic, I. B., R. Varazza, M. Hill, and S. Yu, "Packet switching performance at 10 Gbit/s across a 4x4 optical crosspoint switch matrix", Photonics Technology Letters, 16, 1, 102–104 (2004).

Dods, S., J.P.R. Lacey and R.S. Tucker, "Homodyne crosstalk in WDM ring and bus networks", IEEE Photonics Technology Letters, 9, 9, 1285-1287, (1997)

Donnelly, J.P., J.N. Walpole, G.E. Betts, S.H. Groves, J.D. Woodhouse, F.J. O'Donnell, L.J. Missaggia, R.J. Bailey, A. Napoleone, "High-power 1.3-µm InGaAsP-InP amplifiers with tapered gain regions", Photonics Technology Letters, 8, 11, 1450-1452, (1996)

Dorgeuille, F., B. Mersali, M. Feuillade, S. Sainson, S. Slempkes, M. Foucher, "Monolithic InGaAsP-lnP tapered laser amplifier gate 2 x 2 switch matrix with gain", Photonics Technology Letters, 8, 9, 1178-1180, (1996)

Dorgeuille, F., Lavigne, B., Emery, J.Y., Di Maggio, M., Le Bris, J., Chiaroni, D., Renaud, M., Baucknecht, R., Schneibel, H.P., Graf, C., Melchior, H., " Fast optical amplifier gate array for WDM routing and switching applications",proceedings Optical Fiber Communication Conference, (1998)

Dorgeuille, F., L. Noirie, J.P. Faure, A. Ambrisy, S. Rabaron, F. Boubal, M. Schilling and C. Artigue, "1.28 Tb/s throughput 8 x 8 optical switch based on-arrays of gain-clamped semiconductor optical amplifier gates", proceedings Optical Fiber Communication Conference, 221-223, paper PD18-1/221, (2000)

Dreyer K., C.H. Joyner, J.L. Pleumeekers, T.P. Hessler, P.E. Selbmann, B. Deveaud, B. Dagens and J.Y. Emery, "High gain mode adapted semiconductor optical amplifier with 12.4dBm saturation power at 1550nm", Journal of Lightwave Technology, 20, 4, 718-721, (2002)

Dupertuis, M. A., Pleumeekers, J. L., Hessler, T. P., Selbmann, P. E., Deveaud, B., Dagens, B., and Emery, J. Y., "Extremely fast high-gain and low-current SOA by optical speed-up at transparency", Photonic Technology Letters, 12, 11, 1453–1455, (2000)

Ellis, A. D., D. Malyon and W.A. Stallard, "A novel all electrical scheme for laser amplifier gain control", proceedings European Conference on Optical Communications, 487-490, (1988)

Ellis, A.D., D.M. Patrick, D. Flannery, R.J. Manning, D.A.O. Davies, and D.M. Spirit, "Ultra-high-speed OTDM networks using semiconductor amplifier-based processing nodes", Journal of Lightwave Technology, 13, 5, 761-770, (1995)

Emery J.Y., T. Ducellier, M. Bachmann, P. Doussiere, F. Pommereau, R. Ngo, F. Gaborit, L. Goldstein, G. Laube and J. Barrau, "High performance 1.55um polarisation insensitive semiconductor optical amplifier based on a low tensile strained bulk InGaAsP", Electronics Letters, 33, 12, 1083-1084, (1997)

Eskildsen, L. and P.B. Hansen, "Interferometric noise in lightwave systems with optical preamplifiers", Photonics Technology Letters, 9, 11, 1538-1540, (1997)

Fish, G.A. ,B. Mason, L.A. Coldren and S.P. DenBaars, "Compact 4x4 InGaAsP-InP optical crossconnect with a scalable architecture", Photonics Technology Letters, 10, 9, 1256-1258, (1998)

Gillner, L., C.P. Larsen and M. Gustavsson, "Scalability of optical multiwavelength switching networks: Crosstalk analysis", Journal of Lightwave Technology, 17, 1, 58-67, (1999)

Gini, E.I, G. Guekos, and H. Melchior, "Low loss corner mirrors with 45° deflection angle for integrated optics", Electronics Letters, 28, 5, 499-501 (1992)

Glick, M., M. Dales, D. McAuley, T. Lin, K.A. Williams, R.V. Penty, I.H. White, "SWIFT: A testbed with optically switched data paths for computing applications", proceedings 7th International Conference on Transparent Optical Networks, (2005)

Godefroy A., A le Corre, F. Clerot, S. Salaun, S. Loualiche, J.C. Simon, L. Henry, C. Vaudry, J.C. Keromnes, G. Joulie and P. Lamouler, "1.55um polarisation insensitive optical amplifier with stain balanced superlattice active layer", Photonic Technology Letters, 7, 5, 473-475, (1995)

Goldstein, E.L., L.E. Eskildsen, and A.F. Elrefaie, "Performance implications of component crosstalk in transparent lightwave networks", Photonics Technology Letters, 6, 5, 657-660, (1994)

Goldstein, E.L. and L. Eskildsen, "Scaling limitations in transparent optical networks due to low-level crosstalk", Photonics Technology Letters, 7, 1, 93-94 (1995)

Gripp, J., M. Duelk , J. E. Simsarian , A. Bhardwaj , P. Bernasconi , O. Laznicka and M. Zirngibl, "Optical switch fabrics for ultra-high-capacity IP routers", Journal of Lightwave Technology, 21, 11, 2839-2850, (2003)

Grosskopf, G., R. Ludwig, and H.G. Weber, "Cascaded inline semiconductor laser amplifiers in a coherent optical fiber transmission system", Electronics Letters, 24, 9, 551-552, (1988)

Grubb, S. G., D. F. Welch, D. Perkins, C. Liou, and S. Melle, "OEO versus all-optical networks", proceedings Lasers and Electro-optics Society Annual Meeting, Montreal, invited paper TuH1 (2006)

Gustavsson, M. B. Lagerstrom, L. Thylen, M. Janson, L. Lundgren, A.C. Morner, M. Rask, and B. Stoltz, "Monolithically integrated 4 x 4 InGaAsP/lnP laser amplifier gate switch arrays", Electronics Letters, 28, 24, 2223-2225, (1992)

Gustavsson, M., M. Janson and L. Lundgren, "Digital transmission experiment with monolithic 4 x 4 InGaAsP/InP laser amplifier gate switch array", Electronics Letters, 29, 12, 1083-1085, (1993)

Hamamoto K., K. Komatsu, "Insertion-loss-free 2x2 InGaAsP/lnP optical switch fabricated using bandgap energy controlled selective MOVPE", Electronics Letters, 31, 20, 1779-1781, (1995)

Hemenway, R., R.R. Grzybowski, C. Minkenberg and R. Luijten, "Optical packet switched interconnect for supercomputer applications", OSA Journal of Optical Networking, 3, 12, 900-913, (2004)

Himeno, A., H. Terui, M. Kobayashi, "Loss measurement and analysis of high-silica reflection bending optical waveguides", Journal of Lightwave Technology, 6, 1, 41-46, (1988)

Ho, K.P., S.K. Liaw and C. Lin, "Reduction of semiconductor laser amplifier induced distortion and crosstalk for WDM systems using light injection", Electronics Letters, 32, 24, 2210-2211, (1996)

Hluchyj M.G. and M.J. Karol, "ShuffleNet: An application of generalised perfect shuffles to multihop lightwave networks", Journal Lightwave Technology, 9, 10, 1386-1397, (1991)

Huang, D., T. Sze, A. Landin, R. Lytel, and H. Davidson, "Optical interconnects: out of the box forever?", Journal of Selected Topics in Quantum Electronics, 9, 2, 614-623 (2003)

Inoue, K., "Crosstalk and its power penalty in multichannel transmission due to gain saturation in a semiconductor laser amplifier", Journal of Lightwave Technology, 7, 7, 1118-1124, (1989)

Inoue, K., "Optical filtering technique to suppress waveform distortion induced in a gain-saturated semiconductor optical amplifier", Electronics Letters, 33, 10, 885–886, (1997)

Jennen, J. G. L., R. C. J. Smets, H. de Waardt, G. N. van den Hoven, and A. J. Boot, "4x10Gbit/s NRZ transmission in the 1310 nm window over 80 km of standard single mode fiber using semiconductor optical amplifiers", Proceedings European Conference on Optical Communications, 235 - 236, (1998)

Jeong G. and J.W. Goodman, "Analysis of linear crosstalk in photonic crossbar switches based on on/off gates", Journal of Lightwave Technology, 14, 3, 359-365, (1996)

Kabacinski, W., Nonblocking electronic and photonic switching fabrics, Springer, (2005)

Kakitsuka T., Y. Shibata, M. Itoh, Y. Kadota and Y. Yoshikuni, "Influence of buried structure on polarisation sensitivity in strained bulk semiconductor optical amplifiers", Journal of Quantum Electronics, 38, 1, 85-92, (2002)

Kelly A.E., I.F. Lealman, L.J. Rivers, S.D. Perrin and M. Silver, "Low noise (7.2dB) and high gain (29dB) semiconductor optical amplifier with a single layer AR coating", Electronic Letters, 33, 6, 536-538, (1997)

Kikuchi, N., Y. Shibata, Y. Tohmori, "Monolithically integrated 64-channel WDM channel selector", NTT review , 1, 7, 43-49 (2003)

Kinoshita, S., "Monolithically-integrated SOA gate switch and its application to high-speed switching systems", proceedings Asia Communications and Photonics Conference, ThD2, (2009)

Kim, H.K., Chandrasekhar, S., "Reduction of cross-gain modulation in the semiconductor optical amplifier by using wavelength modulated signal", Photonics Technology Letters", 12, 10, 1412-1414, (2000)

Kuindersma, P.I., G.P.J.M. Cuijpers, J.G.L. Jennen, J.J.E. Reid, L.F. Tiemeijer, H. de Waardt, A.J. Boot, "10Gbit/s RZ transmission at 1309nm over 420km using a chain of multiple quantum well semiconductor optical amplifier modules at 38km intervals", proceedings European Conference on Optical Communications, TuD2.1, (1996)

Larsen, C.P. and M. Gustavsson, "Linear crosstalk in 4x4 semiconductor optical amplifier gate switch matrix", Journal of Lightwave Technology, 15, 10, 1865-1870, (1997)

Lee S.C., R. Varazza, S. Yu, "Optical label processing and 10-Gb/s variable length optical packet switching using a 4 x 4 optical crosspoint switch", Photonics Technology Letters, 17, 5, 1085-1087 (2005)

Lee, S.C., J. Begg, R. Varrazza and S. Yu, "Automatic per-packet dynamic power equalization in a 4 × 4 active coupler-based optical crosspoint packet switch matrix", Photonics Technology Letters, 17, 12, 331- 332 (2005)

Legg P.D., D. Hunter, I. Andonovic and P. Barnsley, "Inter-channel crosstalk phenomena in optical time division multiplexed switching networks", Photonics Technology Letters, 6, 5, 661-663, (1994)

Lemoff, B.E., M.E. Ali, G. Panotopoulos, G.M. Flower, B. Madhavan, A.F.J. Levi, and D.W. Dolfi, "MAUI: Enabling fiber-to-the-processor with parallel multiwavelength optical interconnects", Journal of Lightwave Technology, 22, 9, 2043-2044, (2004)

Lemoff, B.E., M.E. Ali, G. Panotopoulos, E. de Groot, G.M. Flower, G.H. Rankin, A.J. Schmit, K.D. Djordjev, M.R.T. Tan, A. Tandon, W. Gong, R.P. Tella, B. Law, L.-K. Chia, and D.W. Dolfi, "Demonstration of a compact low-power 250-Gb/s parallel-WDM optical interconnect", Photonics Technology Letters, 17, 1, 220-222, (2005)

Liboiron-Ladouceur, O., B. A. Small, and K. Bergman, "The data vortex optical packet switched interconnection network", Journal of Lightwave Technology, 24, 1, 262–270, (2006)

Lin, T., K. A. Williams, R. V. Penty, I. H. White, M. Glick and D. McAuley, "Self-configuring intelligent control for short reach 100Gb/s optical packet routing", proceedings Optical Fiber Communications Conference, Paper OWK5, (2005)

Lin, T., K. A. Williams, R. V. Penty, I. H. White, M. Glick, and D. McAuley, "Performance and scalability of a single-stage SOA switch for 10x10Gb/s wavelength striped packet routing", Photonics Technology Letters, 18, 5, 691-693, (2006)

Lin, T., K. A. Williams, R. V. Penty, I. H. White, M. Glick, "Capacity scaling in a multi-host wavelength-striped SOA-based switch fabric", Journal of Lightwave Technology, 25, 3, 655-663 (2007)

Lindgren, S., M.G. Oberg, J. Andre, S. Nilsson, B. Broberg, B. Holmberg, and L. Backbom, "Loss-compensated optical Y-branch switch in InGaAsP-InP", Journal of Lightwave Technology, 8, 10, 1591-1595 (1990)

Liu, S., Hu, X., Thompson, M.G., Sellin, R.L., Williams, K.A., Penty, R.V., White, I.H., and Kovsh, A.R.:, "Cascaded performance of quantum dot semiconductor optical amplifier in a recirculating loop", Proceedings European Conference on Optical Communications, paper Th4.5.6, (2006)

Liu, Y., E. Tangdiongga, Z. Li, H. de Waardt, A.M.J. Koonen, G.D. Khoe, X.W. Shu, I. Bennion, H.J.S. Dorren, "Error-free 320-Gb/s all-optical wavelength conversion using a single semiconductor optical amplifier", Journal of Lightwave Technology, 25, 1, 103-108, (2007)

Lord, A. and W.A. Stallard, "A laser amplifier model for system optimisation", Optical and Quantum Electronics, 21, 6, 463-470, (1989)

Luijten, R.P. and R. Grzybowski, "The OSMOSIS optical packet switch for supercomputers", proceedings Optical Fiber Communications Conference, OTuF3, (2009)

Manning R., R. Giller, X. Yang, R.P. Webb, D. Cotter, "Faster switching with semiconductor optical amplifiers", Proceedings Photonics in Switching, 145-146, (2007)

Masetti, F., D. Chiaroni, R. Dragnea, R. Robotham, D. Zriny, "High-speed high-capacity packet-switching fabric: a key system for required flexibility and capacity", OSA Journal of Optical Networking , 2, 7, 255-265, (2003)

Maxwell, G., "Low-cost hybrid photonic integrated circuits using passive alignment techniques", proceedings Lasers and Electro-optics Society Annual Meeting, invited paper MJ2, (2006)

Mikkelsen, B. and Raybon, G., "Reduction of crosstalk in semiconductor optical amplifiers by amplifying dispersed WDM signals (7x20Gb/s)", Optical Fiber Communications Conference, ThJ5-I, (2000)

Miller D.A.B., "Optical interconnects to electronic chips", Applied Optics, 49, 25, F59-F70, (2010)

McAuley, D., "Optical Local Area Network", in Computer Systems: Theory, Technology and Applications, A. Herbert and K. Sparck-Jones, Eds. Springer-Verlag, (2003)

Morito K., M. Ekawa, T. Watanabe and Y. Kotaki, "High saturation output power (+17dBm) 1550nm polarisation insensitive semiconductor optical amplifier", proceedings European Conference in Optical Communication, paper 1.3.2, (2000)

Morito, K., Ekawa, M., Watanabe, T., and Kotaki, Y., "High output-power polarization-insensitive semiconductor optical amplifier", Journal of Lightwave Technology, 21, 1, 176-181, (2003)

Morito, K., S. Tanaka, S. Tomabechi and A. Kuramata, "A broadband MQW semiconductor optical amplifier with high saturation output power and low noise figure", Photonics Technology Letters, 17, 5, 974-976, (2005)

Mukai, T., Y. Yamamoto, T. Kimura, "S/N and error rate performance in AlGaAs semiconductor laser preamplifier and linear repeater systems", Transactions Microwave Theory and Techniques, 30, 10, 1548-1556, (1982)

Nagarajan R., and M.K. Smit, "Photonic integration", IEEE Laser and Electro-Optic Society Newsletter, February (2007)

Newkirk, M.A., U. Koren, B.I. Miller, M.D. Chien, M.G. Young, T.L. Koch, G. Raybon, C.A. Burrus, B. Tell, K.F. Brown-Goebeler, "Three-section semiconductor optical amplifier for monitoring of optical gain", Photonics Technology Letters, 4, 11, 1258-1260, (1992)

Nicholes, S.C., M.L. Masanovic, E. Lively, L.A. Coldren and D.J. Blumenthal, "An 8x8 monolithic tuneable optical router (MOTOR) packet forwarding chip", Journal of Lightwave Technology, 28, 4, 641-650, (2010)

Oberg M. G. and N. A. Olsson, "Crosstalk between intensity-modulated wavelength-division multiplexed signals in a semiconductor-laser amplifier," Journal of Quantum Electronics, 24, 1, 52-59, (1988)

Olsson, N.A. "Lightwave systems with optical amplifiers", Journal of Lightwave Technology, 2, 7, 1071-1081, (1989)

Onishchukov, G., V. Lokhnygin, A. Shipulin and P. Reidel, "10Gbit/s transmission over 1500km with semiconductor optical amplifiers", Electronics Letters, 34, 16, 1597-1598, (1998)

Ougazzadeu A., "Atmospheric pressure MOVPE growth of high performance polarisation insensitive strain compensated MQW InGaAsP/InGaAs optical amplifier", Electronics Letters, 31, 15, 1242-1244, (1995)

Patel, R.R., S.W. Bond, M.D. Pocha, M.C. Larson, H.E. Garrett, R.F. Drayton, H.E. Petersen, D.M. Krol, R.J. Deri, and M.E. Lowry, "Multiwavelength parallel optical interconnects for massively parallel processing", Journal of Selected Topics in Quantum Electronics, 9, 2, 657-666, (2003)

Pleumeekers, J. L., Kauer, M., Dreyer, K., Burrus, C., Dentai, A. G., Shunk, S., Leuthold, J., and Joyner, C. H., "Acceleration of gain recovery in semiconductor optical amplifiers by optical injection near transparency", Photonic Technology Letters, 14, 1, 12–14, (2002)

Reid, J. J. E., L. Cucala, M. Settembre, R. C. J. Smets, M. Ferreira, and H. F. Haunstein, "An international field trial at 1.3 μm using an 800 km cascade of semiconductor optical amplifiers", Proceedings European Optical Communications Conference, 567 - 568, (1998)

Renaud, M., M. Bachmann, M. Erman, "Semiconductor optical space switches," Journal of Selected Topics in Quantum Electronics, 2, 2, 277-288 (1996)

Roadmap: International technology roadmap for semiconductors, http://public.itrs.net/, (2005)

Roberts, G.F., Williams, K.A, Penty, R.V., White, I. H, Glick, M., McAuley, D. Kang, D. J. and Blamire, M., "Multi-wavelength data encoding for improved input power dynamic range in semiconductor optical amplifier switches", proceedings European Conference on Optical Communications, (2005)

Rohit, A., K. A. Williams, X. J. M. Leijtens, T. de Vries, Y. S. Oei, M. J. R. Heck, L. M. Augustin, R. Notzel, D. J. Robbins, and M. K. Smit, "Monolithic multi-band nanosecond programmable wavelength router", Photonics Journal, 2, 1, 29-35 (2010)

Ryu, S., Taga H. Yamamoto S., Mochizuki K. Wakabayashi H., "546km, 140Mbit/s FSK coherent transmission experiment through 10 cascaded semiconductor laser amplifiers, Electronics Letters, 25, 25, 1682-1684, (1989)

Sahri, N., D. Prieto , S. Silvestre , D. Keller , F. Pommerau , M. Renaud , O. Rofidal , A. Dupas , F. Dorgeuille and D. Chiaroni, "A highly integrated 32-SOA gates optoelectronic module suitable for IP multi-terabit optical packet routers", Proceedings Optical Fiber Commununications Conference, PD32-1 (2001)

Sato K., H. Toba, "Reduction of mode partition noise by using semiconductor optical amplifiers", Journal of Selected Topics in Quantum Electronics, 7, 2, 328-333, (2001)

Saxtoft C. and P. Chidgey, "Error rate degradation due to switch crosstalk in large modular switched optical networks", IEEE Photonics Technology Letters, 5, 7, 828-831, (1993)

Sasaki, J., H. Hatakeyama, T. Tamanuki, S. Kitamura, M. Yamaguchi, N. Kitamura, T. Shimoda, M. Kitamura, T. Kato, M. Itoh, "Hybrid integrated 4x4 optical matrix switch using self-aligned semiconductor optical amplifier gate arrays and silica planar lightwave circuit", Electronics Letters, 34, 10, 986-987, (1998)

Shacham, A., B.A. Small, O. Liboirin-Ladouceur and K. Bergman, "A fully implemented 12x12 data vortex optical packet switching interconnection network", Journal of Lightwave Technology, 23, 10, 3066-3075, (2005)

Shacham, A. and K. Bergman, "Building ultralow-latency interconnection networks using photonic integration", IEEE Micro, 6-20, (2007)

Shao S.K. and M.S. Kao, "WDM coding for high-capacity lightwave systems", Journal of Lightwave Technology, 12, 1, 137-148, (1994)

Shares, L., J.A. Kash, F.E. Doany, C.L. Schow, C. Schuster, D.M. Kuchta, P.K. Pepeljugoski, J.M. Trewhella, C.W. Baks, R.A. John, L. Shan, Y.H. Kwark, R.A. Budd, P. Chiniwalla, F.R. Libsch, J. Rosner, C.K. Tsang, C.S. Patel, J.D. Schaub, R. Dangel, F. Horst, B.J. Offrein, D. Kucharski, D. Guckenberger, S. Hegde, H. Nyikal, C.-K. Lin, A. Tandon, G.R. Trott, M. Nystrom, D.P. Bour, M.R.T. Tan, and D.W. Dolfi, "Terabus: Terabit/second-class card-level optical interconnect technologies", Journal of Selected Topics in Quantum Electronics, 12, 5, 1032-1044, (2006)

Sherlock G., J.D. Burton, P.J. Fiddyment, P.C. Sully, A.E. Kelly, M.J. Robertson, "Integrated 2x2 optical switch with gain", Electronics Letters, 30, 2, 137-138, (1994)

Smets, R.C.J., J.G.L. Jennen, H. de Waart, B. Teichmann, C. Dorschky, R. Seltz, J.J.E. Reid, L.F. Tiemeijer, P.I. Kuindersma, A.J. Boot, "114km repeaterless, 10Gb/s transmission at 1310nm using an RZ data format", proceedings Optical Fiber Conference, ThH2 (1997)

Soganci, M., T. Tanemura, K. Takeda, M. Zaitsu, M. Takenaka and Y. Nakano, "Monolithic InP 100-port photonic switch", proceedings European Conference on Optical Communications, post-deadline paper, PD1.5, (2010)

Soulage, G., Doussiere, P., Jourdan, A., and Sotom, M., "Clamped gain travelling wave semiconductor optical amplifier as a large dynamic range optical gate", Proceedings European Conference on Optical Communications, Florence, Italy, 451–454, (1994)

Spanke, R. A. and V. E. Beneš, "N-stage planar optical permutation network," Applied Optics, 26, 7, 1226–1229 (1987)

Spiekman, L. H., J. M. Wiesenfeld, A. H. Gnauck, L. D. Garrett, G. N. van den Hoven, T. van Dongen, M. J. H. Sander-Jochem, and J. J. M. Binsma, "Transmission of 8 DWDM channels at 20 Gb/s over 160 km of standard fiber using a cascade of semiconductor optical amplifiers", Photonics Technology Letters, 12, 6, 717-719, (2000)

Spiekman, L.H., A.H. Gnauck, J.M. Wiesenfeld and L.D. Garrett, "DWDM transmission of thirty two 10Gbit/s channels through 160km link using semiconductor optical amplifiers", Electronics Letters, 36, 12, 1046-1047, (2000)

Stabile R. and K.A. Williams, "Low polarisation conversion in whispering gallery mode micro-bends", proceedings European Conference on Integrated Optics, (2010)

Stabile, R., H. Wang, A. Wonfor, K. Wang, R.V. Penty, I.H. White and K. A. Williams, "Multipath routing in a fully scheduled integrated optical switch fabric", proceedings European Conference on Optical Communications, paper We.8.A.6, (2010)

Stubkjaer, K.E., "Semiconductor optical amplifier-based all-optical gates for high-speed optical processing", Journal of Selected Topics in Quantum Electronics, 6, 6, 1428-1435, (2000)

Summerfield, M. A. and R.S. Tucker, , "Frequency-domain model of multiwave mixing in bulk semiconductor optical amplifiers", Journal of Selected Topics in Quantum Electronics, 5, 3, 839-850, (1999)

Sun, Y., A.K. Srivastava, S. Banerjee, J.W. Sulhoff, R. Pan, K. Kantor, R.M. Jopson and A.R. Chraplyvy, "Error free transmission of 32x2.5Gb/s DWDM channels over 125km using cascaded in-line semiconductor optical amplifiers", Electronics Letters, 35, 21, 1863-1865, (1999)

Suzuki, Y., K. Magari, Y. Kondo, Y. Kawaguchi, Y. Kadota, K. Yoshino, "High-gain array of semiconductor optical amplifier integrated with bent spot-size converter (BEND SS-SOA)", Journal of Lightwave Technology 19, 11, 1745-1750, (2001)

Tanaka S., S. Tomabechi, A, Uetake, M. Ekawa, K. Morito, "Highly uniform eight channel SOA-gate array with high saturation output power and low noise figure", Photonics Technology Letters, 19, 16, 1275-1277, (2007)

Tanaka, S., S.H. Jeong, S.Yamazaki, A. Uetake, S. Tomabechi, M. Ekawa, and K. Morito, "Monolithically integrated 8:1 SOA gate switch with large extinction ratio and wide input power dynamic range", Journal of Quantum Electronics, 45, 9, 1155-1162, (2009)

Tanaka S., N. Hatori, A, Uetake, S. Okumura, M. Ekawa, G. Nakagawa and K. Morito, "Compact, very-low-electric-power-consumption (0.84W) 1.3um optical amplifier module using AlGaInAs MQW-SOA", Proceedings European Conference on Optical Communications, Th10D3 (2010)

Tiemeijer, L.F.; Groeneveld, C.M.; "Packaged high gain unidirectional 1300 nm MQW laser amplifiers", proceedings Electronic Components and Technology Conference, 751, (1995)

Tiemeijer L.F., P.J.A. Thijs, T. van Dongen, J.J.M. Binsma and E.J. Jansen, "Polarization resolved, complete characterisation of 1310nm fiber pigtailed multiple-quantum-well optical amplifiers", Photonics Technology Letters, 14, 6, 1524-1533, (1996)

Tiemeijer, L.F., S. Walczyk, A.J.M. Verboven, G.N. van den Hoven, P.J.A. Thijs, T. van Dongen, J.J.M. Binsma, E.J. Jansen, "High-gain 1310 nm semiconductor optical amplifier modules with a built-in amplified signal monitor for optical gain control", Photonics Technology Letters, 9, 3, 309-311, (1997)

Tucker, R.S., "Optical packet switching: A reality check", OSA Journal of Optical Switching and Networking, 5, 2-9, (2008)

van Berlo, W., M. Janson, L. Lungren, A.C. Morner, J. Terlecki, M. Gustavsson, P. Granestrand, P. Svensson, "Polarization-insensitive 4x4 InGaAsP-InP laser amplifier gate switch matrix", Photonics Technology Letters, 7, 11, 1291-1293, (1995)

Varazza R., I.B. Djordjevic and S. Yu, "Active vertical-coupler-based optical crosspoint switch matrix for optical packet-switching applications", Journal of Lightwave Technology, 22, 9, 2034-2042, (2004)

Wang H., K.A. Williams, A. Wonfor, T. de Vries, E. Smallbrugge, Y.S. Oei, M.K. Smit, R. Notzel, S. Liu, R.V. Penty, I.H. White, "Low penalty cascaded operation of a monolithically integrated quantum dot 1x8 port optical switch", Proceedings European Conference on Optical Communications, (2009)

Wang H., Aw E.T., K.A. Williams, A. Wonfor, R.V. Penty, I.H. White, "Lossless multistage SOA switch fabric using high capacity 4x4 switch circuits", Proceedings Optical Fiber Conference (2009)

Wang, H., A. Wonfor, K.A Williams, RV. Penty, I.H. White, "Demonstration of a lossless monolithic 16x16 QW SOA switch", proceedings European Conference in Integrated Optics, (2010)

Wei X., Y. Su, X. Liu, J. Leuthold and S. Chandrasekhar, "10Gb/s RZ-DPSK transmitter using a saturated SOA as a power booster and limiting amplifier", Photonics Technology Letters, 16, 6, 1582-1584, (2004)

White, I.H., K.A. Williams, R.V. Penty, T. Lin, A. Wonfor, E.T. Aw, M. Glick, M. Dales and D. McAuley, "Control architecture for high capacity multistage photonic switch circuits", OSA Journal of Optical Networking, 6, 2, 180-188 (2007)

White, I.H., E.T. Aw, K.A. Williams, H. Wang, A. Wonfor and R.V. Penty, "Scalable optical switches for computing applications", OSA Journal of Optical Networking, Invited paper, 8, 2, 215–224 (2009)

Williams, K.A., R.V. Penty, I.H. White and D. McAuley, "Advantages of gain clamping in semiconductor amplifier crosspoint switches", proceedings Optical Fiber Communication Conference (2002)

Williams, K.A., G.F. Roberts, T. Lin, R.V. Penty, I.H. White, M. Glick and D. McAuley, "Monolithic 2x2 optical switch for wavelength multiplexed interconnects", Journal of Selected Topics in Quantum Electronics, Special issue on integrated optics and optoelectronics, 11, 78-85 (2005)

Williams, K.A., "High capacity switched optical interconnects for low-latency packet routing", IEEE Photonics Society Benelux Symposium, (2006)

Williams, K.A., "Integrated semiconductor optical amplifier based switch fabrics for high-capacity interconnects", OSA Journal of Optical Networking, Invited paper, 6, 2, 189-199 (2007)

Williams, K.A., E.T. Aw, H. Wang, R.V. Penty, I.H. White, "Physical layer modelling of semiconductor optical amplifier based Terabit/second switch fabrics", Numerical Simulation of Optical Devices, Post-deadline paper ThPD5, (2008)

Winzer, P.J., "Modulation and multiplexing in optical communication systems", IEEE Photonics Sociey Newsletter, 23, 1, 4-10, (2009)

Wolfson, D., "Detailed theoretical investigation and comparison of the cascadability of conventional and gain-clamped SOA gates in multiwavelength optical networks", Photonics Technology Letters 11, 11, 1494–1496 (1999)

Wonfor, A., S. Yu , R.V. Penty and I.H. White, "Constant output power control in an optical crosspoint switch allowing enhanced noise performance operation", in European Conference on Optical Communications, 136-137 (2001)

Wu C. and T. Feng, "On a class of multistage interconnection networks", 29, 8, 694-702, (1980)

Yang S. and Y.G. Yao, "Impact of crosstalk induced beat noise on the size of laser amplifier based optical space switch structures", Photonics Technology Letters, 8, 7, 894-896, (1996)

Yoshino, M. and Inoue, K. "Improvement of saturation output power in a semiconductor laser amplifier through pumping light injection", Photonics Technology Letters, 8, 1, 58–59, (1996)

Yu, J. and Jeppesen, P. "Improvement of cascaded semiconductor optical amplifier gates by using holding light injection", Journal of Lightwave Technology, 19, 5, 614–623, (2001)

Zhou J., M. J. O'Mahony, and S.D. Walker, "Analysis of optical crosstalk effects in multi-wavelength switched networks", Photonics Technology Letters, 6, 2, 302-304, (1994)

Zhou J., R. Cadeddu, E. Casaccia and M.J. O'Mahony, "Crosstalk in multiwavelength optical crossconnect networks", Journal of Lightwave Technology, 14, 6, 1423-1435, (1996)

11

Negative Feedback Semiconductor Optical Amplifiers and All-Optical Triode

Yoshinobu Maeda
Kinki University
Japan

1. Introduction

The field of optical communications is moving toward the realization of photonic networks with wavelength division multiplexing (WDM) utilizing the full bandwidth of optical fibers. Conventionally, an erbium-doped fiber amplifier (EDFA) and a semiconductor optical amplifier (SOA) are used for amplifying an optical signal in optical communications. SOAs can contribute to this and offer the key advantages of small size and ease of mass production, but their practical adoption has been largely precluded by their inferior polarization dependence and noise characteristics as compared those of with optical fiber amplifiers doped with erbium and other rare earths. For eliminating noise generated in such amplifiers, the optical signal that is transmitted at a high speed is once converted into an electrical signal, so as to be subjected to noise elimination and signal processing in an electronic circuit, and the processed signal is then reconverted into an optical signal to be transmitted. This incapability to achieve direct processing of an optical signal without its conversion into an electrical signal limits the speed of the optical signal processing. Therefore, there has been demanded a technique which enables an optical signal to be processed without its conversion into an electrical signal. However, in a field of optoelectronics, there have not yet realized high-performance signal amplifiers corresponding to a negative feedback amplifier or an operational amplifier known in a field of electronics. The negative feedback amplifier in electronics is capable of providing an output signal whose gain, waveform and baseline are stabilized without generating large noise. Negative feedback amplification is widely used in electronics and readily enables gain stability and low-noise electric signal amplification, as the existence of negative- and positive- valued entities facilitate its design and implementation. For optical signals, however, the absence of negative-valued entities poses the need for special techniques. One technique for SOA gain stabilization which has been the subject of research and development at many institutions is the use of a clamped-gain SOA (Bachmann et al., 1996), which utilizes a lasing mode generated outside the signal band. An SOA with gain control obtained by an experimental feedback loop system utilizing a bandpass filter (Qureshi et al., 2007), which is conceptually similar to the technique we have proposed, has also been reported (Maeda, 2006).

In the present study, we utilized phase mask interferometry to fabricate an optical fiber filter (a fiber Bragg grating; FBG) having reflection wavelength characteristics specially designed for surrounding light feedback, formed a lens in the optical fiber tip, and coupled the fiber containing the FBG to the SOA, thus constructing a "negative feedback SOA (NF-SOA)", and performed measurements of its bit error rate (BER) in correspondence with the input signal, its noise figure, and other characteristics, which show its noise reduction effect (Maeda et al., 2010).

In previous study, it has been demonstrated that an all-optical triode can be achieved using a tandem wavelength converter employing cross-gain modulation (XGM) in SOAs (Maeda et al., 2003). Basic functions such as switching can be achieved using all optical gates realized by optical nonlinearities in semiconductor materials (Stubkjaer, 2000). The three mainly used schemes to perform their wavelength conversion employing SOAs are based on XGM, cross-phase modulation (XPM), and four-wave mixing (FWM) (Glance et al., 1992; Durhuus et al., 1994; Wiesenfeld, 1996). The XGM scheme has the advantage to be very simple: an input modulated signal and a continuous-wave beam are introduced into the SOA. The input signal saturates the SOA gain and modulates the cw beam inversely at the new wavelength. A large signal dynamic theoretical model was presented for wavelength conversion using XGM in SOA with converted signal feedback (Sun, 2003). The theoretical results predict that the wavelength conversion characteristics can be enhanced significantly with converted signal feedback. We demonstrated a negative feedback optical amplification effect that is capable of providing an output signal whose gain and waveform are stabilized optically using XGM in SOA with amplified spontaneous emission feedback (Maeda, 2006). We have also previously proposed a tandem wavelength converter in the form of an all-optical triode with cross-gain modulation (XGM) in two reflective semiconductor amplifiers (RSOAs), and demonstrated the signal amplifying effect of its three terminals (Maeda et al., 2003). In investigating the cause of an increase in extinction ratio found in the XGM of the RSOAs, we were able to elucidate the negative feedback optical amplification effect and its potential for SOA noise reduction. This effect is due to the feedback to the SOA of spontaneous emission generated in the SOA in response to the input light signal. The spontaneous emission is intensity inverted with respect to the input light signal effected by XGM in the SOA. It can thus be used to dynamically modulate the SOA internal gain in correspondence with the input optical signal, and achieve a noise reducing effect analogous to that of electronic negative feedback amplification.

2. Negative feedback optical amplification effect

Fig. 1 shows the block diagram of the negative feedback optical amplifier. It consists of a semiconductor optical amplifier and an optical add/drop filter, which is equipped with a negative feedback function. It is used the SOA based on ridge waveguide structure InGaAsP/InP MQW material. The composition of the InGaAsP active layer is chosen to have a gain peak wavelength around 1550 nm. The maximum small signal fiber-to-fiber gain is around 15 dB and the output saturation power is approximately 2 mW measured at 1550 nm with a bias current of 250 mA. A tunable laser is used for the input signal, which is modulated by the mean of electro-optic modulators connected to an electrical synthesizer. The input signal is the wavelength of 1550 nm. The modulated input signal is fed into the SOA using an optical coupler. An add/drop filter (spectral half-width: 13 nm) is set at the center wavelength of 1550 nm. The filter is provided to extract an output signal light of the wavelength of 1550 nm and surrounding spontaneous emission L_S having wavelengths (L_S <1543.5 and L_S >1556.5 nm) other than 1550±6.5 nm. Because of the XGM mechanism in the SOA, the spontaneous emission L_s contain an inverted replica of the information carried by the input signal. The output of L_s is fed back and injected together with the input signal into the SOA by using an optical coupler. A variable optical attenuator (VOA) is provided in an optical feedback path. The average output power is measured at the output of the filter using an optical power-meter.

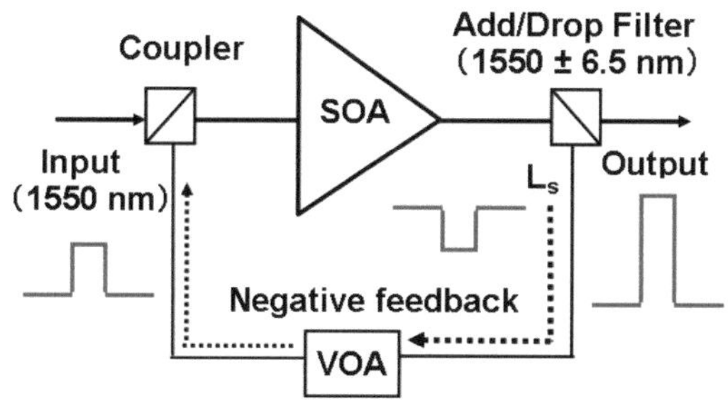

Fig. 1. Block diagram of a negative-feedback semiconductor optical amplifier. VOA: Variable optical attenator.

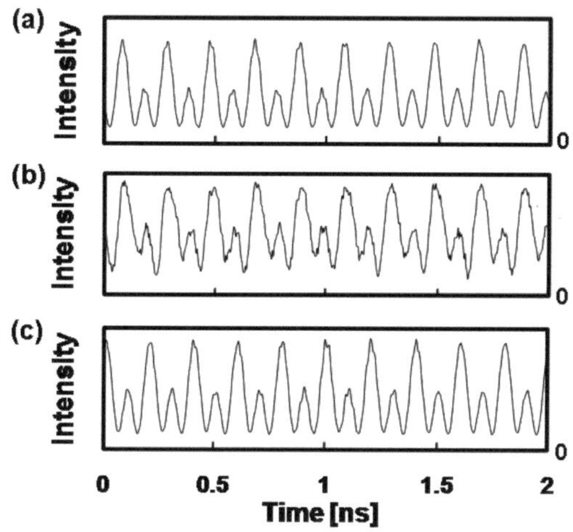

Fig. 2. (a) Input waveform, (b) and (c) Output waveform without and with negative feedback, respectively.

Figs. 2(a), 2(b) and 2(c) show waveforms of the input, the output without negative feedback and the output with negative feedback, respectively. The input average power is approximately 2 mW. They have been measured with a fast photodiode connected to a sampling head oscilloscope. The modulation degree and frequency of the input continuous signal are 80% and 10 GHz, respectively. The modulation degree M is equal to 100 × (P_{max} – P_{min})/(P_{max} + P_{min}) [%], where P_{max} and P_{min} represent the maximum and minimum intensities of the signal, respectively. As is apparent from Figs. 2(b) and 2(c), the output signal was given a higher modulation degree M, a waveform with a higher fidelity and a more stable baseline in the case where the SOA feeding back the spontaneous emission L_s was used with negative feedback, than in the case where the SOA was used without negative feedback. The output average power was around 6.4 mW without negative feedback, as shown in Fig. 2(b). On the other hand, in the SOA with negative feedback, the output average power was approximately 1.9 mW when the negative feedback average power was 0.12 mW, as shown in Fig. 2(c). Therefore, the output signal waveform with negative feedback is remarkably improved over that without negative feedback. Moreover, in the SOA with negative feedback, the distortion of the waveform is extremely small in a wide frequency band of 0.1 - 10 GHz.

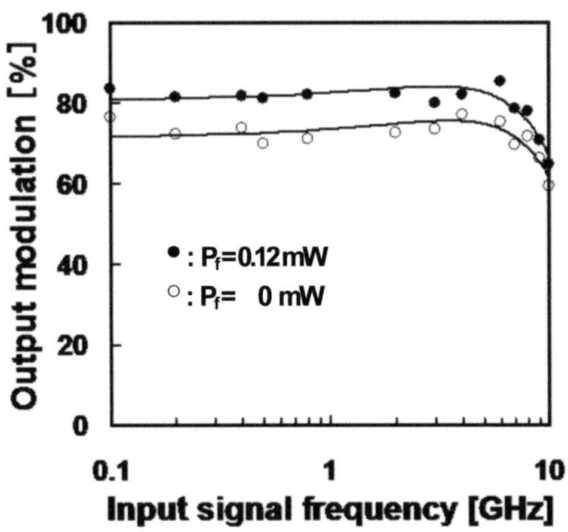

Fig.3. Relationship between the output modulation degrees and the frequency of the input signal.

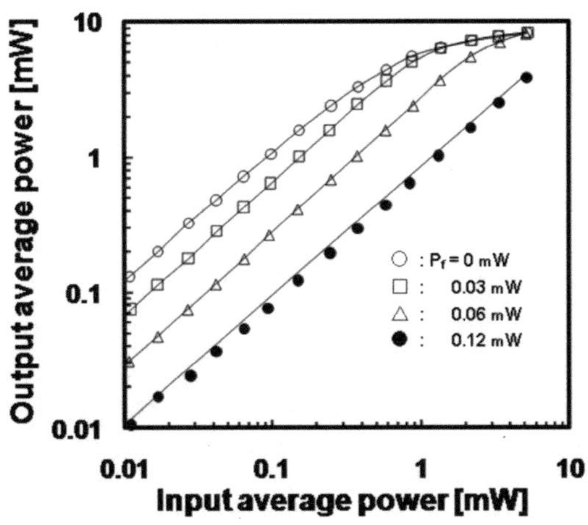

Fig. 4. Relationship between the output average and input average powers for four values of the negative feedback average powers (P_f = 0, 0.03, 0.06, 0.12 mW).

Fig.3 shows the relationship between the output modulation degrees and the frequency of the input signal. The input modulation degree depends on the input signal frequency and decreases relatively at higher frequency due to the characteristics of the electro-optic modulator. The average power of input signal is around 2 mW. The black-dot (●) represents the case of the SOA when the negative feedback average power was around 0.12

mW and the white-dot (○) represents without negative feedback. The output modulation degree (i.e., extinction ratio) with negative feedback is remarkably improved over that without negative feedback in a wide input signal frequency band of 0.1-10 GHz.

Fig. 4 shows the relationship between the output average and input average powers for four values of the negative feedback average powers (P_f = 0, 0.03, 0.06, 0.12 mW). The modulation degree and frequency of the input signal are around 100% and 0.1 GHz, respectively. The input-output characteristic in the SOA with negative feedback has a higher linearity than that without negative feedback. It is also noted that a gain G is defined as $G = 10 \log_{10} (P_{out}/P_{in})$ [dB], where P_{out} and P_{in} represent the respective output and input signal power.

Fig. 5 shows the gain characteristic, i.e., a relationship between the gain and the input signal average power. The gain G with negative feedback is found to be lower than that without negative feedback when the negative feedback average power increases from 0 to 0.12 mW. For P_f = 0.12 mW, the gain remains approximately 0 dB for input signal average powers between 0.01 to 5 mW. In addition, the gain can be adjusted optically between 0 and 11 dB by changing the amount of negative feedback using a variable optical attenuator.

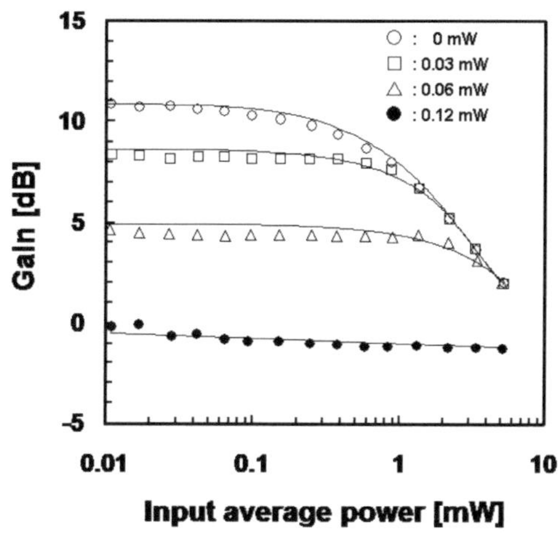

Fig. 5. Relationship between the gain and the input signal average power for four values of the negative feedback average powers (P_f = 0, 0.03, 0.06, 0.12 mW).

In general, since a conventional optical amplifier merely has a simple amplification function (that is almost constant gain), the amplifier disadvantageously amplifies not only the signal but also the noise. Therefore, the waveform and baseline of the output signal cannot be improved basically in relation with the noise, thereby making difficult to achieve an advanced signal processing. For eliminating noise generated in such amplifiers, the optical signal is once converted into an electrical signal, so as to be subjected to noise elimination and signal processing in an electronic circuit, and the processed signal is then reconverted into an optical signal to be transmitted. Fig. 6 shows the concept diagram of a negative feedback optical amplification effect. Figs. 6(a), 6(b) and 6(c) show waveforms of the input, the negative feedback and the gain in SOA, respectively. In the SOA which has a XGM function, spontaneous emission lights which have wavelengths near a wavelength λ_1 of an input light have an intensity varying in response to a variation in the intensity of that input light. Characteristically, the intensity variation of the spontaneous emission lights are

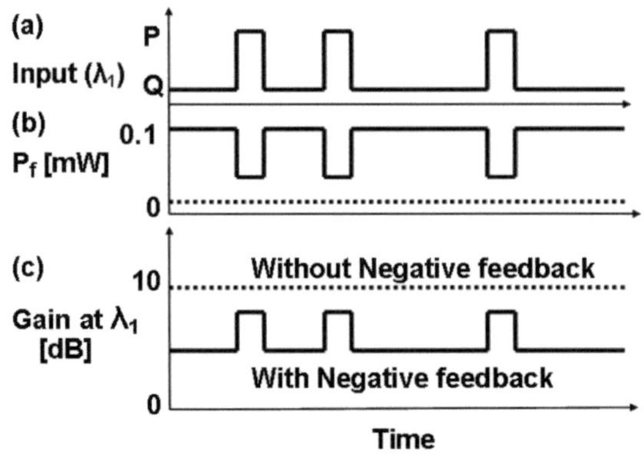

Fig. 6. Concept diagram of a negative feedback optical amplification effect. The straight-line represents the case where the SOA was used with negative feedback, and the dotted line represents the case of the SOA without negative feedback.

inverted with respect to the variation in the input signal, and the spontaneous emission lights are outputted from the SOA, as shown in Fig. 6(b). The straight-line represents the case where the SOA was used with negative feedback, and the dotted line represents the case of the SOA without negative feedback. In the past, it is common that the spontaneous emission lights as the surrounding light having wavelengths other than the wavelength λ_1 are removed by a band-pass filter, since it becomes a factor of noise generation. In such a situation, we found out that a negative feedback optical signal amplification phenomenon in which characteristics of the gain of the SOA is drastically changed by feeding back the separated surrounding light to the SOA, so that the gain is modulated as shown in Fig. 6(c). Therefore, noise reduction is realized all-optically in the SOA, because the output signal waveform with negative feedback is remarkably improved over that without negative feedback, as shown in Fig. 2. Moreover, the baseline of the output signal waveform is suppressed, because the gain in the SOA is low when the power of the input signal is at the low logical level, whereas the output signal is stressed because of the high SOA gain when the input signal power is high, as shown in Fig. 6. In addition, the desired gain was set between 0 and 11 dB by changing the amount of the negative feedback, as shown in Fig. 4. The negative feedback optical amplifier is capable of providing an output signal whose gain is stabilized automatically.

An operational amplifier of the field of electronics has two inputs consisting of a non-inverse input and an inverse input, and is used as a negative feedback circuit for feeding a part of an output voltage of an amplifier, back to the inputs through an external resistance. The operational amplifier is referred to as a non-inverting amplifier where an output is in phase with an input, and is referred to as an inverting amplifier where the phase of the output is delayed by π. The optical amplifier of the present work is physically considered as the optical equivalent of a non-inverting amplifier, since the output is in phase with the input. In addition, the non-inverting amplifier of the electronics is provided a voltage gain of not lower than 1, the gain is 0 dB where the resistance in the feedback path is 0, namely, where the feedback path is provided by a short circuit. The operational amplifier is capable of achieving an analog computing such as summing, differentiating and integrating amplifier. It is therefore no exaggeration to say that the operational amplifier takes change of a major part of an analog electronic circuit today. The optical amplifier of the present work can take charge of an important role in an optical circuit, as the negative feedback or operational amplifiers in electronics.

Therefore, we found out that the negative feedback optical amplification effect is capable of providing an output signal whose gain, waveform and baseline are stabilized automatically. The optical amplifier consists of an InGaAsP/InP semiconductor optical amplifier and an optical add/drop filter, which is equipped with a negative feedback function. In the SOA with negative feedback, the output modulation degree was substantially higher modulation degree and the distortion of the waveform was extremely small in wide frequency band of 0.1 - 10 GHz. The gain in the SOA with negative feedback is suppressed to be lower than that without negative feedback and reaches around 0 dB when the negative feedback power increases. In addition, the desired gain was set between 0 and 11 dB by changing the amount of the negative feedback. The optical amplifier is physically considered as the optical equivalent of a non-inverting amplifier, since the output is in phase with the input. Therefore, the negative feedback optical amplifier of this work can take charge of an important role in an optical circuit, as the negative feedback amplifier in electronics.

3. Negative feedback optical semiconductor amplifier

3.1 Fiber Bragg gratings based on phase mask interferometer

The fiber Bragg grating (FBG) used in the present study is a diffraction grating formed inside an optical fiber. Bragg diffraction gratings are characterized by their reflection of light of certain wavelengths. The FBG is a refractive index modulation grating, with alternating regions of high and low refractive indices in the direction of light propagation. The relation between the grating period Λ and the reflection wavelength (the Bragg wavelength) λ_B is expressed as

$$\lambda_B = 2n_{eff}\Lambda \tag{1}$$

n_{eff} : effective refractive index. The refractive index of the core is raised above that of the clad by adding GeO_2, which induces oxygen-deficient defects in the SiO_2 with an absorption band in the vicinity of 240 nm. A rise in the refractive index occurs under UV irradiation in that vicinity. This has been variously attributed to the Kramers-Kroning relation (the relation between light absorption and change in refractive index) and to the occurrence of defects in the glass structure due to molecular reorientation under UV irradiation. It has also been reported that UV sensitivity can be substantially heightened by pre-treatment with high-pressure hydrogen. In the present study, the change in the optical refractive index was utilized to obtain refractive index modulation in the fiber core as shown in Fig.7.

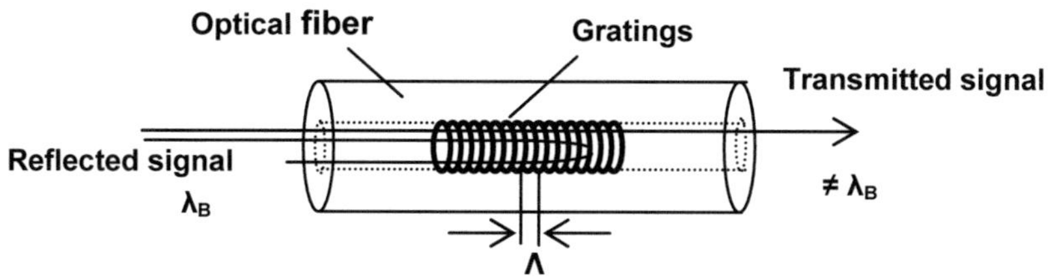

Fig. 7. Drawing of the fiber Bragg grating. Λ : grating period, λ_B : Bragg wavelength.

Phase mask processing is a widely used technique for optical device fabrication. After removal of its covering, the fiber is irradiated from the side (in the circumferential direction) by an intense UV laser beam, which is diffracted by the phase mask and thus forms an interference pattern in the fiber core, resulting in the formation of a refractive index modulation pattern in the core corresponding to the period of the interference pattern. The grating period Λ in the core is 1/2 of the phase mask grating period d, with good reproducibility. A key advantage of the phase mask technique is that it enables the use of a low-coherence laser beam. An alternative technique for the same purpose is the two-light-bundle interference light exposure method. Although it is relatively low in reproducibility, it may be advantageous for multi-grade low-volume production, as it requires no phase mask, which is an expensive consumable, and it enables the use of a broad range of wavelengths in the same optical system. For the production of a chirp grating, however, in which the grating period is gradually changed and the reflection wavelength band is thus broadened, it requires the incorporation of lens systems into the interferometer and presents difficulties relating to adjustment.

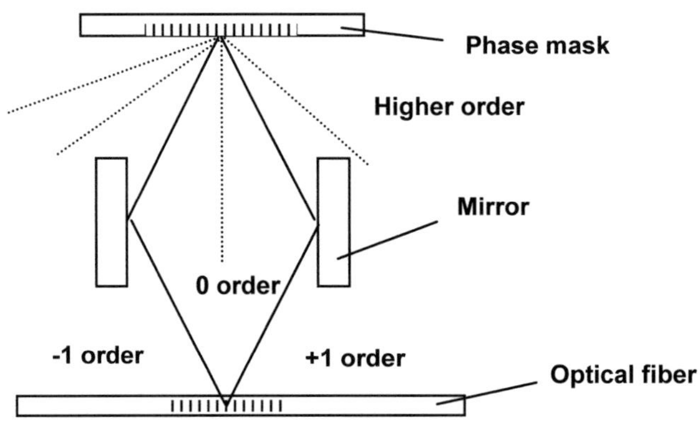

Fig. 8. Phase mask interferometer. The +/− primary light refracted by the phase mask is returned by the mirror.

In the present study, we used the phase mask interference method shown schematically in Fig. 8 as the production/fabrication process. It combines the high reproducibility of the phase mask method and the broad wavelength response of the two-beam interference technique. Only the +/− primary light refracted by the phase mask is returned by the mirror, and the interference fringe period can be controlled by adjusting its angle. It is therefore possible to prevent interference with higher-order diffraction light, and thus reduce passband loss. The key functions required in the fiber grating filter include transmission at the wavelength of the light input to the SOA and feedback to the SOA of a part of the surrounding light generated during amplification. We therefore fabricated the FBG with the transmission and reflection spectra shown in Fig. 9, and with a 92% reflectance ratio. The surrounding light has a certain fixed spread, and it is necessary to provide a spread in the reflection band of the fiber grating. For this purpose, two chirp gratings with slightly different reflection center wavelengths (1548 and 1554 nm) were written in mutually close proximity on the optical fiber. The use of the phase mask interference technique facilitated the selection and control of the reflection center wavelengths, adjustment of the writing positions on the fiber, and other aspects. High stability and reproducibility in the FBG fabrication were also ensured through the use of an original adjustment algorithm and computerized control of the optical system for optical fiber position and mirror angle, unlike the conventional processes which usually depend on visual methods. The tip of the prototype FBG was lensed and coupled to one end of the SOA, thus obtaining a negative-feedback semiconductor optical amplifier (NF-SOA).

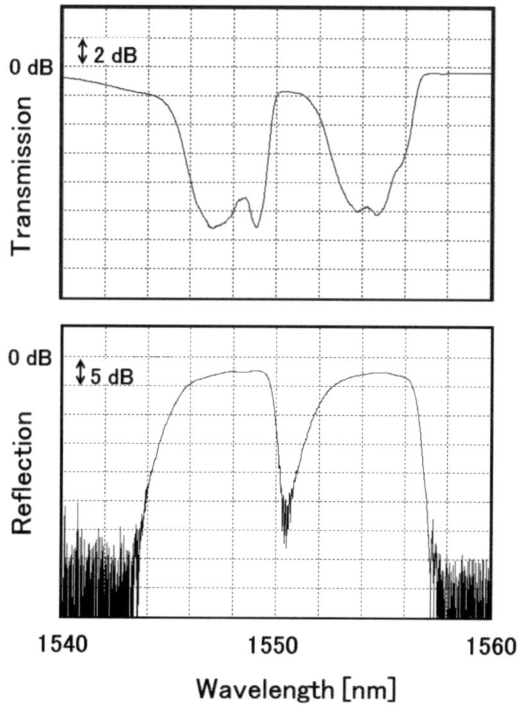

Fig. 9. Transmission and reflection spectrum of the fiber Bragg grating for negative feedback optical amplifier.

3.2 Measurement results of NF-SOA

The composition of the NF-SOA and the measurement system is shown schematically in Fig. 10. The SOA was an InGaAsP-InP ridge semiconductor optical amplifier. The center wavelength of the InGaAsP multi-quantum-well active layer was approximately 1500 nm and the gain at the fiber end was approximately 17 dB with a 250 mA electric current input. As shown by the saturation gain curve in Fig. 11 with the 1550-nm input signal, the gain was constant in the light output power range of −25 to 0 dBm and the saturation power was 7 dBm. Fig. 12 shows the spectrum of the amplified spontaneous emission (ASE) of the SOA, with the characteristics shown in Fig. 9 for the FBG mounted at the output end of the SOA clearly evident in the vicinity of 1550 nm. In the measurement system, the intensity of the wavelength-tunable laser beam was modulated by a lithium niobate (LN) optical modulator. After amplification of the signal by an Er-doped optical fiber amplifier (EDFA), it was passed through a 1-nm bandwidth-tunable filter, and its intensity was then modulated by a variable optical attenuator (VOA) to obtain the input signal. To experimentally verify that it is possible to reduce the output light waveform distortion as an effect of the negative feedback optical amplification, we measured the bit error rate in the transmission of random bits used in optical communication, using an Anritsu PM-1800 signal quality analyzer. The pattern generator provided a continuous signal with an NRZ 2^{31}-1 pseudorandom bit sequence at 10 Gbps.

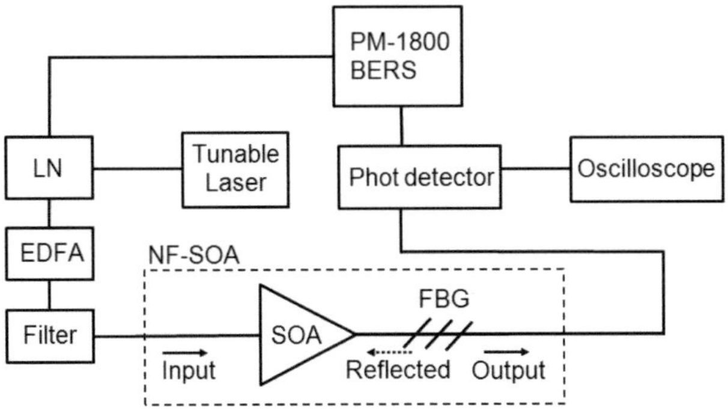

Fig. 10. Block diagram of the negative feedback semiconductor optical amplifier (NF-SOA) and measurement system.

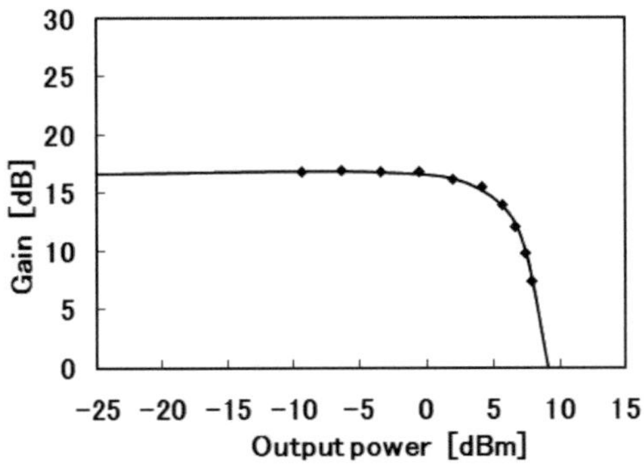

Fig. 11. Characteristic of saturation gain curve for NF-SOA.

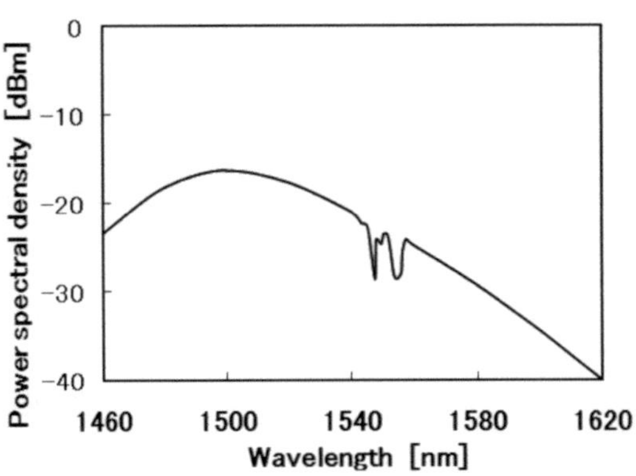

Fig. 12. Amplified spontaneous emission (ASE) spectrum for NF-SOA.

Fig. 13 shows eye-pattern waveforms observed on the oscilloscope, for output light with (a) NF-SOA and (b) SOA, respectively. Figs. 14(a)-14(d) show the eye-pattern waveforms observed on the oscilloscope, for output light with NF-SOA input signals (−7 dBm) at laser beam wavelength settings of 1530, 1544, 1550, and 1558 nm, respectively.

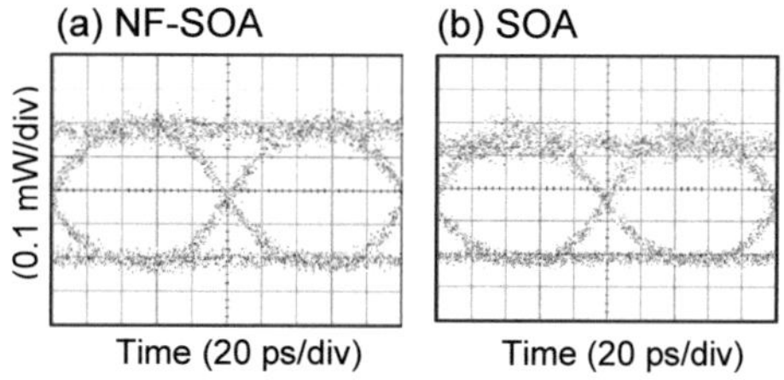

Fig. 13. Eye-pattern waveforms of (a) NF-SOA and (b) SOA at 1550 nm.

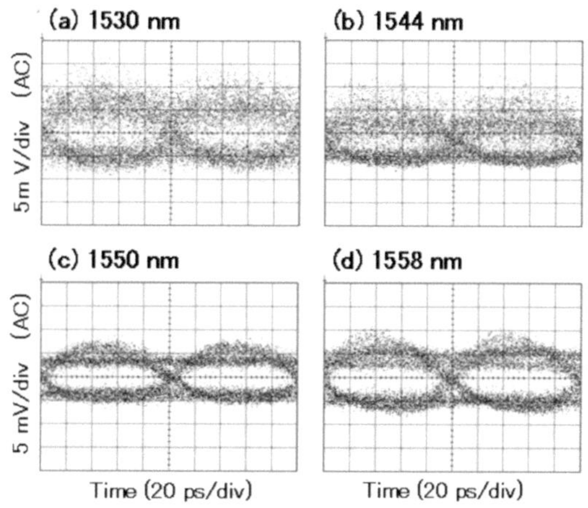

Fig. 14. Eye-pattern waveforms of (a) 1530, (b) 1544, (c) 1550 and (d) 1558 nm for NF-SOA.

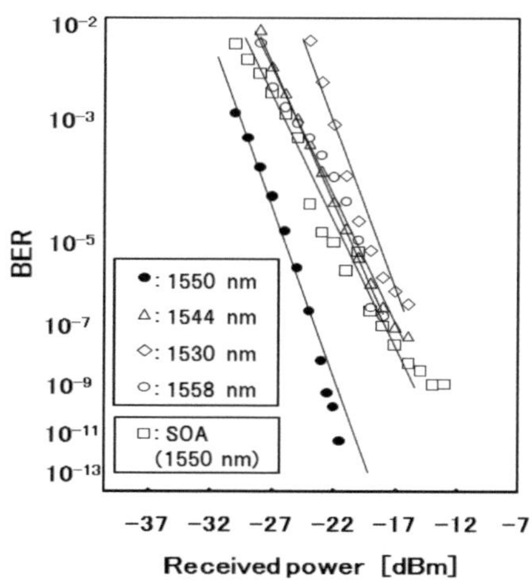

Fig. 15. Measured BER results for the NF-SOA. 1530 (•), 1544 (△), 1550 (◊), and 1558 (○) nm.

The observed aperture ratios were (a) 0.35, (b) 0.37, (c) 0.50, and (d) 0.41, and the waveform at the FBG transmission center wavelength of 1550 nm, as shown in Fig. 14(c), thus exhibited an eye pattern with a better aperture ratio than the other waveforms. Fig. 15 shows the BER measurements for the NF-SOA with laser settings of 1530 nm (•), 1544 nm (∆), 1550 nm (◊), and 1558 (○) nm. As shown by the BER curves, the 1550-nm wavelength resulted in lower BER values than the other wavelengths at the same received power and a power penalty of approximately 1 dB or more. For reference, the measured BER values for the conventional SOA at an optical input wavelength of 1550 nm (□) are also shown. These BER curves demonstrate that the NF-SOA provides lower BER values than the SOA at 1550 nm with the same level of received power. As shown clearly in Fig. 15, the NF-SOA BER curve at 1550 nm achieves a higher negative power penalty than the curve without NF. This indicates that an NF-SOA module can reduce the bit error rate more than an SOA module without NF when random optical signals are transmitted.

Wavelength	G_{max}	G_{min}	PDG	NF
1530 nm	18.3 dB	17.7 dB	0.62 dB	8.3 dB
1544 nm	16.8 dB	16.2 dB	0.63 dB	6.1 dB
1550 nm	17.1 dB	16.0 dB	0.68 dB	5.1 dB
1558 nm	17.2 dB	15.8 dB	0.69 dB	6.2 dB

Table 1. Specification of gain, polarization dependent gain, and noize figure for NF-SOA.

Wavelength	G_{max}	G_{min}	PDG	NF
1530 nm	24.8 dB	24.1 dB	0.78 dB	9.3 dB
1544 nm	22.9 dB	22.7 dB	0.70 dB	8.9 dB
1550 nm	22.0 dB	21.5 dB	0.75 dB	9.0 dB
1558 nm	20.9 dB	20.1 dB	0.76 dB	9.1 dB

Table 2. Specification of gain, polarization dependent gain, and noize figure for SOA.

The characteristics of the NF-SOA and SOA are summarized in Tables 1 and 2, respectively, in terms of maximum and minimum gain (G_{max} and G_{min}), polarization dependent gain (PDG), and noize figure (NF). PDG, the gain of the optical amplifier dependent on the polarization state of the input light signal, is generally low in high-performance amplifiers. The PDG was determined by Muller's method, and the NF was calculated from the S/N ratio of the optical signal (OSNR). As shown by comparison of Tables 1 and 2, the feedback by the FBG of part of the spontaneous emission resulted in a somewhat smaller gain with the NF-SOA than that obtained with the conventional SOA. The NF values of 8.3 dB at 1530 nm, 6.1 dB at 1544 nm, and 6.2 dB at 1558 nm, and particularly 5.1 dB at the FBG transmission center wavelength of 1550 nm obtained with the NF-SOA, however, were all smaller that the value of approximately 9 dB obtained with the conventional SOA at all of these wavelengths, thus clearly demonstrating the noise reduction effect of the negative feedback optical amplification. The PDG values of the NF-SOA (Table 1), moreover, were all approximately 0.1 dB smaller than those of the conventional SOA (Table 2) at all wavelengths, further demonstrating its low polarization dependence and excellent characteristics.

4. All-optical triode

4.1 All-Optical Triode Based on NF-SOA

Fig. 16 shows the block diagram of the optical triode. It consists of two cascaded wavelength converters based on two NF-SOAs. Two tunable lasers are used for the input and the control signals, which are modulated by the mean of electro-optic modulators connected to an electrical synthesizer and a pattern generator respectively. Both input and control signals have the same wavelength equal to 1551 nm. The modulated input signal is fed into the first SOA (SOA-1) using an optical Add/drop filter (central wavelength: 1551 nm, spectral half-width: 0.8 nm). Because of the XGM mechanism in the SOA-1, the FBG-1 reflection beam (Ls) contains an inverted replica of the information carried by the input signal. The modulated output Ls is then injected together with the control signal into the second SOA (SOA-2) by using an optical coupler. A band pass filter (spectral half-width: 0.8 nm) of the central wavelength equal to 1551 nm is placed after the SOA-2. The average output power is measured at the output of the filter. In the second wavelength conversion stage the Ls beam is converted back to the 1551 nm wavelength depending on the power of the control beam.

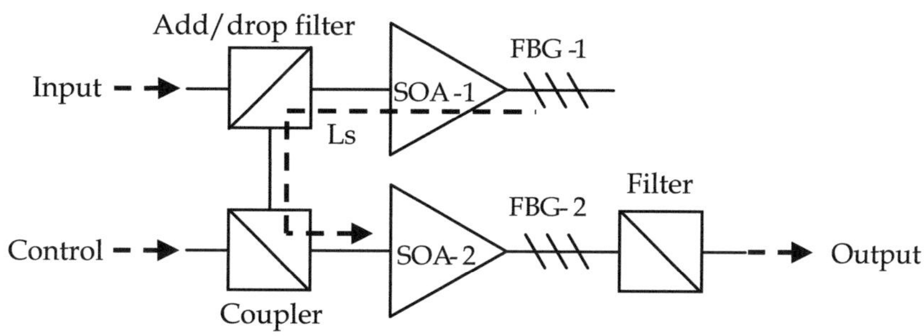

Fig. 16. Block diagram of the all-optical triode. SOA-1 and SOA-2: Semiconductor optical amplifier, FBG-1 and FBG-2: Fiber Bragg grating.

Figs. 17(a) and 17(b) show the input and control signal waveforms, respectively. They have been measured with a fast photodiode connected to a sampling head oscilloscope. The modulation degree and frequency of the input signal are 90% and 1 GHz, respectively. The modulation degree M is equal to $100 \times (P_{max} - P_{min})/(P_{max} + P_{min})$ [%], where P_{max} and P_{min} represent the maximum and minimum intensities of the signal, respectively. Fig. 17(c) shows the output waveforms for four values of the control average power P_c. The information of the input signal is transmitted to the output only when the power of the control signal is at the high logical level, whereas it is blocked when the control power is low. The magnitude of the output waveform is controlled by the power of the control beam: the larger P_c, the higher the peak power at the output.

Fig. 18 shows the dependence of the output power (P_{out}) on the input power (P_{in}) for four values of the control power (P_c). It can be seen that the output power increases when the control power is increased. Fig. 19 shows the dependence of the modulation degree of the output signal on the input signal frequency. By increasing the input signal frequency from 0.1 to 10 GHz, the modulation degree M decreases from 60% to 0%. For frequencies smaller than 2 GHz, M remains approximately the same as the input modulation degree.

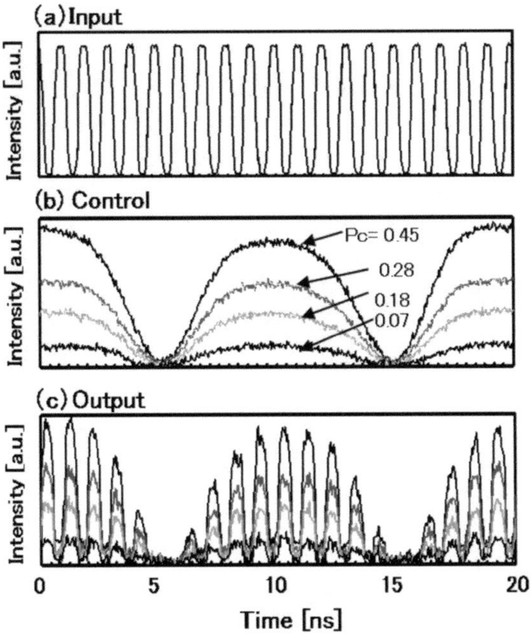

Fig. 17. (a) Input and (b) control signal waveforms. (c) Output waveforms for four values of control average power (Pc=0.45, 0.28, 0.18, 0.07).

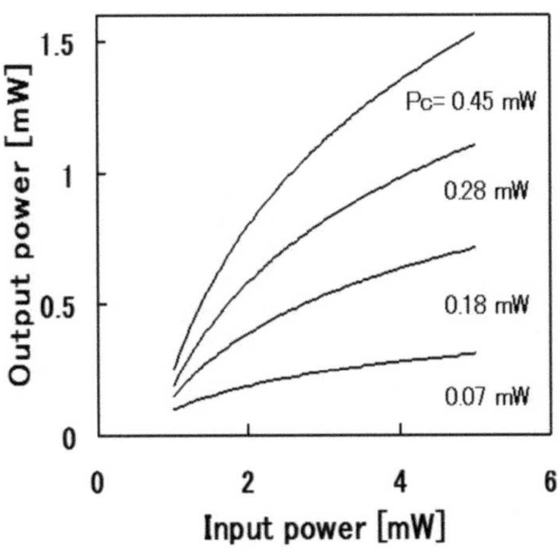

Fig. 18. Output versus input power four for values of the control average power (Pc=0.45, 0.28, 0.18, 0.07).

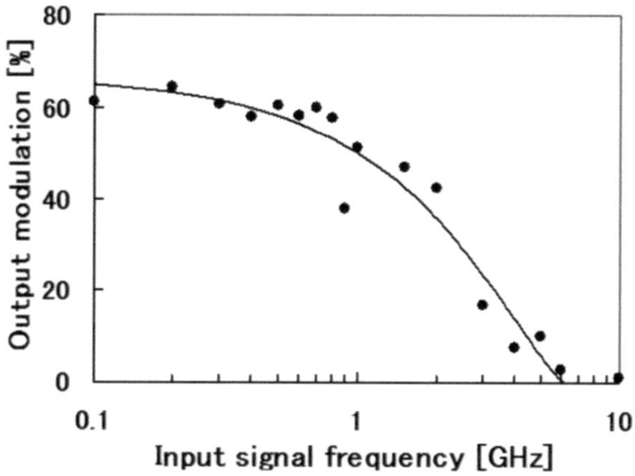

Fig. 19. Output modulation degree versus input signal frequency.

4.2 All-Optical Triode using reflective semiconductor optical amplifiers

Fig. 20 shows the block diagram of the optical triode. It consists of two cascaded wavelength converters based on two RSOAs, which are similar to standard SOAs with one of the facets high-reflection coated. We used two RSOAs based on ridge waveguide structure InGaAsP/InP bulk material. The composition of the InGaAsP active layer is chosen to have a gain peak wavelength around 1555 nm. The small signal fiber-to-fiber gain is around 20 dB and the output saturation power is approximately 2 mW measured at 1555 nm with a bias current of 250 mA. Two tunable lasers are used for the input and the control signals, which are modulated by the mean of electro-optic modulators connected to an electrical synthesizer and a pattern generator respectively. Both input and control signals have the same wavelength equal to 1555 nm. A conventional distributed feedback laser diode is used as bias laser. Its power measured at the point a in Fig. 20 is 0.89 mW (P_b) and the wavelength is 1548 nm. The modulated input signal and the bias beam are fed into the first RSOA (RSOA-1) using an optical coupler and a circulator (C1). The polarization state of both beams is set by maximizing the modulation degree of the bias beam at the RSOA-1 output by using two polarization controllers (P.C.'s).

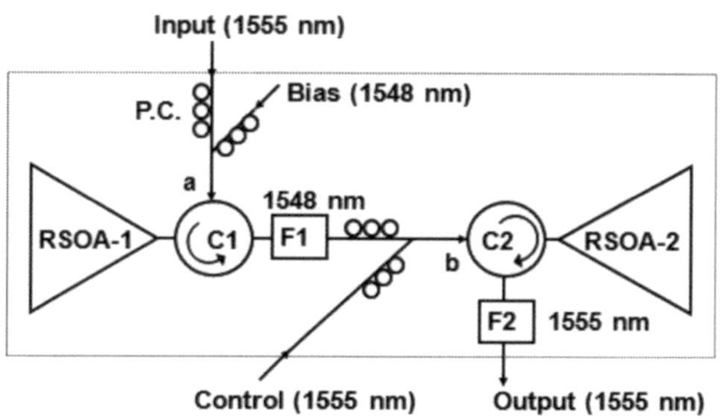

Fig. 20. Block diagram of the optical triode. P.C.: polarization controller, C: circulator, F: tunable band pass filter.

The use of the P.C.'s was necessary because the two employed RSOA are polarization dependent. By using polarization independent devices, the P.C.'s depicted in Fig. 20 are not necessary. A tunable band pass filter at the output of C1 (F1, spectral half-width: 0.6 nm) is set at the center wavelength of 1548 nm to filter away the input signal. Because of the XGM mechanism in the RSOA-1, the bias beam contains an inverted replica of the information carried by the input signal. The modulated output of F1 is then injected together with the control signal into the second RSOA (RSOA-2) by using an optical coupler and a circulator (C2). A second tunable band pass filter (F2, spectral half-width: 1.3 nm) of the central wavelength equal to 1555 nm is placed after the RSOA-2 and it is used to suppress the 1548 nm beam. The average output power is measured at the output of the filter F2. In the second wavelength conversion stage the 1548 nm beam is converted back to the 1555 nm wavelength depending on the power of the control beam.

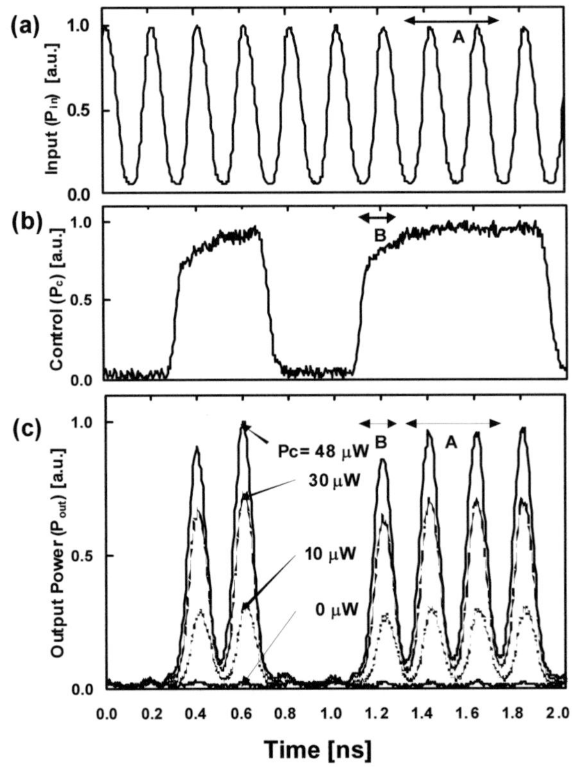

Fig. 21. Input (a) and control (b) signal waveforms. (c) Output waveforms for four values of the control average power (P_c = 0, 10, 30, and 48 μW).

Figs. 21(a) and 21(b) show the input and control signal waveforms, respectively. The average input and control power are 0.86 mW and 48 μW, measured at the points a and b in Fig. 20, respectively. They have been measured with a fast photodiode connected to a sampling head oscilloscope. The modulation degree and frequency of the input signal are 90% and 5 GHz, respectively. The modulation degree M is equal to $100 \times (P_{max} - P_{min})/(P_{max} + P_{min})$ [%], where P_{max} and P_{min} represent the maximum and minimum intensities of the signal, respectively. Fig. 21(c) shows the output waveforms for four values of the control average power P_c. The information of the input signal is transmitted to the output only when the power of the control signal is at the high logical level, whereas it is blocked when the control power is low. The magnitude of the output waveform is controlled by the power of the control beam: the larger P_c, the higher the peak power at the output. The unequal peak output power between the points A and B in Fig. 21(c) is due to a imperfect shape of the control beam as it is evident from Fig. 21(b). Fig. 22 shows the dependence of the output power (P_{out}) on the input power (P_{in}) for four values of the control power (P_c). These transmission characteristics have been obtained by plotting the data from the marked position A in Figs. 21(a) and 21(c) using the time as parameter. For P_c = 0 μW, the output power is very small, less than 0.1 mW even if the input power varies from 0.2 to 1.8 mW. For the other values of P_c, it can be seen that the output power increases when the control power is increased. If we define the output-input gain parameter $g_{tr} = \Delta P_{out} / \Delta P_{in}$, we obtain values of 0.6, 1.1 and 1.6 for P_c = 10, 30, and 48 μW, respectively, at the input power of 1 mW, whereas g_{tr} is almost equal zero, for P_c = 0. The output-control amplification parameter f, defined as $\Delta P_{out} / \Delta P_c$, depends on P_c and P_{in}. For example, when the control peak power increases from 0 to 96 μW, the peak output power increases from 0.1 to 3.3 mW. At the input power of 1.8 mW, f is approximately 33.

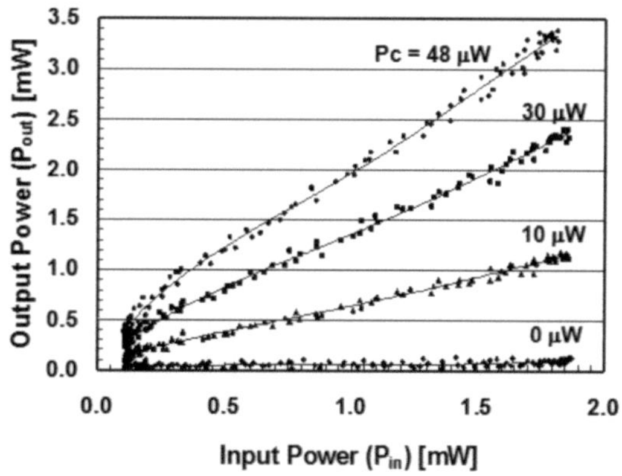

Fig. 22. Output power (P_{out}) versus input power (P_{in}) for four values of the control average power.

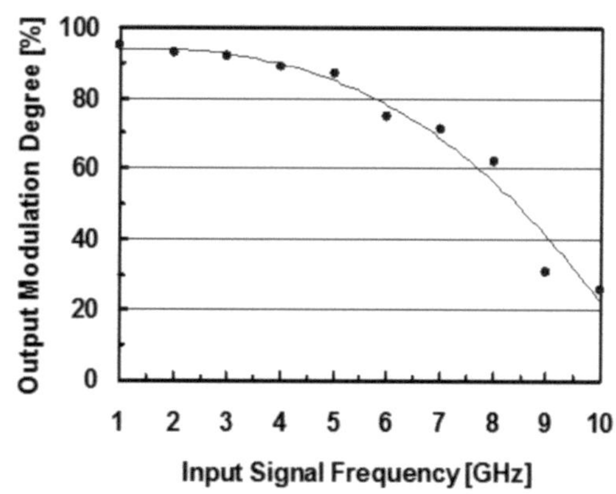

Fig. 23. Output modulation degree versus input signal frequency.

This means that a small change in the control power is able to induce a change 33 times larger at the output. Because of the gain saturation in the SOA, the parameter f decreases when either the input signal power is decreased or the control signal power is increased.
Fig. 23 shows the dependence of the modulation degree of the output signal on the input signal frequency, measured in position marked with A in Fig. 21(c) by using an input modulation degree of approximately 90 %. By increasing the input signal frequency from 1 to 10 GHz, the modulation degree M decreases from 90% to 30%. For frequencies smaller than 5 GHz, M remains approximately the same as the input modulation degree.

5. Discussion

The basic concept of the proposed negative feedback optical amplification effect is shown schematically in Fig. 24.

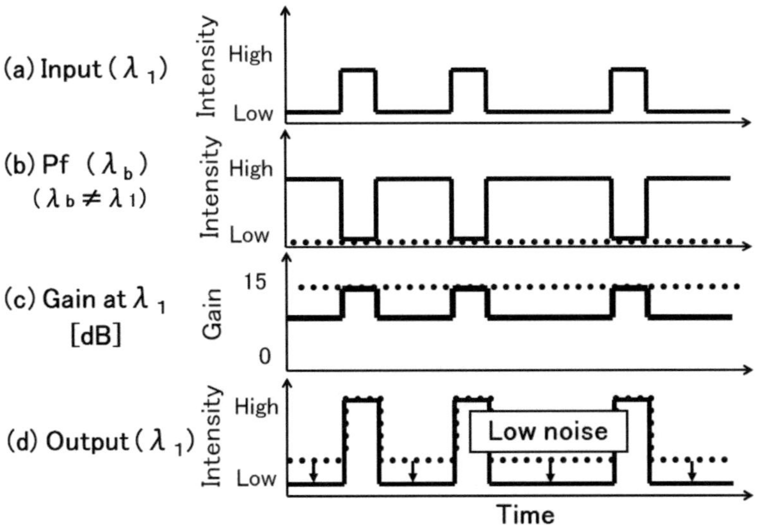

Fig. 24. Concept diagram of a negative feedback optical amplification effect. The straight-line represents the case where the SOA was used with negative feedback, and the dotted line represents the case of the SOA without negative feedback.

When input light of wavelength λ_1 is injected into the SOA as shown in Fig. 24(a), the spontaneous emission at a wavelength λ_b surrounding λ_1, which is generated by the cross-gain modulation within the SOA, undergoes intensity modulation with a reverse phase to that of the input light. This change in intensity is delineated by the solid line in Fig. 24(b). When the resulting P_f (λ_b) is fed back to the SOA as negative feedback light, in the cross-gain modulation within the SOA the gain at wavelength λ_1 is modulated as shown by the solid line in Fig. 24(c). In the absence of feedback light P_f, as shown by the dotted line in Fig. 24(c), the SOA gain is constant. In a constant-gain amplifier, the signal and the noise are both amplified with the same gain and it is therefore impossible in principle to increase the signal to noise ratio (S/N ratio). With the negative feedback P_f, in contrast, the gain at wavelength λ_1 is modulated as shown by the solid line in Fig. 24(c). In an SOA with dynamic gain modulation of this nature, with input light injection as shown in Fig. 24(a), the gain is small for input light of low intensity and the degree of amplification is thus low, and the zero level of the output light is lower than it would be without the negative feedback light P_f. As shown in Fig. 24(d), a high degree of modulation (high extinction ratio) can thus be obtained and it is therefore possible to increase the S/N ratio. Fig. 24 is simply a qualitative illustration of the improvement in the level of output light modulation, but in a previous report (Maeda, 2006), we showed that reduced waveform distortion in the output light and improved modulation can be achieved, as experimentally observed effects of the negative feedback optical amplification.

Positive feedback is performed in optical amplifiers to induce laser oscillation. In electronics, similarly, oscillation tends to occur in an amplifier with positive feedback, and various electronic circuits are utilized as oscillators. Negative feedback is widely used in electronics, as it enables electronic signal amplification with high gain stability and low noise and also makes it possible to stabilize inherently unstable semiconductor elements, for low-noise electric signal amplification. In electronic negative feedback amplification circuits, the degree of amplification is determined by the resistance, which in turn determines the feedback ratio, and effective techniques are essential to avoid temperature dependence in the degree of amplification by transistors, which tend to be highly temperature dependent. In optical amplification, in contrast, the utilization of negative feedback is currently non-existent. For the continuing advance of optoelectronics with optical signals, however, negative feedback optical amplification techniques will presumably be essential. The FBG utilized in the present study is composed primarily of glass, which has a weaker temperature dependence than other semiconductor materials. In negative feedback amplification and other applications, the weak temperature dependence of FBGs in comparison with that of conventional semiconductor SOAs may be expected in principle to facilitate the achievement of systems with increased temperature stability and more specifically the achievement of a superior NF and other optical amplification characteristics through the control of comparatively unstable SOAs by passive FBGs. Detailed experimentation will be required for verification of this possibility.

The improvement of approximately 0.1 dB in PDG observed in the present study with the NF-SOA, as compared with the conventional SOA (Tables 1 and 2), may be attributable to the polarization independent nature of the FBG. As the FBG is not polarization dependent, neither is the light which it returns to the SOA. Although the signal light is not directly influenced by the absence of polarization dependence in the FBG, the influence of polarization on the total SOA gain is presumably reduced by the negative feedback amplification, thus reducing the overall PDG. The data in the present study represents an early stage of experimentation, however, and further data and elucidation will be necessary in relation to the PDG.

It is also necessary to consider the response speed of the negative feedback optical amplifier. A failure in superposition of the signal light and the return light without time delay will result in underutilization of the negative amplification effect or even in an adverse effect. SOA cross-gain modulation is in its principle of operation equivalent to the modulation of a probe light by a signal light. In the signal light input state (with level 0 changed to level 1), the induced emission is the main factor in the negative amplification feedback effect. In the change to the signal light non-input state (with level 1 changed to level 0), however, the spontaneous return of carrier density to the original level becomes the governing factor, and the feedback effect is influenced by the degree of relaxation at approximately 100 ps or higher. For high-speed cross-gain modulation, it is therefore essential to ensure high-speed carrier recovery. The effect of the FBG for negative feedback optical amplification was demonstrated in the present study, but this effect cannot be expected if the input light signal and the negative feedback light signal do not superpose in time. In the prototype fabricated for this study, the separation between the SOA and the FBG was 1 mm or less. Calculations based on the results of this study indicate that, so long as a degree of signal superposition of about 90% can be assumed, the negative feedback effect can be maintained at up to about 10 Gbps, but further investigation will be necessary in this regard.

The present study included an investigation of the dependence of the BER and NF on the wavelength of the input signal. As indicated by Fig. 12, the SOA gain was shown to be wavelength dependent. Because of its origin in spontaneous emission, in contrast, the intensity of the negative feedback light returned by the FBG is independent of the input signal wavelength and it can therefore be considered essentially constant. The intensity of the negative feedback light acting on the input signal determines the degree of negative feedback amplification. The effect of the negative feedback light amplification will therefore be relatively small for input light signals of high-gain wavelength. As shown by the characteristics in Fig. 12 and Tables 1 and 2, an input light signal of 1530 nm showed high gain and the output light signal at that frequency was thus of higher intensity than at the other wavelengths. As the intensity of the negative feedback light returned by the FBG was essentially constant, it is apparently for this reason that the negative feedback light amplification effect was smaller for an input signal of that wavelength than for the other wavelengths of 1544, 1550, and 1558 nm, and the noise reduction effect was accordingly smaller. Conversely, the noise reduction effect was apparently larger with input signals of 1550 and 1558 nm, for which the SOA gain was relatively small, than with the other wavelengths because of the relatively strong negative feedback amplification effect for light of those wavelengths. The results of the measurements also showed the BER and NF to be smaller for the input signal of 1550 nm, which was the FBG center wavelength, than for the input signal of 1558 nm. This may be an effect of the dependence of the speed of propagation through the medium on the wavelength of the light reflected by the FBG. One further aspect which should be noted is the effect of the feedback on the output signal strength, as the measurements showed that the achievement of low noise with the negative feedback light was accompanied by a somewhat lower output (i.e., a somewhat lower degree of amplification). Among the FBG guidelines for practical applications, it will therefore be essential to ensure that the required output will be maintained while obtaining the required negative feedback effect.

In the present study, we have shown that it is possible to achieve negative feedback optical amplification by dynamic SOA gain modulation in correspondence with the input signal. Dynamic gain modulation is obtained by SOA cross-gain modulation, with effective utilization of the spontaneous emission within the SOA. It was shown, moreover, that this can be achieved with a simple, readily fabricated structure consisting of an optical fiber bearing written-in FBGs in addition to the SOA, which is highly advantageous for industrial applications. The results demonstrated, moreover, that by utilizing the NF-SOA it is possible to resolve otherwise intractable problems relating to SOA polarization dependence and noise characteristics. It is therefore expected that the proposed negative optical amplifier will find applications in many areas of optoelectronics, and it is believed that the proposal of this technique as one means of achieving an optical version of feedback amplification technology may be of major significance.

All-optical signal processing is expected to have a wide application in the field of communication and computation due to its capability of handling large bandwidth signals and large information flows. Basic functions such as demultiplexing and switching can be achieved using all optical gates realized by optical nonlinearities in semiconductor materials. For future applications based on all-optical signal processing such as an optical computer, it will be very important to have optical devices showing the same functionality as transistors or triodes in electronics, including the capability of signal amplification. For the triode (i.e. three element tube), small voltage variations on the control grid produce large changes in plate current and voltage.

The proposed scheme shows several features that will be useful for the triode operation. Firstly the transmission gain of the device is directly controlled by the control beam power. Secondly it requires a small control power. The third advantage is that it shows optical signal amplification, which is very important for the cascadability of such devices. Finally, our setup is also flexible regarding the choice of the output wavelength, because we can either use the same wavelength as the input (as done here) or use a different one. Utilizing the advantages mentioned above, the all-optical triode has the potential to be the basis for logic gates and operational amplifiers in digital and analog optoelectronic circuits. This is in analogy with electronics, where devices like the transistors and triodes are the basis of logical gates and operational amplifiers. It is important to note that for the first time the principle of the all-optical triode based on SOAs is demonstrated. At the moment, the proposed scheme has some drawbacks in terms of speed and setup simplicity as compared to the integrated semiconductor devices (e.g. Mach-Zehnder interferometers). To be a practical viable solution to perform all-optical logic gate operation, our approach needs further improvements in terms of speed (for instance by using faster SOAs based on quantum dots).

6. Conclusion

In this study, a FBG for use as a negative feedback optical amplifier was produced by imprinting using the phase mask interference technique, which as a production method combines the high reproducibility of the phase mask technique and the wide wavelength range of the two-beam interference technique. The functions required of the FBG were that it transmit the wavelength of the input light injected into the SOA and simultaneously feed back to the SOA a part of the surrounding light generated in the amplification process. For this purpose, two chirp gratings with slightly different reflection center wavelengths were written onto the optical fiber in close proximity to each other. The use of the phase mask interference technique facilitated selection and control of the center wavelength, adjustment of the fiber write-in positions, and other process functions. The prototype FBG tip was lensed and coupled to one end of the SOA, thus forming a negative feedback semiconductor optical amplifier. In measurement of the bit error rate with a 10 Gbps pseudorandom signal with a wavelength of 1550 nm, which was the same as the FBG transmission center wavelength, the NF-SOA yielded an eye pattern with a higher aperture than those obtained at other frequencies. The BER curves obtained from the measurement results showed that the reduction in BER by the NF-SOA was clearly greater at 1550 nm than at the other wavelengths. At that wavelength, it resulted in a reduction of the noise figure to 5.1 dB. Based on these experimental findings, it was shown possible to reduce optical amplification noise to a low level with a NF-SOA using the FBG.

We also demonstrated an all-optical triode based on a tandem wavelength converter in two NF-SOAs using the same wavelength of input, control and output. This fact can be important for logic applications. The all-optical triode has the input-output transfer function that is similar to the plate voltage-current characteristic of the triode in electronics including the signal amplification function. Because of its characteristics this device can become a key component in future all-optical signal processing and promote the realization of an optical computer.

7. Acknowledgement

This work was supported in part by the Ministry of Education, Culture, Science and Technology of Japan, a Grant-in-Aid （21560048） for Scientific Research (C).

8. References

Maeda, Y.; Tanimoto, H.; Matsuo, T.; Takagi, M. & Nakayama, H. (2010). "Noize reduction effect of negative feedback semiconductor optical amplifiers using fiber Bragg grating based on phase mask interferometer", *Rev. of Laser Engineering,* Vol.38, No.3, pp.219–224.

Maeda, Y. (2006). "Negative feedback optical amplification effect based on cross-gain modulation in semiconductor optical amplifier" *Appl. Phys. Lett.,* Vol.88, No.10, pp.101108-1 - 101108-3.

Maeda, Y. & Occhi L. (2003). "All-optical triode based on a tandem wavelength converter using reflective semiconductor optical amplifiers", *IEEE Photon. Technol. Lett.,* Vol.15, No.2, pp.257-259.

Bachmann, M.; Doussiere, P.; Emery, J. Y.; N'Go, R.; Pommereau, F.; Goldstein, L.; Soulage, G. & Jourdan, A. (1996). "Polarisation-insensitive clamped-gain SOA with integrated spot-size convertor and DBR gratings for WDM applications at 1.55μm wavelength", *Electron. Lett.,* Vol.32, No.22 pp.2076-2078.

Qureshi, K. K.; Tam, H. Y.; Lu, C. & Wai, P. K. A. (2007). "Gain control of semiconductor optical amplifier using a bandpass filter in a feedback loop", *IEEE Photon. Technol. Lett.,* Vol.19, No.18, pp.1401-1403.

Stubkjaer, K. E. (2000). "Semiconductor optical amplifier-based all-optical gates for high-speed optical processing", *IEEE J. Quantum Electron.,* Vol. 6, No. 6, pp. 1428-1435.

Glance, B.; Wiesenfield, J. M.; Koren, U.; Gnauck, A. H.; Presby, H. M. & Jourdan, A. (1992). "High performance optical wavelength shifter", *Electron. Lett.,* Vol.28, No.18, pp.1714-1715.

Wiesenfeld, J. M. (1996). "Gain dynamics and associated nonlinearities in semiconductor optical amplifiers", *J. High speed electronics and systems,* Vol.7, No.1, pp.179-222.

Durhuus, T.; Joergensen, C.; Mikkelsen, B.; Pedersen, R. J. S. & Stubkjaer, K. E. (1994). "All optical wavelength conversion by SOAs in a Mach-Zehnder configuration", *IEEE Photon. Technol. Lett.,* Vol. 6, pp.53-55.

Sun, J. (2003). "Theoretical study on cross-gain modulation wavelength conversion with converted signal feedback"*IEE Proc. Optoelectron.,* Vol. 150, No.6, pp.497-502.

Part 4

Erbium-Doped Amplifiers and Lasers

12

Coherent Radiation Generation and Amplification in Erbium Doped Systems

Sterian Andreea Rodica
University Politehnica of Bucharest, Romania

1. Introduction

The erbium doped fiber systems results into important advantages for information processing and transmission like: possibility of easy integration, highly efficiency and gain, immunity to crosstalk, low noise and high saturation output power (Agrawal, 1995 & 1997; Desurvire, 1995; Sterian, 2006).

During the last years, they have been published many studies carrying out the improving and optimization of the coherent optical systems by "computer experiments". Based on some computational models known in the literature, this chapter proposes to present the main author's results obtained by numerical simulation concerning some coherent optical amplifier and laser systems.

Will be presented firstly the computational model which govern the amplification regime of an uniform doped optical fiber under the form of a system of the nonlinear transport coupled equations, respectively for the signal and for the pumping. This system was used for numerical simulation of the amplification phenomena by a Runge - Kutta type method (Agrawal, 1995; Sterian, 2006; Press et al., 1992).

The study continues with the computational model presentation used for numerical analyses of the laser system doped with Er^{3+} ions, both of the crystal type and of the optical fiber laser type (Pollnau et al., 1996; Maciuc et al., 2001; Sterian & Maciuc, 2003).

The main problems studied by numerical simulation, using these models known in literature are: the amplification, the laser efficiency and threshold for different optical pumping wavelengths, the dependence of the output optical power on the levels life time, the influence of the host materials on the output power and the time dependent phenomena, stability and nonchaotic regime of operation (Maciuc et al., 2001; Pollnau et al., 1994).

We realized the numerical simulation of the erbium doped fiber amplifier concerning the functional and constructive parameters. It was demonstrated that for the Er^{3+} doped fiber laser, the optimum operation condition are obtained for λ = 791 nm, when the upper laser level is directly pumped but in the presence of the „colaser" process which improves two times the laser efficiency for λ = 3 μm.

The Er^{3+} laser system functionment and optimization was studied for different host material, using 3D numerical simulation to take into account the characteristic parameters variations in the range of values resulting from experience; the material selection recommends as efficient materials $LiYF_8$ and $BaYF_8$.

We also put into evidence the existence of a strong interdependence between active medium parameters having important role in the designing of the erbium laser systems with given functional parameters.

We demonstrated the stable, nonchaotic operation of the analyzed laser systems and the modulation performance of them using the communications theory methods.

For the erbium doped fiber laser we explained the complex dynamics of this type of device by simulating the time dependence of the output power correlated with the corresponding changes in the populations of the implied levels.

Another author's numerical simulations refers to nonlinear effects in optical fibers systems (Sanchez et al., 1995; Ninulescu & Sterian, 2005; Ninulescu et al., 2006;). Self - pulsing and chaotic dynamics are studied numerically in the rate equations approximation, based on the ion - pair formation phenomena (Sanchez et al., 1993; Sterian & Ninulescu, 2005; Press et al., 1992).

The developed numerical models concerning the characterization and operation of the EDFA systems and also of the laser systems, both of the "crystal type" or "fiber type" realized in Er^{3+} doped media and the obtained results are consistent with the existing data in the literature.

That was possible due to the valences of the computer experiment method which permits a complex study taking into account parameters intercorrelations by simulating experimental conditions, as have been shown.

The used fourth order Runge - Kutta method for the numerical simulation demonstrates the importance of the "computer experiments" in the designing, improving and optimization of these coherent optical systems for information processing and transmission (Stefanescu et al., 2000, 2002, 2005; Sterian, 2002; Sterian, 2007).

Some new feature of the computer modeled systems and the existence of new situations have been put into evidence, for designers utility in different applications (Petrescu, 2007; Sterian, 2008). Our results are important also for the optimization of the functioning conditions of this kind of devices.

2. Fiber amplifier

2.1 Transport equations for signal and pumping

Let us consider an optical fiber uniformly doped, the concentration of the erbium ions being N_0. The pumping is done with a laser radiation having λ_p wavelength and the pumping power P_p, the absorption cross - section being σ_p^a. The population densities of the atoms on each of the three levels involved in laser process are: $N_1(t,z), N_2(t,z)$ respectively $N_3(t,z)$ which verify the equations:

$$N_3(t,z) \cong 0 \tag{1}$$

$$N_1(t,z) + N_2(t,z) = N_0 . \tag{2}$$

The necessary condition for radiation amplification in this kind of systems is as in the laser case the population inversion.

In the next presentation we refer to the energy levels diagram presented in figure 1 where: σ_s^a is the absorption cross-section for the signal; σ_s^e is the stimulated emission cross-section corresponding to the signal; σ_p^a is the absorption cross-section for the pumping radiation and τ is the relaxation time by spontaneous emission.

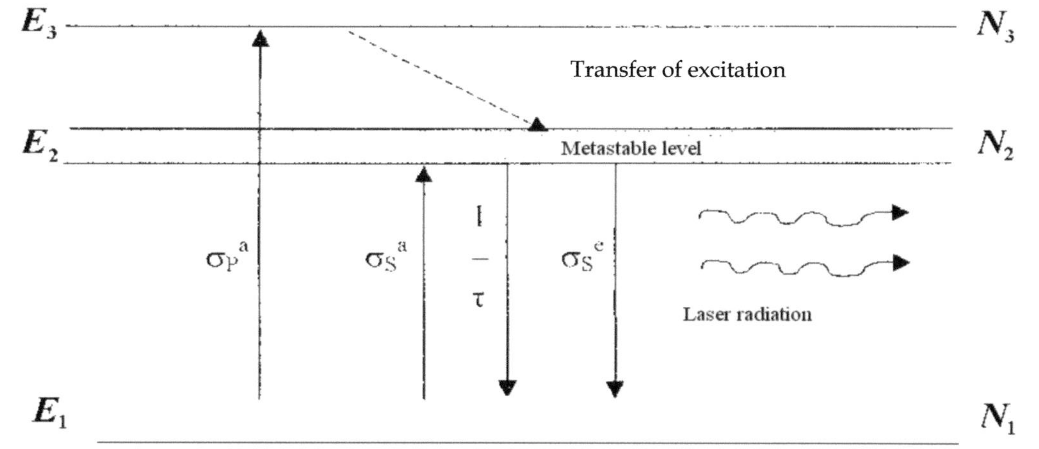

Fig. 1. The diagram of the energy levels involved in radiation amplification

For this system of energy levels on can write three rate equations: one for the population of the E_2 level and two transport equations for the fluxes of the signal and pumping. These rate equations are respectively (Agrawal, 1995):

$$\frac{\partial}{\partial t}N_2(t,z)=\frac{\sigma_p^a N_1(t,z)\cdot I_p(t,z)}{h\nu_p}+\frac{\sigma_s^a N_1(t,z)\cdot I_s(t,z)}{h\nu_s}-\frac{N_2}{\tau}-\frac{\sigma_s^e N_2(t,z)\cdot I_s(t,z)}{h\nu_s}; \quad (3)$$

$$\frac{1}{c}\cdot\frac{\partial}{\partial t}I_p(t,z)=-\frac{\partial}{\partial z}I_p(t,z)-\sigma_p^a\cdot N_1(t,z)\cdot I_p(t,z); \quad (4)$$

$$\frac{1}{c}\cdot\frac{\partial}{\partial t}I_s(t,z)=-\frac{\partial}{\partial z}I_s(t,z)+\sigma_s^e\cdot N_2(t,z)\cdot I_s(t,z)-\sigma_s^a\cdot N_1(t,z)\cdot I_s(t,z); \quad (5)$$

where:

$W_p=\frac{\sigma_p^a\cdot I_p(t,z)}{h\nu_p}$ is the absorption rate for the pumping; $W_s^a=\frac{\sigma_s^a\cdot I_s(t,z)}{h\nu_s}$ - is the absorption rate for the signal; $W_s^e=\frac{\sigma_s^e\cdot I_s(t,z)}{h\nu_s}$ - is the stimulated emission rate; $\frac{1}{\tau}$ - is the spontaneous emission rate; $\sigma_p^a\cdot N_1(t,z)$ - is the rate of pumping diminishing by absorption; $\sigma_s^e\cdot N_2(t,z)$ - rising rate of the signal by stimulated emission and $\sigma_s^a\cdot N_1(t,z)$ - is the rate of signal diminishing by absorption. (It admit that $W_s^a=W_s^e=W_p$).

In the same time the initial condition are:

$$I_p(t,0)=I_p(t) \quad (6)$$

$$I_s(t,0) = I_s(t)\,. \tag{7}$$

If the next conditions are fulfilled:

$$\frac{\partial}{\partial t} N_2(t,z) = 0\,, \tag{8}$$

$$\frac{\partial}{\partial t} I_p(t,0) = \frac{\partial}{\partial t} I_p(t,z) = 0\,, \tag{9}$$

$$\frac{\partial}{\partial t} I_s(t,0) = \frac{\partial}{\partial t} I_s(t,z) = 0, \tag{10}$$

one obtain the steady state equations:

$$\frac{\sigma_p^a N_1(t,z)\cdot I_p(t,z)}{h\nu_p} + \frac{\sigma_s^a N_1(t,z)\cdot I_s(t,z)}{h\nu_s} - \frac{N_2}{\tau} - \frac{\sigma_s^e N_2(t,z)\cdot I_s(t,z)}{h\nu_s} = 0\,, \tag{11}$$

$$\frac{\partial}{\partial z} I_p(t,z) = -\sigma_p^a \cdot N_1(t,z)\cdot I_p(t,z)\,, \tag{12}$$

$$\frac{\partial}{\partial z} I_s(t,z) = \sigma_s^e \cdot N_2(t,z)\cdot I_s(t,z) - \sigma_s^a \cdot N_1(t,z)\cdot I_s(t,z)\,. \tag{13}$$

By eliminating of the populations $N_1(t,z)$ and $N_2(t,z)$, it results the equivalent system of nonlinear coupled equations:

$$\frac{\mathrm{d}I_p}{\mathrm{d}z} = -\sigma_p^a \cdot I_p N_0 \cdot \frac{\dfrac{\sigma_p^a \cdot I_p}{h\nu_p} + \dfrac{\sigma_s^a \cdot I_s}{h\nu_s}}{\dfrac{\sigma_p^a \cdot I_p}{h\nu_p} + \dfrac{\sigma_s^a \cdot I_s}{h\nu_s} + \dfrac{1}{\tau} + \dfrac{\sigma_s^e \cdot I_s}{h\nu_s}}\,, \tag{14}$$

$$\frac{\mathrm{d}I_s}{\mathrm{d}z} = \sigma_s^a \cdot I_s N_0 \cdot \left[\frac{\sigma_s^e + \sigma_s^a}{\sigma_s^a} \cdot \frac{\dfrac{\sigma_p^a \cdot I_p}{h\nu_p} + \dfrac{\sigma_s^a \cdot I_s}{h\nu_s}}{\dfrac{\sigma_p^a \cdot I_p}{h\nu_p} + \dfrac{\sigma_s^a \cdot I_s}{h\nu_s} + \dfrac{1}{\tau} + \dfrac{\sigma_s^e \cdot I_s}{h\nu_s}} - 1 \right]. \tag{15}$$

In the upper equations, there are involved the parameters: $h = 6{,}626\cdot 10^{-34}\,\mathrm{Js}$ - the Planck constant; $c = 2{,}99\cdot 10^{8}\,\mathrm{m/s}$ - the light velocity in vacuum; $\tau = 10^{-2}\,\mathrm{s}$ - the relaxation time for spontaneous emission; $\sigma_p^a = 2\cdot 10^{-16}\,\mathrm{m}^2$ -the absorption cross-section for pumping; $\sigma_s^a = 5\cdot 10^{-16}\,\mathrm{m}^2$ -the absorption cross-section for signal; $\sigma_s^e = 7\cdot 10^{-15}\,\mathrm{m}^2$ -the stimulated emission cross-section for signal; $\lambda_p = 980\cdot 10^{-9}\,\mathrm{m}$ - the pumping radiation wavelength; $\lambda_s = 1550\cdot 10^{-9}\,\mathrm{m}$ - the signal radiation wavelength; L - the amplifier length; $\Delta z = 10^{-3}\,\mathrm{m}$ - the quantization step in the long of the amplifier. We consider also the parameters:

$$\alpha = \left(\frac{hc}{\lambda_p}\right)^{-1}; \beta = \left(\frac{hc}{\lambda_s}\right)^{-1}; \left(\alpha = 4,947 \cdot 10^{18}; \beta = 7,824 \cdot 10^{18}\right).$$

2.2 Numerical simulation

Numerical modeling of the upper rate equations was realized using the MATHLAB programming medium.

The base element of the program was the function *ode 45*, which realize the integration of the right side expressions of the nonlinear coupled equations using Runge - Kutta type methods, for calculation time reducing.

The program was applied for many values of the amplifier length for each of them resulting different sets of results, for the photon fluxes, both for the signal and pumping as well as for the gain coefficients and signal to noise ratio.

From the obtained results by numerical integration of the transport equations, it results that the intensity of the output signal rise with the amplifier length but the pumping diminish in the some time. The calculated gain coefficients of the amplifier have a similar variation as was expected. We observe also the rising of the signal to noise ratio, resulting an improving of the amplifier performances (Sterian, 2006).

The obtained value of the gain coefficient for the signal, of the 40 dB is similar to published values (Agrawal, 1995) So that, the results can be very useful for designers, for example, to calculate the optimum length of the amplifier for maximum efficiency.

3. Laser system in erbium doped active media

3.1 The interaction phenomena and parameters

We analyze the laser systems with Er^{3+} doped active media by particularizing the models and the method of computer simulation for the case of the Er^{3+} continuous wave laser which operate on the 3µm wavelength. This laser system is interesting both from theoretical and practical point of view because the radiation with 3µm wavelength is well absorbed in water.

For this type of laser system don't yet completely are known the interaction mechanisms, in spite of many published works.

Quantitative evaluations by numerical simulations are performed, refering to the representative experimental laser with Er^{3+}:$LiYF_4$, but we analyse also the codoping possibilities of the another host materials: $Y_3Al_5O_{12}$ (YAG), $YAlO_3$, $Y_3Sc_2Al_2O_{12}$ (YSGG) and BaY_2F_8.

The energy level diagram for the Er^{3+}:$LiYF_4$ system and the characteristics processes which interest us in that medium are presented in figure 2.

The energy levels of the Er^{3+} ion include: the ground state in a spectroscopic notation ${}^4I_{15/2}$, the first six excited levels ${}^4I_{13/2}, {}^4I_{11/2}, {}^4I_{9/2}, {}^4F_{9/2}$, the thermally coupled levels ${}^4S_{9/2} + {}^2H_{11/2}$ and the level ${}^4F_{7/2}$.

The possible mechanisms for operation in continuous wave on 3µm of this type of amplifying media are (Pollnau et al., 1996):

a. the depletion of the lower laser level by absorption in excited state (ESA) ${}^4I_{13/2} \rightarrow {}^2H_{11/2}$ for pumping wavelength of 795 nm;

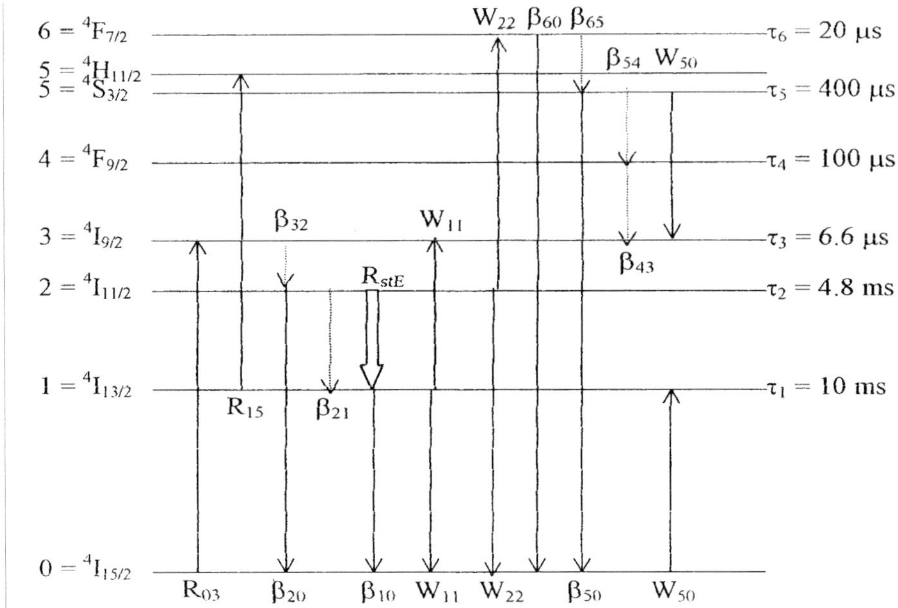

Fig. 2. The energy diagram of the Er^{3+} ion and the characteristic transitions

b. the distribution of levels excitation ${}^4S_{3/2}$ and ${}^2H_{11/2}$ between laser levels due to cross relaxation processes $\left({}^4S_{9/2}+{}^2H_{11/2},{}^4I_{15/2}\right)\rightarrow\left({}^4I_{9/2},{}^4I_{19/2}\right)$ and multiphoton relaxation ${}^4I_{9/2}\rightarrow{}^4I_{11/2}$;

c. the depletion of the lower laser level and enrichment of the upper laser level due to up-conversion processes $\left({}^4H_{13/2},{}^4I_{13/2}\right)\rightarrow\left({}^4I_{15/2},{}^4I_{9/2}\right)$ and multiphoton relaxation ${}^4I_{9/2}\rightarrow{}^4I_{11/2}$;

d. the relatively high lifetime for the upper laser level in combination with low branching ratio of the upper laser level to lower laser level.

These mechanisms, separately considered can't explain satisfactory the complex behavior of the erbium doped system, as has been shown (Pollnau et al., 1996; Maciuc et al., 2001).

That is way it is necessary to put into evidence the most important parameters of the system and to clarify the influence of these non-independent parameters on the amplification conditions as well as the determining the optional conditions of operation.

The levels ${}^4H_{11/2}$ and ${}^4S_{3/2}$ being thermally coupled, will be treated as combined a level, having a Boltzmann type distribution of the populations.

For numerical simulation the parameters of the Er^{3+}:$LiYF_4$ were considered because that medium presents a high efficiency for 3μm continuous wave operation, if the pumping wavelength is $\lambda = 970\,\text{nm}$ on the upper laser level ${}^4I_{9/2}$, or on the level ${}^4I_{11/2}$ in the case of the pumping wavelength $\lambda = 970\,\text{nm}$.

The Active Medium Parameters. Corresponding to the energy levels diagram presented in figure 2, the lifetimes of the implied levels, for low excitations and dopant concentrations have the values: $\tau_1 = 10\,\text{ms}$; $\tau_2 = 4{,}8\,\text{ms}$; $\tau_3 = 6{,}6\,\mu\text{s}$; $\tau_4 = 100\,\mu\text{s}$; $\tau_5 = 400\,\mu\text{s}$ and $\tau_6 = 20\,\mu\text{s}$.

Just the variations of these intrinsic lifetimes due to ion-ion interactions or ESA will be considered in the rate equations.

The radiative transitions on the levels $^4S_{3/2}$ and $^2H_{11/2}$ are calculated taking into account the Boltzmann contributions of these levels for 300K: 0,935 respectively 0,065 for each transition.
The nonradiative transitions are described through the transition rates $A_{i,NR}$ of the level *i*, calculated with formula:

$$A_{i,NR} = \tau_i^{-1} - \sum_{j=0}^{i=-1} A_{ij} , \tag{16}$$

where A_{ij} are the radiative transition rates from level i to level j. In the same time, the branching rations β_{ij} of the level i through the another lower levels are given by:

$$\beta_{ij} = \left(A_{ij} + A_{i,NR}\right)\tau_i^{-1} , \text{for } i - j = 1 \tag{17}$$

respectively:

$$\beta_{ij} = \frac{A_{ij}}{\tau_i^{-1}} , \text{ for } i - j > 1 . \tag{18}$$

The values of the branching ratios have been calculated (Desurvire, 1995; Pollnau et al., 1996).
The considered ion-ion interaction processes are:

$$\begin{aligned}
&\left({}^4I_{13/2}, {}^4I_{13/2}\right) \leftrightarrow \left({}^4I_{15/2}, {}^4I_{9/2}\right) \\
&\left({}^4I_{11/2}, {}^4I_{11/2}\right) \leftrightarrow \left({}^4I_{15/2}, {}^4F_{7/2}\right) \\
&\left({}^4S_{3/2}, {}^2H_{11/2}, {}^4I_{15/2}\right) \leftrightarrow \left({}^4I_{9/2}, {}^4I_{13/2}\right),
\end{aligned} \tag{19}$$

being characterized by the next values of the transition rates:

$$W_{11} = W_{11}^{-1} = 3 \cdot 10^{-23} \text{ m}^3\text{s}^{-1}; W_{22} = W_{22}^{-1} = 1{,}8 \cdot 10^{-23} \text{ m}^3\text{s}^{-1};$$

$$W_{50} = W_{50}^{-1} = 2 \cdot 10^{-23} \text{ m}^3\text{s}^{-1},$$

where the W_{50} parameter take into account the indiscernible character of the corresponding relaxation processes.

The Resonator Parameters. The resonator parameters used in the realized computer experiments are consistent with operational laser systems, as: the crystal length: l = 2 mm; the dopant concentration: $N_0 = 2 \cdot 10^{21} \text{cm}^{-3}$; the pumping wavelength: $\lambda_p = 795 \text{nm}$; we consider for ground state absorption (GSA) $^4I_{15/2} \rightarrow {}^4I_{9/2}$ the cross section $\sigma_{03} = 5 \cdot 10^{-21} \text{cm}^2$ and for excited state absorption (ESA), $^4I_{13/2} \rightarrow {}^4S_{3/2} + {}^2H_{11/2}$, the cross section $\sigma_{15} = 1 \cdot 10^{-20} \text{cm}^2$.
(The ESA contribution of the level $^4I_{11/2}$ was neglected for that wavelength.)
Another considered parameter values are presented in literature, being currently used by researchers.

In the literature (Pollnau et al., 1996; Maciuc et al., 2001). we found also the values of the energy levels populations reported to the dopant concentration and the relative transition rates, for different wavelength used for pumping: $\lambda = 795\,\text{nm}$ and $\lambda = 970\,\text{nm}$.

3.2 Computational model

The presented model, include eight differential equations which describes the population densities of each Er^{3+} ion energy levels presented in figure 2 and the photon laser densities inside the laser cavity.

We take N_i for $i = 1,2,...,6$ to be the population density of the i level and N_0 the population density of the ground state, the photonic density being ϕ .

That model consisting of eight equation system is suitable for crystal laser description (Pollnau et al., 1994). For the fiber laser, the model must be completed with a new field equation to describe the laser emission on $\lambda = 1{,}7\,\mu\text{m}$ between the fifth and the third excited levels.

The rate equations corresponding to energy diagram with seventh levels, for Er^{3+} systems are presented below:

$$\frac{dN_6}{dt} = \sum_{i=0}^{5} R_{i6}N_i - \tau_6^{-1}N_6 + W_{22}\left(N_2^2 - N_0N_6\right); \tag{20}$$

$$\frac{dN_5}{dt} = \sum_{i=0}^{4} R_{i5}N_i - R_{56}N_5 - \tau_5^{-1}N_5 + \beta_{56}\tau_6^{-1}N_6 - W_{50}\left(N_5N_0 - N_3N_1\right) - R_{SE}^{5\to3}; \tag{21}$$

$$\frac{dN_4}{dt} = \sum_{i=0}^{3} R_{i4}N_i - \sum_{j=5}^{6} R_{4j}N_4 - \tau_4^{-1}N_4 + \sum_{i=4}^{6} \beta_{i4}\tau_i^{-1}N_i\,; \tag{22}$$

$$\begin{aligned}\frac{dN_3}{dt} = \sum_{i=0}^{3} R_{i3}N_i - \sum_{j=4}^{6} R_{3j}N_3 - \tau_3^{-1}N_3 + \\ + \sum_{i=4}^{6} \beta_{i3}\tau_i^{-1}N_i + W_{50}\left(N_5N_0 - N_3N_1\right) + W_{11}\left(N_1^2 - N_0N_3\right) + R_{SE}^{5\to3};\end{aligned} \tag{23}$$

$$\begin{aligned}\frac{dN_2}{dt} = \sum_{i=0}^{1} R_{i2}N_i - \sum_{j=3}^{6} R_{2j}N_2 - \tau_2^{-1}N_2 + \\ + \sum_{i=3}^{6} \beta_{i2}\tau_i^{-1}N_i - 2W_{22}\left(N_2^2 - N_0N_6\right) - R_{SE};\end{aligned} \tag{24}$$

$$\begin{aligned}\frac{dN_1}{dt} = R_{01}N_0 - \sum_{j=2}^{6} R_{1j}N_1 - \tau_1^{-1}N_1 + \\ + \sum_{i=2}^{6} \beta_{i1}\tau_i^{-1}N_i + W_{50}\left(N_5N_0 - N_3N_1\right) - 2W_{11}\left(N_1^2 - N_0N_3\right) + R_{SE};\end{aligned} \tag{25}$$

$$\frac{dN_0}{dt} = -\sum_{j=0}^{6} R_{0j}N_0 + \sum_{i=1}^{6} \beta_{i0}\tau_i^{-1}N_i - W_{50}\left(N_5N_0 - N_3N_1\right) + \\ + W_{11}\left(N_1^2 - N_0N_3\right) + W_{22}\left(N_2^2 - N_0N_6\right). \tag{26}$$

$$\frac{d\phi}{dt} = \frac{l}{l_{opt}}\left(\frac{P_l}{P}\gamma_{21}\beta_{21}\tau_2^{-1}N_2 + R_{SE}\right) - \left\{-\ln\left[(1-T)(1-L_r)\right] + 2\kappa l\right\}\frac{c\phi}{2l_{opt}}; \tag{27}$$

$$\frac{d\phi^{5\to3}}{dt} = \frac{l}{l_{opt}}\left(\frac{P_l}{P}\gamma_{53}\beta_{53}\tau_5^{-1}N_5 + R_{SE}^{5\to3}\right) - \left\{-\ln\left[(1-T)(1-L_r)\right] + 2\kappa l\right\}\frac{c\phi^{5\to3}}{2l_{opt}}. \tag{28}$$

A similar models are given in the references (Pollnau et al., 1996; Maciuc et al., 2001,a & b). In the field equations (27) and (28), the parameters $L, T, L_r, \kappa, l_{opt}.P_l / P$ are considered the same for the two type of laser studied. In the equations system (20) ÷ (28) the parameters are: R is the pumping rate from lower levels to the higher ones; τ is the life-times for each corresponding level; W is associated with the transition rates of the ion-ion up-conversion and the corresponding inverse processes; - β_{ij} are the branching ratios of the level i through the other possible levels j; R_{SE} is the stimulated emission rate; l and l_{opt} are the crystal length and the resonator length; γ_{21} is an additional factor for the spontaneous radiative transition fraction between the levels: $^4I_{11/2}$ and $^4I_{13/2}$; P_l / P is the of spontaneous emission power emitted in laser mode; T, L_r, κ, c are the transmission of the output coupling mirror, the scattering losses and the diffraction - reabsorption losses respectively, c being the light speed in vacuum.
The pumping rates depend on the corresponding cross-section and of the other parameters (Maciuc et al., 2001).
The parameters for the lasing in an Er:LiYF$_4$ crystal system are considered the same and for the fiber laser.

3.3 Crystal laser simulation

Laser Efficiency for Different Pumping Wavelength. In the simulation were used for pumping the radiations having $\lambda = 795$, 970 and 1570 nm, which are in resonance with the energy levels in diagram of Er^{3+} ion presented in figure 2.
The pumping radiation for $\lambda = 795\,\text{nm}$ connect the ground state level $^4I_{15}$ with the third excited level $^4I_{9/2}$ and also the second level with the fifth one $\left(^4I_{13/2}, ^4S_{3/2} + ^2I_{11/2}\right)$, processes.
In the case of pumping radiation having $\lambda = 970\,\text{nm}$ the ground state absorption (GSA) corresponds to transition $^4I_{15} \to ^4I_{11/2}$ and excited state absorption (ESA) to transition $^4I_{11/2} \to ^4F_{7/2}$. Similarly the pumping for $\lambda = 1530\,\text{nm}$ determine a single transition GSA that is $^4I_{15/2} \to ^4I_{13/2}$.
The dependence of the output power versus input power for different pumping wavelength (795 nm, 970 nm and 1530 nm) were plotted resulting the functioning thresholds and the slope efficiencies for each situation.
For the crystal laser Er^{3+} doped, the optimum efficiency results for the direct pumping on the upper laser level.

The output power variation with the level lifetimes. The output power variation on the lifetimes for the upper levels having τ_4, τ_5, τ_6 was studied for an input pump power $P_p = 5\text{W}$ and $\lambda_p = 795\text{nm}$.
We found that radiative and nonradiative transitions from the fifth and the sixth levels, improve the population difference for the laser line and determine the raising of the output power of them, the variation of the fourth level lifetime, being without influence for the output power.
The influence of the Er^{3+} ion doped host material on the output power. A three dimensional study was done to investigate the influence in the laser output power due to parameters variations for the host material, using $\lambda_p = 795\text{nm}$.
The relative spontaneous transition rates were considered the same for all simulations.
To determine the host material change influence on the laser output power the next variation scale of the lifetimes have been considered:

$$\tau_1 = (1 \div 15)\text{ms},\ \tau_2 = (0{,}4 \div 9{,}6)\text{ms},\ \tau_3 = (0{,}22 \div 22)\mu\text{s},\ \tau_4 = (3 \div 300)\mu\text{s},\ \tau_5 = (12 \div 1200)\mu\text{s}$$
$$\text{and } \tau_6 = (0{,}6 \div 60)\mu\text{s}.$$

Similarly, the variations of the transition rates corresponding to up-conversion processes for different host materials are considered to span the intervals given bellow:

$$W_{11} = (0{,}1 \div 300)\cdot 10^{-21}\,\text{cm}^3\text{ms}^{-1},\ W_{22} = (1{,}8 \div 180)\cdot 10^{-21}\,\text{cm}^3\text{ms}^{-1},$$
$$W_{50} = (0{,}02 \div 200)\cdot 10^{-21}\,\text{cm}^3\text{ms}^{-1}.$$

For the other parameters used in the numerical simulation the published data was the main source of reference.

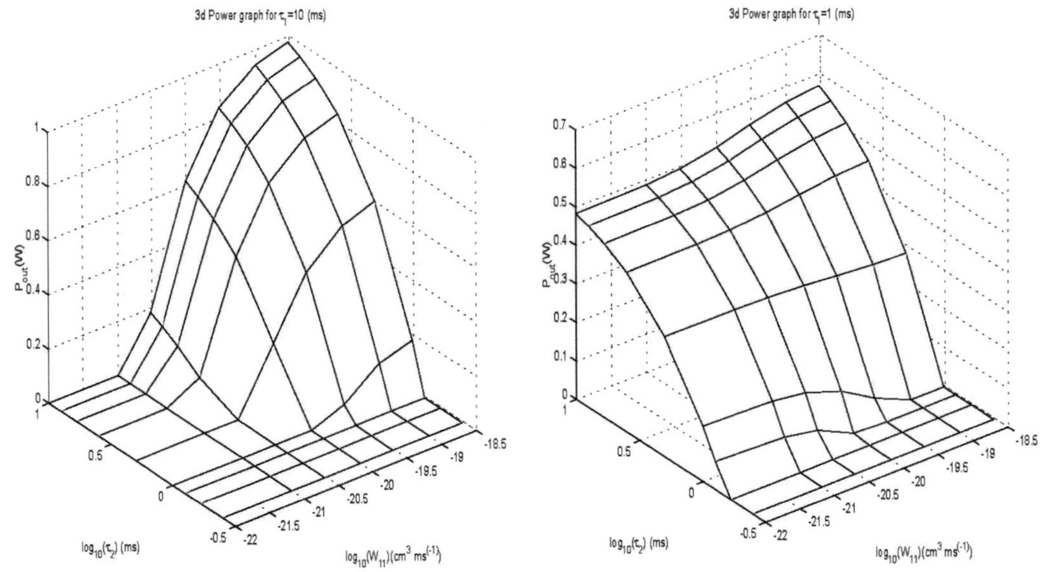

Fig. 3. Output laser power dependence on parameters τ_2 and W_{11} for two values of τ_1

The graphs in figure 3 give the three-dimensional (3D) output power dependence versus lifetime of the higher level τ_2 and the "up-conversion" parameter W_{11} associated to studied transition for two values of the τ_1 parameter: τ_1 =.10 ms and τ_1 = 1 ms. In figure 4 we show the dependence of the output power on the life time of the second excited level, the "up-conversion" parameter W_{11} and "up-conversion parameter W_{22} for two values of this parameter: $W_{22} = 1{,}8 \cdot 10^{-24} m^3 s^{-1}$ and $W_{22} = 1{,}8 \cdot 10^{-22} m^3 s^{-1}$.

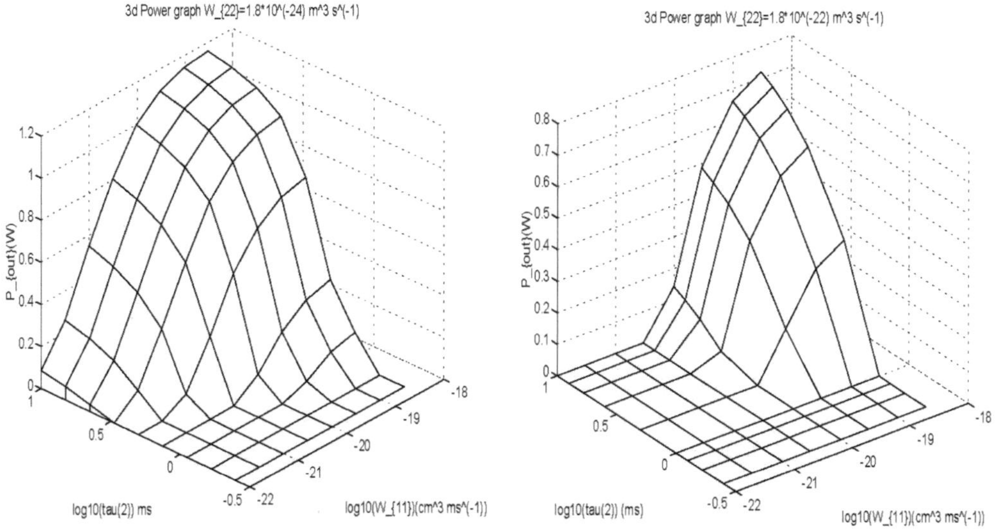

Fig. 4. Output laser power dependence on parameters τ_2 and W_{11} for two values of W_{22}

The graphs in figure 5 give a three-dimensional representation of the function $P_{laser}(\tau_3, W_{11}, \tau_1)$ for two values of the τ_1 parameter: τ_1 =.10 ms and τ_1 = 1 ms.

The three-dimensional study of the parameters variation to increase laser output power showed the role of host material for high laser efficiency, checking that the decisive parameter is the time of life associated with the upper laser level. Selection of the materials having parameters in the areas of variation adopted in the analysis recommend as efficient solutions the fluorides: LiYF4 and BaY2F.

Stable, non-chaotic behavior of the laser systems. A time dependence of the photon density in the cavity of the output power and of the implied level populations in the laser process was analyzed by the input parameters variations that is pumping power and the interaction cross-sections. For the pumping power differently step functions was considered. The analysis represents a satisfactory temporally description of the crystal laser to verify the used computational model.

Our simulation for the time dependence confirm the stability of the continuous wave regime of operation of the crystal laser, after an initial transitory regime of the milliseconds order, which is gradually droped, from the moment we switch on the pump.

This stable non-chaotic behavior is similar for different host materials, the used method not being time prohibitive for such studies.

To understand better the obtained results, we indicate below some of the graphs plotted in that simulation: 3d analysis $P(W_{11}, \tau_1, \tau_2)$ with $\tau_1 = 10\,ms$; 3d analysis $P(\sigma_{15}, \tau_2, W_{11})$ with $\sigma_{15} = 10^{-19}\,cm^2$; 3d analysis $P(W_{50}, \tau_2, W_{22})$ with $W_{22} = 1{,}8 \cdot 10^{-24}\,m^3 s^{-1}$; 3d analysis $P(W_{50}, \tau_1, \tau_3)$ with $\tau_1 = 10\,ms$, etc.

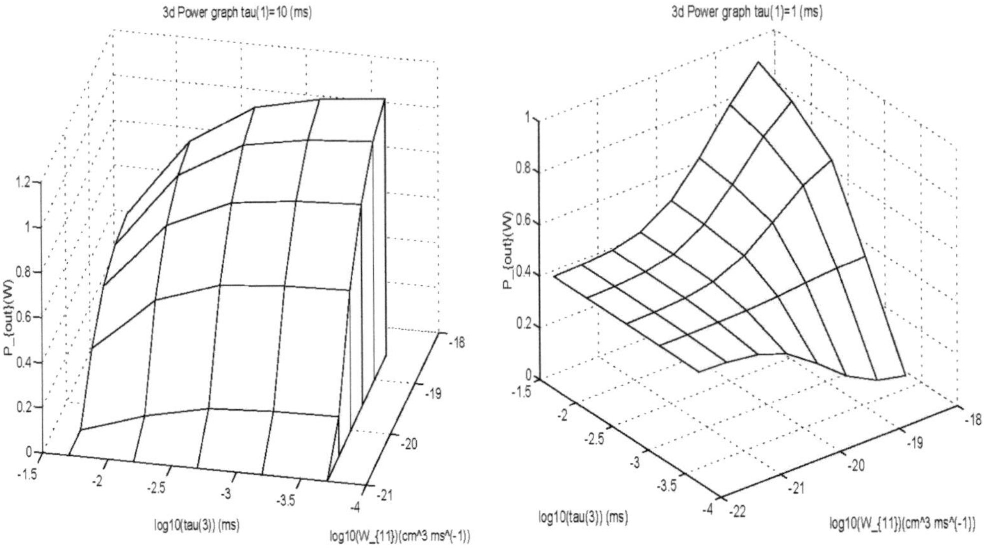

Fig. 5. Output laser power dependence on parameters τ_3 and W_{11} for two values of τ_1

In conclusion, the 3D study of the parameters variations to rise the output laser power put intro evidence the important role of the host materials, the decisively parameter being the lifetime associated with the upper laser level.

By selection, other the parameters variations limits, the most efficiently media are the fluorides: $LiYF_4$ and LiY_2F_8.

In spite of the fact we have analyzed the problems by an original method, the results are consistent with the published data.

A special mention must be made concerning the used "step-size" Runge - Kutta method which is rapidly and don't alterate the results obtained by classical Runge - Kutta method.

In case of 3d analysis we used a 7 order precision and a 6 order stopping criteria.

3.4 Fiber laser simulation

In the fiber laser functioning, were studied almost the same problems as in the crystal laser case, that are:

a. The output power thresholds and efficiencies for different values of the "colaser" process and in the absence of this effect.
b. The relevance and the implications of the "colaser" process, which is specific to fiber laser
c. The dependence of the output power on host material Er^{3+} doped, by variation of the characteristic parameters.
d. The description of the time depended phenomena for the Er^{3+} doped fiber laser, inclusively the population dynamics.

The principal differences between the crystal laser and fiber laser were taken into consideration, the most important being:

- the existence of an extra field equation (Maciuc et al., 2001), which describes the colasing process in the fiber laser;
- the absence of the "up-conversion" processes due to the low concentration of the Er^{3+} dopant.

The role played by the up-conversion in crystal is taken in fiber laser by pumping from the first and second excited level.

The analyzed physical system was the optical fiber with ZBLAN composition, having the next characteristic parameters:

- the dopant concentration: $N_d : 1{,}8 \cdot 10^{19}\,\text{cm}^{-3}$; the amplifier length, $l : 480\,\text{cm}$; the laser mod radius, $r_{\text{mode}} : 3{,}25\,\mu\text{m}$; the pumping wavelength, $\lambda_p : 791\,\text{nm}$; the ground state absorption cross-section, $\sigma_{03} : 4{,}7 \cdot 10^{-22}\,\text{cm}^2$; the excited state absorption cross-section from the level ${}^4I_{13/2}$, $\sigma_{15} : 10^{-21}\,\text{cm}^2$; the excited state absorption cross-section from the level ${}^4I_{11/2}$, $\sigma_{27} : 2 \cdot 10^{-22}\,\text{cm}^2$; the laser wavelength, $\lambda_L : 2{,}71\,\mu\text{m}$; the "colaser" wavelength, $\lambda_{cl} : 1{,}7\,\mu\text{m}$; the emission cross-section, $\sigma_{21} : 5{,}7 \cdot 10^{-21}\,\text{cm}^2$; the "colaser" cross section, $\sigma_{53} : 0{,}5$ or $0{,}1 \cdot 10^{-20}\,\text{cm}^2$; the Boltzmann, b_{14} and $b_{22} : 0{,}113$ respectively 0,2; the mirror transmission *T*: 68%; the optical resonator length, $l_{opt} : 720\,\text{cm}$.

The "colaser" process was studied for three different values of the "colaser" cross section: $\sigma_{53} = 0\,\text{cm}^2; \sigma_{53} = 0{,}5 \cdot 10^{-20}\,\text{cm}^2$ and $\sigma_{53} = 0{,}1 \cdot 10^{-20}\,\text{cm}^2$.

The most important conclusions resulting from the fiber laser analysis are:

- The optimum operating conditions are obtained for $\lambda_p = 791\,\text{nm}$, so that the pumping is realized directly on the upper laser level with the cross - section σ_{03}.
- The presence of the "colaser" process, improves the laser efficiency on $3\,\mu\text{m}$, by a 2 factor in that cascade laser situation. The three - dimensional (3D) analysis shows the determinant role of the τ_2 for laser power similarly to the crystal laser, the parameters σ_{15} and σ_{27} being strong correlated with the laser process, for the high values of the τ_{11}.

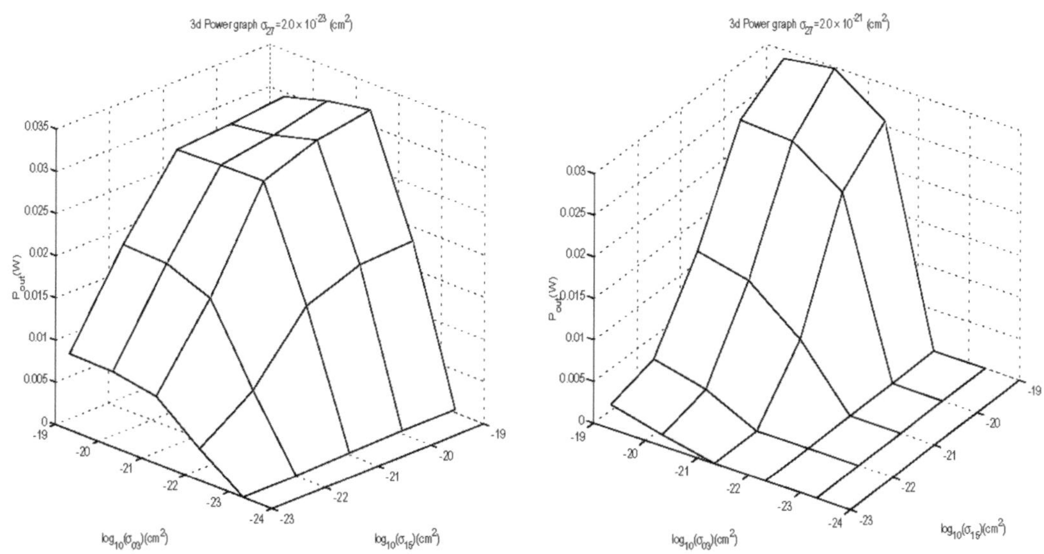

Fig. 6. Dependence of the output laser power both on σ_{15} and σ_{03} cross sections for two values of σ_{27}.

The graphs in Figure 6 shows a strong dependence of the laser power both on σ_{15} and σ_{03} cross sections in terms of the high life time of the lower laser level τ_1 and a sufficiently low value of the effective cross section, $\sigma_{27} = 2 \cdot 10^{-23}\,\text{cm}^2$. Increasing the value of the second process(ESA) (Fig. 7) on obtain the same behavior of the output power with the difference of a rapid increase from zero to a high value of a laser power, followed by a slow saturation.

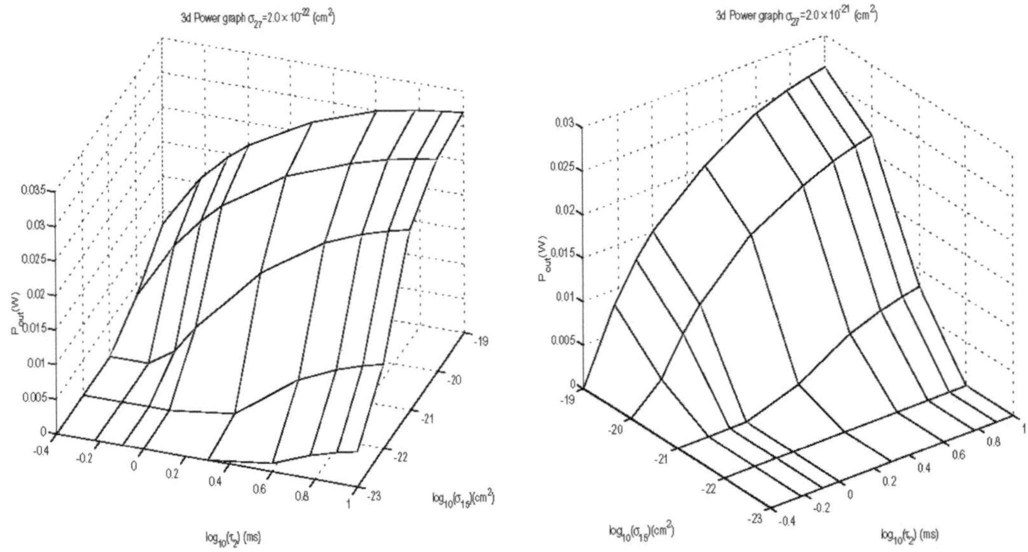

Fig. 7. Dependence of the output laser power both on σ_{15} and τ_2 for two values of σ_{27}

The dependence $P(\sigma_{15},\tau_2,\sigma_{27})$ represented by the graphs in figure 7, leads to the conclusions:

- increasing of τ_2 the laser power increases;
- the rising of σ_{15} is favorable to laser power for $\tau_1 = 10$ ms;
- high values of σ_{27} causes low laser powers, due to the depopulation of the upper laser level. The study of the "colaser" process was made for three different values of effective "colaser" cross-section: $\sigma_{53} = 0\ \text{cm}^2$; $\sigma_{53} = 0{,}5 \cdot 10^{-20}\text{cm}^2$ and $\sigma_{53} = 0{,}1 \cdot 10^{-20}\text{cm}^2$. (Note that ZBLAN fiber optic amplifiers require more pumping power than EDFA's based on silicon dioxide.)

Another important result is represented by the time dependent analysis of the output power and of the level populations, which shows a stable non - chaotic behavior as in the crystal laser case.

All obtained results by numerical analysis are consistent with the data from the literature (Shalibeik, 2007).

4. Nonlinear effects in optical fibers systems.

4.1 The model

In recent years much attention has been paid to the study of nonlinear effects in optical fiber lasers (Agrawal, 1995 & 1997; Desurvire, 1995).

Self-pulsing and chaotic operation (Baker & Gollub, 1990; Abarbanel, 1996) of the EDFLs has been reported in various experimental conditions (Sanchez et al., 1993; Sanchez et al., 1995) including the case of pumping near the laser threshold. We present firstly a model for the single-mode laser taking into account the presence of the erbium ion pairs that act as a saturable absorber.

The nonlinear dynamics of an erbium-doped fiber laser is explained based on a simple model of the ion pairs present in heavily doped fibers. The single-mode laser dynamics is reducible to four nonlinear differential equations. Depending on the ion pair concentration, the pumping level and the photon lifetime in the laser cavity, numerical calculations predicts cw, self-pulsing and sinusoidal dynamics. The regions of these dynamics in the space of the laser parameters are determined.

A modeling of the erbium laser operating around 1.55 μm has been proposed (Sanchez et al., 1993). This considers the amplifying medium as a mixture of isolated erbium ions and erbium ion pairs. For an isolated ion, the laser transition takes place between the energy levels $^4I_{13/2}$ and $^4I_{15/2}$ (Fig. 8). The ion is pumped on some upper energy levels and it fastly relaxes to the level $^4I_{13/2}$. It is noteworthy that energy level $^4I_{9/2}$ is positioned above level $^4I_{13/2}$ at a separation approximately equal to that of the laser transition. Two neighboring ions interact and form an ion pair. The strength of this interaction is small (due to the screening effect of the $4d^{10}$ electrons on the $4f$ electrons) so that the energy levels are practically preserved and the pair energy is the sum of the two ions energy. Because of the quasiresonance of levels $^4I_{9/2}$ and $^4I_{13/2}$ with the laser transition, up-conversion in an ion pair has a significant probability. This is followed by a fast transition back to the $^4I_{13/2}$ level. As a result of these processes, the population inversion decreases by one without the emission of a photon. Thus, the laser effect due to the ion pair is explained based on three ionic levels.

Based on the above picture of the active medium, in the rate equation approximation the laser is described by the population inversion d of the isolated ions, the sum d_+ and the difference d_- of populations of levels 22 and 11, and the normalized laser intensity I, that verify equations (Flohic et al., 1991; Sanchez et al., 1993):

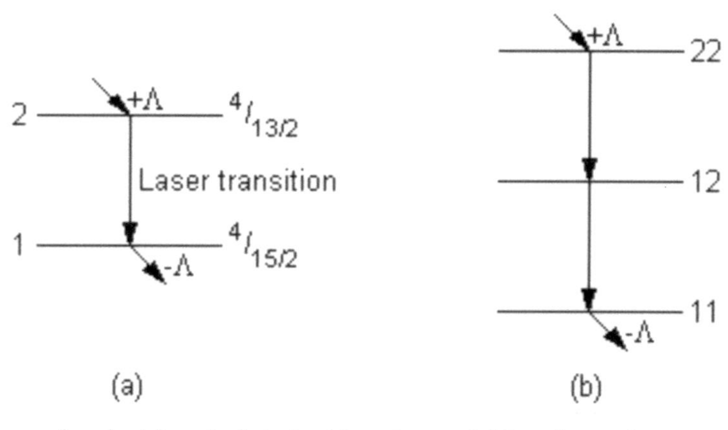

Fig. 8. Laser energy levels: (a) an isolated erbium ion and (b) an ion pair.

$$\dot{d} = \Lambda - a_2(1+d) - 2dI, \quad (1a)$$

$$\dot{d}_+ = a_2(1-d_+) - (a_{22}/2)(d_+ + d_-) + y(2-3d_+)I, \quad (1b)$$

$$\dot{d}_- = \Lambda - a_2(1-d_+) - (a_{22}/2)(d_+ + d_-) - yd_-I, \quad (1c)$$

$$\dot{I} = -I + A(1-2x)dI + Axyd\_I . \qquad (1d)$$

In the above, the time variable is expressed in units of the photon lifetime in the cavity (τ_L). The quantity x is the fraction of ion pairs in the active medium, Λ is the pumping parameter, $a_2 = \tau_L / \tau_2$ and $a_{22} = \tau_L / \tau_{22}$, where τ_2 and τ_{22} are the lifetimes of level 2 and 22, respectively. The other parameters are $y = \sigma'_L / \sigma_L$, where σ_L and σ'_L are the absorption cross-section for the laser transition in isolated ions and pairs, respectively, and $A = \sigma_L N_0 \tau_L$, where N_0 is the erbium concentration. The normalization of the laser intensity I is performed in the form $I = i_L / \sigma_L \tau_L$.

4.2 Stability analysis

Apart from the nonlasing state (zero intensity state), the intensity i_L in a stationary state satisfies the third-order polynomial equation:

$$i_L^3 + c_1 i_L^2 + c_2 i_L + c_3 = 0 , \qquad (5)$$

where:

$$c_1 = Ax(a_{22} - 2a_2)/3 + (a_2 + 2a_{22})/3y + a_2/2 - A(2\Lambda - a_2)/2 , \qquad (6)$$

$$c_2 = x[a_2 A(a_2/6 + a_{22}/6 - \Lambda) + a_{22} A\Lambda / y - a_2 A(a_2 + a_{22})/3y]$$

$$+ a_2(a_2 + 2a_{22})/6y + a_2 a_{22}/3y^2 - A(a_2 + 2a_{22})(2\Lambda - a_2)/6y , \qquad (7)$$

$$c_3 = a_2 Ax(a_2 a_{22} - 2a_2\Lambda - a_{22}\Lambda)/6y + a_2^2 a_{22}/6y^2 - a_2 a_{22} A(1-2x)(2\Lambda - a_2)/6y^2 . \qquad (8)$$

The laser threshold is derived from the condition $c_3 = 0$ that implies:

$$\Lambda_{th} = \frac{1 - Ax(2-y)/(A+1)}{1 - (2 - y/2 - a_2 y / a_{22})x} \Lambda_{th}^0 , \qquad (9)$$

where:

$$\Lambda_{th}^0 = (1 + 1/A) a_2 / 2 \qquad (10)$$

is the pumping parameter at laser threshold in the absence of the ion pairs. Numerical solution of eq. (5) is performed for typical parameters. We take $\tau_2 = 10$ ms, $\tau_{22} = 2$ μs, $N_0 = 5 \times 10^{24}$ m^{-3}, $\sigma_L = 1.6 \times 10^{-16}$ m^2 and $y = 0.2$. The dependence of the threshold pumping level on the concentration of the ion pair for two values of photon lifetime in the cavity is presented in Fig. 9. The increase of the laser threshold due to the presence of the ion pairs limits the use of the fibers in such conditions.

Above threshold there exists one steady-state intensity given by eq. (5), the other two solutions being unphysical (negative). The numerical calculation of the intensity in a significant range of the pumping parameter (Fig.10) gives a laser intensity following a straight line dependence. The influence of the ion pairs is again disadvantageous and manifests in reducing the slope of the characteristics.

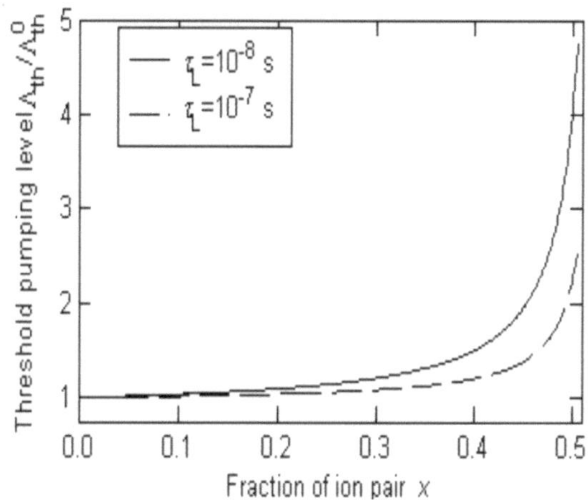

Fig. 9. Threshold pumping parameter vs. ion pair percentage

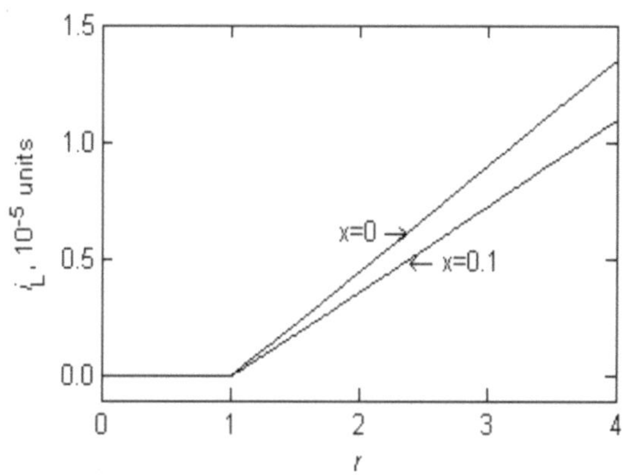

Fig. 10. Laser intensity in the steady state versus the pumping strength

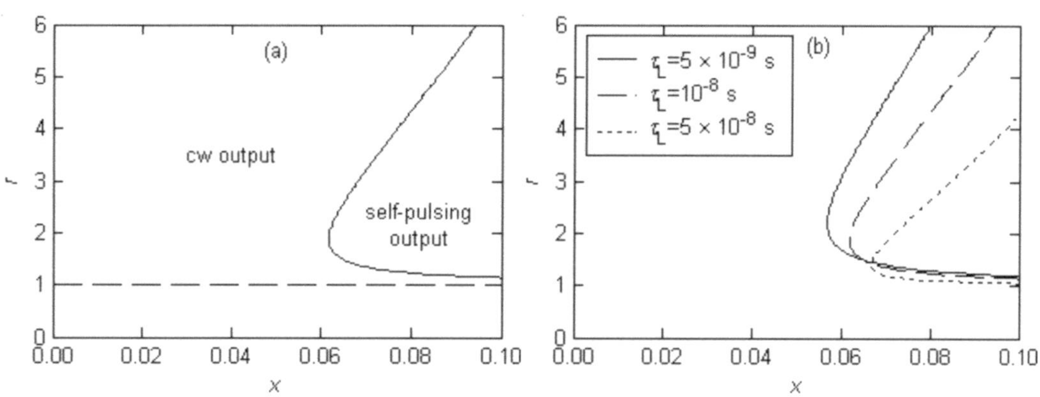

Fig. 11. (a) Calculated stability diagram for EDFL. $\tau_L = 10^{-8}$ s ; (b) The influence of the photon lifetime on the margins of the stability domains.

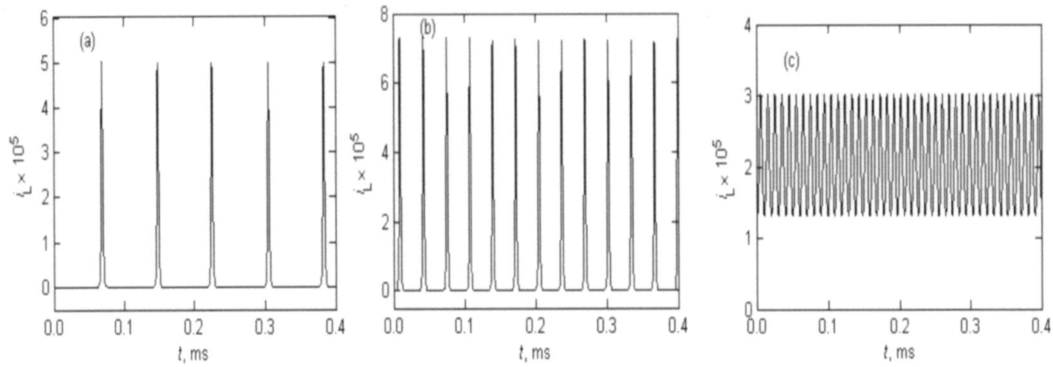

Fig. 12. Long term temporal evolution of laser intensity at the fraction $x = 0.1$ of ion pairs and the pumping level (a) $r = 1.5$, (b) $r = 2.5$, and (c) $r = 6.6$. $\tau_L = 10^{-8}$ s.

Linear stability investigation of the steady states (Crawford, 1991, Press, 1992) reveals the existence of a critical value of the ion pair percentage under which the steady state (cw) solution is stable whatever the pumping level. At larger concentrations of the ion pairs, the laser is in a steady state or self-pulsing, depending on the value of the pumping parameter (Fig. 11.a). The transition from the cw dynamics to a self-pulsing one takes place when two complex conjugate eigenvalues of the linearized system cross the imaginary axis from left to right, i.e., a Hopf bifurcation occurs.(Crawford, 1991). In such conditions the steady state solution becomes unstable and the long-term system evolution settle down on a stable limit cycle in phase space.

The photon lifetime contribution to the stability diagram is presented in Fig. 11.b. This proves that a cavity with low losses makes it possible to preserve the cw dynamics at larger doping levels.

Fig.12 clarifies the quantitative changes of the laser intensity inside the self-pulsing domain. At a fixed value of the ion pair concentration, the increase in pumping gives rise to pulses of a higher repetition rate and close to the bifurcation point the intensity becomes sinusoidal. Besides, pulse amplitude reaches a maximum approximately in the middle of the self-pulsing domain.

4.3 The two-mode erbium-doped fiber laser

A further step in the study of the laser dynamics considers a dual wavelength operation at 1.550 μm and 1.556 μm, treated as a two coupled laser modes (Sanchezet al., 1995).

The nonlinear dynamics of a two-mode fiber laser is explained in terms of a classical two-mode laser model and two additional equations for the ion-pair quantum states.

The laser is described through six coupled differential equations. There are analyzed the steady-states, their dependence on the laser parameters, and the onset of the self-pulsing and chaotic states is pursued.

The two modes of the laser are coupled via a cross-saturation parameter (β). The system is described through the set of coupled equations:

$$\dot{d}_1 = \Lambda - a_2(1 + d_1) - 2d_1(I_1 + \beta I_2)\,, \tag{2a}$$

$$\dot{d}_2 = \gamma\Lambda - a_2(1 + d_2) - 2d_2(\beta I_1 + I_2)\,, \tag{2b}$$

$$\dot{d}_+ = a_2(1-d_+) - (a_{22}/2)(d_+ + d_-) + y(2-3d_+)(I_1+I_2), \tag{2c}$$

$$\dot{d}_- = \Lambda - a_2(1-d_+) - (a_{22}/2)(d_+ + d_-) - yd_-(I_1+I_2), \tag{2d}$$

$$\dot{I}_1 = -I_1 + A(1-2x)(d_1 + \beta d_2)I_1 + Axyd_-I_1, \tag{2e}$$

$$\dot{I}_2 = -I_2 + A(1-2x)(\beta d_1 + d_2)I_2 + Axyd_-I_2. \tag{2f}$$

In the above, $I_{1,2}$ are the normalized intensities and $d_{1,2}$ are the normalized population inversions of the two modes. The supplementary parameter γ takes into account the anisotropy in pumping for the two modes.

System (2) is to be investigated for typical parameters: $\tau_L = 200$ ns, $\tau_2 = 10$ ms, $\tau_{22} = 2$ μs, $N_0 = 5\times10^{18}\,\mathrm{cm}^{-3}$, $\sigma_L = 1.6\times10^{-10}\,\mathrm{cm}^3\mathrm{s}^{-1}$, $y = 0.2$, $\beta = 0.5$ and $\gamma = 0.9$.

There are three types of interesting stationary states: $I_1 = I_2 = 0$ (laser below threshold), ($I_1 > 0$, $I_2 = 0$), i.e., the single-mode laser, and $I_{1,2} > 0$ (the two-mode state). The complete determination requires the solving of a fourth-order polynomial equation for intensity in the single-mode case, and a third-order polynomial equation in the two-mode case. The treatment of the steady-state stability is performed through the linearization of the system (2) around the steady-states and searching of the eigenvalues of a six-order matrix (Crawford,1991; Abarbanel,1996; Ştefănescu,2002). The results are showed in Fig 13(a) for the parameters above and a range of ion-pair concentration in accordance with all practical cases. Here, $\Lambda_{th}^{(1)}$ denotes the pumping parameter at the laser threshold and $\Lambda_{th}^{(2)}$ is the pumping parameter at the switching of the weak mode (mode 2, here). We focus here on the two-mode operation. The pumping strength is usually expressed in the form $r = \Lambda / \Lambda_{th}^{(1)}$.

The steady-state intensities versus the pumping parameter are presented in figure 14. The straight-line dependence is followed for a large range of pumping strengths.

The way to a stable steady-state is always through relaxation oscillations (Fig. 15).

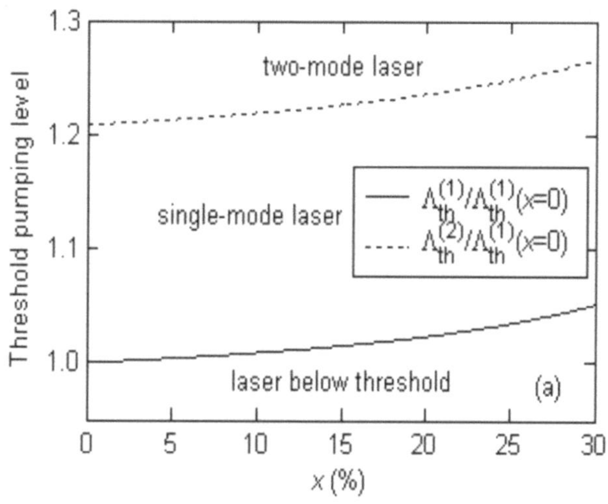

Fig. 13. Threshold pumping parameter for the laser action and two-mode states vs. ion pair percentage

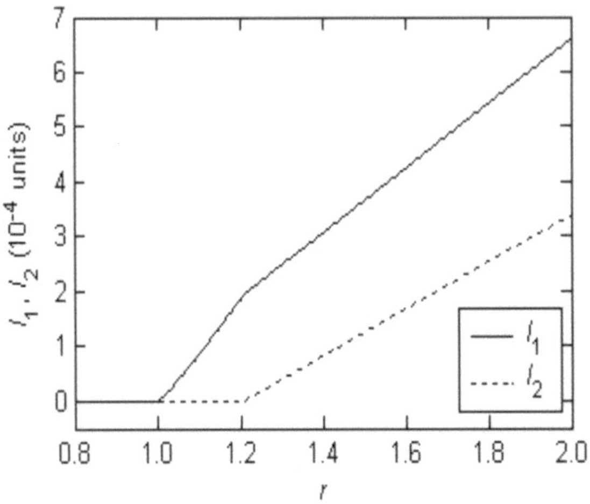

Fig. 14. Steady-state intensities of the two-mode laser versus the pumping parameter, $x = 0.1$

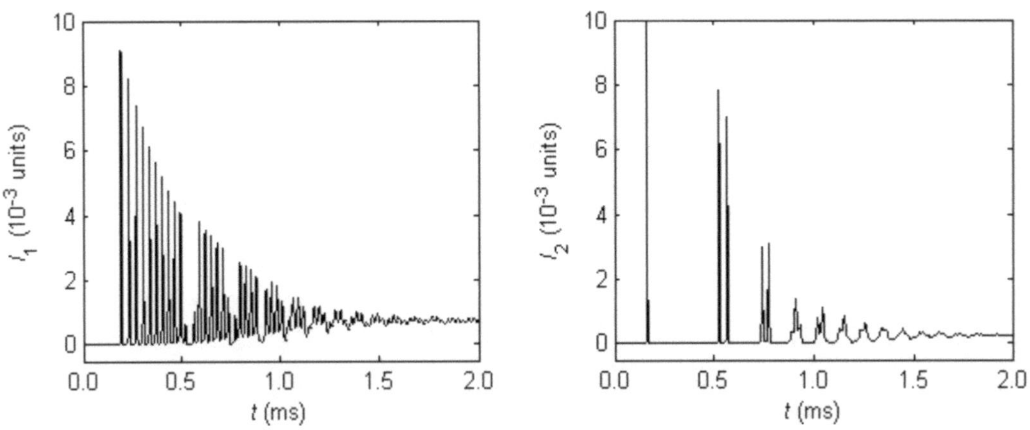

Fig. 15. (a) Transitory regime to a two-mode stationary state. $x = 0.03$ and $r = 1.5$.

A thorough investigation of the laser dynamics in the domain where both modes are active seems to be a hard task. Instead, we try to find out the basics of this. For sufficiently low doping level, the asymptotic dynamics is a steady-state whatever the pumping parameter. For larger values of x, there is a range of self-pulsation dynamics. In Fig.16(a) this is showed by means of a bifurcation diagram, i.e., the maxima of one mode intensity is plotted against the pumping parameter. The dynamics in Fig. 16(a) has also been encountered in the case of the one-mode model (Sanchez et al., 1993) and the change from a stationary stable state to a self-pulsing one at $r \approx 1.8$ is a Hopf bifurcation (Crawford, 1991).
There exists a region of two maximum intensities, i.e., a period doubling. Temporal dynamics in such a case (Fig. 17) exhibits antiphase dynamics of the two laser modes.

A small increase of the ion-pair concentration to the value of 7% [Fig.16(c)] leads to the appearance of a window of a tripled period [Fig. (17c)]. Further increase of the ion-pair concentrations leads to a more complicated bifurcation diagram, including regions with multiple stable states (generalized bistability) and routes to chaotic dynamics such as the period-doubling route or a quasi-periodic one.

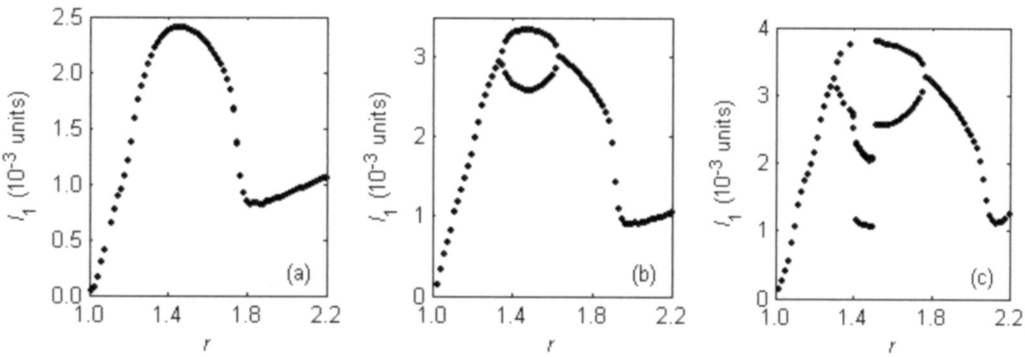

Fig. 16. Bifurcation diagram of the strong mode intensity for relatively small ion-pair percentages: (a) $x = 0.065$, (b) $x = 0.068$, and (c) $x = 0.070$. Figure 16(b) shows the maxima of intensity I_1 at a slightly larger ion-pair concentration.

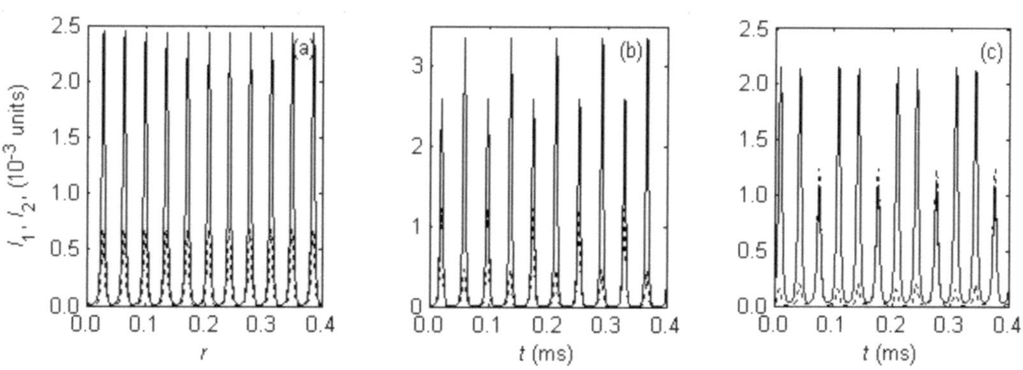

Fig. 17. Long-term temporal evolution of the laser for $r = 1.45$ and ion-pair concentrations in Fig. 6: (a) $x = 0.065$, (b) $x = 0.068$, and (c) $x = 0.070$. Solid line is used for the strong mode (I_1) and dotted line for the weak mode (I_2).

5. Conclusions

The developed numerical models concerning the characterization and operation of the EDFA systems and also of the laser systems, both of the "crystal type" or "fiber type" realized in Er^{3+} doped media and the obtained results are consistent with the existing data in

the literature. Our results put into evidence the existence of the new situations which are important for the optimization of the functioning conditions for this kind of devices.
That was due to the valences of the computer experiment method which make possible a complex study taking into account parameters intercorrelations by simulating experimental conditions, as have been shown.
The erbium ion pairs in a laser fiber can explain the experimentally observed nonlinear dynamics of the system, apart from the intrinsic nonlinearity of a multiple-mode laser. The existence of the erbium ion pairs introduces a supplementary nonlinearity with a saturable absorber action (Sanchez et al.,1995). In the single-mode laser description, as well as in the two-mode laser the ion pairs are responsible for an increase of the laser threshold, a decrease of the laser gain, and self-pulsations. Depending on the laser pumping level, the laser self-pulsations range from an oscillatory form to a well-defined pulse-shape. The nonlinear dynamics is present even close to the laser threshold and does not requires large pumping levels.
The nonlinear dynamics can be avoided by a choice of sufficiently low-doped fibers, such that the average distance between ions is large enough and the interaction between ions is small. It is clear that the limitation to sufficiently low-doped fibers is performed at the expense of using longer fibers.

6. References

Abarbanel, H. (1996). *Analysis of observed chaotic data*, Springer-Verlag, Berlin.

Agrawal, G.P. (1995). *Nonlinear fiber optics*, Academic Press, San Diego.

Agrawal, G.P. (1997). *Optic Communication Systems*, A Wiley - Interscience Publication, J. Wiley and Sons, Inc.,New York.

Baker, G. L. & Gollub, J. P. (1990). , *Chaotic dynamics. An introduction*, Cambridge University Press, Cambridge.

Crawford, J. D. (1991). Introduction to bifurcation theory, *Rev. Mod. Phys.* 63, 991-1037, (1991).

Desurvire, E. (1995). *Erbium – Dopped Fiber Amplifier*, J. Wiley and Sons, Inc., New York.

Le Flohic, M.; Francois, P.L.; Allain, J. Y. ; Sanchez, F. & Stephan, G. (1991). Dynamics of the transient buildup of emission in Nd^{3+} doped fiber lasers, *IEEE J. Quantum Electron.* 27, 1910-1921, (1991).

Maciuc, F. C.; Stere, C. I & Sterian, A.R. (2001). Rate equations for an Erbium laser system, a numerical approach, *Proceedings of SPIE*, Vol. 4430, 136 – 146, The Sixth Conference on Optics ROMOPTO 2000, Bucharest.

Maciuc, F. C.; Stere, C. I & Sterian, A.R. (2001). Time evolution and multiple parameters variations in a time dependent numerical model applied for Er^{+3} laser system, *Proceedings of SPIE*, Vol. 4394, 84 - 89, Laser Physics and Application 10th International School an Quantum Electronics (2001), Varna.

Ninulescu, V.; Sterian, A. R. & Sterian, P. (2006). Dynamics of a two-mode erbium-doped fiber laser, *Proceedings of SPIE*, Vol. 6344, 63440 Q1 - 63440Q6, ALT-05, (Editors: Ivan A. Shcherbakov ş.a.), June (2006), Tianjin.

Ninulescu, V. & Sterian, A. R. (2005). Dynamics of a Two-Level Medium Under the Action of Short Optical Pulses, *Lecture Notes in Computer Science,* LNCS 3482, 635-642, Gervasi, O.et al. (Eds), Springer - Verlag, (2005), Berlin, Heidelberg.

Petrescu, A.; Sterian, A. R. & Sterian, P.E. (2007). Solitons propagation in optical fibers: computer experiments for students training, *Proceedings of Computational Science and Its Applications,* ICCSA 2007, Pt 1, 4705450-461, Editor(s): Gervasi, O; Gavrilova M.L., (2007).

Pollnau, M.; Graf, Th.; Balmer, J. E.; Lüthy, W. & Weber H. P. (1994). Explanation of the cw operation of the Er^{3+} 3-µm crystal laser, *Phys. Rev.* A 49 (5), 3990-3996, (1994).

Pollnau, M.; Spring, R.; Ghisler, Ch.; Wittwer, S.; Luthy, W. & Weber, H.P. (1996). Efficiency of Erbium 3-µm crystal and fiber lasers.*IEEEJ Quantum Electronics,* Vol.32, No. 4, April (1996).

Press, W. H.; Teukolsky, S.A.; Vetterling, W.T. & Flannery, B.P. (1992). *Numerical Recipients in C: The art of Scientific Compunting,* Second Edition, Cambridge University Press, Cambridge.

Sanchez, F.; Le Flohic, M. ; Stephan, G. M. ; Le Boudec, P.& Francois, P.L. (1995). Quasi - periodic route to chaos in erbium-doped fiber laser, *IEEE J. Quantum Electron.* 31, 481- 488, (1995).

Sanchez, F.; Le Boudec, P.; Francois, P.L. & Stephan, G. (1993). Effects of ion pairs on the dynamics of erbium - doped fiber lasers, *Phys. Rev.* A 48, 2220-2229. (1993)

Shalibeik, H. (2007). *Rare-Earth-Doped Fiber Lasers and Amplifiers.* Cuvillier Verlag, 978-3-86727-467-8, Gottingen.

Stefanescu, E.N.; Sterian, A. R. & Sterian, P. E. (2005). Study of the fermion system coupled by electric dipol interaction with the free electromagnetic field, *Proceedings of SPIE,* Vol. 5850, 160-165, Advanced Laser Tehnologies 2004, edited by Anna Giardini, Vitali I. Konov, Vladimir I. Pustavoy, (2005), Rome.

Stefanescu, E.N.; Sterian, P. E. & Sterian, A. R. (2000). Fundamental Interactions in Dissipative Quantum Systems, *Hyperion University Scientific Bulletin,* Vol.1, No 1, 87-92, (2000).

Stefanescu, E.N.; Sterian, P. E. & Sterian, A. R. (2002). The Lindblad dynamics of a Fermi system in a particle dissipative environment, *Proceedings of SPIE,* Vol. 5147, 160-168, ALT 02, (2002), Adelboden.

Sterian, A. R. & Ninulescu, V. (2005). Nonlinear Phenomena in Erbium-Doped Lasers, *Lecture Notes in Computer Science,* LNCS 3482, 643-650, Gervasi, O.et al. (Eds), Springer - Verlag, (2005), Berlin, Heidelberg.

Sterian, A.R. (2006). *Amplificatoare optice,* Editura Printech, ISBN 978-973-718-434-4, Bucuresti.

Sterian, A.R. & Maciuc, F. C. (2003). Numerical model of an EDFA based on rate equations, *Proceedings of SPIE,* Vol.
5226, 74-78, Laser Physics and Application 12th International School an Quantum Electronics (2003), Varna.

Sterian, A.R. (2008). *Mecanica cuantică,* Omnia Univ. S.A.S.T., (ISBN 978-973-134-042-5, 978-973-134-040-1), Braşov.

Sterian, A.R. (2007). Computer modeling of the coherent optical amplifier and laser systems, *Proceedings of Computational Science and Its Applications*, ICCSA 2007, Pt 1, 4705436-449, Editor(s): Gervasi, O; Gavrilova M.L., (2007).

13

Optical Amplifiers from Rare-Earth Co-Doped Glass Waveguides

G. Vijaya Prakash[1], S. Surendra Babu[2] and A. Amarnath Reddy[1]
[1]*Nanophotonics lab, Department of Physics, Indian Institute of Technology, Delhi, New Delhi-110016,*
[2]*Laser Instrumentation Design Centre, Instruments Research & Development Establishment, Dehradun-248 008,*
India

1. Introduction

Optical amplifiers are of potential use in wide variety of optoelectronic and optical communication applications, particularly for Wavelength Division Multiplexing (WDM) to increase the number of channels and transmission capacity in optical network systems. For efficient performance of WDM systems, essential requirements are larger bandwidth, higher output power and flat gain over entire region of operation. Recent research is focused on design and development of fiber/MEMS-compatible optical amplifiers. Some examples of such sources are semiconductor quantum dot light-emitting diodes, super-luminescent diodes, Erbium doped fibre amplifier (EDFA, 1530-1625nm), Erbium doped planar amplifier (EDWAs), Fibre Raman amplifier, Thulium doped fibre amplifier (1460-1510nm). However, for many applications covering the total telecommunication window (1260-1700nm) is highly desirable and as such it is not yet realized. Typical attenuation spectrum for glassy host is shown in Figure 1. Specially the low loss region extending from 1450 to 1600 nm, deemed the 3rd telecommunication window, emerged as the most practical for long haul telecommunication systems. This window has been split into several distinct bands: Short-band (S-band), Centre-band (C-band) and Long-band (L-band). With several generations of development, the transmission rates have increased dramatically so that several Terabits per second data can be transmitted over a single optical fiber at carrier wavelengths near 1550 nm, a principal optical communication window in which propagation losses are minimum. EDFAs are attractive to WDM technology to compensate the losses introduced by WDM systems and hence has grown as a key to upgrade the transmission capacity of the present fiber links. EDFAs are widely used in long-haul fiber optic networks where the fiber losses are limited to 0.2 dB/km, is compensated periodically by placing EDFAs in the transmission link with spacing of up to 100 km. EDFAs make use of trivalent erbium (Er^{3+}) ions to provide the optical amplification at wavelengths near 1550 nm, the long wavelength window dominantly used in optic networks since the fiber losses are found minimum around this wavelength. Light from an external energy source at a wavelength of 980 nm or 1450 nm, coupled along with the information signal, and is passed through the EDFA to excite the Er^{3+} ions in order to produce the optical amplification through stimulated emission of photons at the signal wavelength. Er^{3+} doped waveguides (EDWA) have obvious advantages as short-length devices and can be integrated monolithically or in hybrid form. Thus individual or co-doped rare earth doped glass fibres/waveguide research is of great promise and concentrated mostly on silicate glasses. However, their Amplified Spontaneous Emission (ASE) bandwidth is limited and also intrinsic absorption occurs even in its most pure form and is associated with the vibrational characteristics of the chemical bonds such as silicon-oxygen bonds. Recently, rare earth co-doped fluoride and tellurite glasses have shown considerable ASE with extended bandwidth, where the emission broadness arises due to inter-ion energy transfer. Overall, the specific technological information related to suitability of rare earth doping, and proper choice of glass matrix is still not clear and worthy of pursuit.

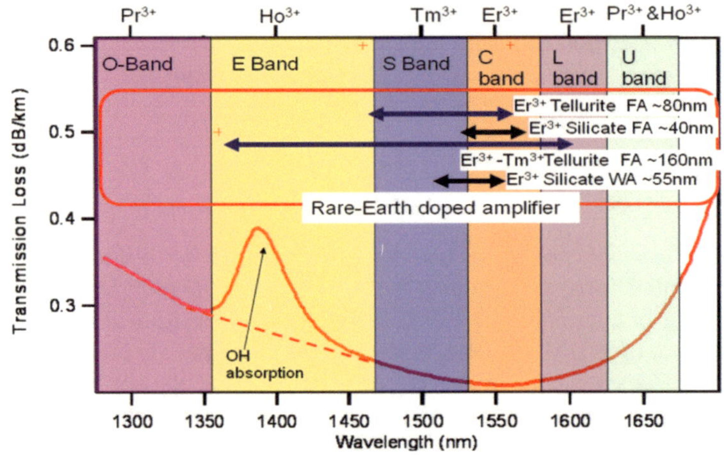

Fig. 1. Loss characteristics of silica fiber and emission bands of some rare earth ions in different glass hosts.

In general, application and utilities of glassy materials are enormous and are governed by the factors like composition, refractive index, dopants/impurities present in the glasses. Demanding physical features of rare earth doped fibers/waveguide devices are (1) high solubility of rare earth for large concentration doping, (2) large temperature gap between the glass transition to crystallization temperature, (3) physical durability such as mechanical and moisture robustness. Moreover, the rare earth emission in glassy matrix is strongly dependent on crystal field effects, local environment where the ion is situated, phonon energies, refractive index and precise details about glass defects (Urbach tails) extended into the band gap. Though silicate glass is inexpensive and most useful matrix for many applications, the low refractive index and strong OH band contributions are major disadvantages. Numerous investigations were made in search for improvised and suitable glass matrix for strong rare earth emission. Among them, phosphate glasses are found to be most suitable for rare earth emission and phosphate glass lasers are already available in the market. However, the knowledge is still largely incomplete due to unusual structural characteristics such as high co-ordinations of the elements present and the large number of chemical elements that are being used in the compositions of these glasses. Our extensive studies on phosphate glasses show wide range of transparency, moisture insensitivity and ability in accepting large amount of rare earths as dopants, and most importantly the compatibility for fibre and waveguides (Vijaya Prakash et al., 2002; Surendra babu et al.,

2007, Pradeesh et al., 2008; Amarnath Reddy et al., 2010). In this chapter we review the essential rare earth spectroscopic details such as absorption analysis through Judd-Ofelt computations, emission lifetimes, gain cross-sections etc,. We further discuss the characteristic features of suitable glass hosts. Finally, we present most common waveguide fabrication techniques and a comparative review of information related to available waveguide devices.

2. Spectral properties of rare earths ions

Optically active rare earth (RE) ions are characterized by their 4f electrons in seven 4f-orbitals. In addition to the most stable electronic configurations (ground state), various configurations with excited higher energy are possible in the thirteen RE ions from Ce^{3+} to Yb^{3+} ion. The energy of these different electronic configurations is separated by the Coulombic spin-orbit and ligand field interactions, resulting in the well-known energy level structures. The energy levels and various absorption and emission transitions of most sought-after Erbium (Er^{3+}) ions are shown in Figure 2A.

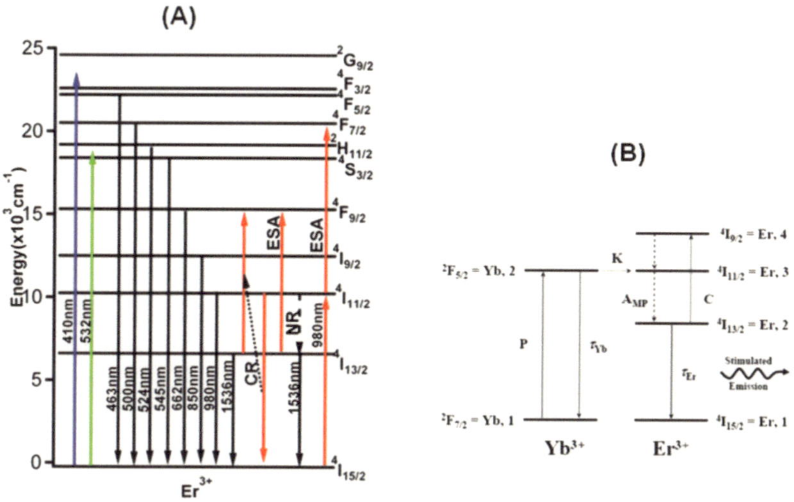

Fig. 2. (A) Partial electronic energy levels and various absorption and emission transitions of Er^{3+} ions and (B) energy transfer between Er^{3+} and Yb^{3+} ions.

The $4f^N$ electron configuration of a rare earth ion is composed of a number of electronic states. The spread of $4f^N$ energy levels arises from various atomic interactions between electrons and can be found by solving the time dependent Schrodinger equation. The weak interaction of the $4f$ electrons with electrons from other ions (in any environment) allows a Hamiltonian to be written for an individual rare earth ion as (Wybourne, 1965).

$$\hat{H} = \hat{H}_{FI} + \hat{H}_{CF} \tag{1}$$

where $\hat{H}_{FI}$ is defined to incorporate the isotropic parts of $\hat{H}$ (including the spherically symmetric part of the $4f$-electron-crystal-field interactions) and $\hat{H}_{CF}$ is defined to represent the non-spherically symmetric components of the even parity crystal-field. The majority of RE optical absorption transitions occur between states which belong to the $4f^N$ configuration.

For intra-*f* shell transitions the initial and final states have the same parity; this would imply that the electric-dipole process is forbidden (Görler-Walrand et. al., 1998). However, in cases where the site symmetry is not a centre of inversion (as usually the case of the glass), the site interaction Hamiltonian can contain odd-parity terms, thereby introducing a degree of electric-dipole strength into the intra-*f* –shell transitions. Using some simplifications, Judd and Ofelt have provided a way to treat conveniently the spectral intensities of intra-configurational transitions of 4*f* ions in solids (Judd, 1962; Ofelt, 1962).
The experimental oscillator strengths of the absorption spectral transitions are calculated from (Reisfeld et al., 1977)

$$f_{\exp} = \frac{2303mc^2}{N\pi e^2} \int \varepsilon(\bar{\nu}) d\bar{\nu} \tag{2}$$

where m and e are the electron mass and charge respectively, c is the velocity of light and $\varepsilon(\bar{\nu})$, the extinction coefficient and is given by $\varepsilon(\bar{\nu}) = \left(\frac{1}{cl}\right)\log\left(\frac{I_0}{I}\right)$ where c is concentration of RE ion in moles/litre, *l* is the optical path length (cm) and $\log(I_0 / I)$ is the absorptivity. According to Judd-Ofelt (JO) theory the oscillator strengths for an induced electric dipole transition can be calculated theoretically from the relation

$$f_{cal} = \sigma \sum T_\lambda \left\langle f^N \psi J \middle| U^\lambda \middle| f^N \psi' J' \right\rangle^2 \tag{3}$$

Where $T_\lambda = \tau_\lambda(2J+1)^{-1}$, σ is the mean energy of the transition in cm^{-1}, T_λ are the adjustable Judd-Ofelt parameters and the U^λ are doubly reduced unit matrix elements. Since these matrix elements are insensitive to ion environment, free ion matrix elements are used from literature. The experimental oscillator strengths were least square fit to obtain the intensity parameters T_λ. These parameters are used in turn to calculate Ω_λ (λ = 2, 4 and 6) using the expression,

$$\Omega_\lambda(cm^2) = \frac{3h}{8\pi^2 mc} \frac{9\eta(2J+1)T_\lambda}{(\eta^2+2)^2} \tag{4}$$

where η is the refractive index of the medium, (2J+1) is the degeneracy of the ground level of the particular ion of interest.
In general, the oscillator strengths and positions of given transitions are sensitive to the local environment of RE ion sites occupied within the glass network. Therefore, JO parameters provide critical information about the nature of bond between RE ions to the surrounding ligands, particularly Ω_2 parameter is sensitive to the local environment of RE ions and is often related to the asymmetry of the coordination structure, the polarizability of ligand ions and nature of bonding. On the other hand the values of Ω_4, and Ω_6 values, depend on bulk properties such as viscosity and dielectric of the media and are also effected by the vibronic transitions of the rare earth ions bound to the ligand atoms (Vijay Prakash et al., 2000).
Accurate knowledge of absorption and emission cross sections is required for a range of applications, such as modelling of optical amplifiers and fiber lasers. While it is not straightforward to measure both of these cross sections accurately, owing to complications such as spectral re-absorption, up-conversion, excited-state absorption etc. The theory of

McCumber (McCumber, 1964) is a powerful tool that permits one of these two cross sections to be calculated if the other is known from measurements. The radiative transition probability between the states ψ J and ψ 'J' is given by

$$A(\psi J,\psi' J') = \frac{64\pi^4 \overline{\upsilon^3}}{3h(2J+1)}\left[\frac{\eta\left(\eta^2+2\right)^2}{9}S_{ed} + \eta^3 S_{md}\right] \tag{5}$$

where S_{ed} and S_{md}, are the electric and magnetic dipole line strengths respectively given by

$$S_{ed} = e^2 \sum_{2,4,6} \Omega_\lambda \left\langle \psi J \left\| U^\lambda \right\| \psi' J' \right\rangle^2 \tag{6}$$

and

$$S_{md} = \frac{e^2 h^2}{16m^2 c^2 \pi^2}\left\langle \psi J \left\| L+2S \right\| \psi' J' \right\rangle^2 \tag{7}$$

The total radiative probability $A_T(\psi J)$, the radiative life time (τ_R), branching ratios(β_R) and stimulated emission cross sections(σ_p) are given by

$$A_T(\psi J) = \sum A(\psi J,\psi' J') \tag{8}$$

$$\beta_R = A(\psi J,\psi' J') / A_T(\psi J) \tag{9}$$

$$\tau_R = (A_T(\psi J))^{-1} \tag{10}$$

and

$$\sigma_p(\lambda_p) = \frac{\lambda_p^4}{8\pi c\eta^2 \Delta\lambda_p} A\left(\psi j,\psi' j'\right) \tag{11}$$

where λ_p and $\Delta\lambda_p$ are average emission wavelength and effective emission band width respectively.

While Er-doped glasses are of significant interest for C-band telecommunication window, in compact systems active material is often co-doped with Ytterbium (Yb^{3+}) to enhance absorption efficiency. To obtain sufficient population inversion, even at low concentration of Er, large amount of Yb^{3+} sensitizing ions is required. Yb → Er migration-assisted non-radiative energy transfer scheme is shown in Figure 2B. For efficient population of the upper $^4I_{13/2}$ laser level of Er^{3+} ions via Yb^{3+} ions, two indispensable conditions have to be fulfilled. The first one is a rapid excitation relaxation from $^4I_{11/2} \rightarrow {}^4I_{13/2}$ of Er^{3+} ion level is necessary. Otherwise reverse energy transfer back to Yb^{3+} ions in combination with the up-conversion processes will compete the upper laser level population process. This may lead to reduction in efficiency, or even inhibit of 1550nm emission completely. Second one is the quantum yield of the upper laser level $^4I_{11/2}$ luminescence should be high. Both conditions are well satisfied in phosphate glasses where Er^{3+} ion luminescent lifetimes are typically τ_1=7.8 - 9.0ms for the $^4I_{13/2}$ level and $\tau_2 \cong$ 1–3 μs for the $^4I_{11/2}$ level (Zhang et al., 2006; Liang et al., 2005)

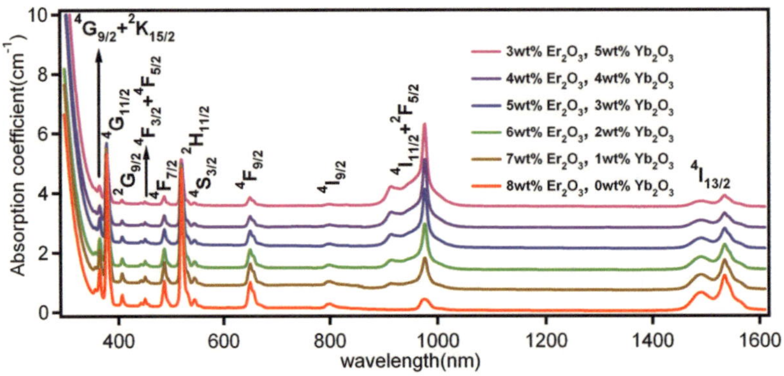

Fig. 3. Typical absorption spectra of Er^{3+}/Yb^{3+} co-doped phosphate (NAP) glasses.

The absorption spectra of Er^{3+}/Yb^{3+} co-doped NAP ($Na_3Al_2P_3O_{12}$) phosphate glass are shown in Figure 3 (Amarnath Reddy et al, 2010). The spectra consist of absorption bands corresponding to the transitions from the ground state $^4I_{15/2}$ of Er^{3+}ions. From the absorption spectral analysis three important JO parameters (Ω_λ = 2, 4 and 6), were obtained. As mentioned before, both covalence and site selectivity of RE ion with non-centro symmetric potential contributes significantly to Ω_2. The variation of Ω_2 with Er^{3+} concentration in various phosphate glasses is shown in Figure 4A and also in Table 1. Also, Ω_2 is strongly dependent on the hypersensitive transitions. Hypersensitivity is related to the covalency through nephelauxetic effect and affects the polarizability of the ligands around the rare earth ions (Kumar et al., 2007). Higher ligand polarizability results in a larger overlap between rare earth ions and ligands orbital, i.e., higher degree of covalency between rare earth ions and the ligands. Hypersensitive nature of $^4I_{15/2} \rightarrow {}^2H_{11/2}$, and $^4G_{11/2}$ transitions of Er^{3+} ions, which is strongly dependent on covalency and site asymmetry, governs the radiative properties of Er^{3+} ions. Though Ω_2 parameter can be an indicator for covalent bonding, in view of hypersensitivity, it is appropriate to look at the variation of the sum of the JO parameters with the oscillator strengths of hypersensitive transitions instead of taking a single parameter. Figure 4B compares the plots of $\Sigma\Omega_\lambda$ (λ= 2, 4 and 6), against oscillator strengths of the hypersensitive transitions of Er^{3+} ions doped in various glass systems. Lower oscillator strengths are observed for ionic systems such as fluorides while highly covalent systems such as phosphate, and borate glasses show higher oscillator strengths.

Electronic polarization of materials is widely regarded as one of the most influencing parameter and many physical, linear and nonlinear optical properties of materials are strongly dependent on it. Duffy, Dimitrov and Sakka correlated many independent linear optical entities to the oxide ion polarizability of single component oxides (Dimitrov & Sakka, 1996; Duffy, 1986). This polarizability approach, predominantly gives the insight into the strong relation between covalent/ ionic nature of materials and other optical parameters, such as optical band gap (E_{opt}) (Vijay Prakash 2000). Recently we have related optical band gaps of various binary, ternary and quaternary oxide glasses to the polarizability (of cation) in terms of $1\text{-}R_m/V_m$,(R_m=molar refractivity and V_m= molar volume) known as covalancy parameter or metallization parameter (Pradeesh etal., 2008). Generally, the covalancy parameter ranges from 0.3 to 0.45 for highly polarisable cation containing oxides (such as Pb^{2+} and Nb^{5+}), while for the alkaline and alkaline-earth (such as Na^+, Li^+) oxides it falls in between 0.5-0.70. Although it is quite complex to correlate optical parameters of the glassy

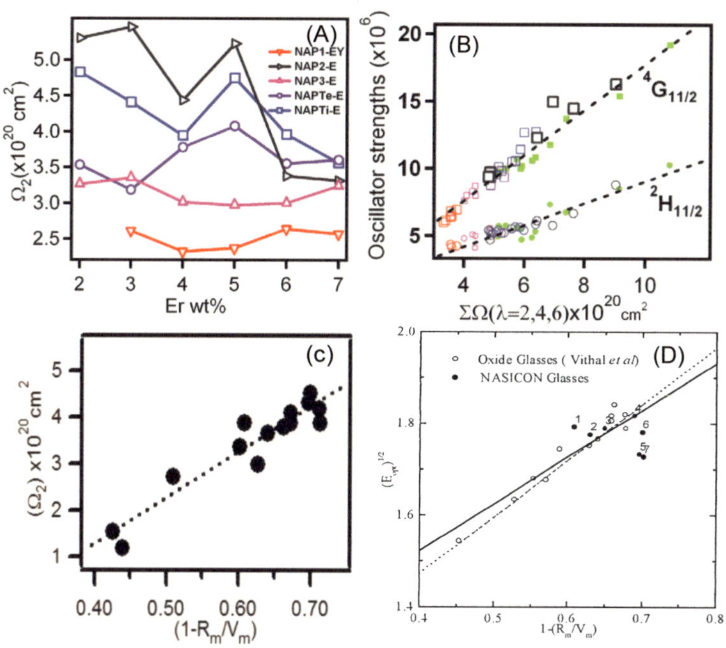

Fig. 4. (A) variation of Ω_2 with Er^{3+} concentration in various phosphate glasses. (B) Oscillator strengths (as open symbols) of various hypersensitive ($^4G_{11/2}$ and $^2H_{11/2}$) and non-hypersensitive ($^4F_{9/2}$ and $^4S_{3/2}$) transitions of Er^{3+} ions with respect to the sum of Judd-Ofelt parameters. Various other glasses are also incorporated (as filled symbols) for comparison (Vijaya Prakash et al., 1999,2000). (C) Ω_2 and (D) E_{opt} values against metallization parameter $(1-R_m/V_m)$ (Pradeesh etal 2008,Vijaya Prakash et al., 1999,2000).

Phosphate glass composition	Ω_2	Ω_4	Ω_6
$92(Na_3Al_2P_3O_{12})$ -(8-x) Al_2O_3-(x)Er_2O_3 x=2	3.26	0.57	0.55
$92(Na_3Al_2P_3O_{12})$ -(8-x) Al_2O_3-(x)Er_2O_3 x=3	3.25	0.54	0.55
$92(Na_3Al_2P_3O_{12})$ -(8-x) Al_2O_3-(x)Er_2O_3 x=4	3.01	0.51	0.56
$92(Na_3Al_2P_3O_{12})$ -(8-x) Al_2O_3-(x)Er_2O_3 x=5	2.97	0.49	0.54
$92(Na_3Al_2P_3O_{12})$ -(8-x) Al_2O_3-(x)Er_2O_3 x=6	2.99	0.84	0.59
$92(Na_3Al_2P_3O_{12})$ -(8-x) Al_2O_3-(x)Er_2O_3 x=7	3.24	0.80	0.71
$92(Na_3Al_2P_3O_{12})$ -(8-x) Al_2O_3-(x)Er_2O_3 x=8	3.53	0.76	0.56
$92(Na_2AlP_5O_{15})$ -(8-x) Al_2O_3-(x)Er_2O_3 x=7	3.08	0.80	0.39
$92(Na_2AlP_5O_{15})$ -(8-x) Al_2O_3-(x)Er_2O_3 x=8	3.39	0.72	0.50
60 $PbCl_2$ - 40P_2O_5, 1wt% Er^{3+}	3.36	0.51	1.51
10Na_2O- 50 $PbCl_2$ - 40P_2O_5, 1wt% Er^{3+}	4.11	0.47	1.16
20Na_2O- 40 $PbCl_2$ - 40P_2O_5, 1wt% Er^{3+}	3.66	0.34	1.86
30Na_2O- 30 $PbCl_2$ - 40P_2O_5, 1wt% Er^{3+}	3.79	0.13	1.21
$Na_3TiZnP_3O_{12}$ 1wt% Er^{3+}	4.94	1.15	5.77
$Na_4AlZnP_3O_{12}$ 1wt% Er^{3+}	3.77	1.34	1.25

Table 1. Judd Ofelt ((Ω_λ = 2, 4 and 6) x 10^{-20}, cm^2) parameters for various phosphate glasses

systems with polarizability of oxide ions, it is interesting to see the influence of cation (such as Na^{+}) polarizability on other independent parameters, such as JO parameters. Since Ω_2 values are known to show dependence on covalent nature of phosphate host, we made a plot between Ω_2 and E_{opt} values against metallization parameter ($1-R_m/V_m$) in Figure 4C&D. Interestingly, the Ω_2 values are monotonically increasing with the metallization parameter. Though with respect to metallization parameters they show more of alkali oxide nature, significantly higher Ω_2 indicates strong covalent nature, could possibly be due to the increased role of phosphate linkage, having more double bonded oxygens (DBOs) coordinated to the rare earth ions.

Further, table 2 gives the most useful physical and spectral parameters, viz., values of density(d, gm/cm^3), Refractive index (n), Judd-Oflet parameters ((Ω_λ, λ=2,4,6)x10^{-20}, cm^2), calculated (τ_{cal}, ms) and measured life times (τ_{exp}, ms), effective line widths ($\Delta\lambda$, nm), absorption (σ_ax10^{-21}, cm^2) and emission cross-sections (σ_ex10^{-21}, cm^2) of Er^{3+} and Er^{3+}/Yb^{3+} doped in phosphate, Tellurite and silicate glasses.

Composition	RE	d/η	Ω_2	Ω_4	Ω_6	τ_{exp}	τ_{cal}	$\Delta\lambda$	σ_a	σ_e	Reference
Phosphate											
$20Na_2O$–$2Al_2O_3$–xEr_2O_3–yYb_2O_3–$30Nb_2O_5$-$15TiO_2$–$30P_2O_5$	Er^{3+}-Yb^{3+}	-/1.84	-	-	-	3.00	-	53	-	-	Bozelli et al., 2010
Schott	Er^{3+}-Yb^{3+}	-/1.52	-	-	-	9.18	9.21	40	-	-	Zhang, et al., 2006
EYDPG	Er^{3+}-Yb^{3+}	-	-	-	-	7.91	9.10	42	6.56	7.20	Liang, et al., 2005
$72P_2O_5$-$8Al_2O_3$-$20Na_2O$	Er^{3+}	-/1.51	6.84	1.94	1.29	-	7.46	-	7.41	8.23	Gangfeng, et al., 2005
WM-1	Er^{3+}-Yb^{3+}	2.83/1.53	6.45	1.56	0.89	7.90	9.10	42	6.56	7.20	Bao - Yu, et al., 2003
$67P_2O_5$-$14Al_2O_3$-$14Li_2O$-$1K_2O$	Er^{3+}-Yb^{3+}	-	7.06	1.70	1.04	8.00	8.62	40	6.4	7.8	Wong, et al., 2002
$29.9Na_2O$-$30.8P_2O_5$-$18.7Nb_2O_5$-$13.9Ga_2O_3$-$6.6Er_2O_3$	Er^{3+}	-/1.71	3.89	1.00	0.55	8.9	3.10	50	-	-	Righini, et al., 2001
IOG-1	Er^{3+}-Yb^{3+}	-/1.52	6.13	1.48	1.12	8.10	8.55	46	4.13	4.62	Veasey, et al., 2000
Tellurite											
$80TeO_2$-$10ZnO$-$10Na_2O$-$20P_2O_5$-$0.5Er_2O_3$	Er^{3+}	4.36/1.95	-	-	-	3.90	-	46	6.99	8.50	Fernandez, et al., 2008
$67TeO_2$-$30P_2O_5$-$1Al_2O_3$-$1.75La_2O_3$-$0.25Er_2O_3$	Er^{3+}	3.75/2.00	3.4	1.0	0.2	4.10	7.90	41	-	6.00	Nandi, et al., 2006
$60TeO_2$–$30WO_3$–$10Na_2O$	Er^{3+}	-	7.13	1.90	0.82	3.40	3.46	52	8.10	9.10	Zhao, et al., 2006
$25WO_3$-$15Na_2O$-$60TeO_2$+(0.05-$2Er_2O_3$)	Er^{3+}	-/2.047	6.7	1.7	1.15	2.80	3.39	62	-	-	Conti, et al., 2004
$8Na_2O$-$27.6Nb_2O_5$-$64.4Te_2O_5$-$1Er_2O_3$	Er^{3+}	4.98/2.12	6.86	1.53	1.12	3.02	2.90	-	8.63	1.02	Lin, et al., 2003
Silicate											
$52.0SiO_2$·$33.0B_2O_3$·$7.7Na_2O$·4.0 CaO·$2.7Al_2O_3$.$0.6CeO_2$	Er^{3+}	1.98/1.46	8.15	1.43	1.22	6.00	9.10	45	-	5.10	Ning, et al., 2006
$61.4SiO_2$–$11.73Na_2O$–$9.23CaO$–$16.67Al_2O_3$–$0.33P_2O_5$-$0.51K_2O$-$0.39Er_2O_3$	Er^{3+}	2.624/ 1.54	7.15	1.95	1.01	6.20	8.43	48	-	7.70	Berneschi, et al., 2005
$61SiO_2$–$12Na_2O$–$3Al_2O_3$–$12LaF_3$–$12PbF_2$	Er^{3+}	3.62/1.66	-	-	-	10.5	-	48	7.50	7.80	Shen, et al.,2004
$73SiO_2$–$14Na_2O$–$11CaO$–$1Al_2O_3$–$0.4P_2O_5$–$0.6K_2O$	Er^{3+}-Yb^{3+},	-/1.53	4.89	0.77	0.50	-	-	-	-	-	Righini, et al., 2001
IOG-10	Er^{3+}-Yb^{3+}	-/1.54	-	-	-	10.2	17.8	32	5.70	5.80	Peters, et al., 1999

Table 2. Important spectroscopic parameters of Er^{3+} and Er^{3+}/Yb^{3+} doped various phosphate, tellurite and silicate glasses.

3. Glasses for rare earth based amplifiers

Improvements in the quantum efficiency of the luminescent levels of RE ions can be achieved by selecting a suitable host material and by modifying the local environment surrounding them (Görler-Walrand et. al., 1998). Such modifications are often achieved by breaking up the structure of atoms surrounding a rare earth ion with other, often larger elements, termed as *network modifiers*. Discussing the role of network modifiers requires an understanding of the basic structure of different glass network formers. Oxide glasses, such as borate, silicate, tellurite and phosphate glasses, have proven to be the appropriate host materials for the development of optoelectronic components. Among the oxide glass hosts, phasephate glasses have attracted much attention due to their high transparency, thermal stability, good RE ion solubility, easy preparation in large scale, shaping and cost effective properties. Let us compare the structural features of phosphate glass with well-known silica and tellurite glass networks.

3.1 Silicate glass

Silica is built from basic structural units, the most common of which is the network former unit, $(SiO_4)^{2-}$. This network former consists of a silicon atom at the centre of a tetrahedron with an O atom bonded to each corner. The basic structure of silica glass, Figure 5A, contains tetrahedral units that are tightly connected by their corners through oxygen atoms (bridging oxygens); these random connections form a 3 dimensional structure similar to that shown in the Figure 5B. The strong electron bond which exists between the silicon and oxygen atoms gives the silica glass its impressive mechanical strength and thermal properties. However, the drawback to the silica glass network is its limited soluability of rare earth ions (<3wt%). One explanation of this phenomenon is that rare earth ions require co-ordination of a sufficiently high number of non-bridging oxygens to screen the electric charge of the ion, whereas highly rigid silica glass network cannot co-ordinate the non-bridging oxygens resulting in a system with a higher enthalpy state. Therefore, rare earth ions tend to share non-bridging oxygens to reduce the excess enthalpy, resulting in the

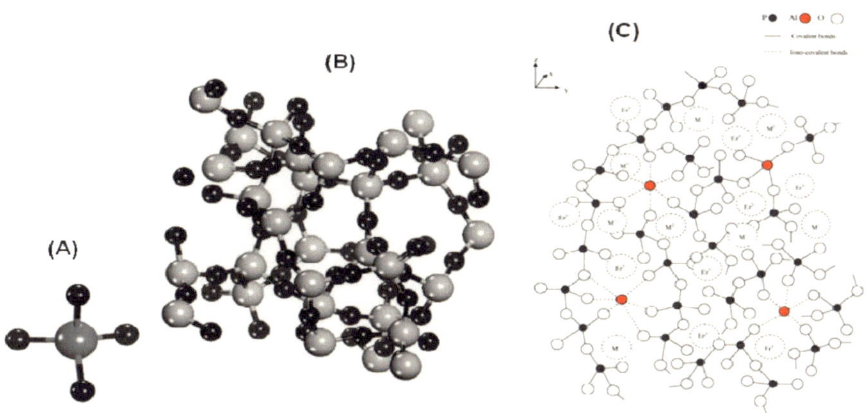

Fig. 5. (A) &(B)Tetrahedral structure of $(SiO_4)^{2-}$, showing 4 oxygen atoms surrounding the central silicon atom and 3 - dimensional structure of SiO_2 , showing the interconnection of the tetrahedral units. (C) Schematic view of the phosphate glass network showing the different type of atoms neighbourhoods.

formation of clusters. To accommodate greater amounts of rare earth ions, network modifiers are required to increase the number of available non-bridging oxygen ions in the silica glass network. Network modifiers such as Na^{3+} and Al^{3+} are often used to facilitate the incorporation of rare earth ions, as their size is substantially greater than the basic network. These modifiers act to break the bridging oxygens to form non-bridging oxygens which can be used to co-ordinate the rare earth ions. Of the network modifiers studied in silica glass to date, Al_2O_3 has shown the most favourable characteristics. Al_2O_3 has also been used to improve the efficiency of Er^{3+} -doped fibre amplifiers by eliminating the quenching effects from the $^4I_{13/2} \rightarrow {}^4I_{15/2}$ transition (Corradi et al., 2006).

3.2 Tellurite glass

Recent work on the structure of tellurite glasses has concluded that the network more closely resembles paratellurite (α-TeO_2), where [TeO_4] units are only linked at their corners (Stavrou et al., 2010). Combining TeO_2 with network modifiers (such as Na_2O) and intermediates (such as ZnO) results in structural modification to chain like structures. In general the tellurite glasses follow the pattern of crystalline α-TeO_2, which are formed by [TeO_4] groups as trigonal bipyramids. Such structural units can progressively form [TeO_{3+1}] and trigonal pyramids [TeO_3] (Surendra babu et al., 2007). Upon network modified, the glass network becomes more open and more non-bridging oxygens are created (Golubeva, 2003; Vijaya Prakash et al., 2001). Tellurite glasses are very promising materials for up-conversion lasers and nonlinear applications (such as photonic crystal fibers) in optics due to some of their important characteristic features such as high refractive index, low phonon maxima and low melting temperatures. Recently, an addition of oxides of heavy metals such as Nb, Pb and W to tellurite glasses is being studied extensively because such additions seem to show remarkable changes in both physical and optical properties of these glasses (Vijaya Prakash et al, 2001).

3.3 Phosphate glass

The structure of phosphate glass consists of random network of phosphorous tetrahedra. In glassy and crystalline phosphate the basic building blocks are PO_4 tetrahedra. In a pure phosphate glass the tetrahedra are linked through three of the oxygens while the fourth oxygen is doubly bonded to the phosphorus atom and does not participate in the network formation. The networks of phosphate glasses can be classified by the oxygen to phosphorus ratio, which sets the number of tetrahedral linkages (through bridging oxygens) between neighboring P-tetrahedra (Vijaya Prakash & Jaganaathan, 1999). When the modifier cations are added to the phosphate glasses, the P=O of phosphate group is unaffected and depolymerization takes place through the breaking linkages only. When a glass modifier (oxides, such as Al_2O_3) is added the network breaks up creating non bridging oxygens in the structure which coordinate the metal ions of the modifier oxide, Figure 5C. With increasing amount of modifier, the number of non-bridging oxygens, per PO_4 unit, will go from zero to three for orthophosphates. At this composition the host structure consists principally of chains of corner linked PO_4 tetrahedra with 2 non-bridging oxygens per tetrahedron. The metal ions of the modifier oxide will not participate in the network but will associate to the non-bridging oxygens (Seneschal et al., 2005).

In recent days NASICON type phosphate glasses (acronym for the crystalline Na-Super-Ionic Conductor, $Na_{1+x}Zr_2P_{3-x}Si_xO_{12}$) has attracted much attention, as they facilitate a large scope for preparing a number of glasses with variation in their constituent metal ions and

compositions. These glasses have the general formula $A_mB_nP_3O_{12}$ where A is an alkali or alkaline earth metal ion and B is one or more metal ions in tri, tetra or pentavalent state. Constituent metal ion variation in these glasses has shown marked changes in Physical, linear and nonlinear optical properties (Mariappan et al, 2005, 2005, et al, Vijaya Prakash et al., 1999,00,01,02). It is also interesting to note that the rare earth ions in NASICON glasses are likely to be located in the sites of A and B, implies that the rare earth oxides are actively involved in glass network than as simple dopants. Morever, due to the presence of alkali ions, the rare earth solubility improves leading to the possibility of using a high concentration of dopants, which is very important for short length optical amplifiers,and further provides suitability for the fabrication of optical wave guide devices by ion exchange. Also chloro/fluorophosphate glasses show potential as hosts for lasers and holographic gratings, specially lead-bearing fluorides are considered to be good candidates for up-conversion studies (Pradeesh et al, 2008, Vijaya Prakash, et al 1999).
Phosphate laser glass is an attractive amplifier material because it combines the required properties of good chemical durability, high gain density, wide bandwidth emission spectrum of erbium, and low up-conversion characteristics (Miniscalco, 1991). Phosphate glass exhibits a high gain density due to a high solubility for rare earth ions. The high ion density results in a significantly short-length optical gain device than silica bsed glass counterparts. For example, Erbium doped silica fiber is typical gain coefficients are about of 2 to 3 dB/m, whereas in phosphate glass waveguide it is about 2 to 3 dB/cm. Erbium-ytterbium doped phosphate glass technology, in particular, has demonstrated a significant capacity for large gain per length coefficients in addition to providing the ability to tailor the absorption by the ytterbium concentration. These combined aspects of the phosphate glass material, and reported results, Table2, support it as a prime candidate for producing compact photonic modules employing gain.

4. Erbium doped waveguide fabrication

As discussed earlier, glass is of particular interest for integrated optics because of relatively low cost, excellent transparency, high optical damage threshold and availability in substantially large sizes. It is rigid and amorphous which makes it easier to produce polarization-insensitive components. Refractive index can be tailored close to that of optical fiber to reduce the coupling losses between the waveguides and optical fibers. There are various waveguide fabrication methods available for waveguides amplifiers such as Ion Implantation of Er ions directly into the pre-fabricated waveguide (Bentini et al., 2008), Thin film techniques (RF Sputtering/PECVD/EBVD) combined with photolithography, reactive ion etching (RIE) and flame hydrolysis (Shmulovich, et. al., 1992; Nakazawa & Ktmuraa, 1992). Composite erbium-doped waveguides (Honkanen et. al., 1992,) and sol-gel based low-cost integrated optoelectronic devices are some of the other interesting developments (Najafi et. al., 1996; Milova et. al., 1997).
Micromachining of glass substrates by high-power femtosecond laser pulses is one of the recent developement in the fabrication of optical channel waveguides, Figure 6C. This technique has an unique advantage in fabricating of 3-D waveguides inside glass substrates, which is not easy from conventional ion-exchange and photolithographic processes. Channel waveguides written using ultrafast lasers in erbium-doped phosphate glasses for integrated amplifiers and lasers operating in the C-band have been already demonstrated (Osellame, 2003).

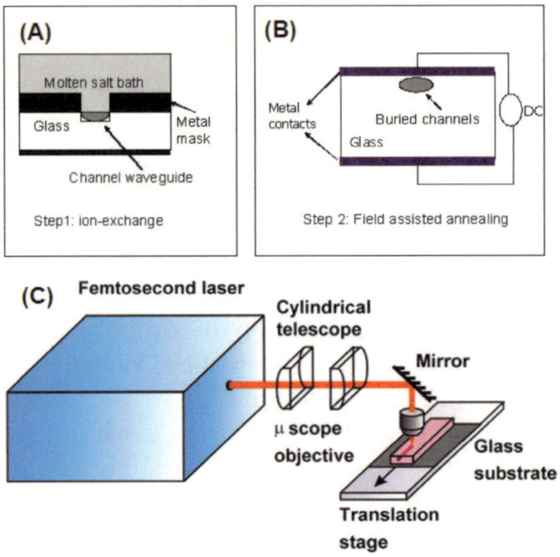

Fig. 6. (A&B) Ion exchange field assisted annealing channel waveguide fabrication (C) Optical setup for femtosecond laser waveguide writing

Among all, ion-exchange has been the most popular technique for rare earth doped waveguides in glass. Molten salt bath has been used as a source of ion-exchange for fabrication of glass waveguides (Ramaswamy & Srivastava, 1988). In this process usually the ion-exchange take place between alkali ions (mostly Na^+) of the glass host and monovalent cations (Cs^+, Rb^+, Li^+, K^+, Ag^+ and Tl^+) from molten salt bath. Alkali containing phosphate glasses are of right choice because of high solubility of rare earth ions without significant reduction in the emission life times. Moreover, many oxide based glasses exhibit poor chemical durability during ion-exchange waveguide fabrication, due to their weak structural network that results into damage of surfaces. Phosphate glasses are considerably stable to thermal and chemical induced fluctuations among other oxides. Ion exchange involves a local change in composition which is brought about by mass transport driven by thermal or electric field gradients or some combination of the two, using lithographically designed mask. Usually, the ion exchange process has no effect on the basic structure of glass network if it is carried out at temperatures well below the softening point of the glass. The ion exchange can be purely a thermal diffusion process, or an electric field assisted diffusion process. The refractive index change and diffusion depth can be controlled easily by proper choice of the exchanged ions and the glass compositions. Table 3 gives ion-exchange conditions for phosphate, tellurite and silicate glass compositions along with refractive index change and diffusion depths. The index profile can be tailored from shallow graded to a step like function with the assistance of electric field. For slightly buried waveguides, field assisted one step or two step ion exchange processes is often used, (Figure 6A&B) (Liu and Pun, 2007). This technique has already well-demonstrated the capacity to form planar and channel waveguides as power splitters, multiplexers, optical amplifiers (EDWAs), with integrated-optical functions with great stability and low losses.

The depth of the waveguide d is related the diffusion time t (time of the ion-exchange process) by the equation $d = \sqrt{D.t}$ where D is the effective diffusion coefficient which depends on the molten salt solution, the glass and the temperature. This kind of diffusion profile indicates that the mobility of the incoming alkali ion (example, Ag^+) is much lower than that of the original ion (example, Na^+) in the glass. Furthermore, the diffusion coefficient shows an Arrhenius temperature dependence:

$$D = D_0 \exp\left[-\frac{E_D}{RT}\right] \tag{12}$$

where D_0 is a fitting constant, E_D is the activation energy, T is the temperature of the bath and R is the universal gas constant (8.314 J/K mol).

A successful ion-exchange process to fabricate stable and low-loss waveguides demands (1) control of host glass composition and identifying suitable processing conditions, (2) increased ASE cross sections with high gain flatness, (3) high refractive index and low-loss glass hosts with a compatibility for efficient emission, and (4) complete understanding of guest-host interaction in rare earth doped glass.

Composition	**Melt composition**	**RE**	**T⁰C**	**Time**	**η**	**Δη**	**d μm**	**D.C. m²/S**	**Reference**
				Phosphate					
$20Na_2O$–(5-x-y)Al_2O_3–xEr_2O_3–yYb_2O_3–$30Nb_2O_5$–$15TiO_2$–$30P_2O_5$	97.33g $NaNO_3$+2.67g $AgNO_3$	Er^{3+}/ Er^{3+}- Yb^{3+}	400	1h	1.836 1.894 1.851	0.007 0.007 0.021	7.50 2.20 7.90	-	Bozelli, et al., 2010
Phosphate IOG-1	$5AgNo_3+95KNO_3$ $4.8AgNo_3+89KNO_3+6.2NaNO_3$	Er^{3+}-Yb^{3+}	345 330	35min 8min	-	0.02	7.00	5.8×10^{-16} 28×10^{-16}	Jose, et al., 2003
				Tellurite					
$25WO_3$-$15Na_2O$-$60TeO_2$+(0.05-$2Er_2O_3$)	$2AgNo_3+43KNO_3+55NaNO_3$	Er^{3+}	330	90min	2.03	0.12	-	6×10^{-10}	Conti, et al., 2004
$12Na_2O$ -$35WO_3$ - $53TeO_2$- $1Er_2O_3$	1.0$AgNO_3$ 49.5$NaNO_3$ 49.5KNO_3	Er^{3+}	300 - 360	5h	2.07	0.10	3.30	-	Sakida et al., 2007
$75TeO_2$-$2GeO_2$-$10Na_2O$-12ZnO-$1Er_2O_3$	$2AgNo_3+49KNO_3+49NaNO_3$	Er^{3+}	250 - 280	3-12h	2.01	0.24	2.21	1.0×10^{-16}	Rivera, et al., 2006
				Silicate					
$73SiO_2$–$14Na_2O$–11CaO–$1Al_2O_3$–$0.4P_2O_5$–$0.6K_2O$ mol%	$0.5AgNo_3+99.5NaNO_3$	Er^{3+}-Yb^{3+}	-	-	1.52	0.044	2.1	6×10^{-9} at1.53	Righini, et al., 2001
MM40 $12Na_2O$+ZnO+MgO+SiO2+$2Er_2O_3$	$100KNO_3$ for $K^+\leftrightarrow Na^+$ $24AgNO_3+50NaNO_3+50KNO_3$ for $Ag^+\leftrightarrow Na^+$	Er^{3+}	375 280	2h 5min	-	0.009 0.085	8.20 3.40	3.12×10^{-9} 12.9×10^{-9}	Salavcoca, et al., 2005
$74.3SiO_2$–$13.4Na_2O$–4.4CaO–$2.8Al_2O_3$–3.2MgO-$0.5K_2O$-1.4BaO	$41KNO_3$-$59Ca(NO_3)_2$	Pure	380	6h	-	0.008	9.1	-	Kosikova et al., 1999

Table 3. Ion-exchange waveguide characteristic parameters, substrate refractive index (n), change in refractive index (Δn), depth (d,μm) of waveguide and diffusion coefficient (DC,m^2/s) of melt composition, temperature (°C) and time used for various phosphate, tellurite and silicate, glass waveguides.

5. Spectroscopic and waveguide characterization

5.1 Prism coupling

Prism coupling technique is widely considered as one of the effective ways to characterise planar optical waveguides. The totally reflecting prism coupler technique, also known as the *m*-line technique, is commonly used to determine the optical properties of thin films. The coupling of an incident laser beam by a prism into a planar waveguide is governed by the incident angle θ of the beam on the prism base. Under total internal reflection conditions, coupling of light into the waveguide would occur via resonant frustrated total reflection, i.e. via evanescent waves in the air layer (Figure 7A). Such coupling occurs only when resonant conditions inside the waveguide are met such phase matching condition. This leads to a finite number of discrete incidences of the light beam, for which the light can be strongly coupled into the waveguide and can be transmitted through the substrate. In the experiment, the resonant coupling of the laser beam into the waveguide is observed through the appearance of dark and bright lines in the reflected beam known as ***m***-lines, Figure 7B. Effective index of each guided mode in a waveguide is calculated by

$$n_{eff} = n_p \sin\left(\theta_p + \sin^{-1}\left(\sin\theta_m / n_p\right)\right) \tag{13}$$

here n_p is the prism index, θp is the angle of prism and θ_m is the synchronised angle. From the measured refractive index profile, one can estimate the information of waveguide, such as the depth, the change in refractive index and the number of modes coupled to the waveguide. The most popular method for refractive-index profiling of planar waveguides is the inverse Wentzel–Kramer–Brillouin (WKB) method, in which the refractive-index profile of a waveguide is defined uniquely by the relationship between the effective index and the mode order, i.e., the effective-index function (Chiang, et. al., 2000).

5.2 Spectroscopic properties

The relative absorption and emission cross sections at both pump and signal wavelengths are obtained from the absorption and emission measurements, as mentioned earlier. For intensity profile of the guided modes and scattering losses, light will be fed through butt-coupling fiber coupled to a tunable laser source through the channels of edge- polished waveguides. The mode field profiles at the signal and pump wavelengths are obtained by near-field IR imaging at the waveguide output facets. The channel waveguide scattering loss are estimated from the Fabry–Perot resonator method (Lee, 1998) and propagation losses can be estimated from the conventional cutback method.

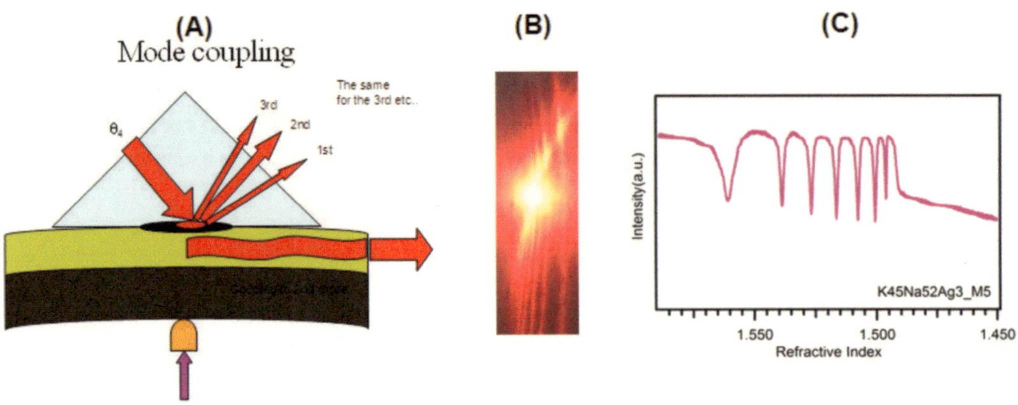

Fig. 7. (A) Schematic representation of prism coupling, (B) photograph of m-line pattern and (C) m-line intensity spectra as a function of effective refractive index.

The small signal gain or optical gain measurements are usually performed from the typical experimental setup showin in Figure 8, consists of a tunable laser (signal) and a semiconductor laser diode (pump) as shown in Figure 8. The signal and pump lasers are combined by a suitable fiber WDM and coupled into the waveguides using a single mode fiber. At the waveguide output facet, the amplified signal will be conveniently separated from the pump in the second fiber WDM coupler and the eventual signal is detected using a detector or optical spectrum analyzer. The signal light intensities from the output of the waveguide with and without pump laser were measured to estimate the internal gain G_{INT} (= signal light power with pump/signal light power without pump).
The optical gain G_O, relative gain G_R (signal enhancement), and net gain G_N of the waveguide amplifier are defined as

$$G_O = 10\log_{10}\left(P_{\text{Sig (pump ON)}}/P_{\text{sig (pump OFF)}}\right) \tag{14}$$

$$G_R = 10\log_{10}\left((P_{\text{Sig (pump ON)}} - P_{ASE})/P_{\text{sig (pump OFF)}}\right) \tag{15}$$

$$G_N = G_R\text{-coupling losses-waveguide losses-RE}^{3+}\text{ absorption losses} \tag{16}$$

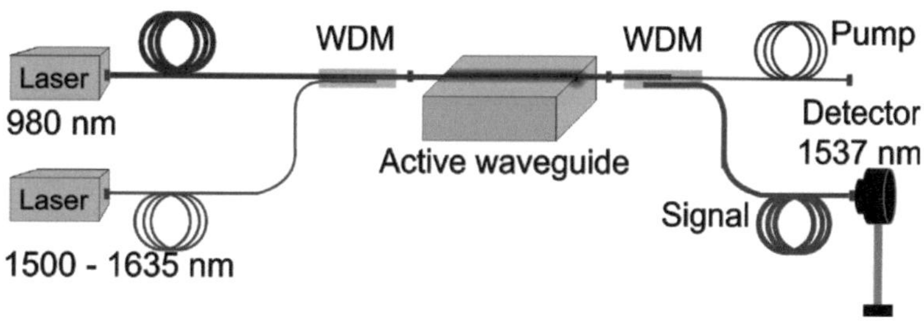

Fig. 8. Experimental setup for Er doped waveguide optical gain measurements

In order to obtain the net gain, it is reasonable to use the following three approaches for a practical waveguide amplifier. The first one is to decrease the waveguide loss by improving the waveguide quality; the second is to increase the coupling efficiency; and the third one is to enhance the pump power, or improve mode confinement in the guide at both pump and signal wavelengths.

6. Conclusion

A brief review of rare earth-doped glass waveguides and their potential application as optical amplifiers is presented. Significance of spectral information, glass composition and rareearth (RE) ion-glass host interaction for engineering waveguide devices, which can potentially useful to design waveguide devices and/or fully integrated photonic structures is discussed. Further, a brief review on fabrication strategies related to waveguides and the influence of the glass composition and other conditions are also presented. Glass-based waveguides thus offer excellent flexibility in fabricating multi-functional optoelectronic devices using cost-effective technologies, having adequate knowledge of glass and rare earth properties.

Table 4 gives the optical gain parameters of various phosphate, tellurite and silicate glass waveguides fabricated from different fabrication methods, ion-exchange (IE), ion exchange filed assisted annealing (IE FAA) and femtosecond laser writing (FSLW). Optical gain parameters includes the coupling losses (CL, dB/facet), propagation losses (α_p, dB), absorption losses (α_a, dB), insertion losses (IL, dB), internal gain (G_{INT}, dB), relative gain (G_R, dB) and net gain (G_N, dB/cm).

Method of fabrication	Channel aperture , laser specification	Length cm	RE	CL dB/facet	α_p dB/cm	α_a dB	IL dB	G_{INT} dB	G_R dB	G_N dB/cm	Gain range (nm)	Reference
Phosphate												
IE,FAA	6μm	1.24 2.72	Er^{3+}-Yb^{3+}	2.20 2.20	2.27 2.22	5.70 28.00	-	- -	7.0	-	1520-1580	Zhang et al., 2006
IE	4-12 μm	3.80	Er^{3+}-Yb^{3+}	-	0.33	-	3.60	-	-	3.65	-	Liang et al.,2005
IE	4-10μm	1.50	Er^{3+}-Yb^{3+}	-	0.80	-	-	-	-	3.30	-	Wong et al., 2002
IE, FAA	6 μm	4.00	Er^{3+}-Yb^{3+}	0.25	0.30	-	12.2	-	-	2.00	1530-1560	Liu. et al, 2007
IE, FAA	6,8 μm	1.20	Er^{3+}-Yb^{3+}	0.32	0.30	5.40	-	-	-	3.40	-	Liu, et al., 2004
FSLW	150fs,500μJ, 1kHz, 270nJ, 885kHz	2.5 2.2	Er^{3+}-Yb^{3+}	2.4 0.1	0.28 0.4	-	5.5 1.2	1.4 7	-	- 2.72	- 1530-1565	Osellame et al., 2008
FSLW	22MHz,350fs, 1μJ	3.70	Er^{3+}-Yb^{3+}	0.25	0.40	-	1.90	-	-	1.97	1530-1580	Valle, et al., 2005
FSLW	166kHz,300fs, 270nJ	2.00	Er^{3+}-Yb^{3+}	0.25	0.80	-	2.10	4.4	-	1.15	1530-1550	Taccheo, et al., 2004
Tellurite												
FSLW	600kHz,350fs, 1.3μJ	2.50	Er^{3+}	0.50	1.35	2.08	4.40	1.25	-	-	1530-1610	Fernandez, et al., 2008
IE	3μm thick	5.00	Er^{3+}	-	8.00	-	-	-	-	-	-	Sakida, et al., 2006
Silicate												
IE	7-13μm	3.50	Er^{3+}-Yb^{3+}	-	8.00		-	2.50	-	-	-	Righini, et al., 2001
IE	3μm	3.00	-	0.76	0.50	-	2.99	-	-	-	-	He, et al., 2008
IE	3μm	1.80	Er^{3+}-Yb^{3+}	0.36	0.25	-	-	-	-	-	-	Peters, et al., 1999
FSLW	1kHz,100fs, 1-90μJ	1.00	Er^{3+}	0.70	0.90	5.5		2.30	-	-	-	Vishnubhatla, et al., 2009
FSLW	600kHz,350fs, 40-150nJ	1.00	Er^{3+}-Yb^{3+}	0.40	0.34	4.34	5.56	1.93	6.1	0.72	1535-1555	Psaila, et al., 2007
FSLW	5kHz,130fs, 0.3μJ	1.00	Er^{3+}	1.20	1.00	6.90	11.2	1.70	8.6	-	-	Thomsom, et al., 2006
FSLW	5kHz,250fs, 0.9μJ	1.90	Er^{3+}	0.80	1.68				2.7		-	Thomson, et al., 2005

Table 4. Various optical gain parameters of Phosphate, tellurite, Silicate glass waveguides

7. Acknowledgments

Authors acknowledge the financial support from Ministry of Information Technology, India through the project on *"Ultra Wideband Optical Sources from rare earth co-doped glass Waveguides-Fabrication and Characterization* "under Photonics switching Multiplexing and Networking (PSMN) programme-Phase II. This work is partially supported by *UK-India Education and Research Initiative* (UKIERI) programme.

8. References

Amarnath Reddy, A.; Sekhar, C.H., Pradeesh, K., Surendra Babu, S. & Vijaya Prakash, G. (2010). *Optical properties of Dy^{3+}-doped Sodium-Aluminium-Phosphate glasses.* Journal of Materials Science. (Accepted)I *0022-2461*

Amarnath Reddy, A.; Surendra Babu, S. Pradeesh, K., Otton, C. J., &Vijaya Prakash, G. (2010). Optical properties of highly Er^{3+} oped Sodium-Aluminium- Phosphate Glasses for broadband 1.5 μm emission. *(un published)*

Bao-Yu, C.; Shi-Long, Z. & Li-Li, H. (2003). Novel Chemically Stable Er^{3+}-Yb^{3+} Codopded Phosphate Glass for Ion-Exchanged Active Waveguide Devices. *chinese Physics Letter* 20, 11, (July, 2003) 2056–2057, 0256-307x

Bentini, G. B.; Chiarini, M., Bianconi, M., Bergamini, F., Castaldini, D., Montanari, G. B., Bogoni, A., Potì, L., Sugliani, S., Nubil, A., De Nicola, P., Gallerani, L., Pennestrì, G. & Petrini, G. (2008). Waveguide formation by ion implantation in Er doped optical materials. *Nuclear Instruments and Methods in Physics Research B,* 266, (2008) 3120–3124, 0168-538x

Berneschi, S.; Bettinelli, M., Brenci, M., DallIgna, R., Conti, N. G., Pelli, S., Profilo, B., Sebastiani, S., Speghini, A. & Righini, G. C. (2006). Optical and spectroscopic properties of soda-lime alumino silicate glasses doped with Er^{3+} and/or Yb^{3+}. *Optical Materials,* 28, (August, 2006) 1271–1275, 0925-3467

Berneschi, S.; Bettinelli, M., Brenci, M., Conti, G., Pelli, S., Sebastiani, S., Siligardi, C., Speghini, A. & Righini, G.C. (2005). Aluminum co-doping of soda-lime silicate glasses:Effect on optical and spectroscopic properties. *Journal of Non-Crystalline Solids,* 351, (July, 2005) 1747–1753, 0022-3093.

Bozelli, J. C.; De Oliveira Nunes, L.A., Sigoli,F.A. & Mazali, I. O. (2010). Erbium and Ytterbium Codoped Titanoniobophosphate Glasses for Ion-Exchange-Based Planar Waveguides. *Journa of American Ceramic Society,* (2010) 1–4, 0002-7820

Chiang, K. S.; Wong, C. L., Cheng, S. Y. & Chan, H. P. (2000). Refractive-Index Profiling of Graded-Index Planar Waveguides from Effective Indexes Measured with Different External Refractive Indexes . *journal of Lightwave Technology,* 18, 10, (October 2000) 1412-1417, 0733-8724

Conti, G. N.; Berneschi, S., Bettinelli, M., Brenci, M., Chen, B., Pelli, S., Speghini, S., Righini,G.C. (2004). Rare earth doped tungsten tellurite glasses and waveguides: fabrication and characterization. *Journal of Non-Crystalline Solids,* 345&346, (October, 2004) 343–348, 0022-3093

Corradi,A. B.; Cannillo, V., Montorsi, M., & Siligardi, C. (2006). Influence of Al_2O_3 addition on thermal and structural properties of erbium doped glasses. *Journal of Materials Science,* 41, 10, (April, 2006) 2811–2819, 0022-2461

Dimitrov, V. & Sakka, S. (1996). Electronic oxide polarizability and optical basicity of simple oxides. *Journal of Applied Physics,* 79, 3, (February, 1996) 1736-1740, 0021-8979

Duffy, J. A. (1986). Chemical bonding in the oxides of the elements: A new appraisal. *Journal of Solid State Chemistry,* 62, (April,1986) 145-157, 0022-2461, 0022-4596

Fernandez, T. T.; Valle, D. G., Osellame, R., Jose, G., Chiodo, N., Jha, A. & Laporta, P. (2008). Active waveguides written by femtosecond laser irradiation in an erbium-doped phospho-tellurite glass. *Optics Express,* 16, 19, (September, 2008) 15198-15205, 1094-4087

Gangfeng,Y.; Zhonghong,J., Zaide,D., Bing,Y., Tingzhao,Y. & Zhouming,F. (2005). Effect of Network Modifiers on Spectroscopic Properties of Erbium-doped Phosphate Glasses. *Journal of Wuhan University of Technology - Material Science Education,* 20, 1, (March, 2005) 60-63, 1000-2413.

Golubeva, O.Y. (2003). Structural Interpretation of the Results of Investigations into the Influence of Residual Water on the Properties of One-Alkali Borate Glasses. *Glass Physics and Chemistry,* 29, 6, (June, 2003) 571–574, 1087-6596

Görler-Walrand, C. & K. Binnemans (1998). *Handbook on the Physics and Chemistry of Rare Earths,* Ed. K. A. Gschneidner Jr. and L. Eyring, North-Holland, Amsterdam, 1998, Vol. 25, pp.101

He, Z.; Li, Y., Li, Y., Zhang, Y., Liu, L. & Xu, L. (2008). Low-loss channel waveguides and Y-splitter formed by ion-exchange in silica-on-silicon. *Optics Express,* 16, 5, (March, 2008) 3172-3177, 1094-4087

Honkanen, S.; Najafi, S.I., Wang, W.J. (1992). Composite rare earth-doped glass waveguides. *Electronics Letters,* 28, 8,(April, 1992) 746-747, 0013-5194

Jose, G.; Sorbello, G., Taccheo, S., Cianci, E., Foglietti, V. & Laporta, P. (2003). Active waveguide devices by Ag–Na ion exchange on erbium–ytterbium doped phosphate glasses. *Journal of Non-Crystalline Solids,* 322, (July, 2003) 256–261, 0022-3093

Judd, B. R. (1962). Optical absorption intensity of rare earth ions. *Physical Review,* 127,3, (March, 1962) 750-761, 0031-899x

Kosıkova, J. & Schrofel, J. (1999). Planar waveguides prepared by K^+–Na^+ field-assisted ion exchange in different types of silicate glass. *Journal of Materials research,* 14, 7, (July, 1999) 3122-3129, 0884-2914

Kumar, G.A.; Riman, R. A., Diaz Torres, L. A., Banerjee, S., Romanelli, M. D., Emge, T. J. & Brennan, J. G. (2007). Near-Infrared Optical Characteristics of Chalcogenide-Bound Nd^{3+} Molecules and Clusters. *Chemistry of Materials,* 19, (May, 2007) 2937-2946, 0897-4756

Lee. C.T. (1998). Nondestructive measurement of separated propagation loss for multimode waveguides. *Applied Physics Letters,* 73,2,(July,1998) 133-135, 0003-6951

Liang, G. J.; Gong-Wang, S., Huan, M. & Li-Li, H., Qu, L. (2005). Gain and Noise Figure of a Double-Pass Waveguide Amplifier Based on Er/Yb-Doped Phosphate Glass. *Chinese Physics Letter,* 22,11, (July, 2005) 2862-2864, 0256-307x

Lin, H.; Jiang, S., Wu, j., Song, F., Peyghambarian, N. & Pun, E.Y.B. (2003). Er^{3+} doped Na_2O–Nb_2O_5–TeO_2 glasses for optical waveguide laser and amplifier. *Journal of Physics D: Applied Physics,* 36,7, (July, 2003) 812–817,0022-3727

Liu, K. & Pun, E. Y. B. (2009). Comparative studies of spectroscopic properties in Er^{3+}–Yb^{3+} codoped phosphate glasses. *Journal of Alloys and Compounds,* 470, (February, 2009) 340–346, 0925-8388

Liu, K. & Pun, E. Y. B. (2007). Modeling and experiments of packaged Er^{3+}–Yb^{3+} co-doped glass waveguide amplifiers. *Optics Communications,* 273, (May, 2007) 413–420, 0030-4018

Liu, K. & Pun, E. Y. B., (2004). K^+-Na^+ ion-exchanged waveguides in Er^{3+}-Yb^{3+} codoped phosphate glasses using field-assisted annealing. *Applied Optics*, 43, 15, (May, 2004) 3179-3184, 0003-6935

Mariappan, C.R.; Govindaraj, G., Rathan, V. S., & Vijaya Prakash, G. (2005). (a)Vitrification of $K_3M_2P_3O_{12}$ (M = B, Al, Bi) NASICON-type materials and electrical relaxation studies. Materials Science and Engineering B, 123, (November, 2005) 63-68, 0921-5107

(b)Preparation, characterization, ac conductivity and permittivity studies on vitreous $M_4AlCdP_3O_{12}$ (M = Li, Na, K) system. Materials Science and Engineering B, 121, (July, 2005) 2-8, 0921-5107

Milova, G.; Najafi, S.I., Skirtach, A.G., Simkin, D.J., Andrews, M.P. (1997). *Proceedings of SPIE*, pp.90-101, San Jose, February, 1977, SPIE, USA

Miniscalco, W.J. (1999). Erbium-Doped Glasses for Fiber Amplifiers at 1500 nm. *Journal of Lightwave Technology*, 9, 2, (February, 1991) 234-250, 0733-8724

Mc Cumber, D. E. (1964). Theory of phonon-Terminated optical masers. *Physical Review*, 134, 2A, (November, 1964) A299-A306, 0031-899x

Nandi, P.; Jose, G., Jayakrishnan, C., Debbarma, S., Chalapathi, K., Alti, K., Dharmadhikari, A. K., Dharmadhikari, J. A. & Mathur, D. (2006). Femtosecond laser written channel waveguides in tellurite glass. *Optics Express* 14, 25, (December, 2006) 12145-12150

Najafi, S.I.; Andrews, M.P., Fardad, M.A., Milova, G., Tahar, T., Coudray, P. (1996). *Proceedings of SPIE*, pp.100-104, Berlin, October, 1996, SPIE, Germany

Nakazawa, M. & Kimura, Y. (1992). Electron-beam vapour-deposited erbium-doped glass waveguide laser at 1.53μm. *Electronics Letters*, 28, 22, (October 1992) 2054-2055, 0013-5194

Ning, N.; lv-yun, Y., Ming-ying, P., Qing-ling, Z., Dan-ping, C., Akai, T. & Kadono, K. (2006). Preparation and spectroscopic properties of Er^{3+}-doped high silica glass fabricated by sintering nanoporous glass. *Materials Letters*, 60, (July, 2006) 1987-1989, 0167-577x

Ofelt, G. S. (1962). Intensities of Crystal Spectra of Rare earth Ions. *Journal of Chemical Physics*, 37, 3, (August, 1962) 511-520, 0021-9606

Osellame, R.; Della Valle, G., Chiodo, N., Taccheo, S., Laporta, P., Svelto, O. & Cerullo, G. (2008). Lasing in femtosecond laser written optical waveguides. *Applied physics* A, 93, 1, (October, 2008) 17-26, 0947-8396

Osellame, R. ; Taccheo, S., Marangoni, M., Ramponi, R., Laporta, P., Polli, D., De Silvestri S. & G. Cerullo (2003). Femtosecond writing of active optical waveguides with astigmatically shaped beams. *Journal of the Optical Society of America B: Optical Physics*, 20, 7, (July, 2003) 1559-1567, 0740-3224

Psaila, N. D.; Thomson, R. R., Bookey, H. T., Kar, A. K., Chiodo, N., Osellame, R., Cerullo, G., Jha, A. & Shen, S. (2007). Er:Yb-doped oxyfluoride silicate glass waveguide amplifier fabricated using femtosecond laser inscription. *Applied Physics Letters*, 90, (January, 2007) 131102, 0003-6951

Peters, P. M.; Funk, D. S., Peskin,A.P., Veasey, D.L., Sanford, N.A., Houde-Walter, S.N. & Hayden, J.S. (1999). Ion-exchanged waveguide lasers in $Er^{3+}Yb^{3+}$ codoped silicate glass. *Applied Optics*, 38, 33, (November, 1999) 6879-6886, 0003-6935

Pradeesh, K.; Oton, C.J., Agotiya, V.K., Raghavendra, M., Vijaya Prakash, G. (2008). Optical properties of Er^{3+} doped alkali chlorophosphate glasses for optical amplifiers. *Optical materials*, 31, (October-December, 2008) 155-160, 0925-3467

Ramaswamy, R.V. & Srivastava, R.(2000). Ion exchanged glass wavewguides: A review. *Journal of Lightwave Technology*, 6, 6, (October 2000) 984-1002, 0733-8724

Reisfeld, R. & Jorgensen, C.K. (2007). *Lasers and excited states of rare earths.* Springer-verlag, 0387083243, Berlin, New York.

Righini, G. C.; Pelli, S., Brenci, M., Ferrari, M., Duverger, C., montagna, M. & dallIgna, R. (2001). Active optical waveguides based on Er- and Er/Yb- doped silicate glasses. *Journal of Non-Crystalline Solids*, 284, (May, 2001) 223-229, 0022-3093

Righini, G.C.; Pelli, S., Fossi, M., Brenci, M., Lipovskii, A.A., Kolobkova, E.V., Speghini, A., Bettinelli, M. (2001). Characterzation of Er-doped sodium-niobium phosphate glasses. Proceedings of SPIE, pp.210-215, 9780819439604, USA, April 2001, SPIE, San Jose

Rivera, V. A. G.; Chillcce, E. F., Rodriguez, E., Cesar, C. L. & Barbosa, L. C. (2006). Planar waveguides by ion exchange in Er^{3+}-doped tellurite glass. *Journal of Non-Crystalline Solids*, 352, (May, 2006) 363–367, 0022-3093

Sakida, S.; Nanba, T. & Miura, Y. (2007). Optical properties of Er^{3+} doped tungsten tellurite glass waveguides by Ag^+-Na^+ ion-exchange. *Optical materials*, 30, (2007) 586–593.

Sakida, S.; Nanba, T. & Miura, Y. (2006). Refractive-index profiles and propagation losses of Er^{3+}-doped tungsten tellurite glass waveguide by Ag^+–Na^+ ion-exchange. *Materials Letters*, 60, (December, 2006) 3413–3415, 0167-577x

Salavcova, L.; Svecova, B., janakova, S., Kolek, O., Míka, M., Spirkova, J. & Langrova, (2005). Planar optical waveguides in newly developeder:silicate glasses: a comparative study of $K^+$and Ag^+ ion exchange. *Ceramics – Silikaty*, 49, 1, (September, 2005) 53-57, 0862-5468

Seneschal, K.; Smektala, F., Bureau, B., Floch, M. L., Jiang, S., Luo, T., Lucas, J. & Peyghambarian, N. (2005). Properties and structure of high erbium doped phosphate glass for short optical fibers amplifiers. *Materials Research Bulletin*, 40, (2005) 1433–1442, 0025-5408

Shen, S. & Jha, A. (2004). The influence of F^- ion doping on the fluorescence ($^4I_{13/2}\rightarrow {}^4I_{15/2}$) line shape broadening in Er^{3+}-doped oxyfluoride silicate glasses. *Optical Materials*, 25, (April, 2004) 321–333, 0925-3467

Shmulovich, J.; Wong, Y.H., Becker, P.C., Bruce, A.J., Adar, R. & Wong, A. (1992) Er^{3+} glass waveguide amplifier at 1.5 μm on silicon. *Electronics Letters*, 28, (June, 1992) 1181-1182, 0013-5194

Stavrou, E.; Tsiantos, C., Tsopouridou, R.D., Kripotou, S., Kontos, A.G., Raptis, C., Capoen, B., Bouazaoui, M., Turrell, S. & Khatir, S. (2010) Raman scattering boson peak and differential scanning calorimetry studies of the glass transition in tellurium–zinc oxide glasses. *Journal of Physics: Condensed Matter*, 22, 19, (May, 2010) 195103, 0953-8984

Surendra Babu, S.; Babu, P., Jayasankar, C.K., Sievers, W., Troster, Th. & Wortmann, G. (2007). Optical absorption of and photoluminescence studies of Eu^{3+} doped phosphate and fluorophosphates glasses. *Journal of Luminescence*, 126, (September, 2007), 109-120, 0022-2313.

Surendra Babu, S.; Jang, K., Cho, E.J., Lee, H. & Jayasankar, C.K. (2007). Thermal, structural and optical properties of Eu^{3+} -doped Zinc-tellurite glasses. *Journal of Physics D: Applied Physics*, 40,18, (August, 2007),5767-5774, 0022-3727.

Taccheo, S.; Della Valle, G., Osellame, R., Cerullo, G., Chiodo, N., Laporta, P., Svelto, O., Killi, A., Morgner, I., Lederer, M. & Kopf, D. (2004). Er:Yb-doped waveguide laser fabricated by femtosecond laser pulses. *Optics Letters*, 29, 22, (November, 2004) 2626-2628, 0146-9592

Thomson, R. R.; Bookey, H. T. , Psaila, N., Campbell, S., Reid, D. T., Shen, S., Jha, A., & Kar, A. K. (2006). Internal Gain From an Erbium-Doped Oxyfluoride-Silicate Glass Waveguide Fabricated Using Femtosecond Waveguide Inscription. *IEEE Photonics Technology Letters*, 18, 14, (July, 2006) 1515-1517, 1041-1135

Thomson, R. R.; Campbell, S., Blewett, I. J., Kar, A. K. & Reid, D. T. (2005). Active waveguide fabrication in erbium-doped oxyfluoride silicate glass using femtosecond pulses. *Applied Physics Letters*, 87, (April, 2005) 121102,0003-6951

Valle, G. D.; Osellame, R., Chiodo, N., Taccheo, S., Cerullo, G., Laporta, P., Killi, A., Morgner, U., Lederer, M. & Kopf, M. (2005). C-band waveguide amplifier produced by femtosecond laser writing. *Optics Express*, 13, 16, (August, 2005) 5976-5983, 1094-4087

Veasey, D.L.; Funk, D.S., Peters, P.M., Sanford, N.A., Obarski, G.E., Fontaine, N., Young, M., Peskin, A.P., Liu, W., Houde-Walter, S.N. & Hayden, J. S. (2000). Yb/Er-codoped and Yb-doped waveguide lasers in phosphate Glass. *Journal of Non-Crystalline Solids*, 263&264, (March, 2000) 369±381, 0022-3093

Vijaya Prakash, G.; Jagannathan, R. & Rao, D.N. (2002). Physical and optical properties of NASICON-type phosphate glasses. *Materials Letters*, 57, (November, 2002) 134-140

Vijaya Prakash, G.; Nachimuthu, P., Vithal, V. & Jagannathan, R. (2002). Rare earth fluorescence in NASICON type phosphate glass, $Na_3TiZnP_3O_{12}$. *Bulletin of Material Science*, 22, 2, (March, 2002) 121-127, 0250-4707

Vijaya Prakash, G.; Rao, D.N. & Bhatnagar, K. (2001). Linear optical properties of niobium- based tellurite glasses. *Solid State Communications*, 119, (June, 2001) 39-44, 0038-1098

Vijaya Prakash, G. (2000). Absorption spectral studies of rare earth ions (Pr^{3+}, Nd^{3+}, Sm^{3+}, Dy^{3+}, Ho^{3+} and Er^{3+}) doped in NASICON type phosphate glass, $Na_4AlZnP_3O_{12}$. *Materials Letters*. 46, (October 2000) 15-20, 0167-577x

Vijaya Prakash, G. & Jagannathan, R. (1999). Fluorescence properties of Eu^{3+} doped lead bearing fluoro-chloro phosphate glasses. *Spectrochimica Acta Part A 55*, (August, 1999) 1799–1808, 1386-1425

Vishnubhatla, K.C; Rao, S. V., Kumar, R. S. S., Osellame, R., Bhaktha, S.N.B., Turrell, S., Chiappini, A., Chiasera, A., Ferrari, M., Mattarelli, M., Montagna, M., Ramponi, R., Righini, G.C. & Rao, D.N. (2009). Femtosecond laser direct writing of gratings and waveguides in high quantum efficiency erbium-doped Baccarat glass. *Journal of Physics D: Applied Physics* 42, 20, (March, 2009) 205106, 0022-3727

Wong, S. F.; Pun, E. Y. B. & Chung, P. S. (2002). Er^{3+}–Yb^{3+} Codoped Phosphate Glass waveguide amplifier using Ag^{+}–Li^{+} ion exchange. *IEEE Photonics Technology Letters,* 14, 1, (January, 2002) 80-82, 1041-1135

Wybourne, B. G. (1965). *Spectroscopic Properties of Rare Earths,* John Wiley, New York

Zhao, S.; Wang, X., Fang, D., Xu, S. & Hu, L. (2006). Spectroscopic properties and thermal stability of Er^{3+}-doped tungsten–tellurite glass for waveguide amplifier application. *Journal of Alloys and Compounds* 424, (November, 2006) 243–246, 0925-8388

Zhang, X.Z.; K. Liu, K., Mu, S. K., Tan, C. Z., Zhang, D., Pun, E.Y.B. & Zhang, D.M. (2006). Er^{3+}–Yb^{3+} co-doped glass waveguide amplifiers using ion-exchange and field assisted annealing. *Optics Communications* 268, (December, 2006) 300–304, 0030-4018

14

Tunable Fibre Lasers Based on Optical Amplifiers and an Opto-VLSI Processor

Feng Xiao[1], Kamal Alameh[1,2] and Yong Tak Lee[2,3]
[1]*Electron Science Research Institute, Edith Cowan University,*
[2]*Department of Nanobio Materials and Electronics,*
Gwangju Institute of Science and Technology,
[3]*Department of Information and Communications,*
Gwangju Institute of Science and Technology,
[1]*Australia*
[2,3]*Korea*

1. Introduction

Tunable lasers are currently used in a wide range of applications such as wavelength-division-multiplexing networks, optical sensors, spectroscopy, wavelength protection, fiber-optic gyroscope, and testing of optical components and instruments. In particular, tunable fiber lasers, which employ optical fiber cavity and wave-guided gain medium, have recently attracted great interest owing to several intrinsic advantages over traditional lasers, namely: (i) easy manufacture without the need of clean rooms and expensive device packaging; (ii) mechanical flexibility and the ability to withstand bending, thus opening the way for numerous applications in biotechnologies and medical instrumentation; (iii) broadband gain spectrum and high energy efficiency, which are very important features for tunable high-power lasers; (iv) high laser beam quality, ensuring their wide potential applications in material processing, printing, marking, cutting and drilling; (v) robustness, because all optical signals are guided within optical fibres, thus eliminating the need for optical alignment; and (vi) narrow line width which is essential for many applications and hard to achieve for their current counterparts. Benefiting greatly from recent developments in fibre communications, fibre lasers are offering a low-cost alternative to the traditional semiconductor or gain dielectric counterparts.
The key features required for tunable fiber lasers include a large number of channels, high--output power, stable operation in power and wavelength, and tuning operation over a wide wavelength range with application-specific wavelength spacing. However, the current fibre lasers, especially multiwavelength fiber lasers, suffer from limited tunability and flexibility, poor stability, and narrow operation ranges.
Two main aspects related to the development of tunable single/multiwavelength fibre lasers are currently under intensive investigation; namely, the wavelength selection (or tuning) mechanism, and the used gain media. While these investigations have resulted in significant progress towards the development of commercially viable products each of the current approaches have particular limitations.

In this Chapter, we discuss the use of Opto-VLSI processors for the development of tunable fiber lasers. Section 2 briefly reviews the recent progress in tuning mechanism for single and multiple wavelength tuning. Section 3 discusses different approaches for implementing the gain media of fiber lasers. In Section 4 a brief background on Opto-VLSI processors is provided. Single-wavelength, multi-wavelength and multiport tunable fiber lasers employing Opto-VLSI processors are presented in Sections 5, 6 and 7, respectively. We conclude this chapter in Section 8.

2. Tuning mechanism

2.1 Single-wavelength tuning

The tunable single-wavelength fiber laser is relatively mature compared to its multiwavelength countparts. However, due to its long cavity, which is able to give narrow linewidths, a fiber laser oscillates on multiple longitudinal modes (A. Bellemare *et al.*, 2001). Mode discrimination by the gain medium is not strong enough to restrict the oscillation to a single longitudinal mode, despite the fact that homogeneous broadening is known to dominate in gain media, such as erbium-doped fibers, at room temperature. Therefore, one or more spectrally discriminating elements or very short cavities should be employed to limit the number of modes over which the laser may effectively oscillate. Nevertheless, this does not ensure the oscillation of one longitudinal mode (Park *et al.*, 1991).

To suppress multimode lasing as well as mode hopping, technologies including the use of cascaded bandpass filters (one narrow and one broad) (Park *et al.*, 1991; Chow *et al.*, 2002) and a passive multiple-ring cavity (C. C. Lee *et al.*, 1998) have commonly been used. More importantly, an unpumped Er^{3+}-doped fiber has been incorporated in a long Fabry–Perot laser cavity as a saturable absorber to establish linewidth narrowing and single-mode operation (Cheng *et al.*, 1995). In addition, the beat-noise of fiber ring lasers, which is primarily in the low-frequency region of about 10 MHz due to the long ring cavity length, has successfully been suppressed by inserting a Fabry–Pérot laser diode (FP-LD). (H. L. Liu *et al.*, 2006).

Several single-wavelength tuning mechanisms have been intensively investigated. Some early works on single-wavelength tuning have been demonstrated using intracavity elements such as gratings or birefringent plates, whose orientation can be changed mechanically. In order to switch the wavelength electrically, the use of bulk electro-optic intracavity tuners, such as liquid crystal cells, has been reported (MOLLIER *et al.*, 1995), where the tuning range of the device was 17 nm and the tuning rate was 8 nm/V.

A tunable single-longitudinal-mode compound-ring Er3+-doped fiber laser has been demonstrated (J. L. Zhang *et al.*, 1996; C. C. Lee *et al.*, 1998), where the laser is fundamentally structured on an all-fiber compound-ring resonator in which a dual-coupler fiber ring is inserted into the main cavity. When combined in tandem with a mode-restricting intracavity tunable bandpass filter, the compound-ring resonator ensures single-longitudinal-mode laser oscillation. The laser can be tuned over much of its 1525 to 1570 nm wavelength tuning range with the short-term linewidth of less than 5 kHz.

Acousto-optic tunable filters (AOTFs) have recently been used as a wavelength selector because of their broad tuning ranges, narrow bandwidths, simple tuning mechanisms, and low drive power requirements (Chang *et al.*, 2001; Kang *et al.*, 2006). A single-frequency tunable fiber ring laser incorporating an all-fiber acousto-optic tunable bandpass filters (AOTBFs) has been characterized in detail and the impact of the AOTBF characteristics on the laser performance has also been examined (Kang *et al.*, 2006). The AOTBF consists of an all-single-mode-fiber acousto-optic tunable filter and a core-mode blocker in the center of the acousto-optic interaction region. Stable single-longitudinal-mode operation has been achieved over the wavelength range of 48 nm with a side mode suppression ratio higher than 50 dB.

The use of tunable fiber Bragg gratings (FBGs) is a simple approach to realize single-wavelength tuning (Song *et al.*, 2001). Recently, a tunable fiber laser based on a tunable phase-shifted linearly chirped FBG (PS-LCFBG) in transmission with ultra-narrow bandwidth of 1.52 kHz has been demonstrated (Li *et al.*, 2008). By thermally tuning the transmissive PS-LCFBG, a tuning range of 15.5 nm was obtained. A tunable L-band fiber laser based on a mechanically induced LPFG into the EDF ring has been demonstrated (Sakata *et al.*, 2009). Wavelength tuning was achieved by adjusting the grating period to shift the rejection band against the ASE spectrum of the ring cavity. Although thermal effects on the laser were controlled, this technique suffers from the environmental disturbances.

Tunable fiber Fabry–Perot filters have widely been utilized as tuning elements in tunable lasers (X. Y. Dong *et al.*, 2003; Zheng *et al.*, 2006; Chien *et al.*, 2005b; Chien *et al.*, 2005a; Fu *et al.*, 2009). A fiber laser utilizing a tunable fiber Fabry–Perot filter as the tuning element has been demonstrated, which had a moderate milli-Watt level power output over almost the whole tuning range from 1530 to 1595 nm with power fluctuations less than 0.15 dB (Fu *et al.*, 2009). High repetition scanning rate of laser operation over the whole tuning range was achieved at rates of up to 200Hz. The limited tuning range of this tunable filter could be extended by incorporating another tunable passband filter. A tunable narrow-bandpass fiber Fabry-Perot (FFP) filter with a bandwidth of 0.2 nm and free spectral range (FSR) of 28 nm combined with six bandpass filters (BPFs) of 20 nm bandwidth has been used as a wide tuning component (X. Y. Dong *et al.*, 2003). Only one of the BPFs was turned on at a time. Since the bandwidth of each BPF was less than the FFP's 28-nm FSR, continuous wavelength tuning was obtained within a 20 nm range by changing the voltage applied to the piezoelectric transducer of the FFP, thus attaining a total tuning range of 120 nm.

Based on the use of a narrow-band tunable microelectromechanical system (MEMS) filter, a wide-band short-cavity length tunable fiber ring laser that can be tuned at high speed from 1520 to 1626 nm has been reported (H. L. Liu *et al.*, 2005a). The optical bandwidth of the MEMS was about 20 pm (2.5 GHz) tuned over 120 nm, from 1630 to 1510 nm by varying the input voltage from 10 to 32 V, respectively. The MEMS filter also exhibited very high scanning speed (greater than 100 000 nm/s). The insertion loss of the MEMS filter at the peak of the passband was less than 1.5 dB, and the out-of-band reflection from both ends of the MEMS filter was as high as 95%.

Lyot filters, which are a type of optical filter that uses birefringence to produce a narrow passband of transmitted wavelengths, have also been used to widely tunable fiber lasers (Zhou *et al.*, 1998; Xu *et al.*, 2002). The semiconductor fiber ring laser utilized the birefringence of optical fibers, especially PM fiber, together with the polarization controller, rotating linear polarizer and SOA, to form polarization-dependent loss and gain mechanisms. For this filter, different wavelengths arrive at the linear polarizer and SOA with different polarization states due to birefringent chromatic dispersion in the cavity, hence, different wavelengths have different losses due to the polarization dependent loss through the linear polarizer. A wavelength with its electrical field vector lined up with the linear polarizer experiences the least loss. The combination of a polarization controller and linear polarizer thus provided a wavelength selection mechanism to generate a tunable laser output. In addition, as the gains of the SOA are different for transverse electrical (TE) fields and transverse magnetic (TM) fields, 3-dB gain difference between TE and TM polarizations were experienced in addition to gain differences due to the SOA spectrum itself. This also provided gain discrimination due to the fact that different wavelengths have different input polarization states at the SOA. Therefore, the combination of the intracavity polarization controller, the polarizer, and the SOA served as wavelength selection mechanisms. Thus, the lasing wavelength is the light wave that repeats its state of polarization in front of both the polarizer and the SOA after one round-trip and whose polarization state lines up with the linear polarizer and TE or TM direction within the SOA whichever has a higher gain.

2.1 Multi-wavelength tuning

Multiwavelength tuning is much more complex compared to single wavelength tuning, and recently it has attracted equal attention from all over the academic society. Several versatile techniques to realize multi-wavelength tuning have been proposed and demonstrated. A spatial mode beating filter has been reported, which can tune a multiwavelength fiber laser by simply incorporating a section of multimode optical fiber into a single-mode fiber ring cavity (Poustie *et al.*, 1994). This combination of two fiber types results in wavelength-dependent filtering action inside the laser cavity arising from the spatial mode beating between the LP01 and LP11 modes in the multimode fiber. The multiwavelength fiber laser can be tuned by controlling the polarization controller in the fiber loop.

Tunable FBGs are still the commonly-used technology for multiwavelenth tuning (Han *et al.*, 2007; Liaw *et al.*, 2007; Alvarez-Chavez *et al.*, 2007; Moon *et al.*, 2005). A spacing-tunable multiwavelength Raman fiber laser with an independently-adjustable channel number, based on a superimposed chirped-fiber Bragg grating (CFBG) and a linear cavity formed by a bandwidth-tunable CFBG reflector, has been demonstrated (X. Y. Dong *et al.*, 2006). Multiwavelength laser operations at room temperature with spacing of 0.3 to 0.6 nm, and channel number of 2 to 10 have been achieved.

Recently a tunable multiwavelength fiber laser based on an all-fiber FP filter which is constructed by a superimposed CFBG has been demonstrated (Han *et al.*, 2007). The FP filter is capable of continuous FSR tenability, through incorporating a specially designed apparatus for induction of the sophisticated bending along the grating. The proposed technique is based on the symmetrical modification of the chirp ratio along the fiber grating attached on a flexible cantilever beam. The FSR of the all-fiber FP filter could be continuously tuned from 0.21 to 0.81 nm with neglect center wavelength shift.

A fiber loop mirror incorporating a piece of polarization maintaining fiber (PMF) has attracted much attention as an optical comb filter due to its intrinsic advantages such as easy fabrication, stability and flexibility (Z. X. Zhang *et al.*, 2009). The PMF-based fiber loop mirror filters can provide various functionalities like both the peak wavelength tunability and FSR tunability, which can be controlled by adjusting the effective length and birefringence of multiple PMF segments depending on the relative phase difference between two orthogonal polarization modes within the loop (Kim *et al.*, 2003; Z. Y. Liu *et al.*, 2008). The comb filter can also be tuned by changing the operating temperature or by applying axial strain to the PMF (Moon *et al.*, 2007). Besides these, an active device such as a phase modulator (Fok *et al.*, 2005) or SOA (K. L. Lee *et al.*, 2004) is also inserted in the fiber loop mirror to tailor the effective birefringence of the loop, enabling continuous shift of the transmission comb spectrum.

Unlike the use of passive single-mode PMFs into the Sagnac loop, a tunable transmission comb filter based on a pumped erbium-ytterbium co-doped polarization maintaining fiber (EYD-PMF) loop mirror has been recently demonstrated (G. Y. Sun *et al.*, 2008). The effective birefringence of the EYD-PMF depends on the power of pump lasers and the polarization state of the traversing signal. Therefore, the comb filter can be tuned by changing the pump power or adjusting a PC adjacent to the EYD-PMF in the loop.

Mach-Zehnder interferometers also offer a tuning mechanism for tunable fiber lasers (H. Dong *et al.*, 2005a). However, they usually suffer from performance instability and difficulty in free spectral range (FSR) control. A multi-wavelength fiber ring laser of tunable channel spacing has been proposed by employing an optical variable delay line (OVDL) in a Mach-Zehnder interferometer (D. R. Chen *et al.*, 2007). The channel spacing of the present multi-wavelength fiber ring laser can be continuously tuned by adjusting the computer-controlled OVDL. Multi-wavelength lasing with standard ITU channel spacing of 25 GHz, 50 GHz and 100 GHz has been demonstrated. Also, a tunable and switchable multiwavelength erbium-doped fiber ring laser based on a modified dual-pass Mach–Zehnder interferometer has been reported (Luo *et al.*, 2009). Through polarization control, the dual-function operation of the channel-spacing tunability and the wavelength interleaving can be achieved. Up to 29 stable lasing lines with 0.4 nm spacing and 14 lasing wavelengths with 0.8 nm spacing in 3 dB bandwidth were obtained.

Wavelength tunability can also be obtained through the use of a thin film tunable filter (Ummy *et al.*, 2009). A widely tunable (30 nm) fiber laser based on two Sagnac loop mirrors and a tunable thin film filter has been demonstrated, where optical power adjustability is accomplished by proper adjustment of each of the loop mirror reflectivity via a polarization controller.

3. Gain medium

3.1 Single gain medium

Many gain media are suitable for single-wavelength lasing in a fiber cavity, including erbium-doped fiber amplifiers (EDFA) (Antoine Bellemare, 2003), semiconductor optical amplifiers (SOA) (Ummy *et al.*, 2009), and hybrid gain media (Yeh&Chi, 2005). SOA-based fiber ring lasers have limited optical signal-to-noise ratio (OSNR), while EDFAs are ideal for single-wavelength tunable fiber ring lasers.

Erbium-doped fiber lasers (EDFLs) have extensively been studied as a very promising solution because EDF offers several advantages over the other gain media, such as high conversion efficiency, low threshold, homogeneous gain property thus high OSNR, narrow linewidth, coverage of the entire (C L)-band, ease of construction, and low cost (X. Y. Dong *et al.*, 2005b; A. Bellemare *et al.*, 2001; Roy *et al.*, 2005). Both theoretical and experimental investigations on how the laser performances are affected by the various parameters in the lasing cavity, such as the intracavity loss, erbium ion clustering, output coupling ratio (or reflectivity of output coupler), and active fiber length, have been carried out based on different theoretical models.

Meanwhile, some improved gain media based on EDF have also been investigated. A ring laser with 106-nm tuning range has been demonstrated based on a La-codoped Bi-EDF with about 84.6 cm long and doped with very high concentration of erbium ions (H. L. Liu *et al.*, 2005a). The erbium concentration in the Bi-EDF is 6470 wt-ppm and the La concentration is 4.4% wt. The peak absorption of the Bi-EDF at 1480 and 1530 nm are 167 and 267 dB/m,

respectively. The La ions extend the distance between Er ions and reduce the concentration quenching significantly. More recently, a Bi2O3-based erbium-doped fiber (BIEDF) ring laser with a 134 nm tunable range has been reported with only 0.2m of BIEDF as a gain media (Ohara&Sugimoto, 2008). The tunable range varies depending on both pump power and BIEDF lengths, and a high optical signal-to-noise ratio of over 70 dB for a 120 nm tunable range has been obtained.
A GeO-doped high-power and widely tunable all-fiber Raman laser using a linear cavity configuration has been demonstrated (Belanger *et al.*, 2008). The RFL was continuously tuned over 60 nm, from 1075 to 1135 nm and provided up to 5 W of Stokes output power for 6.5 W of launched pump power (LPP).

3.2 Multiple gain medium

Different gain mechanisms and media have been used to develop multiwavelength lasers, such as EDFAs (A. Bellemare *et al.*, 2000; X. M. Liu *et al.*, 2005b), SOAs (Pleros *et al.*, 2002), and schemes based on stimulated Raman scattering (SRS) gain (Kim *et al.*, 2003) and stimulated Brillouin scattering (SBS) gain (Nasir *et al.*, 2009). In addition, hybrid gain mechanisms using a combination of the above mechanisms have also been used (Han *et al.*, 2005).
The homogeneous linewidth broadening of the EDF medium limits the narrowest wavelength spacing between adjacent lasing wavelengths to a few nanometers (A. Bellemare *et al.*, 2000). To overcome this limit, various schemes such as cryogenic cooling (Yamashita&Hotate, 1996), frequency shifting (A. Bellemare *et al.*, 2000), careful gain equalization, spatial-spectral multiplexing, polarization anisotropic gain effects (Das&Lit, 2002), polarization-hole burning (J. Q. Sun *et al.*, 2000; Qian *et al.*, 2008), and intracavity four-wave mixing in nonlinear fibers (Tran *et al.*, 2008; Han *et al.*, 2006), have been used, adding more complexity (and therefore, cost) to these multiwavelength tunable lasers.
A nonlinear optical loop mirror (NOLM) based on a highly nonlinear dispersion shifted fiber has been implemented in the ring laser cavity to stabilize the multiwavelength output at room temperature (Tran *et al.*, 2008; Han *et al.*, 2006). Since the energy transfer from the higher-power waves to the lower-power waves is induced by several degenerate four-wave mixing (FWM) processes, the mode competition of the EDF is degraded. Consequently, the homogeneous line broadening of the EDF can dynamically be suppressed, leading to a stable multiwavelength output at room temperature. The FWM effects induced by the highly-nonlinear DSF introduce a dynamic gain flattening, so that the mode competition is suppressed effectively. The lasing wavelengths can be switched individually by two PCs because the nonlinear polarization phenomenon based on the NOLM induces the polarization-dependence loss and the birefringence-induced wavelength-dependent loss in the laser ring cavity, which can determine the total loss of the laser ring cavity.
Stable multiwavelength lasing has been achieved at room temperature with the hybrid gains of a Raman gain medium and an Erbium-doped fiber in a ring structure (D. R. Chen *et al.*, 2007). The multi-wavelength fiber laser employing Raman and EDF gains increases the lasing bandwidth compared with a pure EDF laser and the power conversion efficiency compared with a pure fiber Raman laser. No special fibers are needed in this proposed fiber laser structure.
SOA-based multiwavelength fiber lasers have more advantages over multiwavelength lasers with erbium-doped fiber amplifiers (H. X. Chen, 2005). This is because the SOA has a dominant property of inhomogeneous broadening and can support many wavelength lasing

oscillations simultaneously in the laser cavity, allowing narrower spacing for multiwavelength lasing operation (Qian *et al.*, 2008). A mechanism of multiwavelength generation based on nonlinear polarization rotation in an SOA has been reported (Z. X. Zhang *et al.*, 2009; Z. X. Zhang *et al.*, 2008). The arbitrarily polarized light incident to the SOA can be decomposed as the transverse electric (TE) and transverse magnetic (TM) modes. The modes propagate independently through the SOA, but they have indirect interaction via the carriers. The gain saturation of the TE mode differs from the gain saturation of the TM mode. Hence, the refractive index change of the TE mode also differs from the refractive index change of the TM mode. A phase difference between the two modes builds up as the light propagates through the SOA. A polarization controller (PC) is used to adjust the polarization of the input signal to the SOA so that the phase difference between the modes can be varied. At the following polarization-dependent isolator (PDI), both modes recombine. Another PC is used to adjust the polarization of the SOA output with respect to the orientation of the PDI. The phase difference and the orientation of the two PCs determine the intensity-dependent switch of the combiner. If the polarizations of the two PCs are set appropriately, the transmission of the combined light decreases when the input intensity to the SOA increases, thus suppressing the mode competition for multiwavelength generation.

4. Opto-VLSI processor

Because the limitations of the available underlying tuning mechanisms, current tunable multiwavelength lasers have very poor performances in tunability (can only shift all the wavelength channels over a very narrow wavelength range), flexibility (cannot independently control each wavelength channel), and operation ranges (limited to a single wavelengthrange). Furthermore, the flexibility limitation of conventional wavelength selection mechanisms has prevented researchers from fully understanding the mechanism of multiwavelength lasing in fibre lasers. As a result multiwavelength lasers tend to suffer from poor output power and wavelength stability. Therefore, both the wavelength selection (or tuning) mechanism and the mechanism of multiwavelength lasing are still in intensive investigation.

Advanced Opto-VLSI technology can be incorporated into a novel tuning approach, for the first time ever, to remove all previous disadvantages, thus simultaneously allowing (i) independent tuning of each wavelength channel, (ii) arbitrary control of the power level for each wavelength channel, (iii) the addition and removal of any lasing wavelength without affecting the power levels of other lasing wavelengths, and (iv) the generation of wavelength channels over various operation ranges, where each range may have output at several bands, all achieving (v) sufficient maximum continuous wave power output per channel (>+10dBm), (vi) sufficiently conserved linewidth (i.e. comparable to performance by Agilent of less than +/0.001nm at 1550 nm), (vii) minimal power fluctuations (i.e. ideally 0.01dB over 12 hours). These unique features are far superior to all tuning mechanisms reported in the literature to date, opening an excellent opportunity to develop a novel practical approach to multiwavelength lasing using Opto-VLSI processors.

A reconfigurable Opto-VLSI processor comprises an array of liquid crystal (LC) cells driven by a Very-Large-Scale-Integrated (VLSI) circuit that generates digital holographic diffraction gratings to steer and/or shape optical beams.

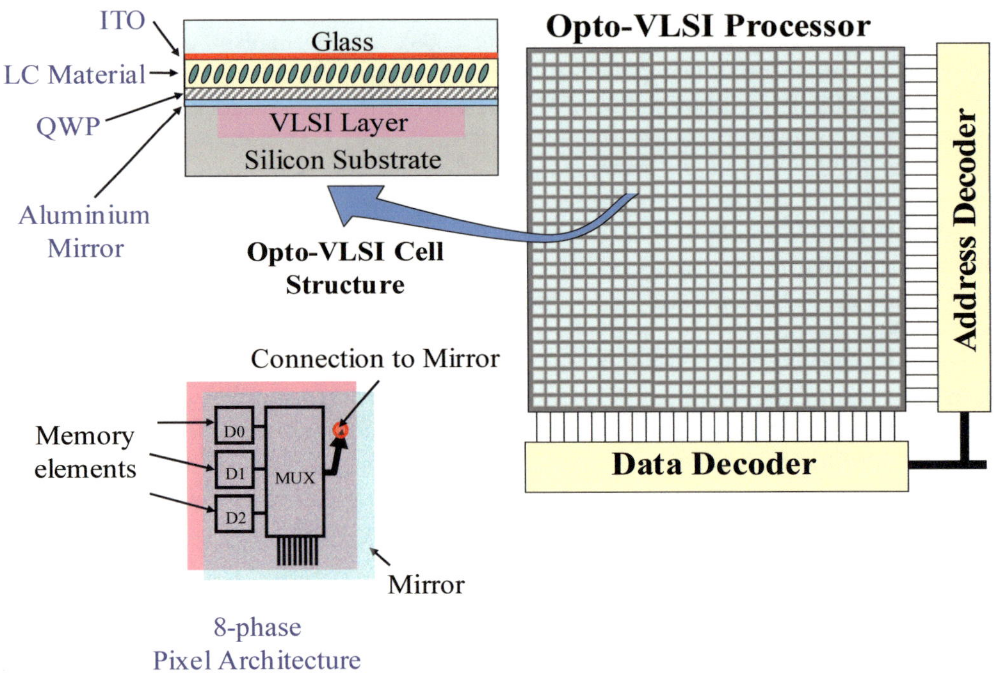

Fig. 1. Typical 8-phase Opto-VLSI processor and LC cell structure design.

Each pixel is assigned a few memory elements that store a digital value, and a multiplexer that selects one of the input voltages and applies it to the aluminium mirror plate. Opto-VLSI processors are electronically controlled, software-configured, polarization independent, cost effective because of the high-volume manufacturing capability of VLSI chips as well as the capability of controlling multiple fiber ports in one compact Opto-VLSI module, and very reliable since beam steering is achieved with no mechanically moving parts. Figure 1 shows a typical layout and a cell design of an 8-phase Opto-VLSI processor. Indium-Tin Oxide (ITO) is used as the transparent electrode, and evaporated aluminium is used as the reflective electrode. The ITO layer is generally grounded and a voltage is applied at the reflective electrode by the VLSI circuit below the LC layer.

Figure 2 illustrates the steering capability of Opto-VLSI processors. For a small incidence angle, the maximum steering angle of the Opto-VLSI processor is given by

$$\theta_{\max} = \frac{\lambda}{M \cdot d} \tag{1}$$

where M is the number of phase levels, d is the pixel size, and λ is the wavelength. For example, a 4-phase Opto-VLSI processor having a pixel size of 5 microns can steer a 1550 nm laser beam by a maximum angle of around ±4°. The maximum diffraction efficiency of an Opto-VLSI processor depends on the number of discrete phase levels that the VLSI can accommodate. The theoretical maximum diffraction efficiency is given by (Dammann, 1979)

$$\eta = \text{sinc}^2\left(\frac{\pi n}{M}\right) \tag{2}$$

where $n = gM + 1$ is the diffraction order ($n = 1$ is the desired order), and g is an integer. Thus an Opto-VLSI processor with binary phase levels can have a maximum diffraction efficiency of 40.5%, while a four phase levels allow for efficiency up to 81%. The higher diffraction orders (which correspond to the cases $g \neq 0$) are usually unwanted crosstalk signals, which must be attenuated or properly routed outside the output ports to maintain a high signal-to-crosstalk performance.

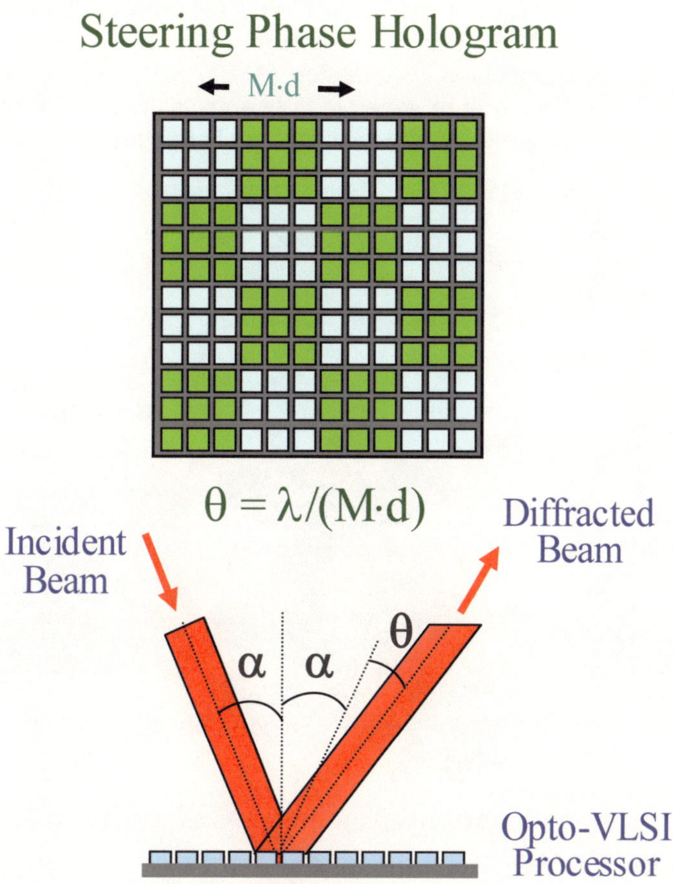

Fig. 2. Steering and multicasting capabilities of an Opto-VLSI processor.

Adaptive optical beam steering can be achieved by reconfiguring the phase hologram uploaded onto the Opto-VLSI processor. Recent advances in low-switching-voltage nematic LC materials and Layer thickness control have allowed the incorporation of a thin quarter-wave-plate (QWP) layer between the LC and the aluminium mirror to accomplish polarization-insensitive multi-phase-level Opto-VLSI processors [25], as shown in Fig. 1. In addition, with current 130nm VLSI fabrication processes, VLSI chips featuring 24mm×24mm active area, maximum switching voltage of 3.0 volts, and pixel size of 5 microns, can be realised. Depositing low-switching-voltage electro-optic materials and QWP over such VLSI chips, can realize a polarization-insensitive Opto-VLSI processor that has a diffraction efficiency of 87% (0.6 dB loss) and a maximum steering angle of more than ±4.0°.

There are several algorithms for the optimization of Opto-VLSI phase holograms to achieve effective beam steering, including simulated annealing and projection methods. In our study, a modified simulated annealing method that can achieve accurate beam steering with low crosstalk is adopted (Yen-Wei Chen *et al.*, 2000).

5. Single wavelength tunable fiber lasers

Recently, a novel tunable fiber laser employing an Opto-VLSI processor has been demonstrated (Xiao *et al.*, 2009). The Opto-VLSI processor is able to arbitrarily select narrowband optical signals from the amplified spontaneous emission (ASE) spectrum of an EDFA and inject them into a recirculating fiber ring to generate laser signals at arbitrary wavelengths. It is motionless and can be tuned electronically over the gain bandwidth of the EDFA.

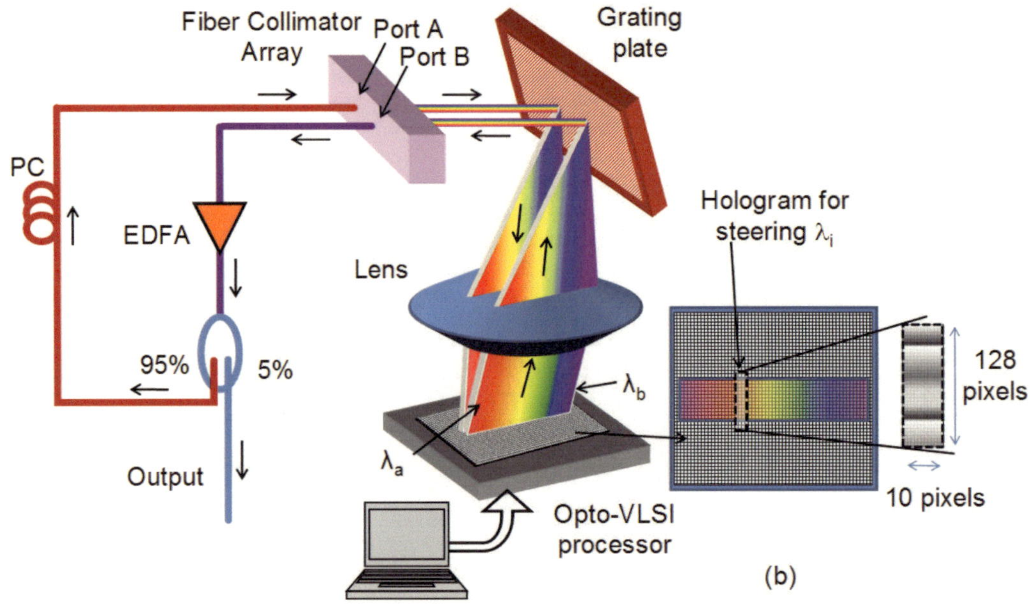

Fig. 3. Opto-VLSI-based tunable fiber laser structure.

Figure 3 shows the Opto-VLSI-based tunable fiber ring laser structure, which consists of an EDFA, an optical coupler, a polarization controller, and a fiber collimator pair (Port A and Port B). 95% of broadband ASE spectrum initially generated by the EDFA is routed to the Opto-VLSI processor through Port A of the fiber collimator array. The polarization controller (PC) is used to align the ASE polarization so that the diffraction efficiency of the Opto-VLSI processor is maximized, and also to enforce single-polarization laser operation. The grating plate demultiplexes the collimated broadband ASE signal along different directions. The lens between the Opto-VLSI processor and the grating plate has a focal length of 10 cm and is placed at 10 cm from the grating plate so that the dispersed ASE wavebands are deflected along the same direction and mapped onto the surface of an Opto-VLSI processor as illustrated in Fig. 3(a). By driving the Opto-VLSI processor with an appropriate steering phase hologram, any waveband of the ASE spectra can be routed to and coupled into Port B of the fiber collimator array (see Fig. 3(b)), and the others are dropped out with dramatic attenuation. The selected wavebands that are coupled into Port B are amplified by the EDFA, leading, after several recirculations, to single-mode laser generation. Therefore, the fiber laser can be tuned by simply uploading appropriate phase holograms that drive the various pixels of the Opto-VLSI processor.

In the experiments, the EDFA was a C-band amplifier having a small signal gain of 14 dB, and a gain spectrum shown in Fig. 4.The EDFA's pump laser was driven with a current of 400 mA. A 256-phase-level 512×512-pixel Opto-VLSI processor of pixel size 15 μm with an insertion loss of about 0.5 dB was used. The spacing between the fiber collimator elements (Port A and Port B) was 3 mm, and the insertion loss and return loss for the two ports were 0.6 dB and 55 dB, respectively. An optical spectrum analyzer with 0.01 nm resolution was used to monitor the laser output power generated at the 5% output port of the optical coupler. The ASE signal was collimated at 0.5 mm diameter, and a blazed grating plate, having 1200 lines/mm and a blazed angle of 70° at 1530 nm, was used to demultiplex the ASE signal and map onto the active window of the Opto-VLSI processor through a lens of focal length 10 cm placed at 10 cm from the grating plate. A Labview software was developed to generate the optimized digital holograms that steer the desired waveband and couple into the collimator Port B.

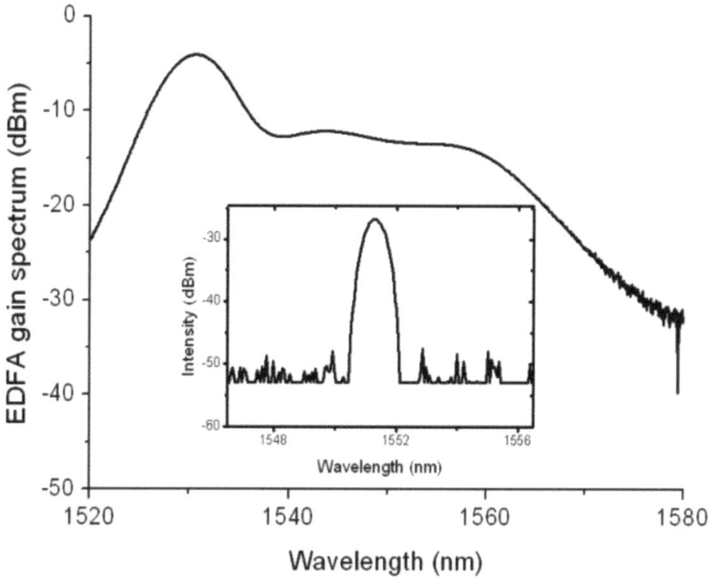

Fig. 4. Amplified spontaneous emission (ASE) noise from the EDFA. The inset is an example of a waveband selected by the Opto-VLSI processor.

When the optical loop was open, the ASE noise signal of the EDFA is shown in Fig. 4. The gain spectrum of the EDFA was linearly mapped along the active window of the Opto-VLSI processor. The inset of Fig. 4 is an example that illustrates the selection and coupling of an arbitrary waveband into Port B by uploading a phase hologram onto the Opto-VLSI processor. The measured total insertion loss from Port A to Port B was around 12 dB, which was mainly due to (i) the lens reflection loss; (ii) the blazed grating loss; and (iii) diffraction loss and insertion loss of the Opto-VLSI processor.

After the optical loop was closed, the Opto-VLSI processor was driven by different phase holograms, each corresponding to a single-mode lasing at a specific wavelength. Each selected waveband experienced a high gain by the EDFA in comparison to the gains experienced by the other ASE wavebands. Figure 5(a) shows the measured outputs of the Opto-VLSI-based fiber laser and demonstrates an excellent tuning capability over the C-band through the generation of 8×512 phase holograms at different position along the active window of the Opto-VLSI processor. The linewidth of the tunable laser was about 0.05 nm, compared to 0.5 nm when the optical loop was open (see the inset in Fig. 4). The measured side-mode suppression ratio (SMSR) was greater than 35 dB and the output power ripple was less than 0.25 dB over the entire C-band. The small ripples in the laser output power levels can be attributed to two main reasons, namely, (i) the EDFA worked in deep saturation, which clips the lasing output power; and (ii) the excellent stability and uniformity of the Opto-VLSI processor in steering and selecting wavebands over the whole C-band.

Figure 5(b) shows the measured laser outputs when fine wavelength tuning was performed by shifting the center of the phase hologram by a single pixel across the active window of the Opto-VLSI processor. The wavelength tuning step was around 0.05 nm. This corresponds to the mapping of 30 nm bandwidth of ASE spectrum of the EDFA across the 512 pixels (each of 15 μm size). Note that the tuning resolution can be made smaller by using an Opto-VLSI processor with a smaller pixel size. Note that the shoulders on both sides of the laser spectrum may be due to self-phase modulation or other nonlinear phenomena arising from a high-level of the output power.

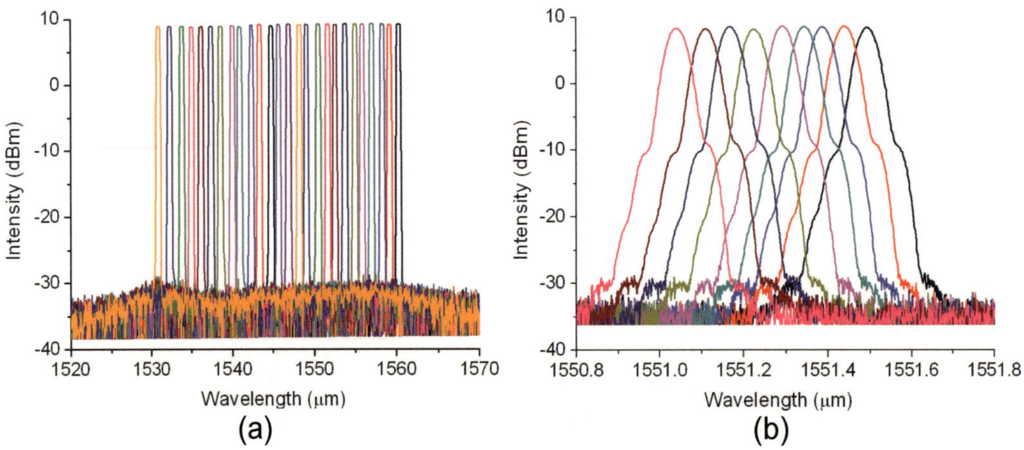

Fig. 5. Measured output intensities of the Opto-VLSI-based fiber laser, (a) Coarse wavelength tuning over C-band, and (b) fine wavelength tuning.

The measured crosstalk between Port A to Port B, defined as the ratio of the unselected ASE signal to the power of the waveband selected by the Opto-VLSI processor, was less than -55 dB. This crosstalk level, which contributes to the background level of the laser output and SMSR, can further be reduced by (i) increasing the spacing between Port A and Port B, (ii) improving the imaging quality of the lens, (iii) increasing the collimated beam diameter, and (iv) reducing the pixel size of the Opto-VLSI processor. The laser exhibited very stable operation at room temperature when it was turned on for different periods of time ranging from a few hours to a few days.

6. Multiple wavelength tunable fiber lasers

Beside its flexible single-wavelength tunable capability, Opto-VLSI processor is also able to offer super multi-wavelength tunability. Furthermore, this tuning mechanism provides two advanced features which cannot be provided by other tuning mechanisms, namely, (i) any of the lasing wavelengths can be tuned simultaneously and independently; (ii) any wavelength channel can be switched on/off without affecting the other wavelength channels, adding an important new function to the current tunable multiwavelength fiber lasers. These novel functions are very attractive for many applications such as optical telecommunications, optical characterization and testing, photonic RF signal processing.

The structure for the Opto-VLSI-based tunable multiwavelength fiber ring laser is the same as the one shown in Fig. 3, with the only difference that an SOA is inserted into the laser cavity to construct a hybrid gain medium which provides stable multiwavelength lasing. By driving the Opto-VLSI processor with an appropriate steering phase hologram, multiple wavebands from the ASE spectra can be steered and coupled into Port 2 of the fiber collimator array, and the others are dropped out with dramatic attenuation. The selected wavebands that are coupled into Port 2 are amplified by the EDFA and SOA, thus forming an optical loop for multiwavelength laser generation. In this way, the fiber laser can be tuned by simply uploading appropriate steering phase holograms that drive the various pixels of the Opto-VLSI processor. The Opto-VLSI-based tunable multiwavelength fiber laser shown in Fig. 3 is able to tune multiple lasers simultaneously and independently. This architecture offers excellent non-inertial tuning flexibility because any waveband within the ASE spectra from the gain media can independently be selected using computer generated holograms.

The principle of the Opto-VLSI-based tunable multiwavelength fiber laser is demonstrated by the experimental setup illustrated in Fig. 3. An EDFA and an SOA, both operating within the wavelength region of C-band, were driven with currents of 400 mA and 300 mA, respectively. A Labview software was specially developed to generate the optimized digital holograms that steer the desired wavebands and couple into Port 2.

Note that the output port was chosen to be immediately after the EDFA, because the homogeneous line broadening of the EDFA results in low background noise when multiple lasing wavelengths are synthesized. A tunable 5-wavelength fiber laser was generated by applying an optimized phase hologram to steer 5 associated wavebands and couple into the fiber ring. Three experimental scenarios were carried out to prove the tunability of the proposed structure, including tuning with equal wavelength spacing, tuning with arbitrary wavelength spacing, and switching on/off wavelength channels.

Figure 6 shows the results of the first scenario, where the Opto-VLSI processor was reconfigured to tune the multiwavelength laser with equal wavelength spacing. A 5-wavelength laser with equal wavelength spacing of 1.14 nm were firstly synthesized using a phase hologram optimized in such a way that the pixel blocks associated with the 5 wavelength channels are uploaded with steering blazed gratings that appropriately steer and couple these channels back into the fiber ring while the other wavelength channels are not steered, thus experiencing dramatic coupling losses. The wavelength spacing of the 5-wavelength laser was tuned from 1.14 nm to 1.68 nm, then 2.22 nm by reconfiguring the phase hologram. The output power for each laser channel was around +6 dBm and the SMSR was better than 30 dB. The laser linewidth was about 0.4 nm, and the power uniformity was less than 0.8 dB. This small measured power uniformity is attributed to the fact that the intensity of each wavelength channel in the cavity can be arbitrarily controlled by adjusting the steering efficiency of the various pixel blocks associated to the lasing channels. In addition, the relatively high laser output power was because of the low insertion loss experienced by each wavelength channel.

The measured minimum wavelength spacing of the multiwavelength laser was around 0.4 nm, which is determined by the pixel size of the Opto-VLSI processor, the focal length of the lens, and the dispersion capability of the grating plate. In the experiments, the 30 nm ASE spectrum was mapped across the 512-pixel surface of the Opto-VLSI processor (each of 15 um size), thus each pixel was occupied by about 0.05 nm of the tuning range, and 8x512-pixel pixel blocks were used to steer the individual wavelength channels, resulting in 0.4 nm channel linewidth (and hence minimum channel spacing), and 0.05 nm tuning resolution for each wavelength channel corresponding to one pixel shift. Note that the wavelength spacing and the tuning resolution can be made smaller by using an Opto-VLSI processor with a smaller pixel size and a larger active window.

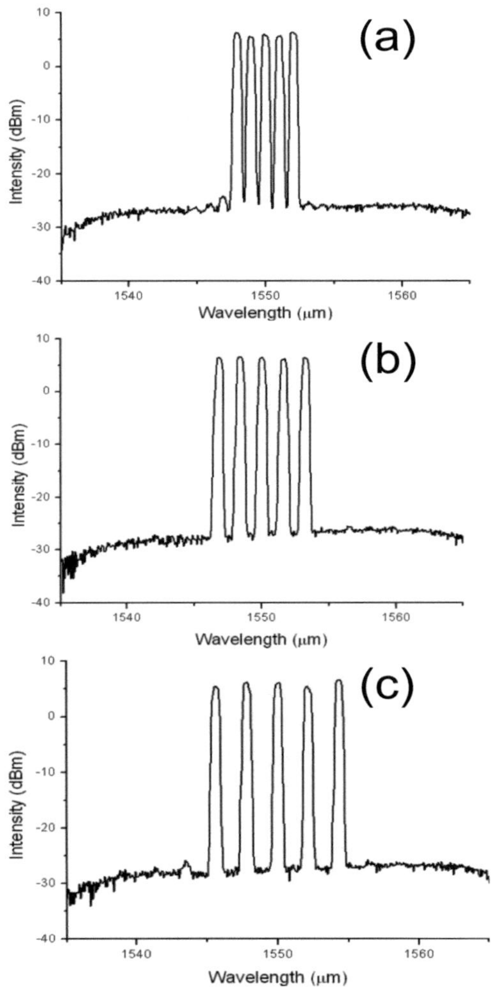

Fig. 6. Tuning with equal wavelength spacing for the Opto-VLSI-based tunable multiwavelength fiber laser.

The second scenario demonstrated that the wavelength spacing of the multiwavelength laser can arbitrary be tuned, as shown in Fig. 7. Starting from the 1.14nm spaced wavelength channels shown in Fig. 6(a), the adjacent wavelength spacings were changed to [1.14 nm, 1.14nm, 1.68 nm, 1.68 nm] as shown in Fig. 7(a). This was accomplished by shifting the steering phase holograms associated to the fourth and fifth wavelength channels by 10 pixels. Similarly, the adjacent wavelength spacing was tuned to [1.14 nm, 1.68nm, 1.68 nm, 1.68 nm] and [1.68 nm, 1.68nm, 1.68 nm, 2.22 nm] as shown in Figs. 7(b) and (c), respectively. Note that when the wavelength spacing of the multiwavelength laser was varied, a negligible change in the output levels for each wavelength and the output power uniformity was observed. However, the other laser characteristics such as output SMSR, laser linewidth, tuning step, did not change.

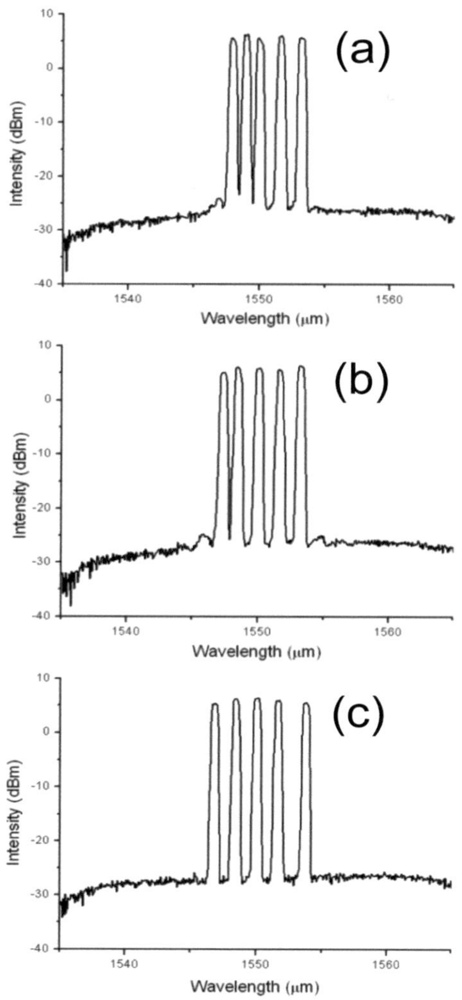

Fig. 7. Tuning with arbitrary wavelength spacing for the Opto-VLSI-based tunable multiwavelength fiber laser.

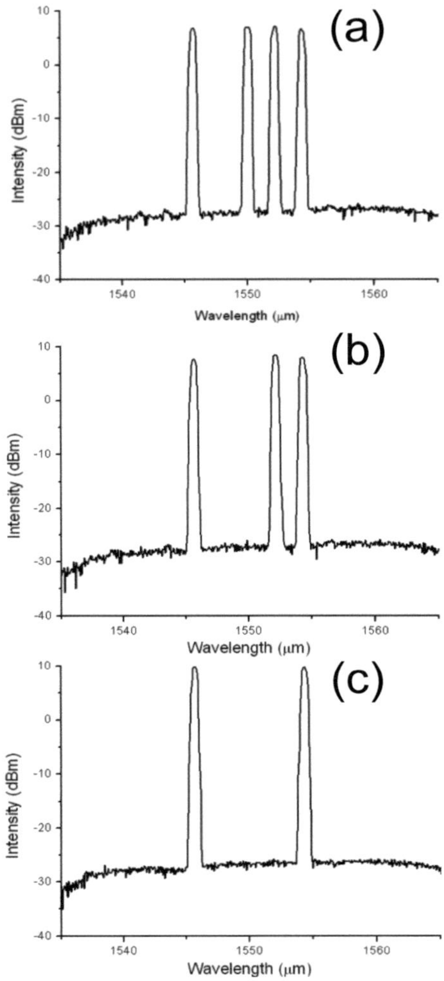

Fig. 8. Switching on/off any wavelength channels.

In the third scenario, we demonstrated that each wavelength channel can independently be switched on/off. Starting from the multiwavelength laser output shown in Fig. 7(c), and by removing the steering phase hologram associated to the second wavelength channel, the latter was switched off and dropped out from the fiber ring while the other channels were kept intact, as shown in Fig. 8(a). Similarly, the third and the fourth wavelength channels were dropped, as illustrated in Figs. 8(b) and (c), by reconfiguring the phase hologram uploaded onto the Opto-VLSI processor. During the switching experiments, the multiwavelength laser characteristics such as the output power level, the power uniformity, laser linewidth, and SMSR were not affected.

The above three scenarios demonstrate the capability of the multiwavelength laser to generate arbitrary wavelength channels via software, leading to significant improvement in flexibility and reconfigurability compared to previously reported tunable multiwavelength laser demonstrators.

Each wavelength channel exhibited very stable operation at room temperature whenever it was turned on for different periods of time ranging from a few hours to a few days. The measured maximum output power fluctuation was less than 0.5 dB for a period of 2-hour observation.

7. Multi-port tunable fiber lasers

In addition to its excellent tunability for both single-wavelength and multi-wavelength lasing, the Opto-VLSI based approach provides a special capability of integrating many tunable single/multi-wavelength fiber lasers into a same tuning system, making it very competitive for commercialization.

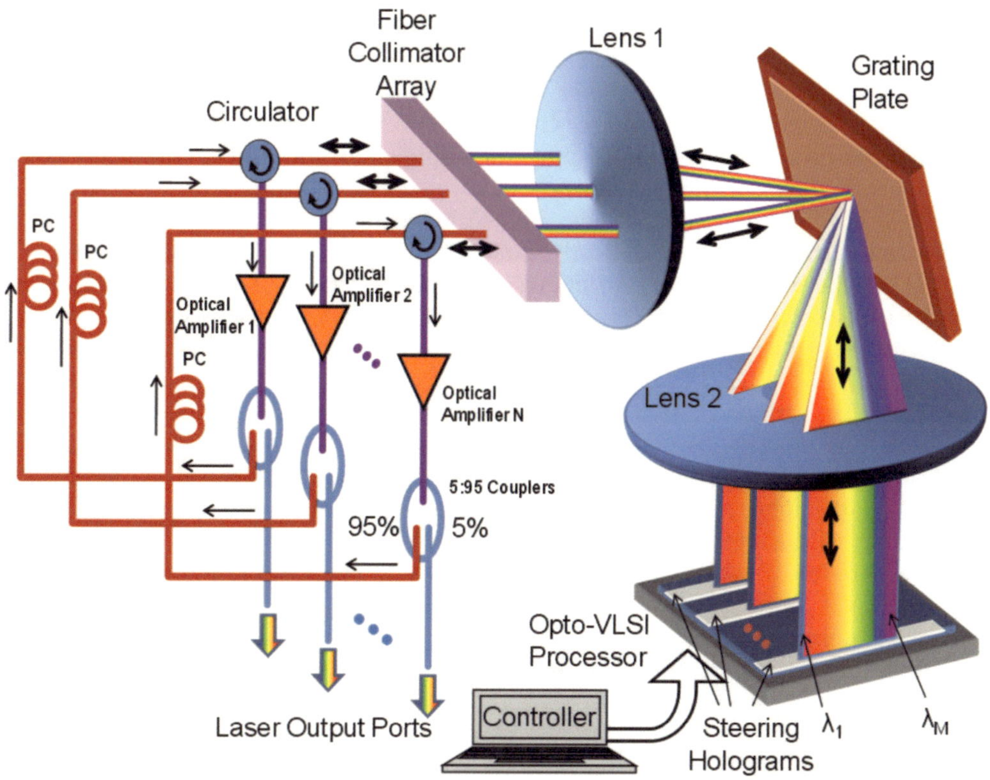

Fig. 9. The proposed multi-port tunable fiber laser structure.

The proposed Opto-VLSI-based multi-port tunable fiber ring laser structure is shown in Fig. 9. It consists of N tunable fiber lasers simultaneously driven by a single Opto-VLSI processor. Each tunable fiber laser employs an optical amplifier, an optical coupler, a polarization controller, a circulator, and one port from a collimator array, as described in Fig. 9. All the broadband ASE signals are directed to the corresponding collimator ports, via their corresponding circulators. A lens (Lens 1) is used between the collimator array and a diffraction grating plate to focus the collimated ASE beams onto a small spot onto the grating plate. The latter demultiplexes all the collimated ASE signals into wavebands (of different center wavelengths) along different directions. Another lens (Lens 2), located in the middle position between the grating plate and the Opto-VLSI processor, is used to collimate the dispersed optical beams in two dimensions and map them onto the surface of a 2-D Opto-VLSI processor, which is partitioned into N rectangular pixel blocks. Each pixel block is assigned to a tunable laser and used to efficiently couple back any part of the ASE spectrum illuminating this pixel block along the incident path into the corresponding collimator port. The selected waveband coupled back into the fiber collimator port is then routed back to the gain medium via the corresponding circulator, thus an optical loop is formed for the single-mode laser generation. Therefore, by uploading the appropriate phase holograms (or blazed grating) that drive all the pixel blocks of the Opto-VLSI processor, N different wavelengths can independently be selected for lasing within the different fiber loops, thus realizing a multiport tunable fiber laser source that can simultaneously generate arbitrary wavelengths at its ports. Note that the N tunable fiber lasers can independent and simultaneously offer lasing in sing wavelength, multi wavelength, or hybrid.

To proof the principle of the proposed Opto-VLSI-based tunable fiber laser, an Opto-VLSI-based 3-wavelength tunable fiber laser was demonstrated using the experimental setup shown in Fig. 9. Each tunable fiber laser channel consists of an EDFA that operates in the C-band, a 1×2 optical coupler with 5/95 power splitting ratio, and a fiber collimator array. A 256-phase-level two-dimensional Opto-VLSI processor having 512×512 pixels with 15 μm pixel size was used to independently and simultaneously select any part of the gain spectrum from each EDFA into the corresponding fiber ring. Two identical lenses of focal length 10 cm were placed at 10 cm from both sides of the grating plate. An optical spectrum analyzer with 0.01 nm resolution was used to monitor the 5% output port of each optical coupler which serves as the output port for each tunable laser channel. The 95% port of each ASE signal was directed to a PC and collimated at about 0.5 mm diameter. A blazed grating, having 1200 lines/mm and a blazed angle of 70° at 1530 nm, was used to demultiplex the three EDFA gain spectra, which were mapped onto the active window of the Opto-VLSI processor by Lens 2. A Labview software was especially developed to generate the optimized digital holograms that steer the desired waveband and couple back into the corresponding collimator for subsequent recirculation in the fiber loop.

The active window of the Opto-VLSI processor was divided into three pixel blocks corresponding to the positions of the three demultiplexed ASE signals, each pixel block dedicated for tuning the wavelength of a fiber laser. Optimized digital phase holograms were applied to the three pixel blocks, so that desired wavebands from the ASE spectra illuminating the Opto-VLSI processor could be selected and coupled back into their fiber rings, leading to simultaneous lasing at specific wavelengths. By changing the position of the phase hologram of each pixel block, the lasing wavelength for each fiber laser could be dynamically and independently tuned. The measured total cavity loss for each channel was around 12 dB, which mainly includes (i) the coupling loss of the associated collimator; (ii) the blazed grating loss; and (iii) the diffraction loss and insertion loss of the Opto-VLSI processor. Note that the total cavity loss influences both the laser output power and the tuning range, as well as the pump current thresholds needed for lasing (60mA in the experiments).

Figure 10 demonstrates the coarse tuning capability of the 3-wavelength Opto-VLSI fiber laser operating over C-band. The measured output laser spectrum for each channel is shown for different optimized phase holograms uploaded onto the Opto-VLSI processor. All the channels could independently and simultaneously be tuned over the whole C-band. Port 1 and Port 2 have an output power level of about 9 dBm with an optical side-mode-suppression-ratio of more than 35 dB. Port 3 has 2 dB less output power because the EDFA's

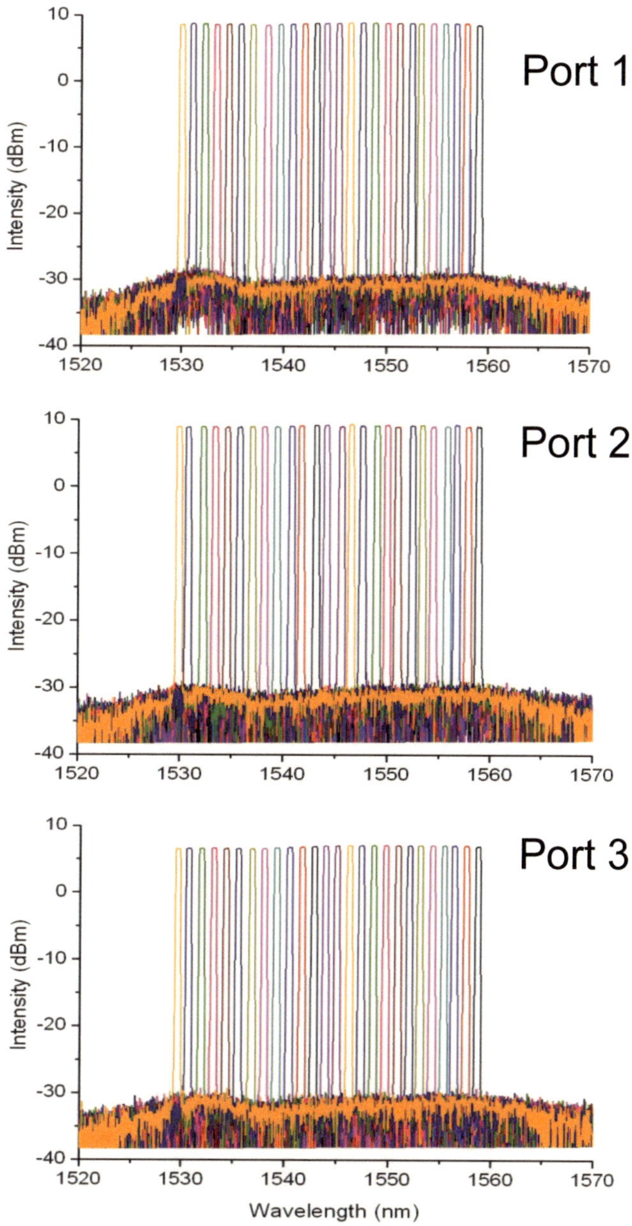

Fig. 10. Measured responses of the Opto-VLSI-based 3-wavelength fiber laser for coarse tuning operation over C-band. These three channels can independently and simultaneously be tuned over the whole C-band.

gain for this channel was intentionally dropped to demonstrate the ability to change the output power level via changing the pump current. The laser output power for each channel has a uniformity of about 0.5 dB over the whole tuning range. Each laser channel exhibited the same performance as described before when only one fiber laser is constructed based on the Opto-VLSI processor.

The maximum output power for the multi-wavelength tunable fiber laser is about 9 dBm. This value is mainly dependent on the gain of the EDFA associated to that channel. Note that the thickness of the liquid crystal layer of the Opto-VLSI processor is very small (several microns), leading to spatial phase-modulation with negligible power loss. For high laser output power levels, the nonlinearity of the LC material could induce unequal phase shifts to the individual pixels of the steering phase hologram, leading to higher coupling loss, which reduces the output laser power. However, properly designed liquid-crystal mixtures can handle optical intensities as high as 700 W/cm2 with negligible nonlinear effects, making the maximum laser output power mainly dependent on the maximum output optical power of the gain medium.

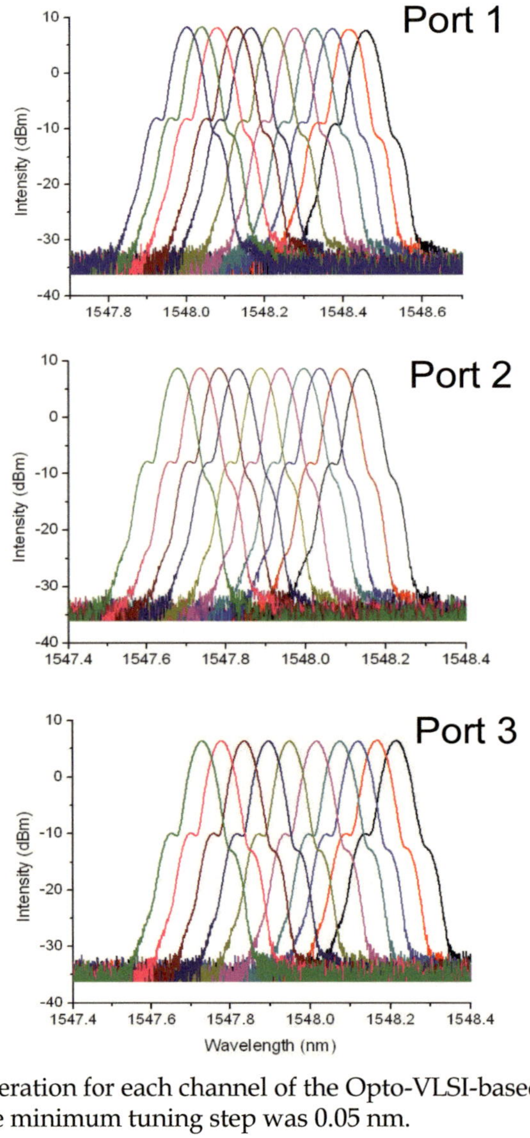

Fig. 11. Fine tuning operation for each channel of the Opto-VLSI-based 3-wavelength tunable fiber laser. The minimum tuning step was 0.05 nm.

The measured laser outputs for fine wavelength tuning operation of the three channels are shown in Fig. 11. By shifting the center of each phase hologram by a single pixel across the active window of the Opto-VLSI processor, the wavelength was tuned by a step of around 0.05 nm for all the three channels. This corresponds to the mapping of 30 nm ASE spectrum of the EDFA of each channel across the 512 pixels (each of 15 μm size). Similarly, the shoulders on both sides of the laser spectrum of each tunable laser channel are due to self-phase modulation or other nonlinear phenomena arising from a high level of the output power, as also shown in Fig. 11(b).

When the output power of each fiber laser is varied via the control of the current driving the pump laser of the EDFA, the other laser characteristics such as output SMSR, laser linewidth, output power uniformity, tuning step, and tuning range were not changed. The pump-independent laser linewidth observation might be due to the limited resolution (0.01 nm) of the OSA we used in the experiments.

Since the Opto-VLSI processor has a broad spectral bandwidth, the multi-port tunable laser structure shown in Fig. 9 could in principle operate over the O-, S-, C- and/or L- bands. Note also that the Opto-VLSI processor used in the experiment was able to achieve wavelength tuning for up to 8 ports independently and simultaneously. This is because each pixel block was about 0.8 mm wide and the active window of the Opto-VLSI active window was 7.6 mm × 7.6 mm.

8. Conclusion

In this chapter, the tuning mechanisms and gain mechanisms for single-wavelength, multi-wavelength tunable fiber lasers have been reviewed. Then the use of optical amplifiers and Opto-VLSI technology to realize a tunable single/multiple wavelength fiber laser and multi-port tunable fiber lasers, has been discussed. The ability of the Opto-VLSI processor to select any part of the gain spectrum from optical amplifiers into desired fiber rings has been demonstrated, leading to many tunable single/multiple wavelength fiber laser sources. We have also experimentally demonstrated the proof-of-principle of tunable fiber lasers capable of generating single and/or multiple wavelengths laser sources with laser linewidth as narrow as 0.05 nm, optical side-mode-suppression-ratio (SMSR) of about 35 dB, as well as outstanding tunability. The demonstrated tunable fiber lasers have excellent stability at room temperature and output power uniformity less than 0.5 dB over the whole C-band. In addition, this tunable fiber laser structure could potentially operate over the O-, S-, C- and/or L- bands.

9. Acknowledgement

We acknowledge the support of the Department of Nano-bio Materials and Electronics, Gwangju Institute of Science and Technology, Republic of Korea, for the development of the tunable laser demonstrator.

10. References

Alvarez-Chavez, J. A., Martinez-Rios, A., Torres-Gomez, I. & Offerhaus, H. L. (2007). Wide wavelength-tuning of a double-clad Yb3+-doped fiber laser based on a fiber Bragg grating array. Laser Physics Letters, Vol. 4, No. 12, pp. 880-883, Issn: 1612-2011.

Belanger, E., Bernier, M., Faucher, D., Cote, D. & Vallee, R. (2008). High-power and widely tunable all-fiber Raman laser. Journal of Lightwave Technology, Vol. 26, No. 9-12, pp. 1696-1701, Issn: 0733-8724.

Bellemare, A. (2003). Continuous-wave silica-based erbium-doped fibre lasers. Progress in Quantum Electronics, Vol. 27, pp. 211-266

Bellemare, A., Karasek, M., Riviere, C., Babin, F., He, G., Roy, V. & Schinn, G. W. (2001). A broadly tunable erbium-doped fiber ring laser: experimentation and modeling. Ieee Journal of Selected Topics in Quantum Electronics, Vol. 7, No. 1, pp. 22-29, Issn: 1077-260X.

Bellemare, A., Karasek, M., Rochette, M., LaRochelle, S. & Tetu, M. (2000). Room temperature multifrequency erbium-doped fiber lasers anchored on the ITU frequency grid. Journal of Lightwave Technology, Vol. 18, No. 6, pp. 825-831, Issn: 0733-8724.

Chang, S. H., Hwang, I. K., Kim, B. Y. & Park, H. G. (2001). Widely tunable single-frequency Er-doped fiber laser with long linear cavity. Ieee Photonics Technology Letters, Vol. 13, No. 4, pp. 287-289, Issn: 1041-1135.

Chen, D. R., Qin, S. & He, S. L. (2007). Channel-spacing-tunable multi-wavelength fiber ring laser with hybrid Raman and erbium-doped fiber gains. Optics Express, Vol. 15, pp. 930-935, Issn: 1094-4087.

Chen, H. X. (2005). Multiwavelength fiber ring lasing by use of a semiconductor optical amplifier. Optics Letters, Vol. 30, No. 6, pp. 619-621, Issn: 0146-9592.

Chen, Y.-W., Yamauchi, S., Wang, N. & Nakao, Z. (2000). A fast kinoform optimization algorithm based on simulated annealing. IEICE Trans. Fundamentals, Vol. E83-A, No. 4, pp. 3

Cheng, Y., Kringlebotn, J. T., Loh, W. H., Laming, R. I. & Payne, D. N. (1995). STABLE SINGLE-FREQUENCY TRAVELING-WAVE FIBER LOOP LASER WITH INTEGRAL SATURABLE-ABSORBER-BASED TRACKING NARROW-BAND-FILTER. Optics Letters, Vol. 20, No. 8, pp. 875-877, Issn: 0146-9592.

Chien, H. C., Yeh, C. H., Lai, K. H., Lee, C. C. & Chi, S. (2005a). Stable and wavelength-tunable erbium-doped fiber double-ring laser in S-band window operation. Optics Communications, Vol. 249, No. 1-3, pp. 261-264, Issn: 0030-4018.

Chien, H. C., Yeh, C. H., Lee, C. C. & Chi, S. (2005b). A tunable and single-frequency S-band erbium fiber laser with saturable-absorber-based autotracking filter. Optics Communications, Vol. 250, No. 1-3, pp. 163-167, Issn: 0030-4018.

Chow, K. K., Shu, C., Mak, M. & Tsang, H. K. (2002). Widely tunable wavelength converter using a double-ring fiber laser with a semiconductor optical amplifier. Ieee Photonics Technology Letters, Vol. 14, No. 10, pp. 1445-1447, Issn: 1041-1135.

Dammann, H. (1979). Spectral characteristics of stepped-phase gratings. Optik, Vol. 53 pp. 9

Das, G. & Lit, J. W. Y. (2002). L-band multiwavelength fiber laser using an elliptical fiber. Ieee Photonics Technology Letters, Vol. 14, No. 5, pp. 606-608, Issn: 1041-1135.

Dong, H., Zhu, G., Wang, Q., Sun, H., Dutta, N. K., Jaques, J. & Piccirilli, A. B. (2005a). Multiwavelength fiber ring laser source based on a delayed interferometer. Ieee Photonics Technology Letters, Vol. 17, No. 2, pp. 303-305, Issn: 1041-1135.

Dong, X. Y., Ngo, N. Q., Shum, P. & Tam, H. Y. (2003). Linear cavity erbium-doped fiber laser with over 100 nm tuning range. Optics Express, Vol. 11, No. 14, pp. 1689-1694, Issn: 1094-4087.

Dong, X. Y., Shum, P., Ngo, N. Q. & Chan, C. C. (2006). Multiwavelength Raman fiber laser with a continuously-tunable spacing. Optics Express, Vol. 14, No. 8, pp. 3288-3293, Issn: 1094-4087.

Dong, X. Y., Shum, P., Ngo, N. Q. & Tam, H. Y. (2005b). Output power characteristics of tunable erbium-doped fiber ring lasers. Journal of Lightwave Technology, Vol. 23, No. 3, pp. 1334-1341, Issn: 0733-8724.

Fok, M. P., Lee, K. L. & Shu, C. (2005). Waveband-switchable SOA ring laser constructed with a phase modulator loop mirror filter. Ieee Photonics Technology Letters, Vol. 17, No. 7, pp. 1393-1395, Issn: 1041-1135.

Fu, Z. H., Yang, D. Z., Ye, W., Kong, J. & Shen, Y. H. (2009). Widely tunable compact erbium-doped fiber ring laser for fiber-optic sensing applications. Optics and Laser Technology, Vol. 41, No. 4, pp. 392-396, Issn: 0030-3992.

Han, Y. G., Dong, X. Y., Kim, C. S., Jeong, M. Y. & Lee, J. H. (2007). Flexible all fiber Fabry-Perot filters based on superimposed chirped fiber Bragg gratings with continuous FSR tunability and its application to a multiwavelength fiber laser. Optics Express, Vol. 15, No. 6, pp. 2921-2926, Issn: 1094-4087.

Han, Y. G., Kim, G., Lee, J. H., Kim, S. H. & Lee, S. B. (2005). Lasing wavelength and spacing switchable multiwavelength fiber laser from 1510 to 1620 nm. Ieee Photonics Technology Letters, Vol. 17, No. 5, pp. 989-991, Issn: 1041-1135.

Han, Y. G., Tran, T. V. A. & Lee, S. B. (2006). Wavelength-spacing tunable multi wavelength erbium-doped fiber laser based on four-wave mixing of dispersion-shifted fiber. Optics Letters, Vol. 31, No. 6, pp. 697-699, Issn: 0146-9592.

Kang, M. S., Lee, M. S., Yong, J. C. & Kim, B. Y. (2006). Characterization of wavelength-tunable single-frequency fiber laser employing acoustooptic tunable filter. Journal of Lightwave Technology, Vol. 24, No. 4, pp. 1812-1823, Issn: 0733-8724.

Kim, C. S., Sova, R. M. & Kang, J. U. (2003). Tunable multi-wavelength all-fiber Raman source using fiber Sagnac loop filter. Optics Communications, Vol. 218, No. 4-6, pp. 291-295, Issn: 0030-4018.

Lee, C. C., Chen, Y. K. & Liaw, S. K. (1998). Single-longitudinal-mode fiber laser with a passive multiple-ring cavity and its application for video transmission. Optics Letters, Vol. 23, No. 5, pp. 358-360

Lee, K. L., Fok, M. P., Wan, S. M. & Shu, C. (2004). Optically controlled Sagnac loop comb filter. Optics Express, Vol. 12, No. 25, pp. 6335-6340

Li, S. Y., Ngo, N. Q. & Zhang, Z. R. (2008). Tunable Fiber Laser With Ultra-Narrow Linewidth Using A Tunable Phase-Shifted Chirped Fiber Grating. Ieee Photonics Technology Letters, Vol. 20, No. 17-20, pp. 1482-1484, Issn: 1041-1135.

Liaw, S. K., Hung, K. L., Lin, Y. T., Chiang, C. C. & Shin, C. S. (2007). C-band continuously tunable lasers using tunable fiber Bragg gratings. Optics and Laser Technology, Vol. 39, No. 6, pp. 1214-1217, Issn: 0030-3992.

Liu, H. L., Tam, H. Y., Chung, W. H., Wai, P. & Sugimoto, N. (2005a). La-codoped bismuth-based erbium-doped fiber ring laser, with 106-nm tuning range. Ieee Photonics Technology Letters, Vol. 17, No. 2, pp. 297-299, Issn: 1041-1135.

Liu, H. L., Tam, H. Y., Chung, W. H., Wai, P. & Sugimoto, N. (2006). Low beat-noise polarized tunable fiber ring laser. Ieee Photonics Technology Letters, Vol. 18, No. 5-8, pp. 706-708, Issn: 1041-1135.

Liu, X. M., Zhou, X. Q., Tang, X. F., Ng, J., Hao, J. Z., Chai, T. Y., Leong, E. & Lu, C. (2005b). Switchable and tunable multiwavelength erbium-doped fiber laser with fiber bragg gratings and photonic crystal fiber. Ieee Photonics Technology Letters, Vol. 17, No. 8, pp. 1626-1628, Issn: 1041-1135.

Liu, Z. Y., Liu, Y. G., Du, J. B., Kai, G. Y. & Dong, X. Y. (2008). Tunable multiwavelength erbium-doped fiber laser with a polarization-maintaining photonic crystal fiber Sagnac loop filter. Laser Physics Letters, Vol. 5, No. 6, pp. 446-448, Issn: 1612-2011.

Luo, A. P., Luo, Z. C. & Xu, W. C. (2009). Tunable and switchable multiwavelength erbium-doped fiber ring laser based on a modified dual-pass Mach-Zehnder interferometer. Optics Letters, Vol. 34, No. 14, pp. 2135-2137, Issn: 0146-9592.

MOLLIER, P., ARMBRUSTER, V., PORTE, H. & GOEDGEBUER, J. P. (1995). ELECTRICALLY TUNABLE ND3+-DOPED FIBER LASER USING NEMATIC LIQUID-CRYSTALS. Electronics Letters, Vol. 31, No. 15, pp. 1248-1250, Issn: 0013-5194.

Moon, D. S., Kim, B. H., Lin, A. X., Sun, G. Y., Han, W. T., Han, Y. G. & Chung, Y. (2007). Tunable multi-wavelength SOA fiber laser based on a Sagnac loop mirror using an elliptical core side-hole fiber. Optics Express, Vol. 15, No. 13, pp. 8371-8376, Issn: 1094-4087.

Moon, D. S., Paek, U. C., Chung, Y. J., Dong, X. Y. & Shum, P. (2005). Multi-wavelength linear-cavity tunable fiber laser using a chirped fiber Bragg grating and a few-mode fiber Bragg grating. Optics Express, Vol. 13, No. 15, pp. 5614-5620, Issn: 1094-4087.

Nasir, M., Yusoff, Z., Al-Mansoori, M. H., Rashid, H. & Choudhury, P. K. (2009). Widely tunable multi-wavelength Brillouin-erbium fiber laser utilizing low SBS threshold photonic crystal fiber. Optics Express, Vol. 17, No. 15, pp. 12829-12834, Issn: 1094-4087.

Ohara, S. & Sugimoto, N. (2008). Bi2O3-based erbium-doped fiber laser with a tunable range over 130 nm. Optics Letters, Vol. 33, No. 11, pp. 1201-1203, Issn: 0146-9592.

Park, N., Dawson, J. W., Vahala, K. J. & Miller, C. (1991). All fiber, low threshold, widely tunable single-frequency, erbium-doped fiber ring laser with a tandem fiber Fabry-Perot filter. Applied Physics Letters, Vol. 59, pp. 2369-2371

Pleros, N., Bintjas, C., Kalyvas, M., Theophilopoulos, G., Yiannopoulos, K., Sygletos, S. & Avramopoulos, H. (2002). Multiwavelength and power equalized SOA laser sources. Ieee Photonics Technology Letters, Vol. 14, No. 5, pp. 693-695, Issn: 1041-1135.

Poustie, A. J., Finlayson, N. & Harper, P. (1994). MULTIWAVELENGTH FIBER LASER USING A SPATIAL MODE BEATING FILTER. Optics Letters, Vol. 19, No. 10, pp. 716-718, Issn: 0146-9592.

Qian, J. R., Su, J. & Hong, L. (2008). A widely tunable dual-wavelength erbium-doped fiber ring laser operating in single longitudinal mode. Optics Communications, Vol. 281, No. 17, pp. 4432-4434, Issn: 0030-4018.

Roy, V., Piche, M., Babin, F. & Schinn, G. W. (2005). Nonlinear wave mixing in a multilongitudinal-mode erbium-doped fiber laser. Optics Express, Vol. 13, No. 18, pp. 6791-6797, Issn: 1094-4087.

Sakata, H., Yoshimi, H. & Otake, Y. (2009). Wavelength tunability of L-band fiber ring lasers using mechanically induced long-period fiber gratings. Optics Communications, Vol. 282, No. 6, pp. 1179-1182, Issn: 0030-4018.

Song, Y. W., Havstad, S. A., Starodubov, D., Xie, Y., Willner, A. E. & Feinberg, J. (2001). 40-nm-wide tunable fiber ring laser with single-mode operation using a highly stretchable FBG. Ieee Photonics Technology Letters, Vol. 13, No. 11, pp. 1167-1169, Issn: 1041-1135.

Sun, G. Y., Moon, D. S., Lin, A. X., Han, W. T. & Chung, Y. J. (2008). Tunable multiwavelength fiber laser using a comb filter based on erbium-ytterbium co-doped polarization maintaining fiber loop mirror. Optics Express, Vol. 16, No. 6, pp. 3652-3658, Issn: 1094-4087.

Sun, J. Q., Qiu, J. L. & Huang, D. X. (2000). Multiwavelength erbium-doped fiber lasers exploiting polarization hole burning. Optics Communications, Vol. 182, No. 1-3, pp. 193-197, Issn: 0030-4018.

Tran, T. V. A., Lee, K., Lee, S. B. & Han, Y. G. (2008). Switchable multiwavelength erbium doped fiber laser based on a nonlinear optical loop mirror incorporating multiple fiber Bragg gratings. Optics Express, Vol. 16, No. 3, pp. 1460-1465, Issn: 1094-4087.

Ummy, M. A., Madamopoulos, N., Lama, P. & Dorsinville, R. (2009). Dual Sagnac loop mirror SOA-based widely tunable dual-output port fiber laser. Optics Express, Vol. 17, No. 17, pp. 14495-14501, Issn: 1094-4087.

Xiao, F., Alameh, K. & Lee, T. (2009). Opto-VLSI-based tunable single-mode fiber laser. Optics Express, Vol. 17, No. 21, pp. 18676-18680, Issn: 1094-4087.

Xu, L., Wang, B. C., Baby, V., Glesk, I. & Prucnal, P. R. (2002). Optical spectral bistability in a semiconductor fiber ring laser through gain saturation in an SOA. Ieee Photonics Technology Letters, Vol. 14, No. 2, pp. 149-151, Issn: 1041-1135.

Yamashita, S. & Hotate, K. (1996). Multiwavelength erbium-doped fibre laser using intracavity etalon and cooled by liquid nitrogen. Electronics Letters, Vol. 32, No. 14, pp. 1298-1299, Issn: 0013-5194.

Yeh, C. H. & Chi, S. (2005). A broadband fiber ring laser technique with stable and tunable signal-frequency operation. Optics Express, Vol. 13, No. 14, pp. 5240-5244, Issn: 1094-4087.

Zhang, J. L., Yue, C. Y., Schinn, G. W., Clements, W. R. L. & Lit, J. W. Y. (1996). Stable single-mode compound-ring erbium-doped fiber laser. Journal of Lightwave Technology, Vol. 14, No. 1, pp. 104-109, Issn: 0733-8724.

Zhang, Z. X., Wu, J., Xu, K., Hong, X. B. & Lin, J. T. (2009). Tunable multiwavelength SOA fiber laser with ultra-narrow wavelength spacing based on nonlinear polarization rotation. Optics Express, Vol. 17, No. 19, pp. 17200-17205, Issn: 1094-4087.

Zhang, Z. X., Zhan, L., Xu, K., Wu, J., Xia, Y. X. & Lin, J. T. (2008). Multiwavelength fiber laser with fine adjustment, based on nonlinear polarization rotation and birefringence fiber filter. Optics Letters, Vol. 33, No. 4, pp. 324-326, Issn: 0146-9592.

Zheng, L., Vaillancourt, J., Armiento, C. & Lu, X. J. (2006). Thermo-optically tunable fiber ring laser without any mechanical moving parts. Optical Engineering, Vol. 45, No. 7, Issn: 0091-3286.

Zhou, D. Y., Prucnal, P. R. & Glesk, I. (1998). A widely tunable narrow linewidth semiconductor fiber ring laser. Ieee Photonics Technology Letters, Vol. 10, No. 6, pp. 781-783, Issn: 1041-1135.

15

Equivalent Circuit Models for Optical Amplifiers

Jau-Ji Jou[1] and Cheng-Kuang Liu[2]
[1]*National Kaohsiung University of Applied Sciences*
[2]*National Taiwan University of Science and Technology*
Taiwan

1. Introduction

Electrical equivalent circuit models for optical components are useful as they allow existing, well-developed circuit simulators to be used in design and analysis of optoelectronic devices. A circuit simulator also allows integration with electrical components (package parasitic, laser driver circuit, etc.). Equivalent circuit models were developed and investigated for some optoelectronic circuit elements, including p-i-n diodes, laser diodes, and waveguide modulators (Bononi et al., 1997; Chen et al., 2000; Desai et al., 1993; Jou et al., 2002; Mortazy & Moravvej-Farshi, 2005; Tsou & Pulfrey, 1997).
The features of erbium-doped fiber amplifiers (EDFAs) are continuously investigated because of their great importance in optical communication systems. In order to design and analyze the characteristics of EDFAs, it is essential to have an accurate model. A dynamic model of EDFAs is helpful to understand the transient behavior in networks. The EDFA dynamics can also be used to monitor information in optical networks (Murakami et al., 1996; Shimizu et al., 1993). In this chapter, using a new circuit model for EDFAs, the static and dynamic characteristics of EDFAs can be analyzed conveniently through the aid of a SPICE simulator. The dc gain, amplified spontaneous emission (ASE) spectrum, frequency response and transient analysis of EDFAs can be simulated.
Semiconductor optical amplifiers (SOAs) are also important components for optical networks. They are very attractive for their wide gain spectrum, and capability of integration with other devices. In the linear regime, they can be used for both booster and in-line amplifiers (O'Mahony, 1988; Settembre et al., 1997; Simon, 1987). Also, much research activities have been done on all-optical signal processing with SOAs (Danielsen et al., 1998; Durhuus et al., 1996). Laser diodes (LDs) are similar devices to SOAs, and they are also the key components for various applications ranging from high-end and high-speed (i.e. fiber communications, and compact-disc players) to low-end and low-speed (i.e. laser pointers, and laser displays) systems. In this chapter, a new unified equivalent circuit model for SOAs and LDs is also presented.

2. Equivalent circuit model for erbium-doped fiber amplifiers

Sun et al. (Sun et al., 1996) derived a nonlinear ordinary differential equation to describe EDFA dynamics. Then, Bononi, Rusch, and Tancevski (Bononi et al., 1997) developed an equivalent circuit model to study EDFA dynamics. Based on this equation, Novak and

Gieske (Novak & Gieske, 2002) also presented a MATLAB Simulink model of EDFA. However, most EDFA models (Barnard et al., 1994; Freeman & Conradi, 1993; Giles et al., 1989; Novak & Gieske, 2002; Novak & Moesle, 2002) didn't take the ASE into account. Some models or methods of EDFA analysis had been presented with ASE (Araci & Kahraman, 2003; Burgmeier et al., 1998; Ko et al., 1994; Wu & Lowery, 1998), but a complex numerical computation was involved in a model or the ASE was simply taken as an independent light source.

Thus, in this section, the Bononi-Rusch-Tancevski model is extended to develop a new equivalent circuit model of EDFAs including ASE. Through the aid of a SPICE simulator, it is convenient to implement the circuit model and to analyze accurately the static and dynamic features of EDFAs.

2.1 Circuit model of EDFA including ASE

Considering a co-pumped two-level EDFA system, it is assumed that the excited-state absorption and the wavelength dependence of group velocity (υ_g) can be ignored. Let the optical beams propagate in z-direction through an EDF of length L. The rate equation and the propagating equations of photon fluxes in time frame can be simplified by transforming to a retarded-time frame moving with υ_g. These equations are shown as

$$\left(\frac{1}{\tau}+\frac{\partial}{\partial t}\right)N_2(z,t)=\sum_{k=p,\,s,\,A}\left[N_t\sigma_k^a-N_2(z,t)\sigma_k^{ae}\right]\Gamma_k P_k(z,t) \tag{1}$$

$$\frac{\partial P_k(z,t)}{\partial z}=\left[\sigma_k^{ae}N_2(z,t)-\sigma_k^a N_t\right]\Gamma_k P_k(z,t) \tag{2}$$

$$\pm\frac{\partial P_{a,l}^{\pm}(z,t)}{\partial z}=\left[\sigma_l^{ae}N_2(z,t)-\sigma_l^a N_t\right]\Gamma_l P_{a,l}^{\pm}(z,t)+\frac{2\Gamma_l\sigma_l^e\Delta\nu_{a,l}}{A}N_2(z,t) \tag{3}$$

where $P_k=P_k'/(h\nu_k A)$, P_k' is the power of the kth optical beam, ν_k is the optical frequency, h is Planck's constant; N_t is the erbium density in the fiber core of effective area A; Γ_k is the overlap factor of the kth beam; τ is the fluorescence lifetime of the metastable level; $\sigma_k^{ae}=\sigma_k^a+\sigma_k^e$, σ_k^a and σ_k^e are respectively the absorption and emission cross sections at the wavelength λ_k; k = p, s, A represent the pump beam (p), signal beam (s), and ASE (A); $P_A(z,t)=P_A^+(z,t)+P_A^-(z,t)=\sum_{l=1}^{m}\left[P_{a,l}^+(z,t)+P_{a,l}^-(z,t)\right]$, m is the number of frequency slots used in the ASE subdivision, $P_{a,l}^+$ and $P_{a,l}^-$ represent the forward and backward ASE fluxes within a frequency slot of width $\Delta\nu_{a,l}$, centered at optical frequency $\nu_{a,l}$ (wavelength $\lambda_{a,l}$). It is noted that s may be replaced by multichannel signals s(1), s(2), ..., and s(M).

By Eqs. (1)-(3), the equations can be obtained

$$\left(\frac{1}{\tau_a}+\frac{d}{dt}\right)\bar{N}_2(t)=\sum_{k=s,p,\,A}\left[P_k^{in}(t)-P_k^{out}(t)\right] \tag{4}$$

$$P_{s,p}(z=L,t)=P_{s,p}(z=0,t)G_{s,p}(t) \tag{5}$$

$$P_{a,l}^{\pm}(z=\frac{L\pm L}{2},t)\approx P_{a,l}^{\pm}(z=\frac{L\mp L}{2},t)\left[1+\frac{2\Gamma_l\sigma_l^e\Delta\nu_{a,l}}{A}\int_0^L\frac{N_2(z,t)}{P_{a,l}^{\pm}(z,t)}dz\right]G_l(t) \tag{6}$$

where $\tau_a^{-1}=\tau^{-1}-\sum_{l=1}^{m}4\Gamma_l\sigma_l^e\Delta\nu_{a,l}/A$, $\bar{N}_2(t)=\int_0^L N_2(z,t)dz$, $P_{s,p}^{in}=P_{s,p}(z=0,t)$, $P_{s,p}^{out}=P_{s,p}(z=L,t)$,

$P_A^{in}(t)=\sum_{l=1}^{m}\left[P_{a,l}^{+}(z=0,t)+P_{a,l}^{-}(z=L,t)\right]$, $P_A^{out}(t)=\sum_{l=1}^{m}\left[P_{a,l}^{+}(z=L,t)+P_{a,l}^{-}(z=0,t)\right]$, and $G_k(t)=\exp\left\{\Gamma_k\left[\sigma_k^{ae}\bar{N}_2(t)-\sigma_k^a N_t L\right]\right\}$. For simplicity, let an approximation make in this model: $P_{a,l}^{\pm}(z,t)\approx$ constant in Eq. (6) and write

$$P_{a,l}^{\pm}(z=\frac{L\pm L}{2},t)\approx\left[P_{a,l}^{\pm}(z=\frac{L\mp L}{2},t)+\frac{2\Gamma_l\sigma_l^e\Delta\nu_{a,l}}{A}\bar{N}_2(t)\right]G_l(t) \tag{7}$$

In general, the forward ASE remains constant at moderate pump power if the high-gain EDF length is not too long (around 4m in the case of (Pederson et al., 1990)). The forward ASE grows with pump power if the EDF fiber is long. Moreover, for a long EDF fiber (>10m in this case), the growth (or attenuation) of forward ASE along fiber length can not be ignored if the pump power is large (or small). A subdivision of EDF into small segments is necessary in case of long fiber. A similar conclusion holds for the backward ASE. The validity of the approximation of constant ASE power along the EDF will be shown in next subsection.

Subdividing the EDF into n segments with lengths L_i, i = 1, 2, …, n, an equivalent circuit model of EDFA including ASE contributions is developed for Eqs. (4), (5), and (7), as shown in Fig. 1, where $V_{N2,i}=\bar{N}_{2,i}(t)$; the subscript i in the $P_{s(M),i}^{in(out)}$, $P_{p,i}^{in(out)}$, $P_{A,i}^{\pm in(out)}$, or $P_{SE,i}^{\pm in(out)}$ represents the number of EDF segments; $I_{total,i}^{in(out)}=\sum\left(P_{s(M),i}^{in(out)}+P_{p,i}^{in(out)}+P_{A,i}^{\pm in(out)}\right)$; M is the

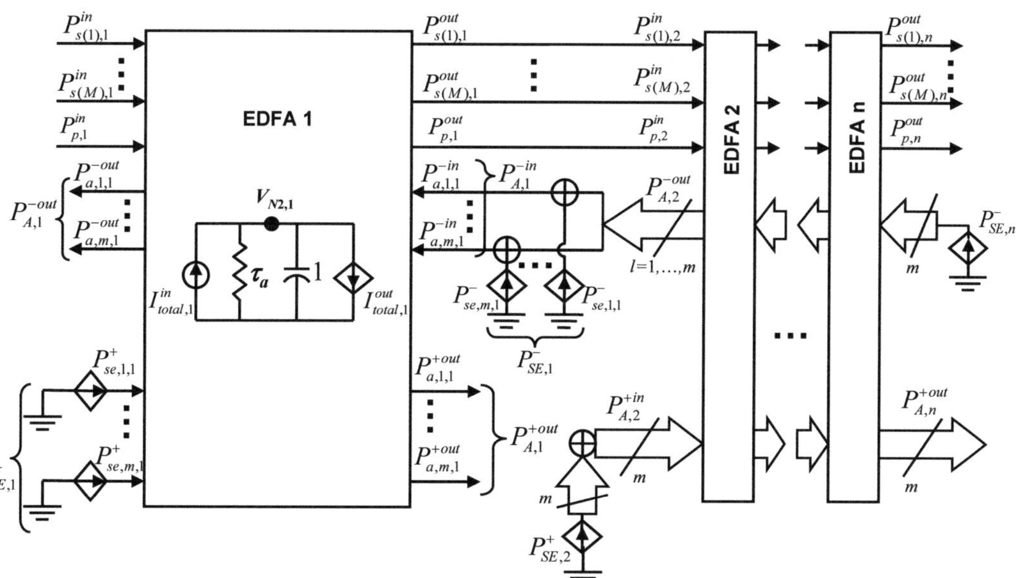

Fig. 1. Equivalent circuit model of EDFA including ASE

number of channels in a multichannel system; $P_{A,i}^{+in} = P_{A,i-1}^{+out} + P_{SE,i}^{+} = \sum_{l=1}^{m}\left(P_{a,l,i-1}^{+out} + P_{se,l,i}^{+}\right)$ with $P_{A,0}^{+out} = 0$; $P_{A,i}^{-in} = P_{A,i+1}^{-out} + P_{SE,i}^{-} = \sum_{l=1}^{m}\left(P_{a,l,i+1}^{-out} + P_{se,l,i}^{-}\right)$ with $P_{A,n+1}^{-out} = 0$; and $P_{se,l,i}^{\pm} = 2\Gamma_l \sigma_l^e \Delta\nu_{a,l} \bar{N}_{2,i} / A$.

2.2 Static gain of EDFA

A forward 980nm-pump EDFA with 12m EDF length and 50μW input signal power at 1558nm are considered. Other parameters used in these simulations are: $N_T = 7.7\times10^{24}m^{-3}$, τ= 10ms, $A = 2.5\times10^{-11}m^2$, $\sigma_s^a = 2.4\times10^{-25}m^2$, $\sigma_s^e = 3.8\times10^{-25}m^2$, $\sigma_p^a = 2.0\times10^{-25}m^2$, and NA = 0.18. These parameters are obtained from manufacturing data or the fitting of experimental gain (Jou et al., 2000; Lai et al., 1999). The dc gain is shown in Fig. 2, as a function of pump power. The square keys represent experimental data (Lai et al., 1999). As a comparison, the numerical calculation is from Eqs. (1)-(3). The numerical results calculated with ASE are shown as the filled circle keys. The EDF length is subdivided into 1200 segments in this numerical computation. The dash curves represent the circuit-model simulation without ASE (Bononi et al., 1997). The solid curve represents the simulation result using this equivalent circuit model. In the circuit models, the parameters used are: the EDF segment number n = 5, the centered wavelength of ASE light λ_a = 1540nm, the bandwidth $\Delta\nu_a$ = 5.06 THz (from 1520nm to 1560nm), and the average cross-section $\sigma_a^e = 5.6\times10^{-25}m^2$ and $\sigma_a^a = 4.8\times10^{-25}m^2$.

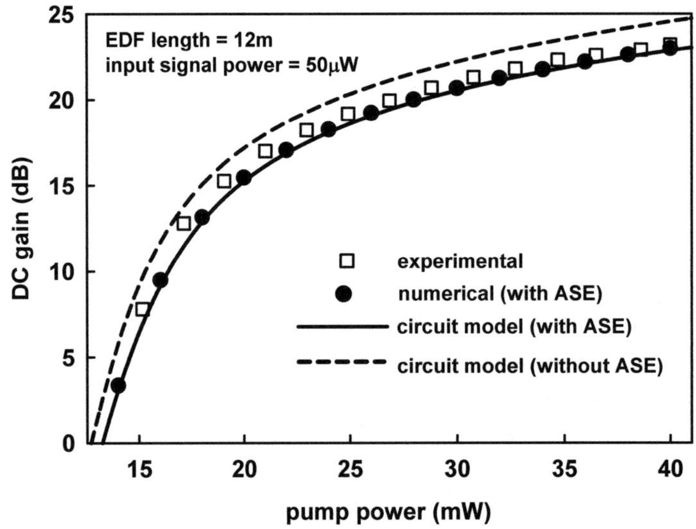

Fig. 2. Measured and simulated dc gain versus pump power for EDFA

Because of the ASE influences on EDFA, the dc gain without ASE is higher than the numerical and circuit-model simulations with ASE. Similar result was also reported by (Novak & Moesle, 2002). In Fig. 2, it is observed that the simulation without ASE results in a dc gain of 2dB lager than the experimental data when the pump power is around 25mW. A good agreement between the measured data and the simulation using this circuit model is obtained. The result of this circuit-model simulation is also in very good agreement with the numerical computation.

Generally, if the EDF is subdivided into more segments, the numerical computation results would be more accurate (Yu & Fan, 1999). However, the circuit model of EDFA without ASE should not be subdivided (Bononi et al., 1997), but the EDF should be subdivided for including ASE in this model. The simulated dc gains of the EDFA are shown in Fig. 3(a), using different number of EDF segments. The dc gain deviation is about 0.2dB between the numerical result using 1200 EDF segments and the circuit-model simulation using 3 EDF segments. Using 5 EDF segments, this simulation can be very close to the numerical result. In Fig. 3(b), the total forward and backward ASE power versus pump power can be shown. It is observed that the forward ASE power is lower than the backward one, since a forward pump EDFA is considered here.

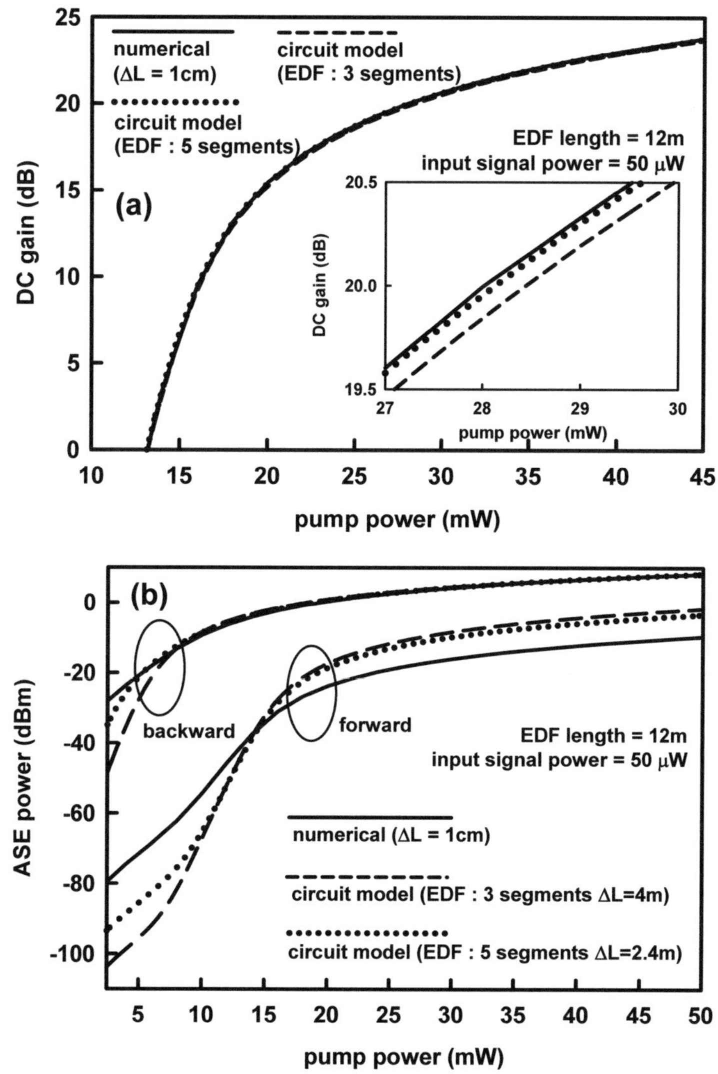

Fig. 3. Static characteristics of EDFA for the different number of subdivided EDF segments (a) DC gain versus pump power (b) ASE power versus pump power

The ASE power of circuit-model simulation is slightly higher than the numerical computation. This makes the dc gain of this circuit-model simulation slightly lower, as expected. When the pump power is around 13mW, the dc gain is around 0dB and the ASE power of circuit-model simulation is in excellent agreement with the numerical computation. The forward ASE power from this circuit model is overestimated when the pump power is above 13mW. This overestimation is believed to be due to this approximation of constant $P^{\pm}_{a,l}(z,t)$ in Eq. (6) and the fact that the forward ASE grows with EDF fiber length in a longer fiber under some given pump power. It is noted that $P^{\pm}_{se,l} = 2\Gamma_l \sigma^e_l \Delta\nu_{a,l} \bar{N}_2(t)/A$ appears in Eq. (7). The use of larger $P^{\pm}_{se,l}$ in this circuit-model simulation can lead to the overestimation of forward ASE, especially in the first and last EDF segments. An improvement in the overestimation is observed using 5-segment EDF instead of 3-segment one, but the model with 5-segment EDF would be more complex. Therefore, the tradeoff between the calculation accuracy and the circuit model complexity should be cautiously considered. Below 13 mW pump power, an underestimation of forward ASE can be found in Fig. 3(b). It is also believed to come from this approximation of constant $P^{\pm}_{a,l}(z,t)$ in Eq. (6) and the attenuation of ASE along EDF fiber. It is noted that the smaller $P^{\pm}_{se,l}$ appears in Eq. (7). A similar conclusion can be reached for the backward ASE. However, the agreement between the backward ASE of this circuit model and the numerical computation is better owing to considering a forward pumped EDFA.

2.3 ASE spectrum of EDFA

An ASE spectrum can also be simulated through this circuit model. It is considered that there are not any input signal beams, the pump power is 55mW, the span of ASE spectrum is from 1520nm to 1560nm, and the cross sections in Fig. 2 of (Desurvire & Simpson, 1989) are used in this simulation. In Figs. 4(a) and 4(b), the numerical results of forward and backward ASE spectra with 40 ASE slots (1nm wavelength spacing) are shown as the filled circle keys. Using this circuit model with 1, 3, and 5 ASE slots, the simulation results are shown as the dotted, dash, and solid curves respectively. The stepped ASE spectrum with 5 ASE slots is coarsely similar to the numerical calculation. Although the stepped ASE spectra using 1 and 3 ASE slots are unlike with the numerical calculation, each total ASE power of those spectra is very similar. If the ASE is subdivided into more slots in this circuit-model simulation, the ASE spectrum would be more accurate, but the model would also become more complex. However, the rough spectra of circuit-model simulations are still worthwhile in the estimation of EDFA's characteristics. In Fig. 4(a), it can also be obtained that the forward ASE power of this circuit model simulation is slightly higher than that of the numerical computation, as the result in Fig. 3(b).

2.4 Frequency response of EDFA

Using this circuit model with ASE, the frequency response of EDFA is also analyzed. The 10% signal power modulation index is used in these ac analyses. When the input pump power is 40mW, the ac gain and phase responses are shown in Figs. 5(a) and 5(b), as a function of frequency. The numerical computation is in a good agreement with this circuit-model simulation and the ac gain response shows a high-pass characteristic (Freeman & Conradi, 1993; Liu et al., 1995; Novak & Gieske, 2002; Novak & Moesle, 2002). When the modulation frequency is below 100Hz, the maximum ac gain deviation can be near 10dB between the simulation results with and without ASE. The peak phase of simulation

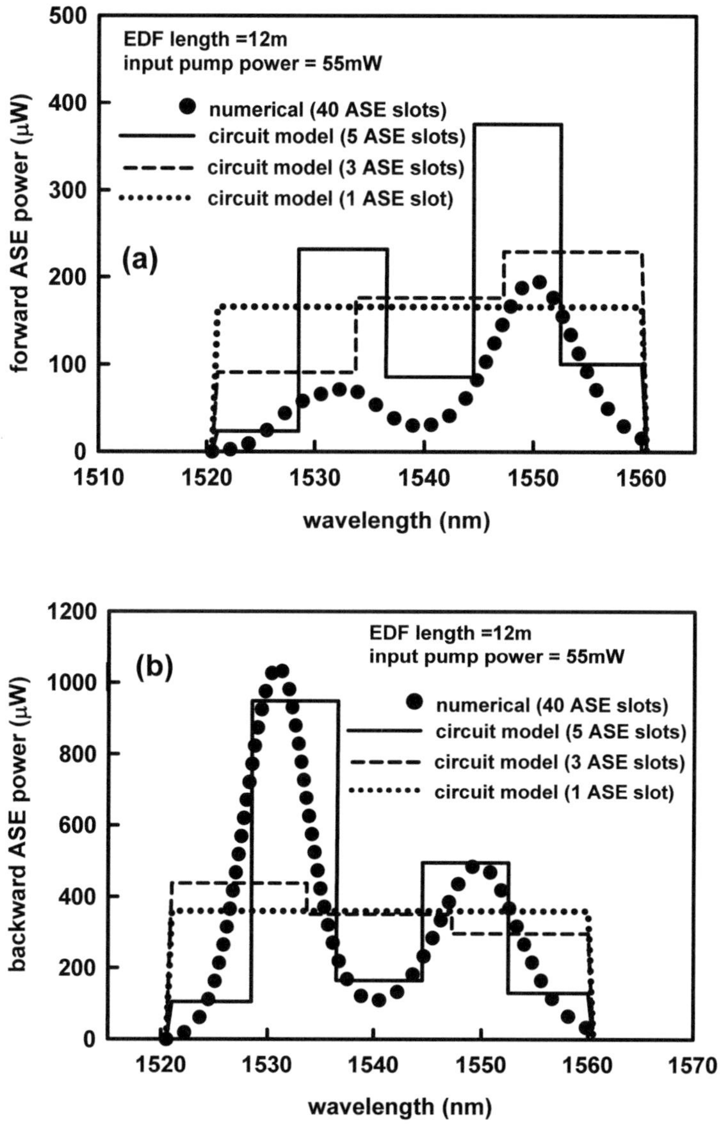

Fig. 4. ASE spectra of EDFA (a) Forward ASE spectrum (b) Backward ASE spectrum

without ASE is about 70 degree at 100Hz. With ASE it is about 30 degree at 230Hz. The deviation of peak phase between the results with and without ASE is quite large. The peak phase position shifts to higher frequency when the simulation includes ASE. The influence of ASE on frequency response of EDFA is significant in low modulation frequency region (below 100Hz). The dc gain of the numerical computation is slightly higher than that of the circuit-model simulation in Fig. 3(a), so the ac gain of the numerical computation is also slightly higher than that of this simulation in low modulation frequency region.

After the numerical computation of transient analysis from Eqs. (1)-(3), the ac gain and phase can be estimated or calculated as functions of frequency. This computing process is more complex. However, using the circuit model, the frequency response of EDFA can be obtained easily and rapidly by the frequency-sweep analysis command in a SPICE simulator.

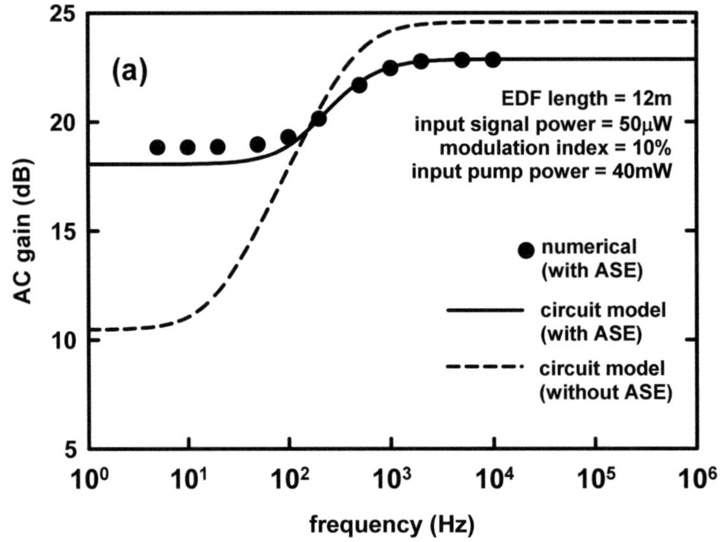

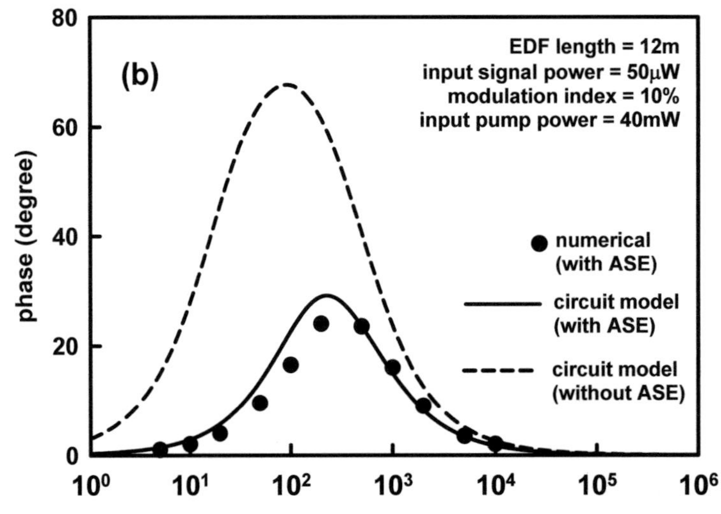

Fig. 5. Frequency responses of EDFA (a) AC gain response (b) Phase response

2.5 Transient analysis of EDFA

Using this circuit model with ASE, the transient response of EDFA dynamics can also be obtained readily through the aid of a SPICE simulator. An eight-channel EDFA system with 35mW pump power at 980nm is considered and seven out of eight channels are added and dropped. These seven channels are represented by a 1558nm signal, while the surviving channel by a 1530nm signal, as shown in Fig. 6(a). (This system can also be regarded as a two-channel system.) $\sigma_s^a = 8.5\times10^{-25}m^2$ and $\sigma_s^e = 8.1\times10^{-25}m^2$ at 1530nm are used in this simulation. Each channel has 10μW power input to the EDFA. In Fig. 6(b), it is shown that the 1558nm output signal has a large power excursion (Giles et al., 1989; Ko et al., 1994; Wu & Lowery, 1998), when the input signals of seven channels are added simultaneously. According to the above-mentioned results, the dc gain is lower and the ac gain deviation between high and low frequency is smaller, because of the ASE influence on EDFA. Therefore, the power excursion of the transient response is lower in this simulation with ASE, as shown by the features in Fig. 6(b). A good agreement between the numerical computation and this circuit-model simulation is obtained.

When the 1558nm signal is added or dropped, the output power of 1530nm surviving channel can decrease or increase due to the effect of cross talk, as shown in Fig. 6(c). Because of the ASE influence on EDFA, the 1530nm output power of the simulation without ASE is higher, and that of the numerical computation is also slightly higher than that of the circuit-model simulation with ASE. A larger difference between the result of the numerical computation and this circuit-model simulation can be observed when the 1558nm signal is

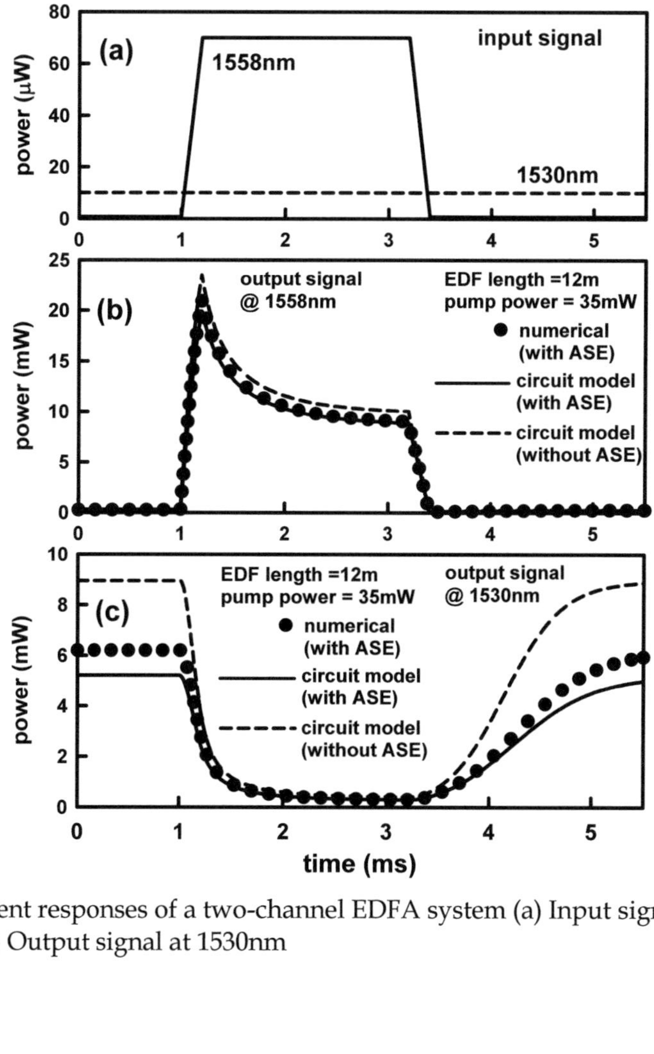

Fig. 6. Transient responses of a two-channel EDFA system (a) Input signals (b) Output signal at 1558nm (c) Output signal at 1530nm

dropped. However, the agreement between this circuit model and the numerical computation is better when the 1558nm signal is added.
To include the forward and backward ASE in a transient analysis, the iteration process in the two-point boundary-value problem gets complex. The numerical simulation requires a large amount of computation time and dynamic data storage, and the convergence problem is not easy to deal with. However, using the circuit model, the transient response of EDFA can be obtained conveniently through the aid of a SPICE simulator. Setting a given tolerance criteria and the number of iterations in a SPICE simulator, the convergence problem can be solved easily and the computer time can also be reduced (Avant!, 2001).

3. Unified circuit model for semiconductor optical amplifiers and laser diodes

There are two primary types of SOAs: traveling wave (TW) SOAs and Fabry-Perot (FP) SOAs. The principle of TW-SOA and FP-SOA is identical, i.e. intrinsic stimulated light amplification. The difference between TW-SOA and FP-SOA is reflectivity of cavity facets. The internal reflectivity of FP-SOA is higher than TW-SOA. Actually, an FP-SOA can be regarded as a FP-LD that is biased below the threshold current. The active layer of an SOA has a positive medium gain but not large enough for laser emission.
Equivalent circuit models have been separately reported for LDs (Lu et al., 1995; Rossi et al., 1998; Tsou & Pulfrey, 1997) and SOAs (Chen et al., 2000; Chu & Ghafouri-Shiraz, 1994; Sharaiha & Guegan, 2000). However, the principles of SOAs and LDs are extremely similar. In this section, a unified equivalent circuit model is presented for SOAs and LDs.

3.1 Circuit model of SOA and LD

Schematic diagrams of a TW-SOA and an FP-SOA are shown in Fig.7 (a) and (b), respectively. TW-SOAs are of a very low internal reflectivity and the incident light is amplified in single pass. FP-SOAs are of a higher reflectivity and incident light can be bounced back and forth within the cavity, resulting in resonance amplification. A basic LD structure is similar to an FP-SOA, but it doesn't need any incident light.
Assume that the nonradiative recombination and carrier leakage rate can be neglected; the wavelength dependence of group velocity (υ_g) can be ignored; any transport time for carriers to reach the active region is not considered. The rate equation for carrier density N, and the continuity equations for signal photon $N^{\pm}_p$ and ASE photon $N^{\pm}_{sp}$ propagating in ±z-direction can be written as (Coldren & Corzine, 1995)

$$\frac{\partial N}{\partial t'}=\frac{\eta_i I}{qV}-R_{sp}(N)-\upsilon_g\left[g_m(N_p^+ + N_p^-)+\langle g_m\rangle\left(N_{sp}^+ + N_{sp}^-\right)\right] \tag{8}$$

$$\frac{1}{\upsilon_g}\frac{\partial N_p^{\pm}}{\partial t'}\pm\frac{\partial N_p^{\pm}}{\partial z'}=\Gamma g_m N_p^{\pm}-\alpha_i N_p^{\pm} \tag{9}$$

$$\frac{1}{\upsilon_g}\frac{\partial N_{sp}^{\pm}}{\partial t'}\pm\frac{\partial N_{sp}^{\pm}}{\partial z'}=\Gamma\langle g_m\rangle N_{sp}^{\pm}-\langle\alpha_i\rangle N_{sp}^{\pm}+\Gamma\frac{\beta_{sp}}{\upsilon_g}R_{sp} \tag{10}$$

where η_i is the internal quantum efficiency, I is the injected current, V is the volume of the active region, Γ is the confinement factor, $R_{sp}(N)$ is the spontaneous emission rate,

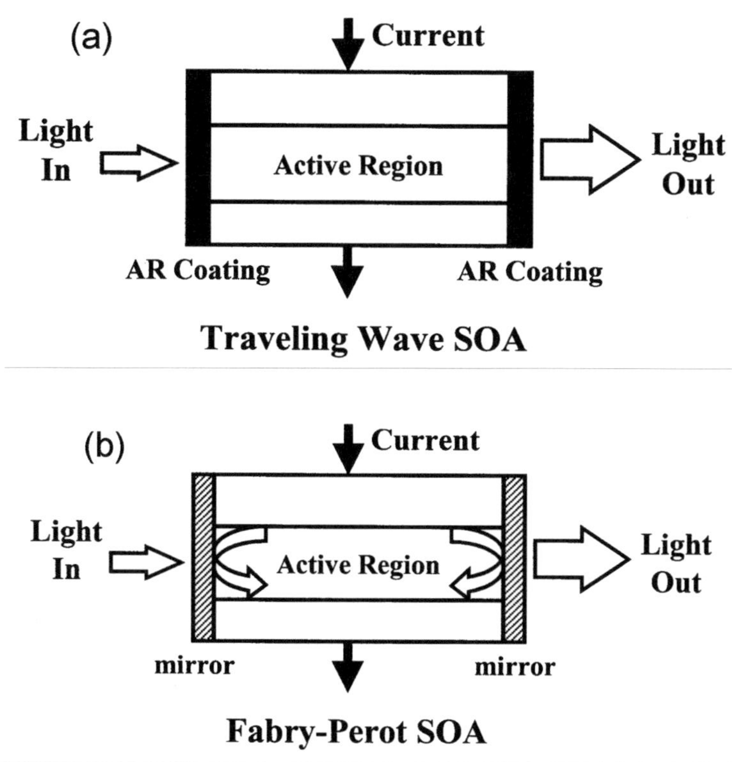

Fig. 7. Schematic diagrams of (a) TW-SOA and (b) FP-SOA

$g_m(\approx a(N-N_0))$ is the material gain, a is the differential gain, N_0 is the transparency carrier density, α_i is the internal loss, and β_{sp} is the spontaneous emission factor. $\langle g_m \rangle$ and $\langle \alpha_i \rangle$ are the spectral average material gain and internal loss over ASE spectrum, respectively. The equations can be simplified by transforming to a retarded-time frame moving with velocity υ_g, $t = t' - z/\upsilon_g$, and $z = z'$. Eqs. (9) and (10) can be written as

$$\pm\frac{\partial N_p^\pm}{\partial z} = \Gamma g_m N_p^\pm - \alpha_i N_p^\pm \tag{11}$$

$$\pm\frac{\partial N_{sp}^\pm}{\partial z} = \Gamma\langle g_m\rangle N_{sp}^\pm - \langle\alpha_i\rangle N_{sp}^\pm + \Gamma\frac{\beta_{sp}}{\upsilon_g}R_{sp} \tag{12}$$

Eqs. (11) and (12) are integrated over z from 0 to the length of active region L. The photon $N_p^\pm$ and $N_{sp}^\pm$ can be transformed into the signal power $P_s^\pm$ and the ASE average power $P_{sp}^\pm$, respectively. Then, by Eqs. (8), (11), and (12), these equations are shown as

$$\frac{\partial N}{\partial t} \approx \frac{\eta_i I}{qV} - R_{sp} - \frac{\Gamma g_m}{h\nu_s A_s g_s}\left(\frac{\partial P_s^+}{\partial z} - \frac{\partial P_s^-}{\partial z}\right) - \frac{\Gamma\langle g_m\rangle}{h\nu_s A_s\langle g_s\rangle}\left(\frac{\partial P_{sp}^+}{\partial z} - \frac{\partial P_{sp}^-}{\partial z}\right) \tag{13}$$

$$P_s^{\pm}(z=\frac{L\pm L}{2})=P_s^{\pm}(z=\frac{L\mp L}{2})\exp\left(\int_0^L g_s dz\right) \tag{14}$$

$$P_{sp}^{\pm}(z=\frac{L\pm L}{2})=P_{sp}^{\pm}(z=\frac{L\mp L}{2})\exp\left(\int_0^L \langle g_s\rangle dz+\beta_{sp}h\nu_{sp}A_s\int_0^L \frac{R_{sp}}{P_{sp}^{\pm}}dz\right) \tag{15}$$

where g_s = (Γg_m - α_i), h is the Plank's constant, ν_s is the signal frequency, ν_{sp} is the ASE central frequency, A_s is the cross-section area of the active region in z-direction. Assume that the distribution of carrier density N approximates a constant in z-direction and $P_{sp}^{\pm}(z)\approx P_{sp}^{\pm}(z=\frac{L\mp L}{2})\exp(\langle g_s\rangle z)$. Eq. (13) is integrated over z from 0 to L. These equations are shown as

$$\begin{aligned} q\frac{\partial N_T}{\partial t} \approx \eta_i I - qVR_{sp} &- \frac{q\Gamma g_m}{h\nu_s g_s}\left[\left(P_s^{+out}+P_s^{-out}\right)-\left(P_s^{+in}+P_s^{-in}\right)\right] \\ &- \frac{q\Gamma\langle g_m\rangle}{h\nu_{sp}\langle g_s\rangle}\left[\left(P_{sp}^{+out}+P_{sp}^{-out}\right)-\left(P_{sp}^{+in}+P_{sp}^{-in}\right)\right] \end{aligned} \tag{16}$$

$$P_s^{\pm out} \approx P_s^{\pm in}G_s \tag{17}$$

$$P_{sp}^{\pm out} \approx \left(P_{sp}^{\pm in}+\beta_{sp}h\nu_{sp}A_sR_{sp}\frac{\langle G_s\rangle-1}{\langle g_s\rangle\langle G_s\rangle}\right)\langle G_s\rangle=\left(P_{sp}^{\pm in}+P_{sp0}\right)\langle G_s\rangle \tag{18}$$

where N_T is the total carriers in the active region, $G_s=\exp(g_sL)$, $P_{s(sp)}^{\pm in}=P_{s(sp)}^{\pm}(z=\frac{L\mp L}{2})$, and $P_{s(sp)}^{\pm out}=P_{s(sp)}^{\pm}(z=\frac{L\pm L}{2})$.

The principles of FP-SOA, TW-SOA, and LD are identical, but their boundary conditions are different. The boundary conditions can be considered: 1) FP-SOA: at z = 0, $P_s^{+in}=\left(\sqrt{(1-R)P_{signal}^{in}}+\sqrt{RP_s^{-out}}\right)^2$, where P_{signal}^{in} is the incident signal light power, and $P_{sp}^{+in}=RP_{sp}^{-out}$; at z = L, $P_{s(sp)}^{-in}=RP_{s(sp)}^{+out}$, and $P_{signal}^{out}=(1-R)P_s^{+out}$, where P_{signal}^{out} is the amplified signal output power. 2) TW-SOA: the reflectivity of facets R = 0, $P_s^{-in(out)}=0$, $P_s^{+in}=P_{signal}^{in}$, and $P_{signal}^{out}=P_s^{+out}$. 3) Laser: no incident signal, $P_s^{\pm in(out)}=0$, $P_{sp}^{\pm in}=RP_{sp}^{\mp out}$, and $P_{signal}^{\pm out}=(1-R)P_{sp}^{\pm out}$, where $P_{signal}^{\pm out}$ is laser output power in ±z-direction. A unified equivalent circuit model of SOA and LD is developed for the Eqs. (16)-(18), as shown in Fig. 8, where V_{Nt} = qVR_{sp}, $E_{Nt}(V_{Nt})$ = $N_T(R_{sp})$, $P_{s,total}^{in(out)}=g_k\left(P_s^{+in(out)}+P_s^{-in(out)}\right)$, $P_{sp,total}^{in(out)}=\langle g_k\rangle\left(P_{sp}^{+in(out)}+P_{sp}^{-in(out)}\right)$, and $g_k=q\Gamma g_m/(h\nu_s g_s)$. Lossless transmission lines are employed with time delay $\Delta = L/\upsilon_g$ from light beams propagating. Using this circuit model and suitable boundary conditions, the performances of SOA and LD can be analyzed and simulated.

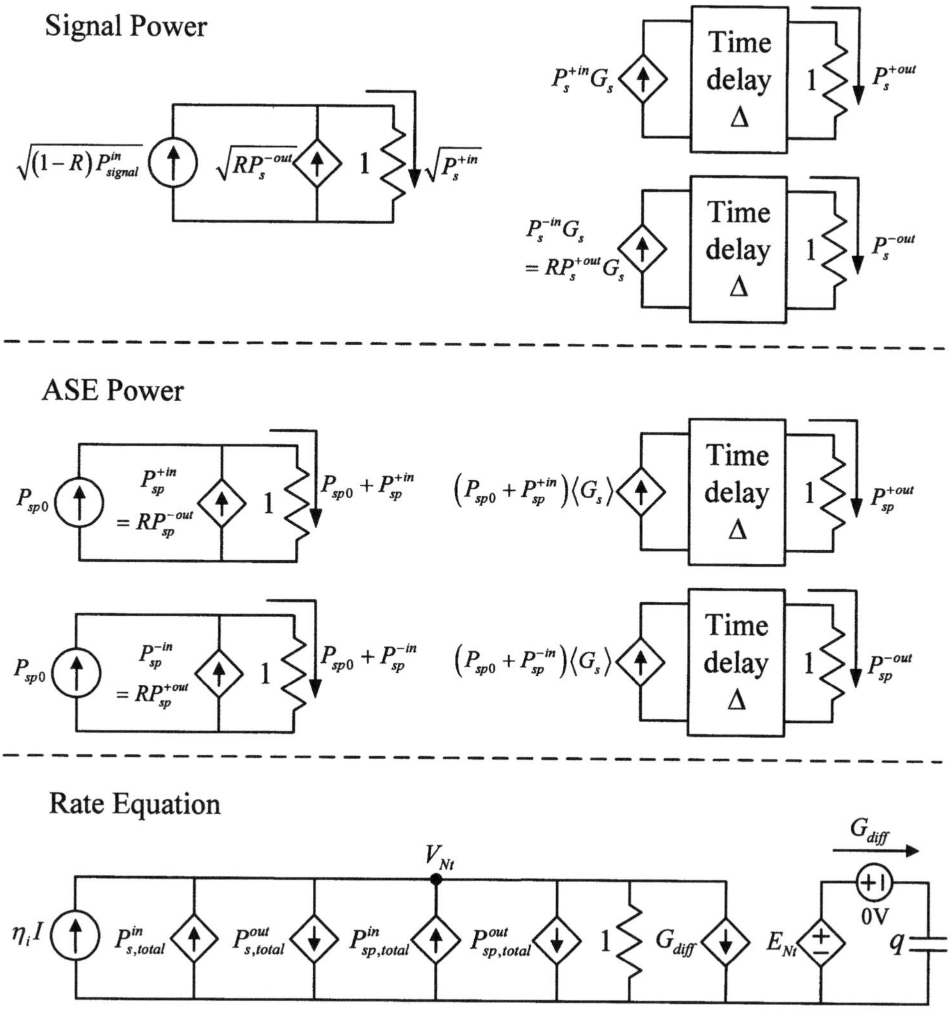

Fig. 8. A unified equivalent circuit model of SOA and LD

3.2 Model validation and analysis of SOA

To demonstrate the validity of this model, the gain against input signal power is simulated for SOAs. First, it is considered the TW-SOA having $R_{sp} = N/\tau_{sp} = (N_T/V)/\tau_{sp}$, τ_{sp} is the spontaneous carrier lifetime. With τ_{sp} = 4ns, η_i = 1, R = 0, Γ = 0.5, a = 5×10^{-16}cm^2, N_0 = 10^{18}cm^{-3}, L = 200μm, A_s = 0.3μm^2, V = 60μm^3, $h\nu_s$ = 0.8eV, υ_g = 0.75×10^8m/s, α_i = 0cm^{-1}, and β_{sp} = 0 (Adams et al., 1985), the results of the TW-SOA without ASE are shown in Fig. 9. It can be shown as expected that the gain of TW-SOAs becomes higher when the higher current injects or the lower signal light power inputs to the SOA. In Fig. 9, with the 3mA injection current and the -60dBm to -20dBm input signal power, the gain of TW-SOAs is fixed about 5.4dB. The influence of input signal power on gain becomes obvious when the higher current injects. These simulations are shown as the solid curves. The results of Ref. (Adams et al., 1985) are shown as the circle keys. A good agreement between this simulation and Ref. (Adams et al., 1985) is observed.

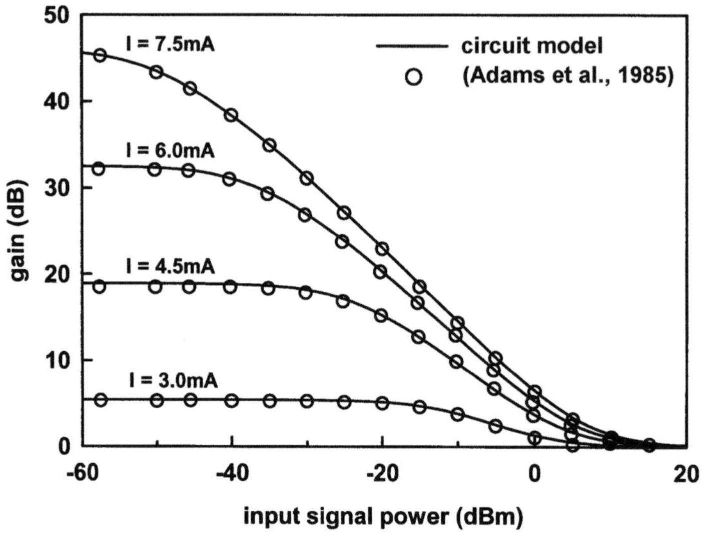

Fig. 9. Gain against input signal power for TW-SOA

Next, it is considered the FP-SOA having $R_{sp} = BN^2 = B(N_T/V)^2$, B is the bimolecular recombination coefficient. The same parameters are used in the first example except R = 0.01, B = 10^{-10}cm^3/s, α_i =25cm^{-1}, β_{sp} = 10^{-4} (Adams et al., 1985), and the threshold current $I_{th} \approx 3.93$mA can be obtained. FP-SOAs must be biased below the threshold current. In Fig. 4, the $0.99I_{th}$, $0.95I_{th}$ and $0.9I_{th}$ injection currents are used, and the results of the FP-SOA without and with ASE are shown in Fig. 10(a) and (b), respectively. Similar to the TW-SOA, the gain of FP-SOAs becomes higher when the injection current more tends to the threshold current or the lower signal light power inputs to the SOA. In Fig. 10(b), it can be observed that the degeneration of the gain of FP-SOAs is influenced by ASE, it becomes more obvious when the injection current tends to the threshold current. These simulations (solid curves) are in good agreement with Ref. (Adams et al., 1985) (circle keys). The agreement can indicate the validity of this model for TW-SOA and FP-SOA.

In this simplified model shown above, the non-uniformity of carrier density is neglected. In fact, the carrier density is non-uniform along the SOA active region. In this simplified SOA model, the rate equation for spatially averaged values of carrier density is used, and the simulation results of gain are slightly high in comparison with a real SOA (Giuliani & D'Alessandro, 2000). However, the simplified model can be helpful for the coarse definition of SOA's parameters. More accurate results can be obtained by the method of cascading, i.e., by subdividing the SOA into many longitudinal sections and using a simplified model with uniform carrier density for each section (Giuliani & D'Alessandro, 2000). In this simulation, the SOA can be also divided into many sections, and the circuit model as Fig. 8 can be used for each section, as done in the multi-section circuit model of fiber lasers (Liu et al., 2002). However, the tradeoff between the result accuracy and the model complexity should be considered.

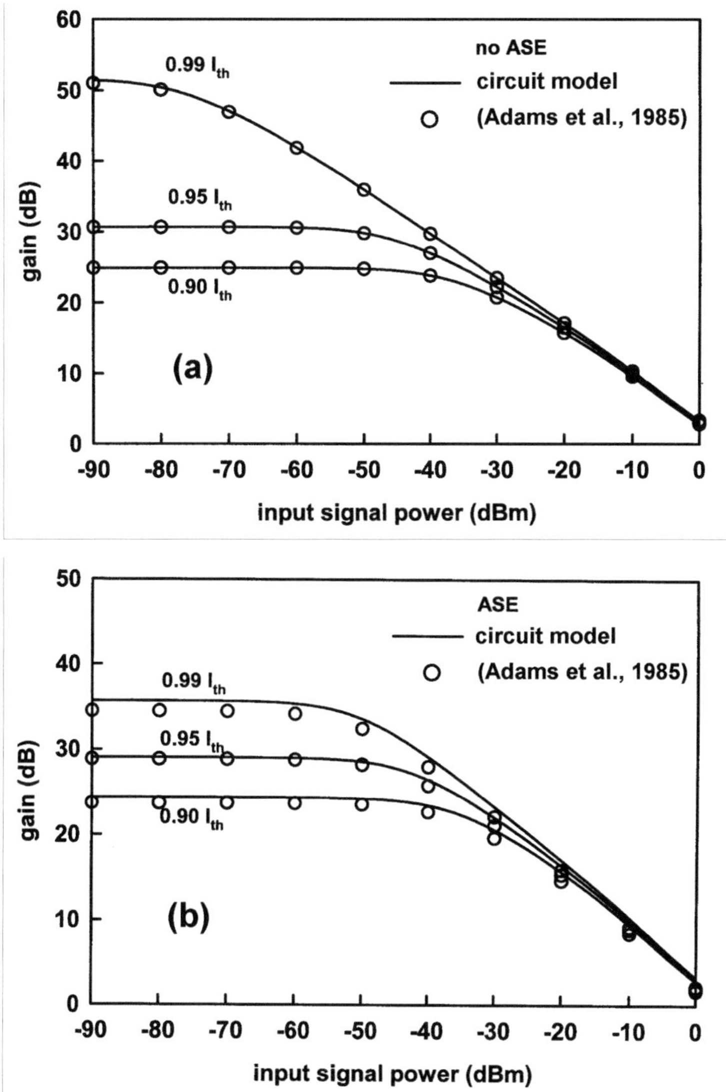

Fig. 10. Gain against input signal power for FP-SOA (a) without ASE and (b) with ASE

3.3 Static and dynamic analysis of laser diode

Then, the characteristics of LD are also analyzed by this model. The parameters used are the same as those of the above example for the FP-SOA. The light output power against injection dc current is simulated for laser diodes having different reflectivity of facets and the results are shown in Fig. 11(a). The higher reflectivity of facets is, and the lower threshold current of the laser is. The threshold current can be expressed as (Liu, 1996),

$$I_{th} = \frac{1}{\eta_i} qVR_{sp}(N_{th}) \tag{19}$$

where $N_{th} = N_0 + \alpha_{tot}/(\Gamma a)$ and $\alpha_{tot} = \alpha_i - (\ln R)/L$. The threshold current versus reflectivity of facets is shown in Fig. 11(b). The circle keys and the solid curve represent, respectively, these simulations and the results of Eq. (19), and the results of the both methods are agreeable.

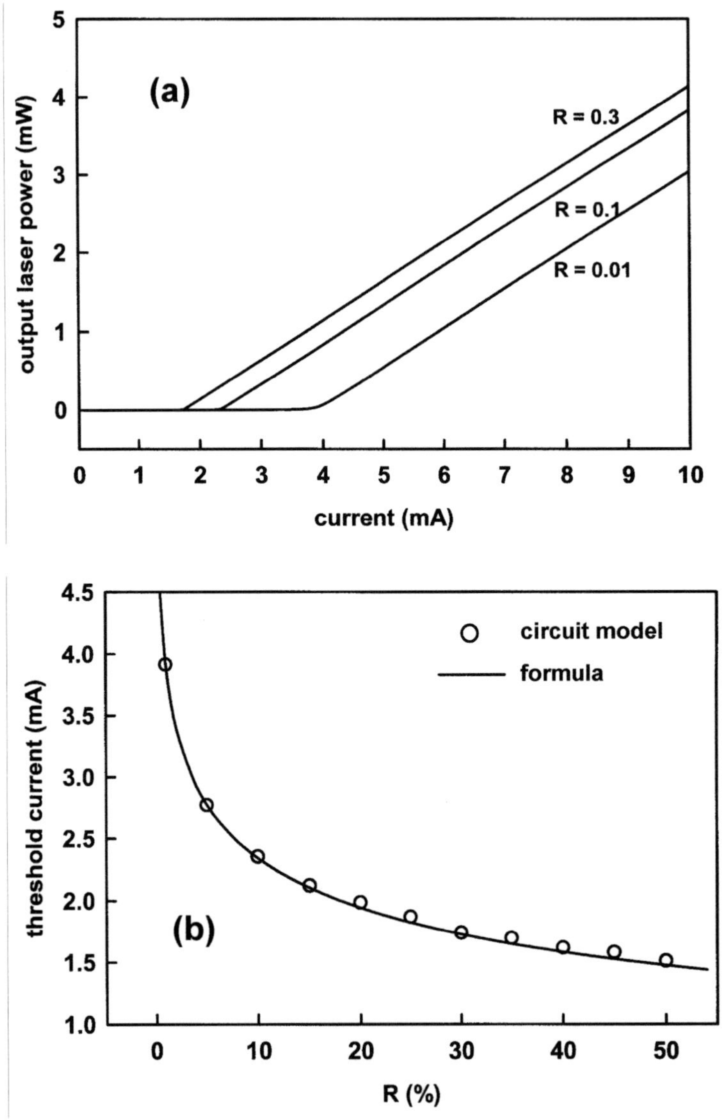

Fig. 11. (a) L-I curve of LDs and (b) the threshold current vs. reflectivity of facets

In Fig. 12 (a), the frequency responses of the laser diode are shown using this circuit model. The 10% current modulation index is used in these ac analyses. The responses have different peak values at different bias currents. When the bias current becomes higher, the peak frequency becomes also higher. The peak frequency can be written as (Liu, 1996),

$$f_p = \frac{1}{2\pi}\left[\upsilon_g \alpha_{tot} \frac{R_{sp}(N_{th})}{N_{th}} \frac{I/I_{th} - 1}{1 - N_0/N_{th}}\right]^{0.5} \tag{20}$$

The peak frequency as a function of bias current is shown in Fig. 12(b). The circle keys exhibit these simulations and the results of Eq. (20) are represented by the solid curve. These simulations are in good agreement with the results of this formula.

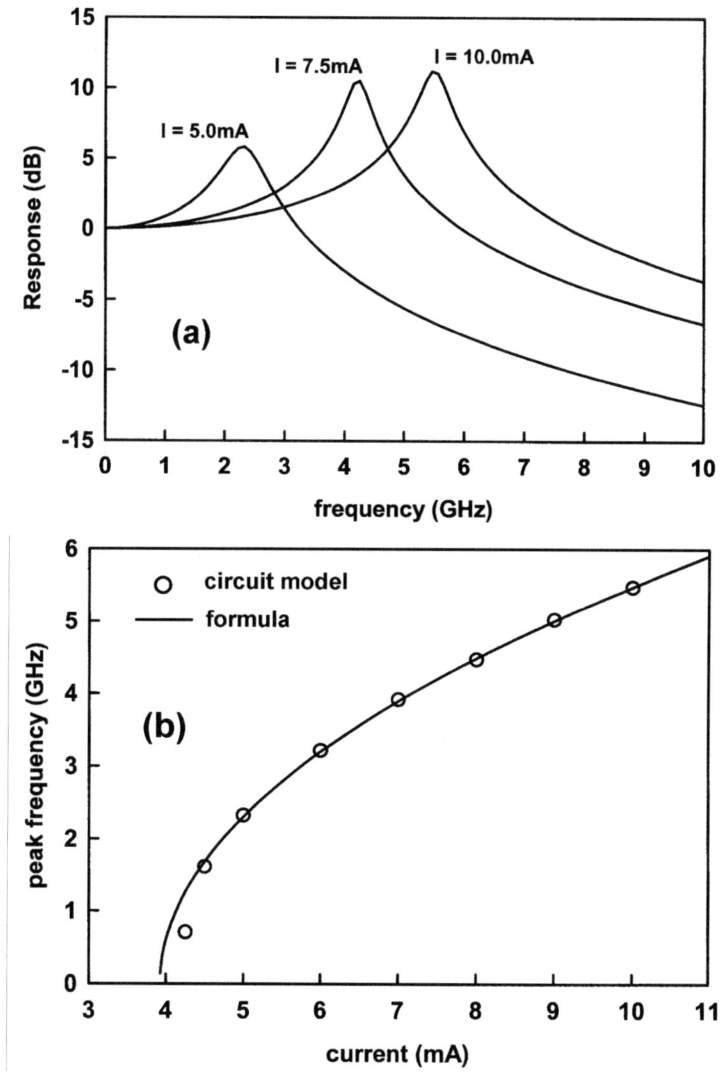

Fig. 12. (a) Frequency responses of LDs and (b) the peak frequency vs. bias current

The transient responses of laser diodes are analyzed during the switching of current. The feature of relaxation oscillation and turn-on delay can be observed in this simulation, as shown in Fig. 13(a). When the turn-on current is higher, the relaxation frequency becomes higher and turn-on delay time is shorter. The turn-on delay time and the relaxation frequency can be formularized as (Liu, 1996),

$$t_d = \frac{N_{th}}{R_{sp}(N_{th})} \frac{\eta_i I/(qV)}{\eta_i I/(qV) - R_{sp}(N_{th})} \tag{21}$$

$$f_r = \frac{1}{2\pi}\sqrt{(2\pi f_p)^2 - \alpha^2} \tag{22}$$

where $\alpha = \frac{1}{2}\frac{dR_{sp}}{dN}\bigg|_{N=N_{th}} + \frac{1}{2}\upsilon_g \alpha_{tot}(2\pi f_p)^2$. As shown in Fig. 13(b), the circle keys and the square keys represent these simulations for the turn-on delay time and the relaxation frequency, respectively. The solid curves are the results of Eqs. (21) and (22). These simulations are in agreement with the results of the formulas.

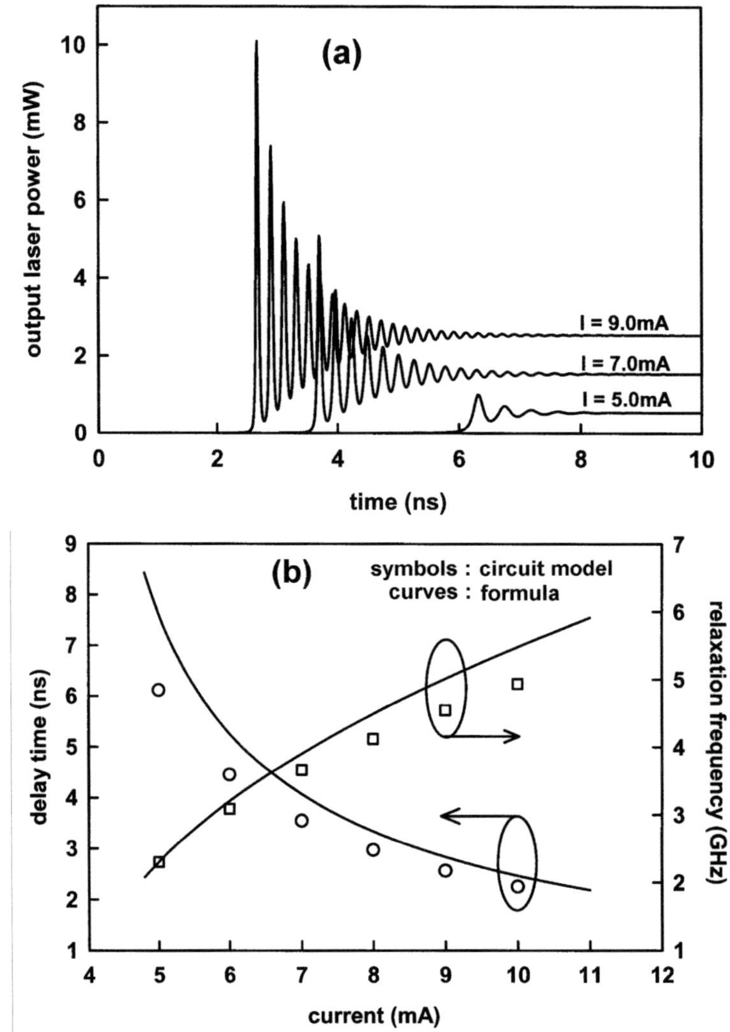

Fig. 13. (a) Transient responses of LDs, and (b) the turn-on delay time and the relaxation frequency vs. bias current

4. Conclusions

An equivalent circuit model of EDFA including ASE is developed. A good agreement between the measured data and this simulation is obtained in the static gain analysis of EDFA. A rough ASE spectrum of EDFA can also be simulated. Furthermore, these simulation results are also in agreement with the numerical computations in the dynamic gain analysis of EDFA. The transient response analyses of a two-channel EDFA system using this circuit model have been demonstrated. Moreover, this circuit model can also be extended to simulate the multichannel, backward-pump, or bidirectional-pump EDFA systems. Besides, this approach could also be applied to develop other numerical model or Simulink model of EDFAs including ASE.

A unified equivalent circuit model of semiconductor optical amplifiers and laser diodes is proposed. The model has been verified by analyzing 1) the gain against input signal power in FP-SOAs and TW-SOAs and 2) the L-I curve, small signal response, and pulse response in laser diodes. The simulation results of this model showed a good agreement with the published results.

Through the aid of SPICE circuit simulator, it is convenient to implement these circuit models. These equivalent circuit models of optical amplifiers can be also extended to include other effects or devices such as modulators, fibers, multiplexers, and parasitic components. The circuit models may be of great value for integrated circuit designers requiring an equivalent circuit model for the amplifiers of the optical communication systems in order to simulate accurately the mixed photonic/electronic modules.

5. References

Adams, M. J.; Collins, J. V. & Henning, I. D. (1985). Analysis of semiconductor laser optical amplifier. *IEE Proc. J. Optoelectron.*, Vol. 132, No. 1, Feb. 1985, pp. 58-63, ISSN 0267-3932

Araci, I. E. & Kahraman, G. (2003). Performance failure analysis of EDFA cascades in optical DWDM packet-switched networks. *J. Lightwave Technol.*, Vol. 21, No. 5, May 2003, pp. 1156-1163, ISSN 0733-8724

Avant! (2001). *Star-Hspice Manual Release 2001.2*, Avant! Corporation

Barnard, C.; Myslinski, P.; Chrostowski, J. & Kavehrad, M. (1994). Analytical model for rare-earth-doped fiber amplifiers and lasers. *IEEE J. Quantum Electron.*, Vol. 30, No. 8, Aug. 1994, pp. 1817-1830, ISSN 0018-9197

Bononi, A.; Rusch, L. A. & Tancevski, L. (1997). Simple dynamic model of fibre amplifiers and equivalent electrical circuit. *Electron. Lett.*, Vol. 33, No. 22, Oct. 1997, pp.1887-1888, ISSN 0013-5194

Burgmeier, J.; Cords, A.; Marz, R.; Schaffer, C. & Stummer, B. (1998). A block box model of EDFAs operating in WDM systems. *J. Lightwave Technol.*, Vol. 16, No. 7, Jul. 1998, pp. 1271-1275, ISSN 0733-8724

Chen, W.; Wang, A.; Zhang, Y., Liu; C. & Liu, S. (2000). Circuit model for traveling wave semiconductor laser. *Solid-State Electron.*, Vol. 44, No. 6, Jun. 2000, pp. 1009-1012, ISSN 0038-1101

Chu, C. Y. J. & Ghafouri-Shiraz, H. (1994). Equivalent circuit theory of spontaneous emission power in semiconductor laser optical amplifiers. *J. Lightwave Technol.*, Vol. 12, No. 5, May 1994, pp. 760-767, ISSN 0733-8724

Coldren, L. A. & Corzine, S. W. (1995). *Diode Lasers and Photonic Integrated Circuit*, John Wiley & Sons, ISBN 978-0471118756, New York

Danielsen, S. L.; Hansen, P. B. & Stubkjaer, K. E. (1998). Wavelength Conversion in Optical Packet Switching. *J. Lightwave Technol.*, Vol. 16, No. 12, Dec. 1998, pp. 2095-2108, ISSN 0733-8724

Desai, N. R.; Hoang, K. V. & Sonek, G. J. (1993). Applications of PSPICE simulation software to the study of optoelectronic integrated circuits and devices. *IEEE Trans. Educ.*, Vol. 36, NO. 4, Nov. 1993, pp. 357-362, ISSN 0018-9359

Desurvire, E. & Simpson, J. R. (1989). Amplification of spontaneous emission in erbium-doped single-mode fibers. *J. Lightwave Technol.*, Vol. 7, No. 5, May 1989, pp. 835-845, ISSN 0733-8724

Durhuus, T.; Mikkelsen, B.; Joergensen, C.; Danielsen, S. L. & Stubkjaer, K. E. (1996). All-optical wavelength conversion by semiconductor optical amplifiers. *J. Lightwave Technol.*, Vol. 14, No. 6, Jun. 1996, pp. 942-954, ISSN 0733-8724

Freeman, J. & Conradi, J. (1993). Gain modulation response of erbium-doped fiber amplifiers. *IEEE Photon. Technol. Lett.*, Vol. 5, No. 2, Feb. 1993, pp. 224-226, ISSN 1041-1135

Giles, C. R.; Desurvire, E. & Simpson, J. R. (1989). Transient gain and cross talk in erbium-doped fiber amplifiers. *Opt. Lett.*, Vol. 14, No. 16, Aug. 1989, pp. 880-882, ISSN 0146-9592

Giuliani, G. & D'Alessandro, D. (2000). Noise analysis of conventional and gain-clamped semiconductor optical amplifiers. *J. Lightwave Technol.*, Vol. 18, No. 9, Sep. 2000, pp. 1256-1263, ISSN 0733-8724

Jou, J.-J.; Lai, F.-S.; Chen, B.-H. & Liu, C.-K. (2000). On-line extraction of parameters in erbium-doped fiber amplifiers. *J. Chinese Ins. Eng.*, Vol. 23, No. 5, Sep. 2000, pp. 615-623, ISSN 0253-3839

Jou, J.-J.; Liu, C.-K.; Hsiao, C.-M.; Lin, H.-H. & Lee, H.-C. (2002). Time-delay circuit model of high-speed p-i-n photodiodes. *IEEE Photon. Technol. Lett.*, Vol. 14, No. 4, Apr. 2002, pp. 525-527, ISSN 1041-1135

Ko, K. Y.; Demokan, M. S. & Tam, H. Y. (1994). Transient analysis of erbium-doped fiber amplifiers. *IEEE Photon. Technol. Lett.*, Vol. 6, No. 12, Dec. 1994, pp. 1436-1438, ISSN 1041-1135

Lai, F.-S.; Jou, J.-J. & Liu, C.-K. (1999). Indicator of amplified spontaneous emission in erbium doped fiber amplifiers. *Elecrton. Lett.*, Vol. 35, No. 7, Apr. 1999, pp. 587-588, ISSN 0013-5194

Liu, C.-K.; Jou, J.-J. & Lai, F.-S. (1995). Second-order harmonic distortion and optimal fiber length in erbium-doped fiber amplifiers. *IEEE Photon. Technol. Lett.*, Vol. 7, No. 12, Dec. 1995, pp. 1412-1414, ISSN 1041-1135

Liu, C.-K.; Jou, J.-J.; Liaw, S.-K. & Lee, H.-C. (2002). Computer-aided analysis of transients in fiber lasers and gain-clamped fiber amplifiers in ring and line configurations through a circuit simulator. *Opt. Commun.*, Vol. 209, No. 4-6, Aug. 2002, pp. 427-436, ISSN 0030-4018

Liu, M. M. K. (1996). *Principles and Applications of Optical Communications.* Richard D. Irwin, ISBN 978-0256164152, Chicago

Lu, M. F.; Deng, J.-S.; Juang, C.; Jou, M. J. & Lee, B. J. (1995). Equivalent circuit model of quantum-well lasers. *IEEE J. Quantum Electron.*, Vol. 31, No. 8, Aug. 1995, pp. 1418-1422, ISSN 0018-9197

Mortazy, E. & Moravvej-Farshi, M. K. (2005). A new model for optical communication systems. *Opt. Fiber Technol.*, Vol. 11, No. 1, Jan. 2005, pp. 69-80, ISSN 1068-5200

Murakami, M.; Imai, T. & Aoyama, M. (1996). A remote supervisory system based on subcarrier overmodulation for submarine optical amplifier systems. *J. Lightwave Technol.*, Vol. 14, No. 5, May 1996, pp. 671-677, ISSN 0733-8724

Novak, S. & Gieske, R. (2002). Simulink model for EDFA dynamics applied to gain modulation. *J. Lightwave Technol.*, Vol. 20, No. 6, Jun. 2002, pp. 986-992, ISSN 0733-8724

Novak, S. & Moesle, A. (2002). Analytic model for gain modulation in EDFAs. *J. Lightwave Technol.*, Vol. 20, No. 6, Jun. 2002, pp. 975-985, ISSN 0733-8724

O'Mahony, M. J. (1988). Semiconductor laser optical amplifiers for use in future fiber systems. *J. Lightwave Technol.*, Vol. 6, No. 4, Apr. 1988, pp. 1556-1562, ISSN 0733-8724

Pederson, B.; Dybdal, K.; Hansen, C. D.; Bjarklev, A.; Povlsen, J. H.; Vendeltorp-Pommer, H. & Larsen, C. C. (1990). Detailed theoretical and experimental investigation of high-gain erbium-doped. *IEEE Photon. Technol. Lett.*, Vol. 2, No.12, Dec. 1990, pp. 863-865, ISSN 1041-1135

Rossi, G.; Paoletti, R. & Meliga, M. (1998). SPICE simulation for analysis and design of fast 1.55μm MQW laser diodes. *J. Lightwave Technol.*, Vol. 16, No. 8, Aug. 1998, pp. 1509-1516, ISSN 0733-8724

Settembre, M.; Matera, F.; Hagele, V.; Gabitov, I.; Mattheus, A. W. & Turitsyn, S. K. (1997). Cascaded optical communication systems with in-line semiconductor optical amplifiers. *J. Lightwave Technol.*, Vol. 15, No. 6, Jun. 1997, pp. 962-967, ISSN 0733-8724

Sharaiha, A. & Guegan, M. (2000). Equivalent circuit model for multi-electrode semiconductor optical amplifiers and analysis of inline photodetection in bidirectional transmissions. *J. Lightwave Technol.*, Vol. 18, No. 5, May 2000, pp. 700-707, ISSN 0733-8724

Shimizu, K.; Mizuochi, T. & Kitayama, T. (1993). Supervisory signal transmission experiments over 10000 km by modulated ASE of EDFAs. *Electron. Lett.*, Vol. 29, No. 12, Jun. 1993, pp. 1081-1083, ISSN 0013-5194

Simon, J. C. (1987). GaInAsP Semiconductor laser amplifier for single-mode optical fiber communications., *J. Lightwave Technol.*, Vol. 5, No. 9, Sep. 1987, pp. 1286-1295, ISSN 0733-8724

Sun, Y.; Luo, G.; Zyskind, J. L.; Saleh, A. A. M.; Srivastave, A. K. & Sulhoff, J. W. (1996). Model for gain dynamics in erbium-doped fibre amplifiers. *Electron. Lett.*, Vol. 32, No. 16, Aug. 1996, pp. 1490-1491, ISSN 0013-5194

Tsou, B. P. C. & Pulfrey, D. L. (1997). A versatile SPICE model for quantum-well lasers. *IEEE J. Quantum Electron.*, Vol. 33, No. 2, Feb. 1997, pp. 246-254, ISSN 0018-9197

Wu, A. W. T. & Lowery, A. J. (1998). Efficient multiwavelength dynamic model for erbium-doped fiber amplifier. *IEEE J. Quantum Electron.*, Vol. 34, No. 8, Aug. 1998, pp. 1325-1331, ISSN 0018-9197

Yu, Q. & Fan, C. (1999). Simple dynamic model of all-optical gain-clamped erbium-doped fiber amplifiers. *J. Lightwave Technol.*, Vol. 17, No. 7, Jul. 1999, pp. 1166-1171, ISSN 0733-8724

Part 5

Other Amplifier Mechanisms

16

Dual-Wavelength Pumped Dispersion-Compensating Fibre Raman Amplifiers

André Brückmann[1], Guido Boyen[1], Paul Urquhart[2], Amaia Legarrea Imízcoz[2], Nuria Miguel Zamora[2], Bruno Bristiel[3] and Juan Mir Pieras[3]

[1]Hochschule Niederrhein
[2]Universidad Pública de Navarra
[3]Télécom Bretagne
[1]Germany
[2]Spain
[3]France

1. Introduction

Fibre Raman amplifiers (FRAs) use optical pumping to provide low-noise gain in fibre waveguides by means of stimulated Raman scattering (SRS). They can be operated over a range of telecommunications windows, from below 1300 nm to beyond 1650 nm, often with broader spectra than those of erbium doped fibre amplifiers (EDFAs). The gain medium can be transmission fibre or dispersion compensating fibre (DCF). DCF-based Raman amplifiers simultaneously boost the propagating signals and compensate for accumulated chromatic dispersion, thereby fulfilling a dual role (Bromage, 2004, Urquhart et al., 2007).

Dispersion compensating Raman amplifiers (DCRAs) normally consist of modules incorporating several kilometres of DCF plus up to around twelve pumps at different wavelengths (Islam, 2004; Namiki et al., 2005), usually launched contra-directionally with respect to the signals, as illustrated in Fig. 1. The Raman gain is often several decibels above the transparency condition of the DCF medium to mitigate the loss of associated passive components. A single pump excites a gain profile with a full width at half height of ~7 GHz but it is far from spectrally uniform, rendering it unsuitable for wavelength division multiplexed (WDM) communications. Gain flattening is thus required and it is normally achieved by the multiple pumps. Complicated optical interactions occur within the fibre, in which power is coupled from the pumps to the signals, from one pump to another and from one signal to another. Additionally, there are the noise processes of amplified spontaneous Raman scattering and amplified distributed Rayleigh backscattering, which can be sufficiently powerful to contribute to the gain saturation. Nevertheless, by carefully optimising the launched powers, the desired spectral equalisation can be achieved.

Multi-wavelength pumped DCF modules have been used to provide gain bandwidths that exceed 100 nm with uniformities of better than 0.3 nm but they are complicated sub-systems (Giltrelli and Santagiustina, 2004; Namiki et al., 2004; Neto et al, 2009). Wavelength-stabilised pump lasers are expensive and the resulting gain spectra are sensitive to the precise values of the launched powers. Sophisticated simulation software with advanced

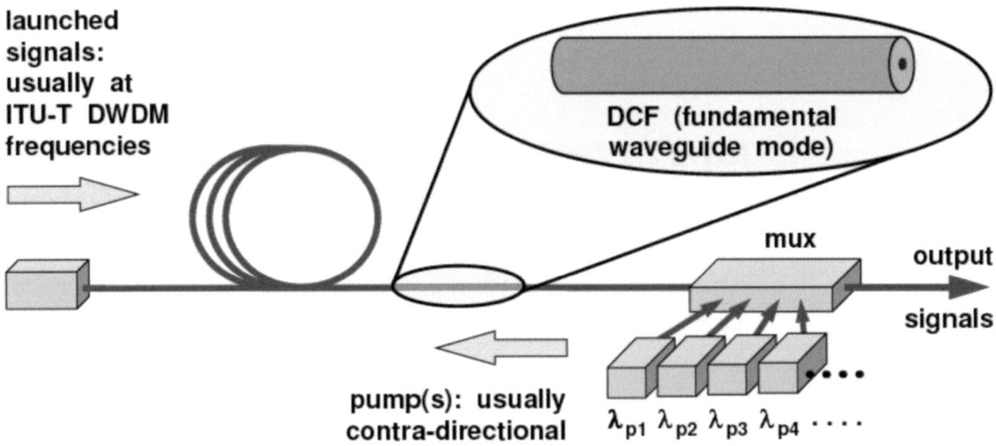

Fig. 1. Dispersion compensating fibre Raman amplifier with contra-directional multi-wavelength pumping. mux = pump-signal wavelength multiplexer.

optimisation algorithms is required to predict the best operating conditions. However, gain uniformity is perturbed by small changes in the power of any of the waves propagating in the fibre. Therefore, the possibility of, for example, the failure of a few channels, the addition of fibre splices elsewhere in the network or electrical power feed fluctuations to the pumps requires that there be continuous monitoring and re-optimisation.

The aim of this chapter is to present simulation results for a simpler and cheaper strategy for gain-equalised DCRAs and to understand its limitations. They are pumped with only two backward-propagating wavelengths (Koch et al., 1999) to obtain very broad spectra and then a customised gain equalising filter (GEF) provides profile uniformity comparable to the multi-wavelength strategy outlined above. Such amplifiers are relatively simple, offering application in cost-constrained networks, such as shorter regional links and in the metropolitan area, where large numbers of WDM channels are being deployed. We describe how they can amplify over 100 channels on the 100 GHz ITU-T dense WDM grid (ITU-T, 2002) with acceptable noise performance and achieve spectral equalisation of under 0.4 dB in typical operation. Moreover, they can tolerate growth in the number of channels, without necessarily having to change filter specifications. We have designed customised thin film transmission filters with spectral profiles specifically for this role and we explain their encouraging operational flexibility.

2. Overview of fibre Raman amplifiers

The SRS, upon which Raman amplification is based, is an inelastic scattering process, in which a pump wave, of frequency ν_p, surrenders energy to the medium through which it passes. The wave causes the medium's molecules to vibrate and any propagating signal at a lower frequency ν_s then receives energy from these excited molecules, producing additional photons at ν_s that are in phase with those of the signal; the result is amplification. An FRA can be provided, as shown in Fig. 1, by launching one or more pump waves into the same fibre as the signal(s). In this way, the signal(s) experience gain during transit in the fibre.

FRAs are "non-resonant"; in contrast to EDFAs, their operation does not depend on electronic energy levels. The non-resonant nature of SRS permits amplification over all spectral regions where the fibre does not exhibit high loss, merely by the provision of one or more pump lasers of suitable wavelength and power. A single optical pump provides a gain profile that is characteristic of the fibre's glass constituents in the form of a spectrum of frequency shifts from ν_p to *lower* frequencies (i.e. longer wavelengths). The peak shift, which is material dependent, is commonly ~13 THz from ν_p. A profile for one reported DCF design is included in Fig. 2, from which a key feature for this chapter is evident: the gain is not at all spectrally uniform (Miyamoto et al., 2002; Namiki et al., 2005). The best pumping efficiencies are achieved by using fibre types with a small effective area (A_{eff}) to maximise the power concentration. This fact favours DCF as a Raman gain medium because in most designs A_{eff} is 15–25 μm², which is about a quarter of the value of many transmission fibres.

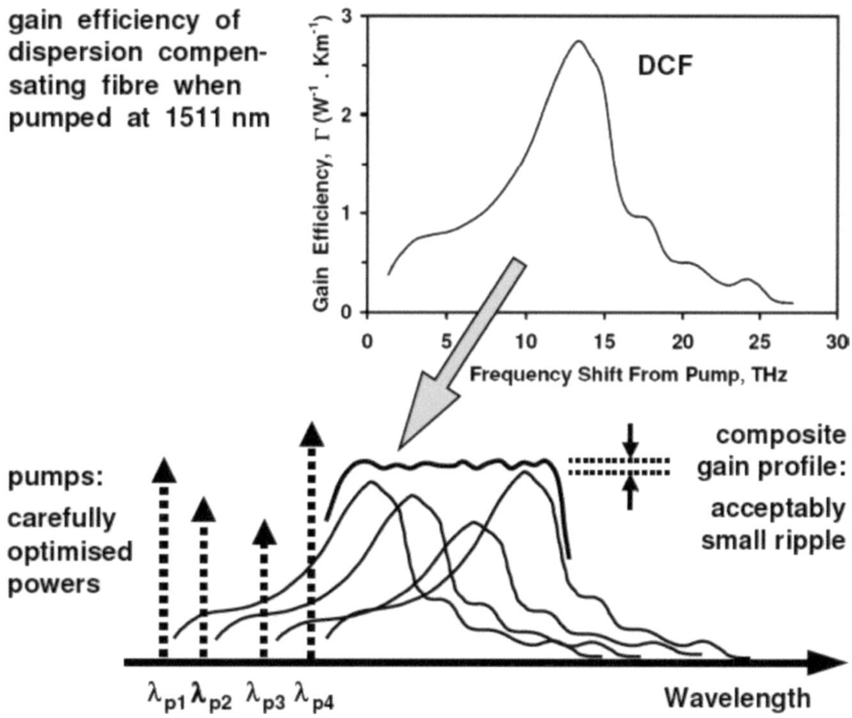

Fig. 2. Multi-wavelength pumping method of gain equalisation of a DCRA. The gain profile on the top right is adapted from Namiki et al., 2005. Other features are schematic.

Figure 2 shows how multi-wavelength pumping can provide spectral gain flattening. Every propagating pump contributes a gain profile and then a wide overall bandwidth of acceptable uniformity is obtained by launching several pumps of suitably optimised powers and wavelengths. However, as stated in Section 1, many interactions contribute to the amplification. Predicting the correct powers with only two pumps is reasonably straightforward using trial and error or by a simple systematic search procedure (as we have done). However, the effort becomes ever greater and the sensitivity to launched powers grows as the number of pumps is increased. Advanced optimisation algorithms are thus used to achieve gain flattening over a wide bandwidth (Cui et al., 2004; Miyamoto et al., 2002; Neto et al., 2009; Zhou et al., 2006).

Noise adversely affects all communications systems and in digital operation it increases the probability of bit errors (Urquhart, 2008). There are three main noise processes in FRAs: amplified spontaneous Raman scattering (often called amplified spontaneous emission, ASE), Rayleigh backscattering (RBS) and relative intensity noise (RIN) transfer. ASE is often the most prominent one and results from "spontaneous" Raman scattering, which occurs in a pumped fibre, irrespective of the presence of signal photons. Spontaneously scattered photons, which are created all along the fibre, encounter further excited (vibrating) molecules, caused by the presence of the pump, and they are amplified. The ASE power grows bi-directionally, sometimes reaching significant magnitudes with respect to the signal. It is broad bandwidth and unpolarised and it is transmitted to the detectors along with the signals, where it reduces the optical signal-to-noise ratio (SNR).

Rayleigh scattering results from microscopic random fluctuations in the glass's refractive index, which exist even in high quality fibres (Bromage et al., 2004; Jiang et al., 2007a). Variations that happen to be $\lambda/4$ for any of the guided waves provoke weak reflections that add in phase, creating a distributed reflector. The pump, signal and bi-directional ASE waves are all reflected but, unlike SRS, the process is "elastic" and so there are *no* frequency shifts. Rayleigh scattered waves are themselves reflected, causing double scattering. As the backscattered waves progress in the fibre they experience amplification, due to the presence of pump photons, becoming reasonably powerful. RBS enhances the ASE power and it causes time-delayed replicas of the signals to be incident on the detectors. In either case, the consequence is a reduction of the SNR.

Normally, the pump lasers are continuous wave but, owing to the oscillatory interactions within their semiconductor active media, they always exhibit random high frequency temporal power fluctuations, known as RIN. Raman gain occurs within a silicate fibre on a sub-picosecond time scale and so it is almost instantaneous. Consequently, when the pumps and signals travel in the same direction in the fibre, the random fluctuations of the pumps are directly transferred to the signals and the effect is amplified within the gain medium. Fortunately, there is a simple means to reduce the problem significantly, which is contra-directional pumping, as shown in Fig. 1. The waves then pass through each other, providing good time averaging during transit. Throughout this chapter, we assume the use of such pump schemes and so RIN transfer is ignored in our analysis.

3. Theory

The optical fibre gain medium, such as in Fig. 1, has a length coordinate z, which ranges from 0, at the signal input, to L, at the pump launch point. It is specified by a Raman material gain coefficient, g (W.m^{-1}), an effective area, A_{eff} (m^2), a loss coefficient, α (m^{-1}) and a Rayleigh scattering coefficient γ (m^{-1}). All of these parameters depend on the fibre's glass composition and waveguide design and they vary with wavelength (Jiang et al., 2007b). When modelling FRAs it is convenient to define a gain efficiency, Γ (W$^{-1}\cdot$km^{-1}) for any two interacting frequencies ν_i and ν_j (corresponding to wavelengths λ_i and λ_j):

$$\Gamma(\nu_i, \nu_j) = g_{i,j} / K A_{eff} \; ; \; \nu_i < \nu_j \tag{1}$$

The constant K accounts for the polarisation states of the two interacting waves and in most circumstances, where there is good randomisation, K = 2. A plot of Γ, which applies when λ_p is 1511 nm (Namiki et al., 2005), is included in Fig. 2. The values presented can be scaled for

another *pump* wavelength, $\lambda_{p\text{-}new}$ in nm, using $\Gamma(\lambda_{p\text{-}new}) = (\lambda_{p\text{-}new}/1511)\,\Gamma$. Plots of $\alpha(\lambda)$ and $\gamma(\lambda)$ for the same DCF type are also presented by Namiki and Emori, 2005.
The signal and pump waves enter the amplifier at z = 0 and z = L, respectively and broad band bi-directional ASE grows throughout the fibre. Raman interactions occur between any pair of guided waves (be they pump, signal or ASE) regardless of their wavelengths, direction of propagation or whether they are modulated to vary in time. Power transfer is from the higher frequency (shorter wavelength) waves to the lower frequency (longer wavelength) ones. Moreover, all guided waves are subject to both loss and RBS. The amplifier is modelled by establishing sets of bi-directional differential equations to account for all of these influences.
The format that we use in Sections 5 - 8 is stated in Equation (2), the derivation of which is outside the scope of this chapter, but details are in Agrawal, 2005 and Islam et al., 2004.

$$\begin{aligned}\pm\frac{dP_i^{\pm}}{dz} &= +P_i^{\pm}\sum_{j=1}^{i-1}\Gamma(\nu_j,\nu_i)\cdot P_j + p_{0i}\sum_{j=1}^{i-1}\Gamma(\nu_j,\nu_i)\cdot P_j\cdot H(\nu_j-\nu_i)\\ &- P_i^{\pm}\sum_{j=i+1}^{N}(\nu_i/\nu_j)\cdot\Gamma(\nu_j,\nu_i)\cdot P_j - P_i^{\pm}\sum_{j=i+1}^{N}(\nu_i/\nu_j)\cdot\Gamma(\nu_i,\nu_j)\cdot 2p_{0j}\cdot H(\nu_i-\nu_j)\\ &- \alpha(\nu_i)P_i^{\pm} + \gamma(\nu_i)P_i^{\mp}\,,\end{aligned} \tag{2}$$

where $P_{i,j} = [\,P_j^+ + P_j^-\,]$. Equation (2) describes the evolution of all of the propagating waves: the pumps, the signals plus a segmentation of the bidirectional ASE into N bands of width $\Delta\nu$ that are sufficiently narrow to provide realistic spectral resolution but not so numerous as to create excessive computation. The length-independent terms $p_{0i,j}$ are the "spontaneous scattering equivalent input powers", given by $p_{0i,j} = 2h\nu_{i,j}\,\Delta\nu$, where h is Planck's constant and the factor of 2 accounts for the fibre's orthogonal polarisation modes. The factor of 2 within the fourth term on the right of Equation (2) results from the two directions of the ASE. The function H, called the "ASE thermal factor" or "Bose-Einstein factor", quantifies the temperature variation of the spontaneous scattering, (Lewis et al., 1999):

$$H(\nu_j-\nu_i) = 1 + \{\exp[\,h\cdot(\nu_j-\nu_i)/k_B T\,] - 1\}^{-1}, \tag{3}$$

where k_B is the Boltzman constant and T is the temperature (Kelvin).
The amplifier's "net" (or "input-output") gain, G_{net} is specified as the ratio of signal powers at one of the channel wavelengths, $G_{net} = P(\lambda_s, z = L)/P(\lambda_s, z = 0)$, and it is converted to decibels when necessary. Another definition, the "on-off" gain, is important for distributed amplification (Urquhart, 2008) but it is not used here. Equation (2) is a comprehensive model, incorporating many phenomena, at the cost of sacrificing physical insight. With this limitation in mind, we have derived an approximate analytical formula for dual wavelength pumping and it is stated as an appendix in Section 11.
The noise performance of an optical amplifier is specified by its "noise figure" (NF), which should be minimised when other considerations, such as cost, permit. When regarded as a black box, Raman (or other phase-insensitive) amplifiers exert three main influences: they amplify the signal(s), amplify the incident noise and add noise of their own. The NF quantifies the amplifier's deterioration of the optical SNR. It is the ratio of the optical SNR at input and output of the amplifier, as measured within a *narrow* optical bandwidth $\Delta\nu$, and the format that we use is:

$$NF = \frac{1}{G_{net}} \left(1 + \frac{P^{+}_{ASE}}{h\nu_s \, \Delta\nu} + \frac{P^{+}_{RBS}}{h\nu_s \, B} \right) \tag{4}$$

P^{+}_{ASE} is the forward propagating ASE power within $\Delta\nu$, imposed by a filter at the detector that is centred on signal frequency ν_s. P^{+}_{RBS} is the Rayleigh backscattered power incident on the detector within a bandwidth B (in Hz) defined by an *electrical* filter. Care is required to account for the states of polarisation in the RBS contribution; Bromage et al., 2004 explains the need for a factor of 5/9 in the equations. The literature on distributed Raman amplifiers often refers to an "effective noise figure" (Agrawal, 2005) but it is not used here.
Amplifiers are commonly concatenated or used together with loss elements, such as passive spans of fibre and optical components. Their total noise figure can be determined from the "Friis cascade formula" (Desurvire, 1994). For two elements with net gains G_1 and G_2, having noise figures NF_1 and NF_2, respectively,

$$NF_{total} = NF_1 + (NF_2 - 1)/G_1 \ , \tag{5}$$

where all terms are linear (*not* decibel) quantities. A passive loss element can be assigned a "gain" in the range $0 < G_{loss} < 1$ but it produces neither ASE nor significant RBS noise. Thus, by Equation (4), it has a noise figure of $1/G_{loss}$, which is greater than unity. A key property of Equation (5) is that it is non-commutative; the order of its constituents matters. The total noise figure for a loss element that precedes an FRA is $NF_{total} = NF_{FRA}/G_{loss}$. In contrast, if the loss follows the FRA, Equation (5) gives $NF_{total} = NF_{FRA} + (1 - G_{loss})/(G_{loss} \cdot G_{FRA})$. Substitution of some typical values, such as $G_{loss} = ½$, $G_{FRA} = 10$ and $NF_{FRA} = 6$, reveals that the lowest total noise figure is obtained by placing the passive loss *after* the fibre Raman amplifier. A loss element in front of it attenuates the signal plus any input noise. Thereafter, the amplifier provides gain and (importantly) it adds noise in the process. However, if the loss follows the amplifier, the amplified signal plus the added noise are attenuated equally by the loss element. For this reason, the GEF and associated components that we consider are located after the amplifying fibre, further details of which are in Section 4.
Guided waves in optical fibres are subject to nonlinear optical interactions, such as self-phase modulation, cross-phase modulation and four-wave mixing, all of which increase the bit error rate. Normally the effects are not problematical in a few kilometres of fibre but DCFs have small core diameters and high germania-content glasses, both of which enhance the nonlinear optical processes (Boskovic et al., 1996). Moreover, the Raman gain maintains a higher signal power along the fibre, which is also detrimental in this respect. The key effects that cause nonlinear crosstalk in optical systems all depend on a nonlinear phase shift, which is determined by a path-averaged power at each signal wavelength λ_s:

$$\Phi_{NL}(\lambda_s) = \frac{2\pi n_2}{\lambda_s A_{eff}} \int_{z=0}^{L} P_s^{+}(z) \cdot dz \ , \tag{6}$$

where n_2 is the nonlinear refractive index ($m^2.W^{-1}$) of the core of the DCF. A typical value of the ratio n_2/A_{eff} for DCF is 1.55×10^{-9} W^{-1}, which is higher than for standard single mode transmission fibre, mainly due to the difference in effective areas. The value of Φ_{NL} (radians) should be as low as possible and so it provides a means to compare FRAs in different operating regimes. Equation (6) is used in Section 8 in this comparative manner.

4. Gain spectrum equalisation

Our simple method to obtain gain equalisation is illustrated in Fig. 3, in which there are only two pumps plus a customised GEF. The filter is electrically passive, requiring neither power feeds nor control circuits. It is placed after the DCF to minimise the impact on the overall noise figure and lessen the disruption in the event of post-installation upgrades. The optical isolator prevents reflections from the GEF re-entering the amplifier and causing multipath interference noise. The two pumps provide a straightforward means to obtain a broad composite gain profile that has two peaks with a central minimum between them. We refer to the gain at this central minimum as the "baseline". The GEF is then designed to suppress the gain above the baseline by presenting a wavelength-dependent passive loss, as shown in Fig. 3. Thus the filter's transfer function, T(λ) is the inverse of the gain curve above the baseline. The gains are in decibels but T(λ) is linear:

$$G_{\text{baseline}}^{(\text{dB})}(\lambda) = T(\lambda) \cdot G_{\text{DCF}}^{(\text{dB})}(\lambda), \tag{7}$$

The GFFs that we simulate are thin film interference filters, composed of alternating layers of high and low refractive index dielectrics deposited on a transparent substrate (Macleod, 2010). Two favoured vitreous film materials for operation in the S-, C- and L- bands are silica (SiO_2) and tantalum pentoxide (Ta_2O_5), which have refractive indices at 1550 nm of ~1.465 and ~2.065, respectively. We assume a substrate refractive index of 1.55 and that the entry medium is air. Thin film filters are a versatile and mature technology with dependable design and fabrication methodologies that allow excellent quality control and good production yields. Low cost and robust packaging is available with single mode fibre pigtails and low insertion losses. Moreover, thin films are one of the most thermally insensitive filter types, making them ideal for outdoor applications (Takahishi, 1995).

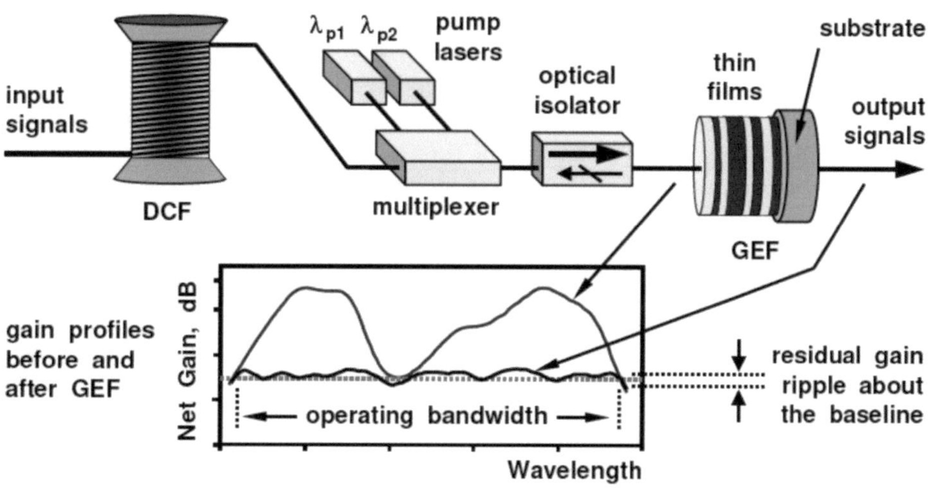

Fig. 3. The use of a passive thin film gain equalisation filter (GEF) to provide spectral flattening of a dual-wavelength pumped dispersion compensating fibre Raman amplifier. The thin film structure is not to scale and is shown unpackaged.

Light incident on the boundary between two films encounters a refractive index discontinuity, causing partial transmission and partial reflection. It also undergoes thickness and refractive index dependent phase changes in transmitting within each film. Filter simulation normally uses a matrix methodology with complex electric fields to account for the infinite number of coherent superpositions that occur during transit through a multi-layer film stack. The end result is a filter transfer function, normally expressed as the wavelength variation of the transmittance, $T(\lambda)$, the *real* function used in Equation (7).

The matrix formulation of thin film optics predicts $T(\lambda)$ for a user-defined film stack but it does not do the opposite and specify the stack structure to create a user-defined $T(\lambda)$. For that we need to supplement the matrix calculations with an optimisation procedure. The approach that we used is called the "needle method" (Tikhonravov et al., 1966; Sullivan and Dobrowolski, 1996; Thelen et al., 2001), which is a numerical technique that inserts a small needle into an empty or a given starting structure. The needle is a layer of a refractive index which is lower than that of the surrounding material. After its insertion, a merit function is calculated and the algorithm changes the position of the needle. (The merit function is commonly a root mean square difference between the target value of $T(\lambda)$ and the value obtained.) The algorithm repeats the needle placement until all positions have been evaluated and it then fixes the needle at the one with the lowest merit value. Many needles must be inserted until an acceptable design is found. Often a needle is placed adjacent to another one, which indicates that the thickness of the existing needle is insufficient and that a thicker layer yields a better merit value. Some implementations combine those needles into one layer; others refine the thicknesses after every insertion.

We performed the algorithm by using needles of different thickness to achieve an almost refined result at the end of each insertion step. This is a simple approach which achieves *desirable* sub-optimal designs. Takashashi, 1995 explains how sub-optimal designs of thin film filters are less sensitive to fabrication errors and temperature fluctuations. Depending on the accuracy chosen for incrementing the thickness and position, there are k calculations for the thickness and for each thickness there are m calculations for every position. We therefore had to compute $k \cdot m$ merit values for each needle, which can be very time consuming. However, because all of these calculations were based on the same TFF structure, we could subdivide them into small packets of parameters that were distributed to many computers, allowing us to reduce the time to obtain a result and/or to use a smaller step size to find a more optimised TFF. Nevertheless, even with a network of 20 standard desktop PCs, each synthesis could take several days of run time.

It is instructive to contrast the computer processing requirements of the two gain equalisation strategies: multiple pumps (Fig. 1) or dual pumps plus a static GEF (Fig. 3). Multi-wavelength pumping imposes more demanding pump power optimisation, which can be in real time to adapt to revised network requirements. In the GEF strategy the heavy computation is performed during the filter's development but it is not in real time. After GEF installation, any adjustment of the amplifier's pumps is relatively straightforward because there are only two to control. (Our approximate analytical model in Section 11 provides an insight into this aspect.) In effect, the computational burden is thus shifted from amplifier operation to the GEF production. A comparison of the two strategies is outside the scope of this chapter. However, a key consideration is the flexibility to adapt to changing amplifier operating requirements: does a sub-optimal filter synthesis leave sufficient latitude for network upgrading? This theme is addressed Sections 6 and 7.

5. Small signal and saturation regimes with a GEF

We start by simulating the net gain form 10 km of DCF and the results are plotted in Fig. 4. DCF spans are determined primarily by the transmission system's chromatic dispersion (Grüner-Nielsen et al., 2006). Our choice of 10 km is slightly longer than usual to provide pessimistic values of unpumped losses and a more challenging gain profile to flatten. The (contra-directional) pumps were at λ_{p1} = 1435 nm and λ_{p2} = 1485 nm to provide amplification across all of the communications C-band plus parts of the S- and L- bands.

In order to obtain small signal behaviour, we launched 151 channels separated by 100 GHz (over an extension of the ITU-T DWDM grid) with powers of -35 dBm per channel. Our total launched signal power was therefore 47.8 μW and the resulting signal wavelength range was between 1490 and 1611 nm. The 115 exiting channels from 1514 to 1606 nm are plotted as Curve a on Fig. 4. The pump powers, specified in Table 1, were chosen to ensure two peaks of equal magnitude and a minimum between them that is as close as possible to 3 dB, which was our selected baseline gain. The difference between the baseline and the gain peaks is the "excursion" and in Curve a it is 2.93 dB, which applies at ~1534 nm and ~1590 nm. We chose a baseline of 3 dB as it would be useful to overcome additional losses of associated components, such as the multiplexer and isolator shown in Fig. 3. Alternatively, suitably selected pump powers enable other values, as addressed in Section 7. The analytical model that we report in Section 11 gives further insight into the small signal limit.

We turn to saturated operation. In Curve b in Fig. 4 we used the same pump powers as in the small signal simulation but we launched 115 signals with the higher powers of -3 dBm/channel, causing two main effects: pump depletion and signal-to-signal Raman

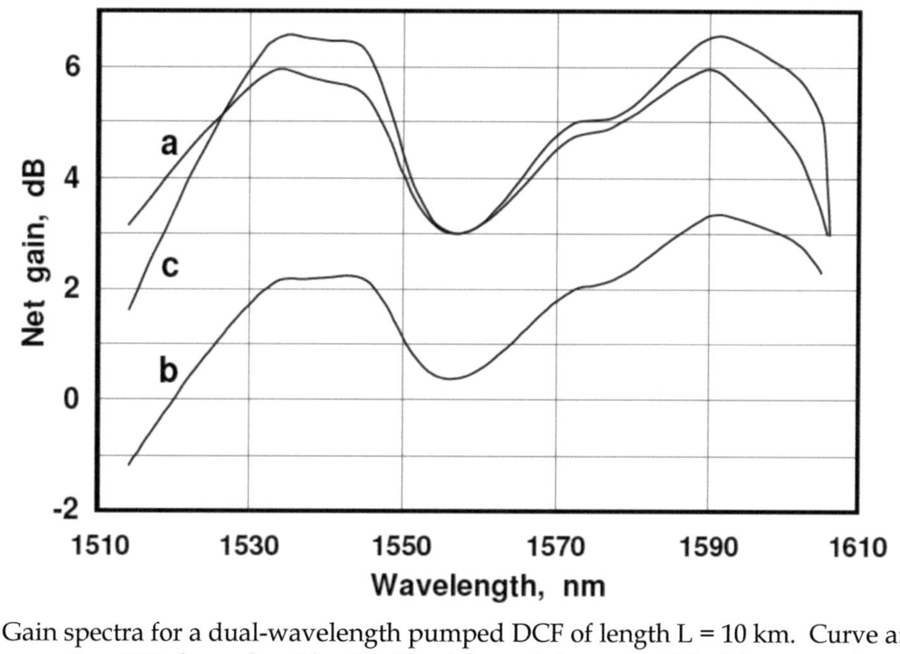

Fig. 4. Gain spectra for a dual-wavelength pumped DCF of length L = 10 km. Curve a: small signal operation, 151 channels with – 35 dBm/channel (but only 115 of them are plotted). Curves b and c: 115 channels spaced 100 GHz with –3 dBm/channel. The pump powers are stated in Table 1.

Operating mode	Channel power, dBm	Power λ_{1435}, mW	Power λ_{1485}, mW
Small signal	-35	173.3	89.1
Saturating signal, no pump readjustment	-3	173.3	89.1
Saturating signal with pump readjustment	-3	373.2	129.3

Table 1. Pump powers for different operation modes for Figs. 4 and 5.

interactions. Consequently, the gains are lower and there is an overall tilt, the most obvious manifestation of which is that the long wavelength peak is higher than the short wavelength peak. Curve c of Fig. 4 also uses 115 saturating signals of -3 dBm/channel but the pump powers have been readjusted to ensure two equal peaks and a central minimum of 3 dB. As stated in Table 1, higher pump powers were required to achieve this condition. In Curve c the operating bandwidth is about 88 nm and the gain excursion is now 3.56 dB.
The amplifier's noise figures are plotted in Fig. 5, where the curves correspond directly to those of Fig. 4. Small signal operation provides the lowest NF, followed by the saturating signals with unadjusted pump powers, followed by the saturated signals with readjusted pump powers. All curves show higher values at shorter wavelengths, which results mainly from the thermal factor that influences the ASE spectrum. In general, the NF depends on the signal and pump powers plus the powers of any other propagating waves, such as the ASE and Rayleigh scattering, within the fibre gain profiles. In common with EDFAs and other amplifier types, lower signal and ASE powers cause light or negligible saturation and lower noise figures. There are two peaks on Curve c associated with the two gain peaks on Fig. 4. By re-performing the calculation with $\gamma(\lambda) = 0$ to eliminate all distributed reflections, we have found that they are caused largely by Rayleigh backscattering of the signals and ASE.

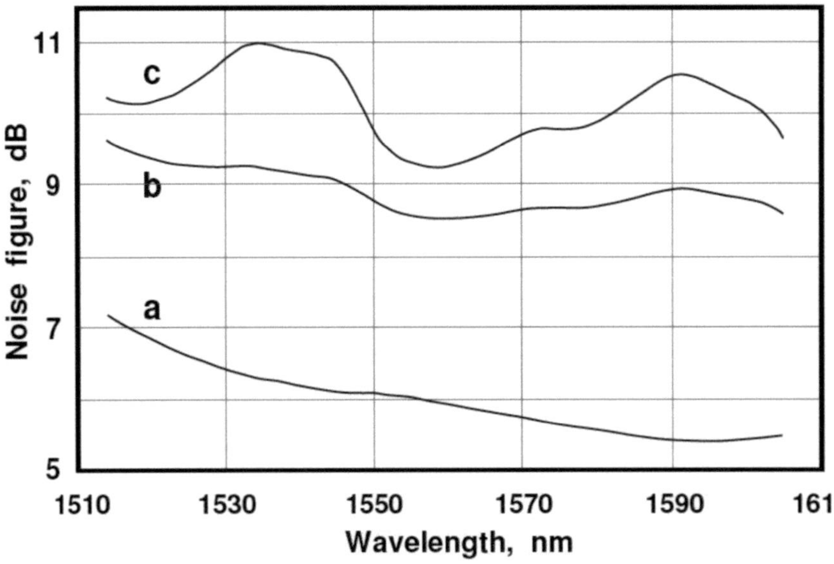

Fig. 5. Noise figure spectra for a dual-wavelength pumped DCF of length L = 10 km. Curve a: small signal operation, 151 channels with –35 dBm/channel (but only 115 of them are plotted). Curves b and c: 115 channels spaced 100 GHz with –3 dBm/channel. The pump powers are stated in Table 1.

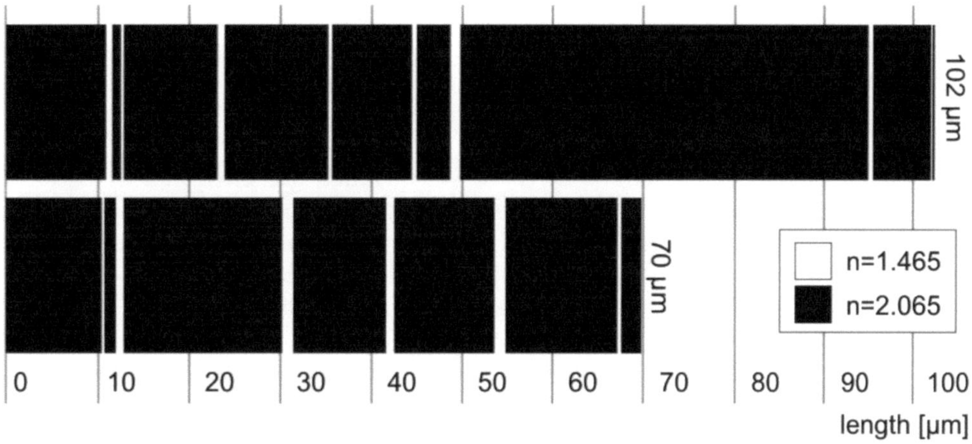

Fig. 6. Thin film structures formed from layers of SiO_2 (refractive index 1.465) in Ta_2O_5 (refractive index 2.065). Top: fifteen-layer filter to flatten Curve a in Fig. 4. Bottom: fourteen-layer filter to flatten Curve c in Fig. 4. The finer layers are not easily seen.

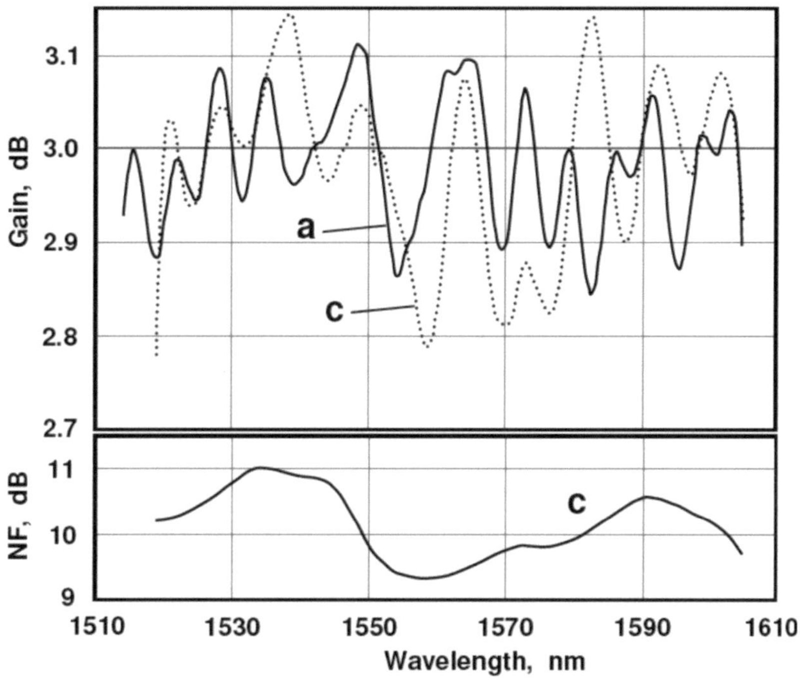

Fig. 7. Gain (top) and noise figure (bottom) spectra for a 10 km DCF Raman amplifier plus GEFs. Curve a: small launched signals and the corresponding GEF. Curves c: saturating signals with readjusted pump powers and the corresponding GEF. The pump powers are stated in Table 1 and the filter designs are illustrated in Fig. 6.

An ideal gain profile would be a horizontal line at the 3 dB baseline over a wide range. Accordingly, we now turn to the inclusion of a GEF with this aim, as described in Section 4. We have designed filters for two cases, bearing in mind that the profile to be flattened depends upon the pumping and saturation details. We constrained each SiO_2 layer to be no thinner than 10 nm. The film structures that we obtained are depicted, with statements of their total stack thicknesses and numbers of needles, in Figs. 6 (top) for small signal operation and in Fig. 6 (bottom) for saturating signals with readjusted pumps. Some of the layers are very fine and are not easily illustrated.

Net gain and noise figure spectra with the GEFs included are presented in Fig. 7. Gain Curve a is for small signal operation with the small-signal optimised filter (top structure on Fig. 6), whereas gain Curve c is for saturated operation with pump power readjustment and its corresponding filter (lower structure on Fig. 6). The peak-to-peak ripples on Fig. 7 are 0.29 dB and 0.37 dB for Curves a and c, respectively. These values compare well to reported results using a larger number of pumps but no GEF (Namiki et al., 2001; Giltrelli et al., 2004; Cui et al., 2004). The NF curve in Fig. 7 is for saturated operation with the corresponding filter included, as determined by the Friis formula (Equation (5) with a passive element that follows the gain). The result differs little from Curve c of Fig. 5, which is not surprising; losses that come after the gain region in a compound amplifier exert negligible influence on its total noise figure.

We set the NF values on Fig. 7 in context by comparing them with what can be achieved using one C-band EDFA with saturating channels. We assume that the lowest loss of a passive 10 km DCF module (including one splice between non-identical fibres) is 5.5 dB and that the NF for a good gain-saturated commercial EDFA that follows it is 6.5 dB. Thus, the NF for the DCF+EDFA combination is 12 dB at best. Alternatively, by bringing the gain forward into the DCF via Raman amplification, Fig. 7 shows that NF_{total}, which includes the GEF and other passive components, is never worse than 11 dB. A Raman-pumped DCF with a thin film GEF therefore provides better noise performance than the combined DCF+EDFA unit, doing so across a gain bandwidth that is more than twice as wide.

6. Different channel numbers with one passive GEF

Several questions remain to assess the suitability of thin film GEFs and we now address two of them. The first follows from the variability of the Raman gain profile according to the degree of gain saturation and the associated pump powers. Does a static GEF that is optimised for amplifiers with a certain number of saturating channels remain useful when the system capacity is augmented by adding more channels? Our second question is: Is the final gain flattening tolerant to manufacture thickness errors in the thin film filters?

We started by simulating a DCRA without GEFs. As before, the pumps were backward propagating with wavelengths λ_{p1} = 1435 nm and λ_{p2} = 1485 nm but the fibre length was 11.43 km. In every case the channel powers were –3 dBm and they all lay on ITU-T grid lines. We chose the pump powers so that the internal losses of the DCF were compensated and, as before, there was an additional margin of at least 3 dB over its operating range to accommodate losses of any additional passive components.

Various channel wavelength plans can be used according to the evolution of capacity demand but we simulated a simple one in which the overall bandwidth is constant and the inter-channel spacing is reduced. Figure 8 shows gain and noise figure spectra for 30, 40, 60 and 120 launched channels. In each case we adjusted the pump powers to ensure a gain profile with two equal peaks and a baseline of 3 dB. It is clear that the gain excursion increases with the number of channels (and the associated pump power adjustment). There is also a slight migration of the operating band at the 3 dB baseline to longer wavelengths,

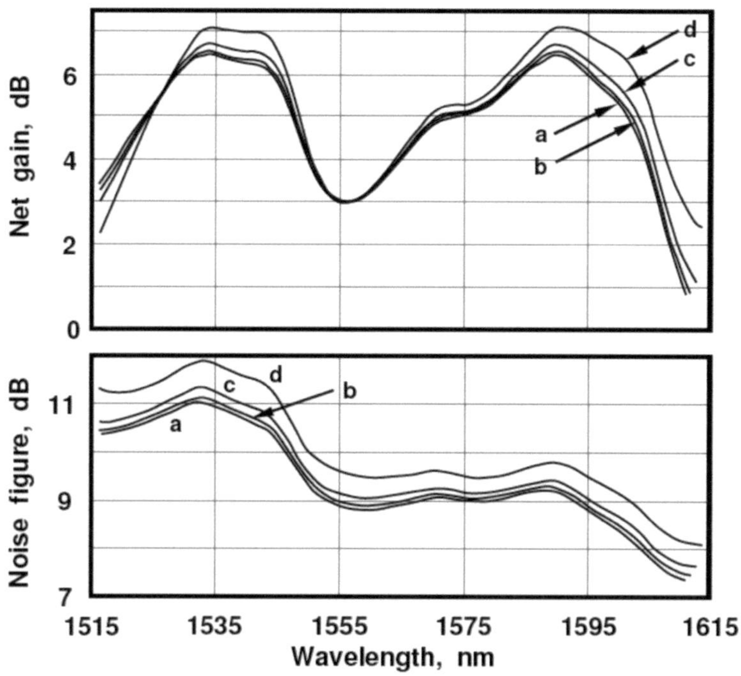

Fig. 8. Spectral variation of gain (top) and noise figure (bottom) for a 11.43 km DCF fibre Raman amplifier. The number and channel spacings are (a) 30 channels at 400 GHz, (b) 40 channels at 300 GHz, (c) 60 channels at 200 GHz and (d) 120 channels at 100 GHz. The corresponding launched pump powers at 1435 and 1485 nm are (a) 232.9 and 97.8 mW, (b) 250.5 and 101.7 mW, (c) 288.3 and 109.0 mW and (d) 410.5 and 133.1 mW.

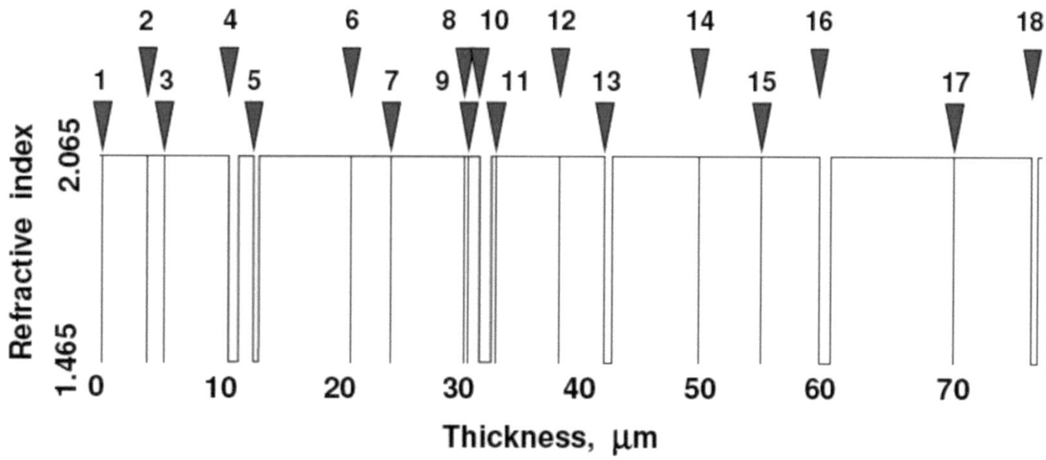

Fig. 9. The thin film design for gain equalization when operating with 60 channels spaced at 200 GHz intervals (corresponding to Curves c of Fig. 8). The arrows indicate the start positions of the eighteen SiO_2 layers.

owing to the asymmetric Raman gain profile of silica-germania glass. The noise figure curves of Fig. 8 exhibit slightly degraded performance with increasing channel numbers, and hence greater pump depletion, at shorter length coordinates within the gain medium.

We have synthesised another thin-film GEF design using the needle method, the details of which are shown in Fig. 9 (in a different pictorial format). In searching for solutions we accepted a sub-optimal film stack. Raman amplifiers exhibit a diversity of gain profiles and we aimed for simple but versatile structures. The filter was targeted at operation with 60 channels, each of –3 dBm launched power and 200 GHz spacing, in which the pumps were selected to ensure a gain profile as in Curve c of Fig. 8. Our filter design procedure predicted 18 SiO_2 layers in a Ta_2O_5 stack. The total stack thickness was 78 μm and we constrained each layer to be no thinner than 10 nm.

Figure 10 depicts the gain spectra when the GEF is included. When 60 channels of 200 GHz spacing are launched, the GEF operates according to its design criteria and the resulting peak-to-peak gain ripple is 0.30 dB over the range 1517 - 1607 nm. If there are only 30 channels with 400 GHz spacing, the filter no longer satisfies its ideal design condition and the peak-to-peak ripple is 0.38 dB within the range 1520 - 1599 nm. This small sacrifice in flattening and bandwidth is tolerable in many situations. Alternatively, if the channel density is increased above the GEF design target to 100 GHz spacing, there is a greater deterioration in performance and the peak-to-peak ripple is 0.55 dB within the range 1524 - 1596 nm. Nevertheless, such performance can be acceptable when the signals pass through only a small number of amplifying stages.

As argued in Section 3, a loss element that is located after a gain medium exerts little influence on the total noise figure. We have computed the noise figures for the whole assembly (in which the GEF is the one in Fig. 9) using the Friis cascade formula of Equation (5) and found that the resulting curves always lie within 0.15 dB of their counterparts in the bottom graph of Fig. 8.

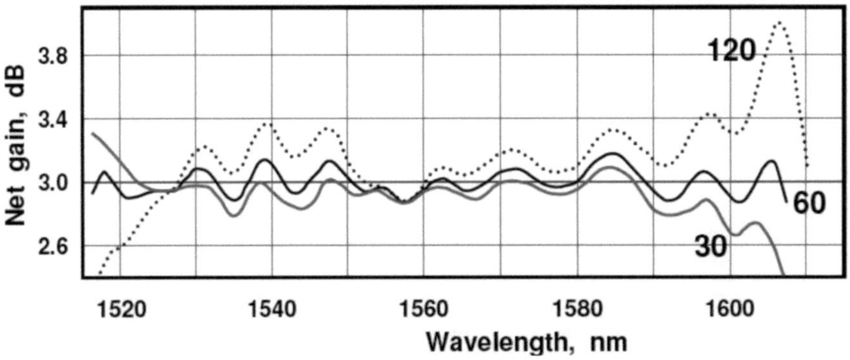

Fig. 10. Spectral variation of gain when the GEF design in Fig. 9 is placed after the DCF. The channel number and spacing are 30 channels at 400 GHz, 60 channels at 200 GHz and 120 channels at 100 GHz. The launched pump powers are as in Fig. 8.

Errors in film thickness during fabrication lead to filters with spectral characteristics that differ from their design target and so cause a deterioration in the amplifier's gain flatness. The errors can be in the form of relative or absolute film thicknesses (Thelen et al., 2002). We have applied a random number generator to simulate relative and absolute errors of 0.25% and 1.5 nm, respectively in the layers of the filter depicted in Fig. 9. (These values significantly exceed the ones reported in Verly et al., 2002 and Thelen et al., 2002.) In this way we have simulated 100 fabricated filters for each of the two types of errors and we then selected the *worst* of each batch. Figure 11 shows the gain variation with the application of the two worst filters in which 60 channels were launched and it can be seen that the peak-to-peak gain ripples increase to 0.37 and 0.39 dB for the relative and absolute errors, respectively. Such values are tolerable in many applications.

In this section we have shown that, when there is freedom to readjust the launched pump powers, a filter that is designed for use with a certain number of saturating signals can still be applicable as the network is upgraded by adding channels. We have also studied the influence of film thickness fabrication errors on gain flattening and demonstrated that the performance of our sub-optimal GEF designs remains viable even with reasonably large error values.

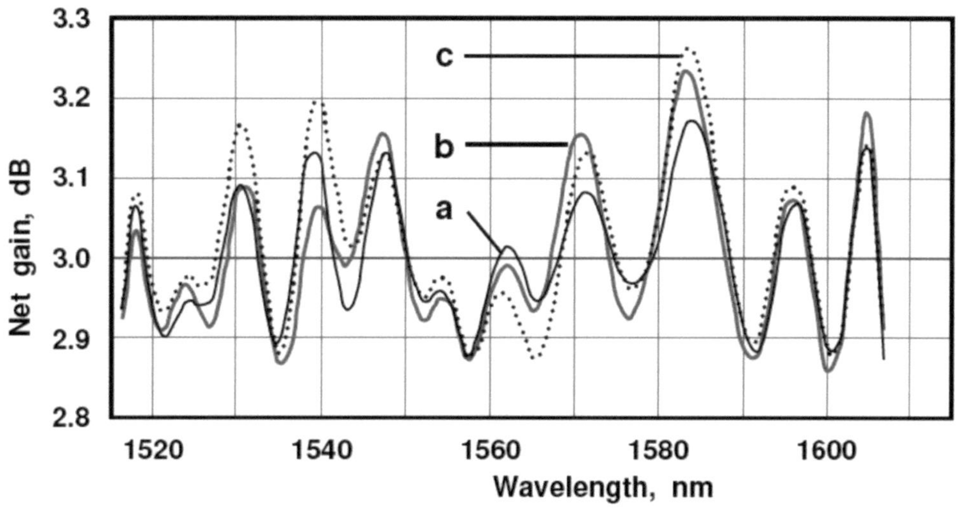

Fig. 11. Gain equalised profile for 60 channels spaced at 200 GHz: (a) no fabrication errors, (b) relative film thickness errors of 0.25%, (c) absolute film thickness errors of 0.15 nm.

7. Variation of base-line gain with one static GEF

During their operational lifetime, optical networks are often upgraded and repaired, leading to revised optical power budgets. The re-routing of cables, re-assignment of channels and inclusion of new components are common and the amplifiers in the network should ideally be sufficiently flexible to accommodate such changes without replacement. To this end, we now consider whether a DCRA can continue to provide spectrally flat performance with a new baseline gain, without having to change its GEF.

First, we modelled an 8 km DCF-based discrete Raman amplifier without a GEF. There were two backward propagating pumps (λ_{p1} = 1440 nm and λ_{p2} = 1490 nm) and 116 channels lying on ITU grid lines with 100 GHz spacing were launched, each with a power of –4 dBm. The results are presented in Fig. 12. As before, the pump powers were adjusted to provide a gain profile with two equal peaks and a local minimum between them. The local minima are at 3, 4, 5, 6 and 7 dB and so they define a sequence of baseline gains. All of the curves are very similar in shape, but the gain excursions increase slightly from bottom to top; the highest being 4.97 dB when the baseline is at 7 dB.

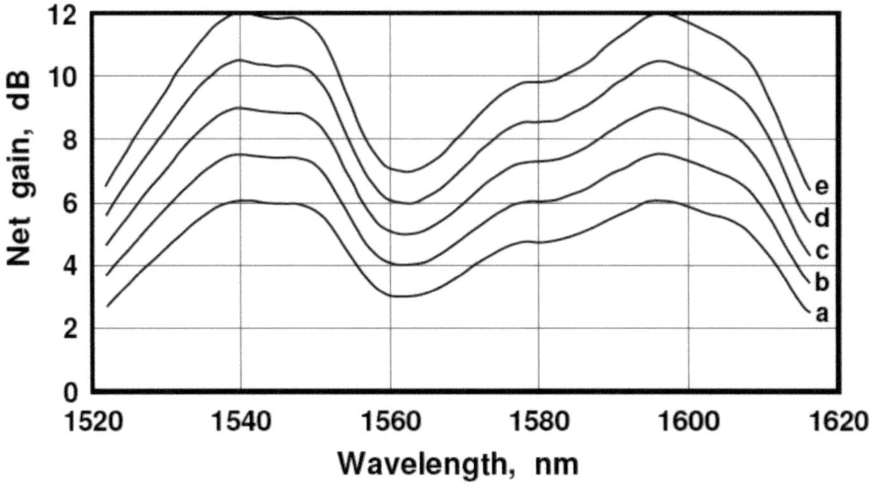

Fig. 12. Input-output gain as a function of wavelength for the following launched pump powers, P (λ_{p1}), P (λ_{p2}) in mW: (a) 281.7 and 120.5; (b) 356.3 and 142.0; (c) 448.5 and 166.1; (d) 565.4 and 194.1; (e) 716.3 and 227.8. The DCF length is 8 km.

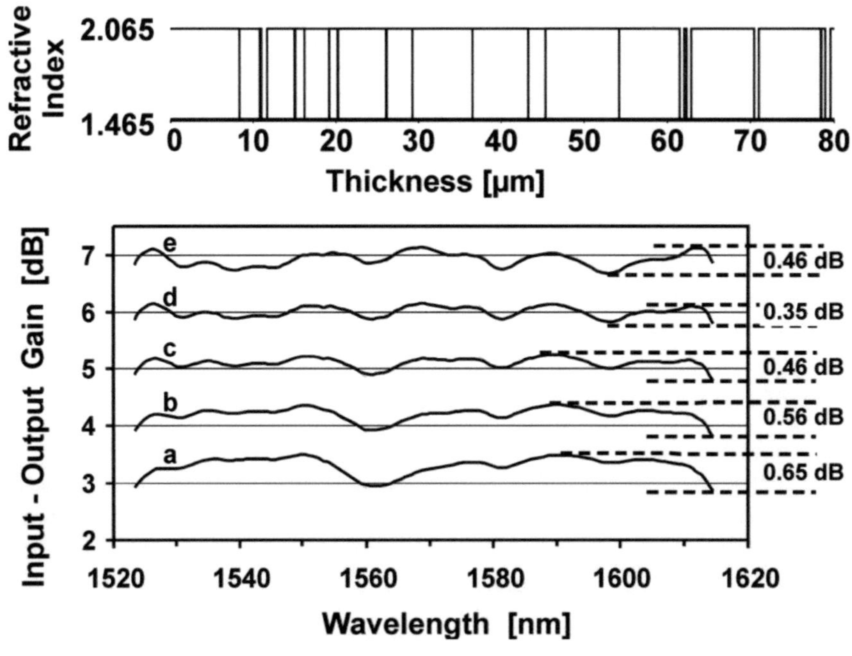

Fig. 13. Top: Thin film filter design to achieve spectral equalization of the amplifier gain Curve d of Fig. 12. Bottom: Spectral gain variations when the filter is applied. The peak-to-peak gain ripples are marked on the right. For curves a to e, the pump powers are the same as the corresponding curves in Fig. 12.

We have designed a GEF to flatten Curve d of Fig. 12 (where the baseline gain is 6 dB). Our aim was to apply it to all of the curves and compare how well it performs. The top of Fig. 13 shows its film structure, which is 18 SiO_2 layers in a Ta_2O_5 stack, with a total thickness of 80 μm. When this GEF is included in the amplifier, we obtained the gain profiles depicted in the lower part of Fig. 13. The curves are similar to each other, in which the quoted peak-to-peak gain ripples apply across the entire range of 1523 to 1614 nm. Clearly, the best flattening is achieved when the GEF was applied to the gain profile for which it was designed but acceptable performance is evident from other curves. The importance of Fig. 13 is that, provided the pump powers can be readjusted appropriately, the static GEF need not necessarily be replaced if ever the baseline gain must be varied in response to changing network operating requirements.

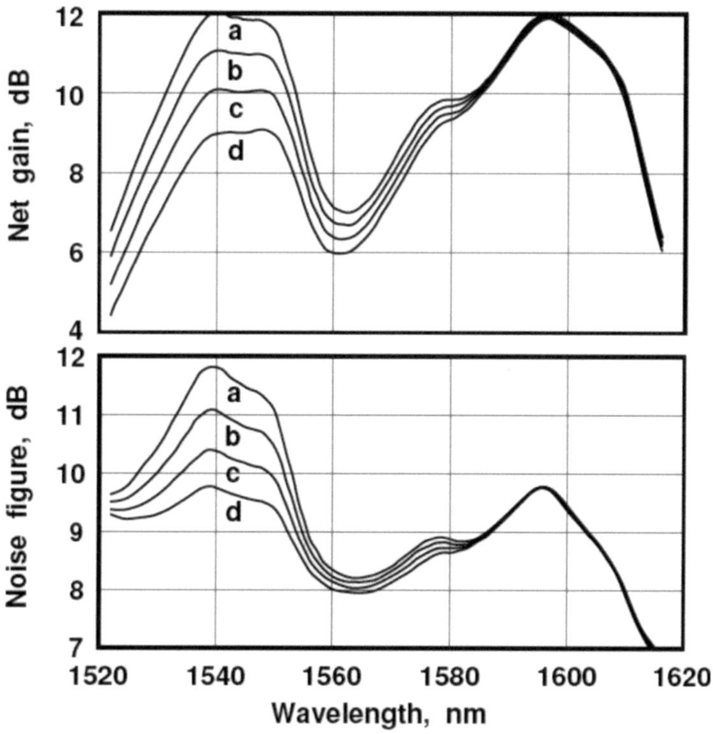

Fig. 14. Top: Gain variation with wavelength, Bottom: Noise figure variation with wavelength. Launched pump powers, $P(\lambda_{p1})$, $P(\lambda_{p2})$ in mW: (a) 716.3 and 227.8; (b) 642.0 and 241.0; (c) 569.5 and 255.1; (d) 496.5 and 271.8. Other amplifier characteristics are as in Fig. 12.

8. Reduction of the noise figure

The results in Sections 5 - 7 were mainly about the quality of gain spectral uniformity; achieving low noise figures was a secondary concern. Although we argued in Section 5 that our amplification strategy compares favourably with the use of an EDFA located after a passive span of DCF, further improvements are desirable. In particular, Figs. 5 and 8 both show two unwanted features in the noise figure spectrum: (i) performance is worst at short wavelengths because of the thermal dependence of ASE, as stated in Equation (3), and (ii) two peaks appear in saturated operation, owing to Rayleigh backscattering. Some WDM channels might therefore fail to meet the network's bit error rate requirements (Urquhart, 2008). In this section we show that, by the appropriate adjustment of the two pumps and

synthesising a filter to match, the noise figures can be improved without unacceptably deteriorating the amplifier's performance in other respects.
Gain and noise figure profiles are plotted in Fig. 14, in which the fibre span, pump wavelengths and channel input powers and spacing remain as in Section 7. The top curve on each graph (designated "a") corresponds directly to Curve e on Fig. 12, where the launched pump powers ensure a baseline gain of 7 dB and equal gain peaks. Curve a of the bottom plot of Fig. 14 shows that the corresponding noise figure spectrum has a main peak of 11.8 dB at 1539 nm and a secondary peak of 9.7 dB at 1596 nm.

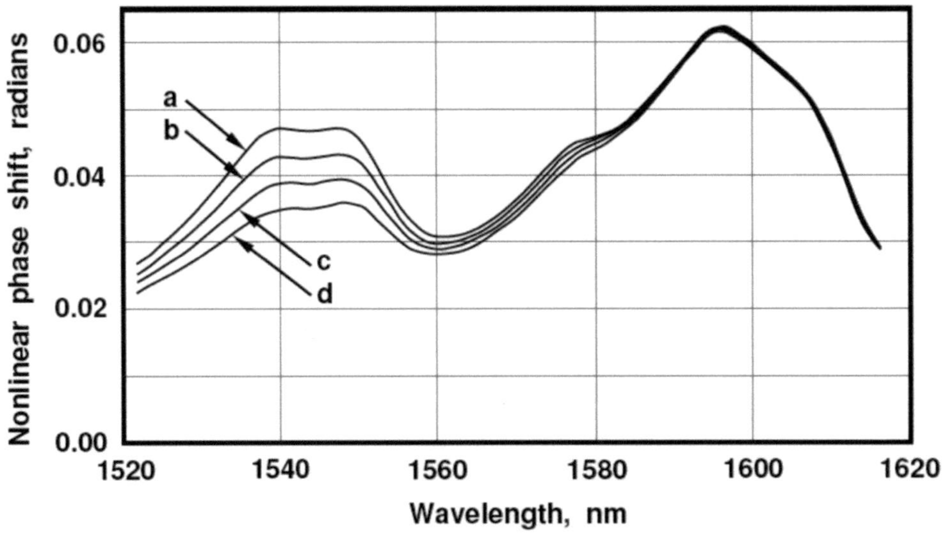

Fig. 15. Variation of nonlinear phase shift with wavelength. All of the pump powers and fibre characteristics are as in Fig. 14.

In Curves b, c and d of Fig. 14, we relaxed our previous requirement that there be two identical gain maxima and a baseline at 7 dB. Instead, we concentrated on the noise figure spectrum with the aim of selecting pump powers that progressively suppress the peak at 1538 nm to achieve better uniformity. In Curve d the chosen pump powers ensure that the two noise figure peaks are equal, having a value of 9.7 dB, which is clearly beneficial for the shorter wavelength channels. The method used was to reduce the power at λ_{p1} and increase it at λ_{p2}. When this was done, there was a significant reduction of the signal DRB near the short wavelength gain peak and a slight lowering of the ASE at all shorter signal wavelengths. The pump-to-pump SRS was also reduced and so the power at λ_{p2} had to be increased to compensate and thus achieve a revised baseline gain of 6 dB. However, as Curve d in the upper plot of Fig. 14 shows, there was a corresponding loss of symmetry in the gain profile and the question is: does it matter?
One possible concern is that an asymmetrical gain profile might provoke unacceptable nonlinear cross-talk within the DCF at the longer signal wavelengths. According to Equation (6), nonlinear phase shift is proportional to the integral of the signal powers over the fibre length coordinate. We need to ensure that it is not intolerably enhanced by varying the ratio of pump powers to achieve an asymmetric gain profile. Figure 15 is a plot of the nonlinear phase shift. It displays two peaks that are associated with the gain peaks on Fig. 14 because

the path averaged signal power is high when the gain is high. Curves b, c and d of Fig. 15 show that by progressively changing the ratio of powers at λ_{p1} and λ_{p2} the short wavelength peak is depressed but the long wavelength peak is largely unaffected. In Curve a much of the long wavelength gain results from the guided wave at λ_{p2} being amplified by that at λ_{p1}. Lowering the power at λ_{p1} reduces the Raman pump-to-pump interaction. Therefore, pump-to-pump interactions are less important contributors in the other curves, where the launched power at λ_{p2} is increased in a manner that compensates for the effect.

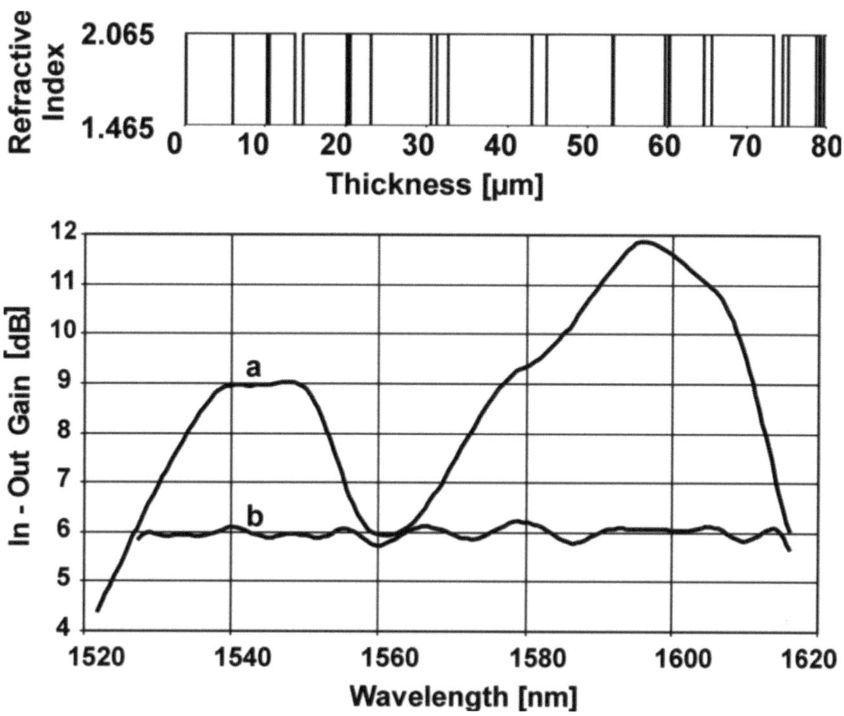

Fig. 16. Top: Thin film filter design (with 23 SiO_2 layers) to achieve spectral equalisation of the gain Curve d of Fig. 14. Bottom: The spectral gain variation before (Curve a) and after (Curve b) the filter is applied.

We have synthesised a GEF design to equalise the distorted gain profile (Curve d of Fig. 14). Figure 16 shows the result after applying the filter to the amplifier, together with the target function for which it was designed; a peak-to-peak gain ripple of 0.49 dB over the range 1527–1615 nm was achieved. The performance should be compared with that of Curve d of Fig. 13, where the baseline gain is also 6 dB. The 0.49 dB ripple in Fig. 16 is slightly worse but it is an acceptable value for intermediate haul DWDM systems.

In summary, when the pump powers are adjusted to obtain tilted gains, the noise figure and the nonlinear phase shift are both reduced at shorter wavelengths but there is no decrease of the nonlinear phase shift at longer signal wavelengths. We have designed a GEF to compensate for the asymmetric profile, providing acceptably uniform amplification across a wider spectrum than a combined C- and L-band EDFA. Pumping to achieve improvements in the short wavelength noise figures at the expense of an asymmetric gain profile is therefore beneficial.

9. Conclusion

We have presented simulations of a spectral equalisation strategy for dispersion compensating fibre Raman amplifiers. Flat and broad bandwidth gain profiles can be obtained by launching only two contra-directional pumps into dispersion compensating fibre (DCF) and following it with a gain equalisation filter (GEF) based on dielectric thin film technology. To this end, we have optimised such filters using a trusted technique called the needle method. When a GEF is included, the amplifier can provide good gain profile uniformities for over 100 channels in both small signal and gain saturation.

The strategy explained in this chapter is versatile, which is partly because our filters were "sub-optimal", meaning that we avoided highly refined design solutions that perform well only within specific circumstances. For example, sub-optimal film structures can provide good spectral equalisation even when imperfect fabrication causes thickness errors. Moreover, by retaining the flexibility to adjust the two launched pump powers, the spectral flatness continues to be acceptable in revised operating conditions, without necessarily having to change the GEF. In particular, we have shown that the amplifier performance remains viable when the number of channels is increased or when the baseline gain must be varied to accommodate new demands of the optical network.

The total noise figure of the amplifier is least when the GEF is placed after the DCF gain medium, as we have done. Moreover, it is possible to achieve further improvements to the noise performance by pumping for tilted (asymmetric) gains and designing a GEF for such operation to obtain good profile uniformity. We have argued that this approach can be used without incurring unacceptable additional nonlinear crosstalk.

The gain equalisation strategy that we describe in this chapter uses passive filters, which are low cost, high reliability components of low thermal sensitivity. Moreover, with only two pump lasers to manage, it reduces the demands on the control software, and hence network management, and we have derived an approximate analytical model to illustrate this feature. Application can be in cost-sensitive but high capacity systems where the signals pass through a cascade of dispersion-compensating amplifier stages.

10. Acknowledgement

Guido Boyen and André Brückmann each received an EU Erasmus scholarship. Project funding was by Spanish Ministry of Education and Science project TEC2010-20224-C02-01.

11. Appendix: Approximate analytical model of dual-wavelength pumping

We present a theory to predict the dual-wavelength pumped gain spectrum when the channels are contra-directional and low power. For ease of calculation, we assume that the two pumps are launched at z = 0. Their mutual interaction is described by:

$$\frac{dP_{p1}}{dz} = -\Lambda\,\Gamma_{12}\,P_{p1}\,P_{p2} - \alpha_p\,P_{p1} \tag{8}$$

$$\frac{dP_{p2}}{dz} = +\Gamma_{12}\,P_{p1}\,P_{p2} - \alpha_p\,P_{p2}\ , \tag{9}$$

where $\Lambda = \lambda_{p2}/\lambda_{p1}$ and the losses are assumed equal ($\alpha_{p1} = \alpha_{p2} = \alpha_p$). Equations (8) and (9) include saturation by transfer of power from one pump to the other with gain coefficient Γ_{12}

from Equation (1). However, they do not account for ASE or depletion by the propagating signal(s). They yield analytical solutions and to do this we transform the coordinates by defining the power ratio $R(z) = P_{p1}/P_{p2}$ and the weighted sum $S(z) = (P_{p1} + \Lambda P_{p2})$, giving:

$$\frac{dS}{dz} = -\alpha_p S \tag{10}$$

$$\frac{dR}{dz} = -\Gamma_{12} R S \cdot \tag{11}$$

The initial conditions are $S(z = 0) = S_0 = [P_{p1}(0) + \Lambda P_{p2}(0)]$ and $R(z = 0) = R_0 = P_{p1}(0)/P_{p2}(0)$. Equations (10) and (11) are then easily solved to provide:

$$S(L) = S_0 \exp(-\alpha_p L) \tag{12}$$

$$R(L) = R_0 \exp(-\Gamma_{12} S_0 L_{eff}) . \tag{13}$$

L_{eff} is identified as the effective length term, which is commonly used in Raman amplifier theory: $L_{eff} = [1 - \exp(-\alpha_p L)]/\alpha_p$. When necessary, Equations (12) and (13) are easily converted back to power coordinates using $P_{p1} = SR/(R + \Lambda)$ and $P_{p2} = S/(R + \Lambda)$.

We now introduce a propagating signal of tuneable wavelength λ_s and we assume that it is not sufficiently powerful to deplete the pumps. The signal is contra-directional and therefore it is launched at $z = L$ with power P_{in}. The fibre loss at the signal wavelength is α_s and in general $\alpha_s \neq \alpha_p$. We also require the Raman gain coefficients for the pump-signal interactions, Γ_{1s} and Γ_{2s}, defined in Equation (1).

When the signal is included, we have three differential equations describing the optical powers propagating in the fibre. The first two are Equations (8) and (9), without any modifications and subject to the same boundary conditions as before. The third one is

$$\frac{dP_s}{dz} = -\Gamma_{1s} P_{p1} P_s - \Gamma_{2s} P_{p2} P_s + \alpha_s P_s \tag{14}$$

Equation (14) can be solved using the previous quantities R(z) and S(z). There are three terms on the right hand side to be integrated. The first one can be transformed using R and S, together with Equation (11):

$$\begin{aligned} \int_{z'=L}^{z} P_{p1}(z')\,dz' &= \int_{z'=L}^{z} \frac{S(z')R(z')}{(R(z') + \Lambda)} . dz' \\ &= -\left(\frac{1}{\Gamma_{12}}\right) \int_{z'=L}^{z} \frac{dR(z')/dz'}{(R(z') + \Lambda)} . dz' = -\left(\frac{1}{\Gamma_{12}}\right) . \ln\left(\frac{R(z) + \Lambda}{R(L) + \Lambda}\right) \end{aligned} \tag{15}$$

The second term on the right of Equation (14) can be integrated with the aid of another defined real variable, I(z), which is the inverse of R(z): $I(z) = 1/R(z)$. In this way, an alternative statement of Equation (11) is $dI/dz = + \Gamma_{12} S I$. Then, the second integration becomes:

$$\begin{aligned} \int_{z'=L}^{z} P_{p2}(z')\,dz' &= \int_{z'=L}^{z} \frac{S(z')I(z')}{(1 + \Lambda I(z'))} . dz' = \left(\frac{1}{\Gamma_{12}}\right) \int_{z'=L}^{z} \frac{dI(z')/dz'}{(1 + \Lambda I(z'))} . dz' \\ &= \left(\frac{1}{\Lambda \Gamma_{12}}\right) . \ln\left(\frac{1 + \Lambda / R(z)}{1 + \Lambda / R(L)}\right) \end{aligned} \tag{16}$$

The third integration from the right of Equation (14) is merely a loss term.
By combining all three integrations, Equation (14) can be solved to give the length-dependence of the signal power and thus the net (input-output) amplifier gain:

$$G_{net} = \frac{P_s(z=0,\lambda_s)}{P_{in}} = \left(\frac{R_0 + \Lambda}{R(L) + \Lambda}\right)^{\frac{\Gamma_{1s}(\lambda_s)}{\Gamma_{12}}} \cdot \left(\frac{1 + \Lambda/R(L)}{1 + \Lambda/R_0}\right)^{\frac{\Gamma_{2s}(\lambda_s)}{\Lambda\Gamma_{12}}} \cdot \exp[-\alpha_s L], \quad (17)$$

where the wavelength dependence is provided through the terms $\Gamma_{1s}(\lambda)$ and $\Gamma_{2s}(\lambda)$. (In general, α_p and α_s also vary with wavelength.) Equation (13) allows us to determine R(L) as a function of the inputs S_0 and R_0. Owing to our choice of length coordinate, the amplified signal exits the fibre at z = 0. However, $P_{p1}(0)$ and $P_{p2}(0)$ are known values and G_{net} is a relation between input and output signal powers, which applies irrespective of the point of launch, provided that it is opposite that of the pumps.
Equation (17) is plotted for some typical small signal operating conditions in Fig. 17, where the DCF is the same as in Fig. 4. The pump wavelength separation is 30, 40 and 50 nm in Curves a, b and c, respectively, so as to provide ever wider operating profiles. All of the pump powers, which are specified in Table 2, were selected for a baseline gain of 3 dB and two peaks of equal magnitude. Curve c is, in fact, two sets of points, which largely overlap. The solid line was plotted from Equation (17) by scanning λ_s across a range; the dotted line is merely a reproduction of Curve a from Fig. 4, which was obtained by numerical solution of Equation (2) with 151 simultaneous low power launched signals.
The two sets of points that constitute Curve c are very close but Table 2 reveals that slightly different pump powers were required. (However, the total power, $[P(\lambda_{p1}) + P(\lambda_{p2})]$, is the same within computational error.) We attribute the small difference to two effects: (i) the absence of ASE and (to a lesser extent) RBS from the analytical model and (ii) our assumption of equal pump losses ($\alpha_{p1} = \alpha_{p2} = \alpha_p$). Although the launched signals might be very small, the spontaneous scattering is amplified bi-directionally, causing some depletion of the pumps. Moreover, the choice of α_p is a compromise that becomes less satisfactory as the wavelength difference ($\lambda_{p2} - \lambda_{p1}$) is increased. Nevertheless, the strong correlation between the two sets of points in Curve c indicates that the analytical model can provide a good approximation to the required pump powers and a physical insight into the gain mechanism in the small signal limit. It can thus be used to establish initial performance estimates prior to finding accurate numerical solutions.

Curve	**λ_{p1} & λ_{p2}, nm**	**$P(\lambda_{p1})$, mW**	**$P(\lambda_{p2})$, mW**	**Bandwidth at baseline, nm**	**Excursion, dB**
a	1445 & 1475	79.6	86.9	41.6	0.6
b	1440 & 1480	118.9	94.3	71.0	1.6
c (solid)	1435 & 1485	160.8	101.5	92.4 ± 0.5	2.9 ± 0.05
c (dotted)		173.3	89.1		

Table 2. Data for plotting Fig. 17 and the resulting bandwidths and gain excursions.

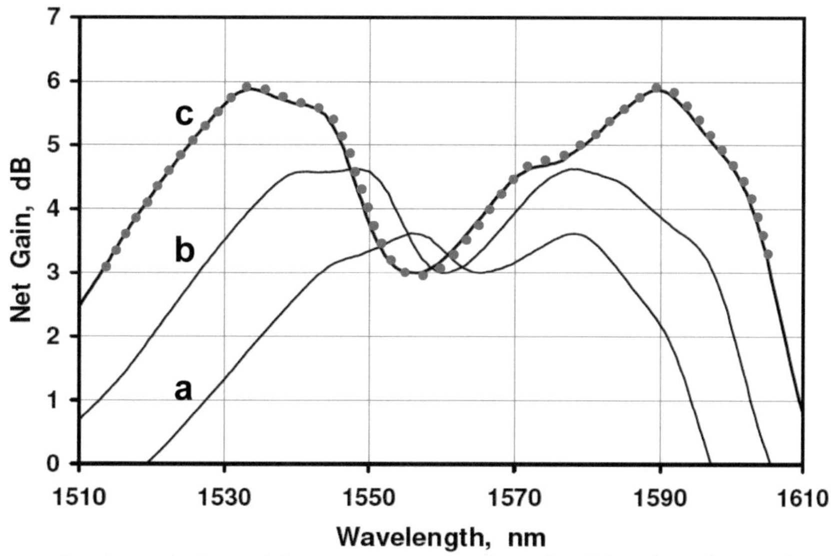

Fig. 17. Spectral gain variation with two pump wavelengths. Fibre length is 10 km and the pump loss coefficient α_p is the average of $\alpha(\lambda_{p1})$ and $\alpha(\lambda_{p2})$. Other data are in Table 2.

12. References

G.P. Agrawal (2005), Theory of Fibre Amplifiers, Chapter 2 in *Raman Amplification in Fiber Optical Communication Systems*, C. Headly and G.P. Agrawal (Editors), Academic Press, ISBN 978-0-120445066.

A. Boskovic, S.V. Chernikov, J.R. Taylor, L. Grüner-Nielsen, O.A. Levring (1996), Direct Continuous-Wave Measurement of n_2 in Various Types of Telecommunication Fiber at 1.55 μm, *Optics Letters*, 21 (24), 1966-1968.

J. Bromage (2004), Raman Amplification for Fiber Communications Systems, *IEEE Journal of Lightwave Technology*, 22 (1), 79–93.

J. Bromage, P.J. Winzer, R.-J. Essiambre (2004) Multiple Path Interference and its Impact on System Design, Chapter 15 in M.N. Islam (Ed.), *Raman Amplifiers for Telecommunications 2: Sub-Systems and Systems*, Springer. ISBN 978-0387406565.

S. Cui, J. Liu, X. Ma (2004), A Novel Efficient Optimal Design Method for Gain-Flattened Multiwavelength Pumped Fiber Raman Amplifier, *IEEE Photonics Technology Letters*, 16 (11), 2451–2453.

E. Desurvire (1994) Erbium Doped Fiber Amplifiers: Principles and Applications, Wiley. ISBN 978-0471266273.

M. Giltrelli, M. Santagiustina (2004), Semianalytical Approach to the Gain Ripple Minimization in Multiple Pump Fiber Raman Amplifiers, *IEEE Photonics Technology Letters*, 16 (11), 2454–2456.

L. Grüner-Nielsen, Y. Qian, P.B. Gaarde (2006), Dispersion Compensating Fibers For Raman Applications, *Journal of Optical Fiber Communications*, 3, 61-89.

ITU-T Recommendation G.694.1 (2002-06), Spectral Grids for WDM Applications: DWDM Frequency Grid, available from www.itu.int/rec/T-REC-G.694.1-200206-I.

M.N. Islam, C. DeWilde, A. Kuditcher (2004), Wideband Raman Amplifiers, Chapter 14 in *Raman Amplifiers for Telecommunications*, Vol. 2, M.N. Islam (Editor), Springer, ISBN 978-0-387406565.

S. Jiang, B. Bristiel, Y. Jaouen, P. Gallion, E. Pincemin (2007a), Bit-Error-Rate Evaluation of the Distributed Raman Amplified Transmission Systems in the Presence of Double Rayleigh Backscattering Noise, *IEEE Photonics Technology Letters*, 19 (7), 468-470.

S. Jiang, B. Bristiel, Y. Jaouen, P. Gallion, E. Pincemin, S. Capouilliet (2007b), Full Characterisation of Modern Transmission Fibers for Raman-Amplified Communication Systems, *Optics Express*, 15 (8), 4883-4892.

F. Koch, S.A.E. Lewis, S.V. Chernikov, J.R. Taylor, V. Grubsky, D.S. Starodubov (2000), Broadband Gain Flattened Raman Amplifier to Extend Operation in the Third Telecommunication Window, Proceedings, *Conference on Optical Fiber Communications*, paper FF-3, ISBN 978-0780359529.

S.A.E. Lewis, S.V. Chernikov, J.R. Taylor (1999), Temperature-Dependent Gain and Noise in Fiber Raman Amplifiers, *Optics Letters*, 24 (24), 1823-1825.

H.A. Macleod (2010), *Thin Film Optical Filters*, 4th Edition, CRC Press, see especially Chapters 2 and 3, ISBN 978-1420073027.

T. Miyamoto, T. Tsuzaki, T. Okuno, M. Kakui, M. Hirano, M. Onishi, M. Shigematsu (2002), Raman Amplification over 100-nm Bandwidth with Dispersion and Dispersion Slope Compensation for Conventional Single Mode Fiber, *Conference on Optical Fiber Communications*, Anaheim, USA, Paper TuJ7, ISBN 1-55752-701-6.

S. Namiki, Y. Emori, (2001), Ultrabroad-band Raman Amplifiers Pumped and Gain-Equalized by Wavelength-Division-Multiplexed High-Power Laser Diodes, *IEEE Journal of Selected Topics in Quantum Electronics*, 7 (1), 3–16.

S. Namiki, Y. Emori, A. Oguri (2005), Discrete Raman Amplifiers, Chapter 4 in *Raman Amplification in Fiber Optical Communication Systems*, C. Headley, G.P. Agrawal (Editors), Elsevier Academic Press, ISBN 978-0-120445066.

S. Namiki, N. Tsukiji, Y. Emori (2004), Pump Laser Diodes and WDM Pumping, Chapter 5 of M.N. Islam, *Raman Amplifiers for Telecommunications 1: Physical Principles*, M.N. Islam (Editor), Springer, New York. ISBN 978-0-387007519.

B. Neto, C. Reis, A. Teixeira, P.S. Andre, N. Wada (2009), proceedings, *SBMO/IEEE Microwave and Optoelectronics Conference*, Boston USA, 687-689. ISBN 978-1-4244-5356-6.

B.T. Sullivan, J.A. Dobrowolski (1996), Implementation of a Numerical Needle Method for Thin-Film Design, *Applied Optics*, 35 (28), 5484-5492.

H. Takashashi (1995), Temperature Stability of Thin-Film Narrow Bandpass Filters Produced by Ion-Assisted Deposition, *Applied Optics*, 34 (4), 667-675.

A. Thelen, M. Tilsch, A.V. Tikhonravov, M.K. Trubetskov, U. Brauneck (2002), Topical Meeting on Optical Interference Coatings (OIC 2001): Design Contest Results, *Applied Optics*, 41 (16), 3022-3038.

A.V. Tikhonravov, M.K. Trubetskov, G.W. DeBell (1996), Application of the Needle Optimization Technique to the Design of Optical Coatings, *Applied Optics*, 35 (28), 5493–5508.

P. Urquhart (2008), Optical Fiber Transmission, Chapter 6 in G. Hill (editor), *The Cable and Telecommunications Professionals' Reference* Vol. 2, 3rd Ed., Focal Press, ISBN 978-0-240-80748-5.

P. Urquhart, O. García López, G. Boyen, A. Brückmann (2007), Optical Amplifiers for Telecommunications, Proceedings, *IEEE Symposium on Intelligent Signal Processing, WISP 2007*, Alcala de Henares, 3-5 October 2007, ISBN 978-1-4244-0830-6.

P.G. Verly (2002), Design of a Robust Thin-Film Interference Filter for Erbium-Doped Fiber Amplifier Gain Equalization, *Applied Optics*, 41 (16), 3092–3096.

X. Zhou, M. Feuer, M. Birk (2006), A Simple Feed-Forward Control Algorithm for Fast Dynamic Gain Profile Control in a Multi-Wavelength Forward Pumped Fibre Raman Amplifier, *IEEE Photonics Technology Letters*, 18 (9), 1004-1006.

17

Fiber-Bragg-Grating Based Optical Amplifiers

Shien-Kuei Liaw[1], Kuang-Yu Hsu[2],
Kuei-Chu Hsu[3] and Peng-Chun Peng[4]
[1]National Taiwan University of Science and Technology
[2]National Taiwan University
[3]National Central University
[4]National Taipei University of Technology
Taiwan

1. Introduction

In this chapter, we propose several schemes for fiber amplifiers which are all using fiber Bragg gratings (FBGs) as the key elements for their advantages of better uniformity, higher contrast ratio and lower cost. Several applications using FBG-based fiber amplifiers are also introduced. Therefore, this chapter initiates the overview of FBGs characteristics in Sec. 1, then addresses the FBGs playing solo-function roles in the fiber amplifiers in Sec. 2. The examples are like gain equalizer, dispersion compensator, signal reflector, pump reflector and other purposes. Next, Sec. 3 addresses the FBGs playing multiple-function roles in the hybrid fiber amplifiers. It is based on the construction of a C-band erbium-doped fiber amplifier (EDFA) and an L-band Raman fiber amplifier (RFA). The serial type, parallel type, bridge type and bidirectional C + L band hybrid amplifiers using single-wavelength pump laser diodes are proposed. Dispersion management and gain equalization among the C+L-band channels are realized simultaneously. Pump reflectors and double-pass schemes are used to increase the slope efficiency. Signal power variation among the channels are reduced by varying the reflectivities of these corresponding FBGs. Sec. 4 deals with FBG-based optical network devices with built-in fiber amplifier. Such networks devices include the reconfigurable optical add-drop multiplexer (ROADM) either in unidirectional or bidirectional transmission scheme, and the optical cross-connect (OXC) device is also proposed. Power compensation are realized with a bulit-in fiber amplifier. After various kinds of FBG-based fiber amplifiers are discussed, we summarize this chapter and conclude their potential utilization within optical communications and optical sensing.

2. Overview of FBGs characteristics

2.1 Introduction to Fiber Bragg Gratings

The fiber Bragg gratings (FBGs) research can be dating back to Ken Hill in 1978, and they were initially fabricated using a visible laser propagating along the fiber core. Then holographic technique was demonstrated using the interference pattern of ultraviolet laser light to create the periodic structure of the Bragg grating. Due to all-fiber geometry, FBGs have found important applications such as gain equalizer, dispersion compensator, pump reflector and other purposes as the telecommunication industry grew, as shown in Fig. 1

(Kashyap, 2010). FBGs are reflective type filters with slightly periodic refractive index modulation running along fiber axis in the fiber core. Incident wavelengths are reflected when the Bragg conditions ($\lambda_B=n_{eff}\Lambda$, where λ_B is the Bragg wavelength, n_{eff} is the effective modal index, and Λ is the grating period) are satisfied, and otherwise they are transmitted. The typical grating period is around 0.5 μm, the reflection bandwidth is around 0.2 nm, the reflectivity is larger than 99 % (>20 dB), and the insertion loss is less than 0.1 dB for reflective applications at the C-band optical communication window. Environmental stability for fiber gratings was originally a big issue but can be controlled now by a suitable annealing process and an appropriate package.

FBGs are usually fabricated by inscribing the periodical intensity of UV lights onto the photo-sensitive fiber core to induce the permanent periodical refractive index change. The photosensitivity of the fiber core is mainly caused by formation of color center, or densification and increase in tension. Various laser light sources have been used to induce refractive index changes in optical fibers. The commonly used pulse lasers are KrF (248 nm), ArF (193 nm), and Ti:Sapphire (800 nm), while the commonly used continuous wave laser is the frequency-doubled Ar-Ion laser (244 nm). Under the intensities of 100-1000 mJ/cm^2, the amount of induced refractive index change in germanium doped optical fibers is around 10^{-5}-10^{-3}. Higher index changes can be achieved by hydrogen loading in high pressure (Hill et. al., 1997).

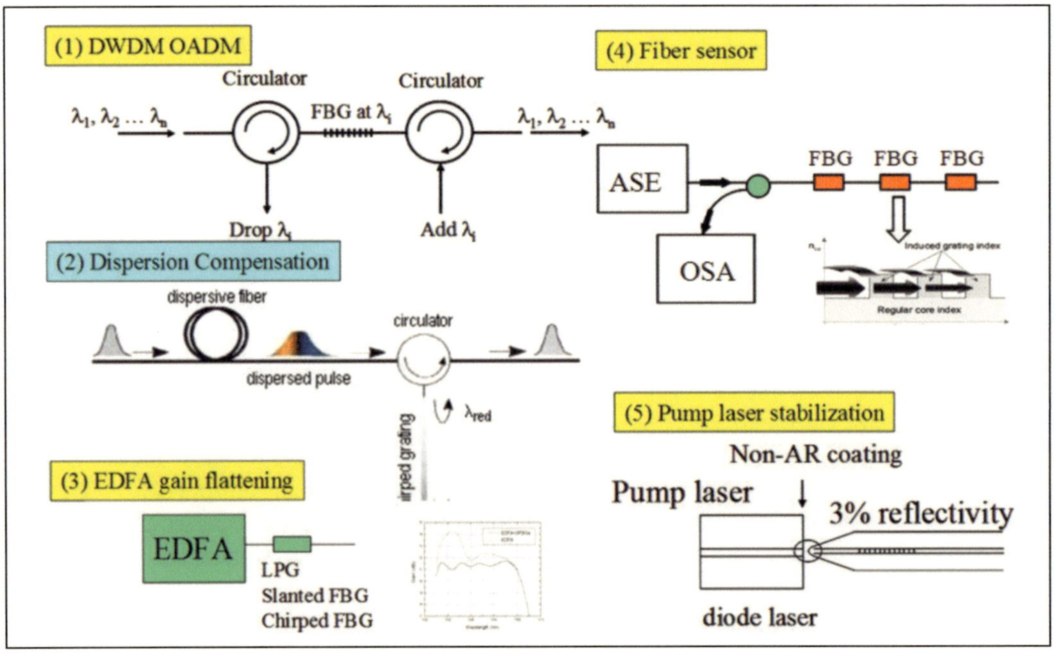

Fig. 1. Various applications of fiber Bragg gratings. (Kashyap, 2010)

Establishing the easy-realized FBGs technology is promising and very useful in various photonic industries. Several fabrication schemes have been proposed for FBG inscription by forming interference light fringes; including the common-used phase mask method and holographic method. In the phase mask technique, the two first diffraction orders of UV beam interfere to form periodic intensity distribution that is half the period of the phase mask, and the zero order beam is totally suppressed. The advantages of the phase mask approach are the easy alignment, low stability requirement, and low coherence laser source requirement. Its drawback, which is the advantage of the holographic approach, is the lack of flexible wavelength tuning capability and the limitation of the grating length. However, the highly environmental requirement is exactly the drawback of the holographic approach. Fiber gratings have various kinds of grating structures. For a phase-shifted FBG, a π phase shift is inserted into the center of the exposure fiber grating during the fabrication process, and there is a narrow transmission peak within the stop-band due to the resonance caused by the π phase shift. Chirped fiber grating in general has a non-uniform period along fiber length, and its phase information can also be contributed from the dc-index change, phase shift and period change. Chirped gratings can be used as dispersion compensators, for they are designed to introduce a time delay as a function of wavelength, so that different wavelength is reflected at a different grating location to achieve wavelength-dependent group delay. Fiber grating with periods as few hundred micrometers is called long period gratings (LPGs). They are transmission gratings, which couple light from forward-propagating guided modes to the forward-propagating cladding modes and the radiation field. LPGs are particularly useful for equalizing the gain of optical fiber amplifiers, and they are also good sensors with operation relies on resonance wavelength shift corresponding to environmental perturbations of strain and temperature.

2.2 Theorem and mathematical model

Both the coupled mode equation model and the transfer matrix analysis are commonly used to describe the relation between the filter spectrum and grating structure (Kogelnik et al., 1972), (Yamada et al., 1975). Consider a fiber grating with small spatial index perturbation $\overline{\delta n_{eff}}$ and the period Λ along the axis coordinate z. Two counter-propagating waves in the optical fiber are denoted as R, and S. Hence the resultant wave coupling can be derived as

$$\begin{aligned} \frac{dR}{dz} &= i\sigma R(z) + i\kappa S(z), \\ \frac{dS}{dz} &= -i\sigma S(z) + i\kappa^* R(z), \end{aligned} \tag{1}$$

where

$$\begin{aligned} \sigma &= \frac{2\pi}{\lambda}\overline{\delta n_{eff}}, \\ \kappa = \kappa^* &= \frac{\pi}{\lambda} v\overline{\delta n_{eff}}. \end{aligned} \tag{2}$$

Here v is confinement factor. For LPGs, phase matching condition occurs between co-directional coupling waves, and the coupling constant should accordingly be modified as the mode overlap area varies between different cladding modes. The amplitude reflection coefficient, the reflectivity and the transmission ratio are denoted as ρ, r, t respectively and are defined according to the following equations,

$$\rho = \frac{S(\frac{-L}{2})}{R(\frac{-L}{2})},$$
$$r = |\rho|^2, \tag{3}$$
$$t = 1 - r.$$

The transfer matrix method is a simple way to analyzing complex grating structures by dividing the gratings into small sections with constant period and uniform refractive index modulation. The transform matrix yields the following relationship between the reflected wave u and transmitted wave v,

$$\begin{bmatrix} \mathrm{u}(L) \\ \mathrm{v}(L) \end{bmatrix} = T_N \cdot T_{N-1} \cdot \ldots\ldots \cdot T_1 \begin{bmatrix} u(0) \\ v(0) \end{bmatrix} = T \begin{bmatrix} u(0) \\ v(0) \end{bmatrix}$$
$$= \begin{bmatrix} T_{11} & T_{12} \\ T_{21} & T_{22} \end{bmatrix} \begin{bmatrix} u(0) \\ v(0) \end{bmatrix}, \tag{4}$$

where u(0) and u(L) represent the input and output forward-propagating waves, v(0) and v(L) are the input and output backward-propagating waves, and L is the length of the grating. The matrix T is a function of the refractive index modulation Δn, and the matrices T_1, T_2, ...,T_N are governed by the parameters of every grating section. The matrix product of T_{11}, T_{12}, T_{21}, T_{22} forms the final transform matrix. The transmission ratio of the grating can be calculated by

$$t(\delta) = 1 / T_{11} \tag{5}$$

where δ is the frequency detuning. Inverse methods such as layer-peeling method or evolutionary programming synthesis can find complex coupling coefficient of a FBG from the reflection spectrum (Lee et al., 2002).

2.3 Fiber grating fabrication technology development and applications

The uniform FBG reflection spectrum possesses apparent side-lobes and thus the FBG refractive index envelopes are usually apodized to be of a gaussian or cosine square shape in order to diminish the side-lobes. The quasi-periodic structure on the long wavelength side originates from the resonance between the abrupt index change of the two ends and can be suppressed by apodizing the index profile. On the other hand, Fabry-Perot resonance between peripheral sections of the grating with apodization cause quasi-periodic structures of the reflection spectrum in shorter wavelengths, which can be reduced by keeping the refractive index constant along the fiber length (pure apodization, see Fig. 2). To keep average refractive index the same throughout the length of the grating, pure-apodization method is used to maintain the dose of the UV radiation the same throughout the fiber length but the fringe pattern is gradually altering (Chuang et al., 2004). Conventional method to achieve pure-apodization relies on double UV exposure. The first exposure is to imprint the interference pattern onto the fiber core, followed by the second scan to keep the total doze along the entire grating length unchanged. FBGs as narrowband filters have important applications in single-longitudinal mode fiber lasers and DWDM systems. The required high sidelobe suppression ratio is achieved by pure apodization, while the spectral shape of narrow and flat-top bandwidth with high reflectivity is achieved by slight index difference and long grating length. Several procedures that can realize long and complex FBG structures have been developed, however, the accumulative position reading errors have caused significant difficulties on the fabrication of long-length fiber Bragg gratings. For advanced realization of long-length FBGs, real-time side-diffraction position monitoring scheme for fabricating long FBGs was proposed (see Fig. 3), and the overlapped FBG sections can be connected section-by-section without obvious phase errors. (Hsu et al., 2005).

FBGs are critical components in fiber-optic communication and fiber sensor applications. FBGs are commonly used as spectral filters, feedback mirrors in erbium-doped amplifiers, fiber lasers and semiconductor diode lasers, and add-drop multiplexers in optical communication network. Narrow linewidth (bandwidth less than the cavity mode spacing) makes FBG a good choice perfectly suited for stabilizing the wavelength of semiconductor lasers and fiber lasers as feedback mirrors to stabilize the frequency and attain single-frequency operation. The use of narrowband FBGs for add/drop multiplexers can also help extracting a single wavelength from the fiber without disturbing other wavelengths thus can achieve high optical data rates. A demultiplexer can be achieved by cascading multiple drop sections of the OADM, where each drop element uses a FBG set to the wavelength to be

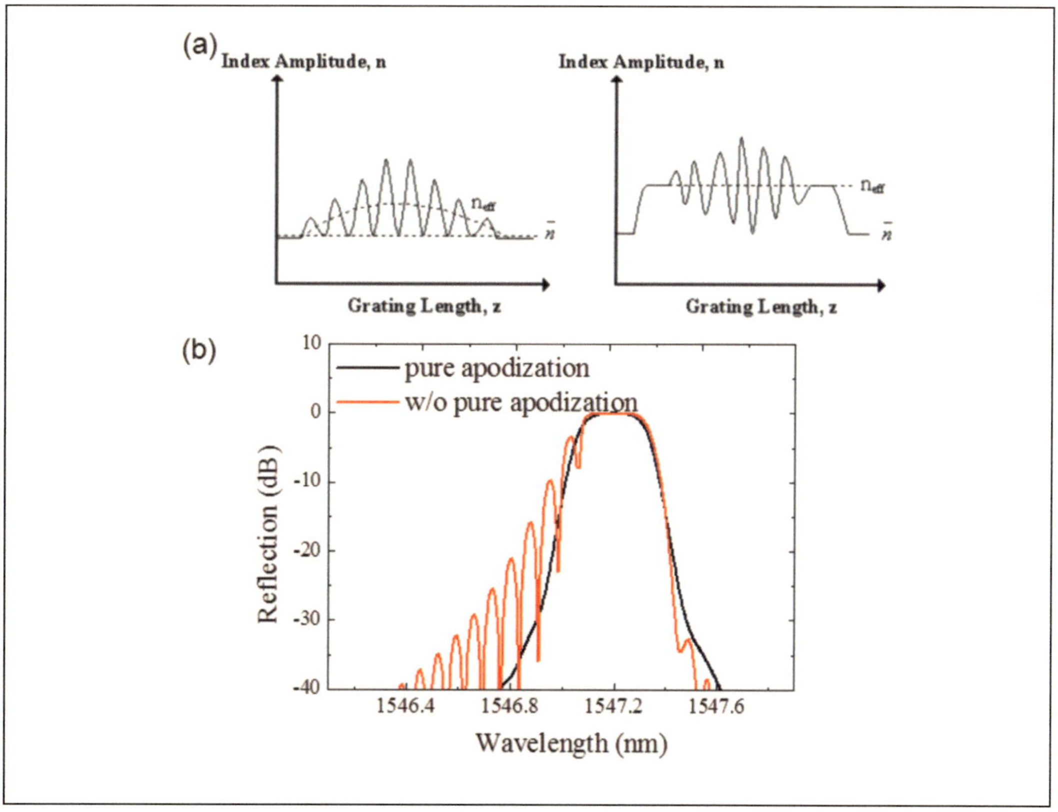

Fig. 2. Pure apodization of Gaussian apodize (a) refractive index profile (b) spectrum with and without pure apodization.

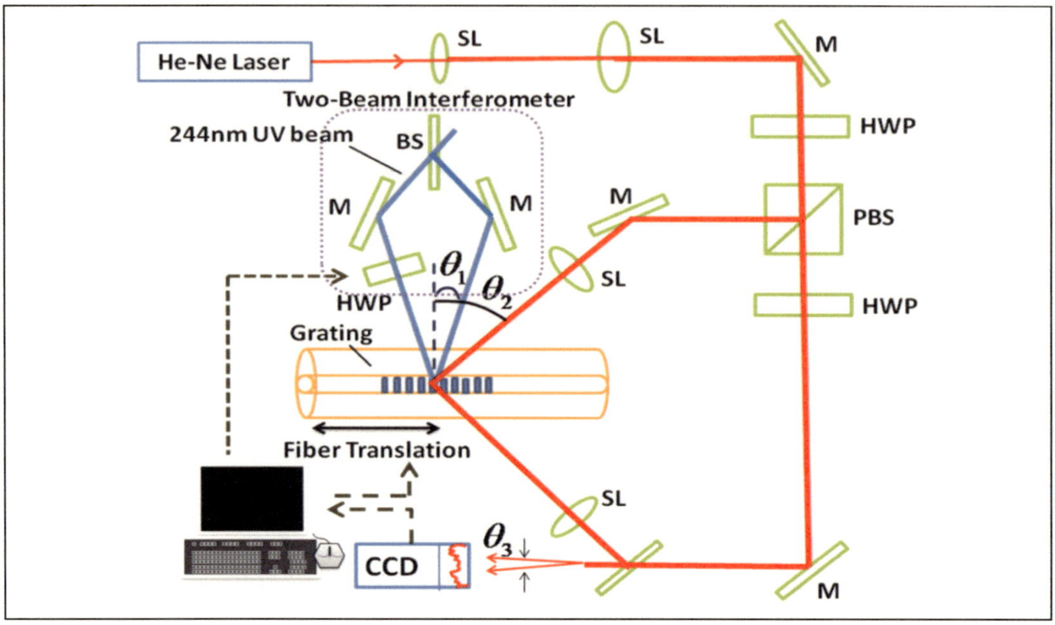

Fig. 3. Interferometric side-diffraction position monitoring technique for writing long fiber Bragg gratings. (Hsu et al., 2005)

demultiplexed. Conversely, a multiplexer can be achieved by cascading multiple add sections of the OADM. FBG demultiplexers and OADMs can also be tunable. Narrow-band FBGs at two ends of rare-earth-doped fibers form Fabry-Perot laser cavities as DFB (distributed feedback) lasers that support single-longitudinal mode operation (Qiu et al., 2005). DBR (distributed Bragg reflector) fiber laser is obtained by putting a π-phase-shifted grating on the rare-earth-doped fibers, so that the grating is treated as a narrow-band transmission filter. In high-power fiber laser systems, the high and low reflectors are mission-critical elements that have a significant impact on the system's performance and reliability. Semiconductor diode laser with short cavity length results a stable single-frequency operation; and the output is coupled into an optical fiber with low reflectivity FBGs (2-4%) incorporated in output fiber end under external feedback mechanism to efficiently suppress mode hopping and reduce output noise (Archambault et al., 1997).

For the applications in EDFAs, FBGs are quite useful for gain-flattening, pump reflection and wavelength stabilization. To maintain a reasonable amount of population inversion in the gain medium, a counter-propagating amplifier configuration is used for optimum power conversion efficiency, and the use of a broad, highly reflecting FBG is needed to double pass the pump light in the amplifier. Tilted FBGs and LPGs with proper designs can couple the guided modes into the cladding to attain flattened EDFA gain spectrum. Another method of fiber amplifier gain equalization is obtained by appropriate choice of individual FBG loss within the gain bandwidth. Furthermore, the center wavelength can be fine-tuned by adding stress on the FBGs to change its period, and the wideband tunability of FBGs widely broadens the application area (Liaw et al., 2008). Besides, chirped fiber gratings as dispersion compensators are widely applied in optical communication systems to compensate chromatic dispersion, or compensate anomalous or normal dispersion caused by the nonlinear effects for pulses propagating in the fiber.

3. FBGs play as solo-function role in a fiber amplifier

The fiber Bragg gratings have been widely used in optical amplifier design for achieving various functions. FBG acts as solo function including fixed or dynamic gain equalization, the dispersion compensation, and the signal and pump reflectors are introduced in this section.

3.1 Fixed and dynamic gain equalization

The gain equalization of the EDFA in a multi-channel wavelength division multiplex (WDM) system can be realized by using LPG (Vensarkar et al., 1996). The unwanted power is coupled from the guided mode to the cladding modes through the following phase matching condition:

$$n_{co} - n_{cl}^{m} = \frac{\lambda}{\Lambda^{m}} \tag{6}$$

where n_{co} and n_{cl}^{m} are the effective core mode index and the cladding mode index, respectively. m is the order of the cladding mode and λ is the signal wavelength in free space. Λ^{m} is the grating pitch that attains the phase matching criteria for coupling the core mode into the m-th cladding mode. Since the index difference between the core mode and the cladding mode is very small, the typical pitch of the long-period grating is in the order of several hundreds of micrometers. Arbitrary spectral shape can be realized by cascading several LPG with appropriate resonance wavelengths and grating strengths. The transmission spectrum of the gain-flattening filter using two cascaded LPG is shown in Fig. 4. (Vensarkar et al., 1996) The flatness is within 0.2 dB over a 25 ~ 30 nm bandwidth.
In the re-configurable add-drop multiplexer system, the power of the add-drop channel changes. Such power variations among channels lead to substantial differences in the signal powers and the signal-to-noise ratios. Thus, the dynamic gain equalization for the fiber

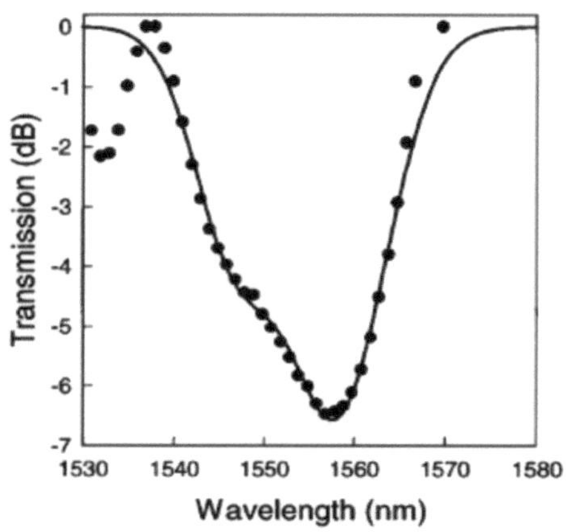

Fig. 4. Transmission spectrum of the gain-flattening long-period fiber grating. Filled circles: inverted erbium fiber spectrum; solid curve: transmission spectrum of two cascaded long-period fiber grating. (Vensarkar et al., 1996)

amplifier is necessary. The most typical structure is to de-multiplex the channels and insert the variable optical attenuators before the multiplexer (Shehadeh et al., 1995). However, the accumulated component loss is usually large so another fiber amplifier is required to compensate the loss. An acousto-optic tunable filter can also achieve the dynamic equalization (Kim et al., 1998). However, the control of the appropriate RF signal is quite complicated. The strain-tunable FBGs has been proposed for dynamic equalization of the EDFA (Liaw et al., 1999). An FBG is actually a wavelength-selective optical attenuator. By detuning the Bragg wavelength from its original wavelength, the FBG becomes a wavelength-selective tunable optical attenuator. The strain-tunable FBGs are placed after the EDFA in either the transmission or reflection structure with an optical circulator. Four structures of the dynamic equalized EDFA are shown in Fig. 5.

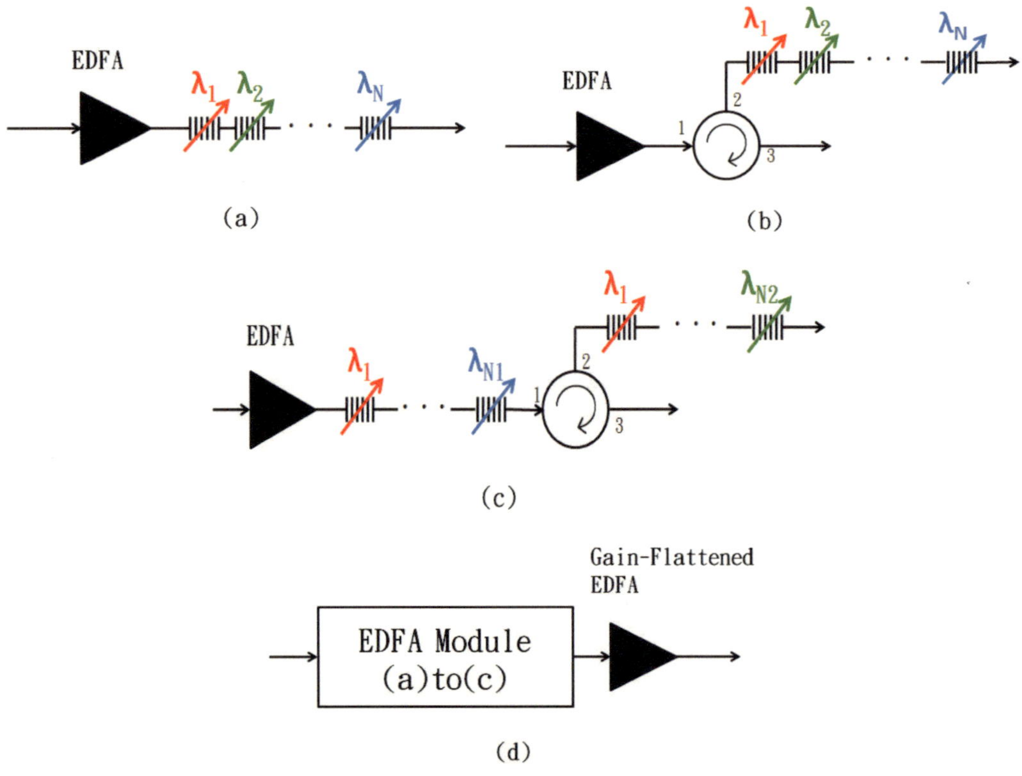

Fig. 5. Schematic diagrams of the dynamic equalized EDFA using strain-tunable FBGs. (a) pass-through structure (b) reflection structure (c) hybrid structure (4) high output power structure. (Liaw et al., 1999)

As no multiplexer and de-multiplexer pair is used, the channel loss is reduced and no optical post-amplifier is required. By stretching or compressing the FBG, the Bragg wavelength is shifted so the reflectance or the transmittance is changed for a specific channel wavelength. The wavelength shift $\Delta\lambda$ is related to the applied longitudinal strain ε as:

$$\Delta\lambda = \lambda(1-p_e)\varepsilon \quad (7)$$

where p_e is the photoelastic coefficient of the fiber. The applied strain can be controlled with high precision by using a piezoelectric transducer. The spectra of a train-tunable FBG with and without applying strain are shown in Fig. 6. The Bragg wavelength is shifted from 1555.4 nm to 1556.5 nm. The reflectivity of the FBG is over 99% and the 10- and 20-dB bandwidths are 0.25 and 0.6 nm, respectively. The dynamic range of the strain-tunable FBG between the two tuning points is as large as 20 dB and is enough for most system applications.

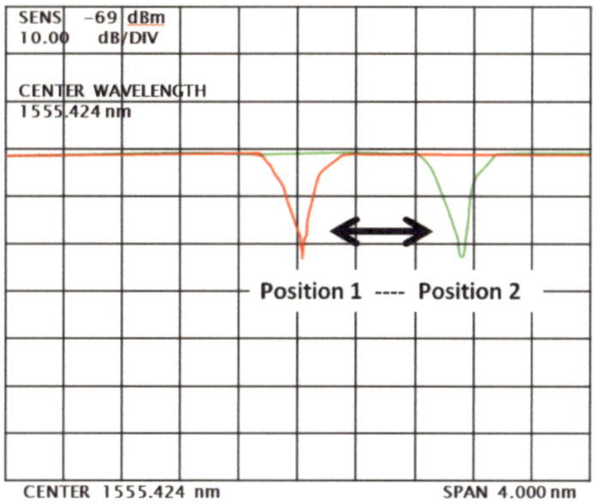

Fig. 6. Transmission spectrum of a strain-tunable FBG. The Bragg wavelength is 1555.4 nm without applied strain (position 1) and 1556.5 nm with applied strain (position 2). (Liaw et al., 1999)

The measured individual channel spectra of a five-channel equalized EDFA module is demonstrated in Figure 7. Figure. 7(a) shows the signals before the FBG chain. The power variation between the input channels is as high as 11 dB. Figure 7(b) is the transmission spectrum of the cascaded strain tunable FBGs. Figure 7(c) shows the output signals after the FBG chain. The power variation between channels is less than 0.3 dB after equalization.

3.2 Dispersion compensation

The chirped FBG is an alternative to the conventional dispersion compensation fiber (DCF) to compensate the dispersion in the optical fiber transmission link. (Hill et al., 1994) The DCF is a long section of fiber with significant loss and high non-linearities due to its small core diameter. The chirped FBG is a compact, all-fiber device with a short interaction length and low non-linearities. The period Λ of a chirped FBG is non-constant. The chirp parameter is expressed as: $d\lambda_D/dz$. $\lambda_D \equiv 2n_{co}\Lambda$ is the designed wavelength for Bragg scattering. The dispersion of a linearly chirped FBG can be estimated as:

$$D \approx 100\left(\frac{d\lambda_D}{dz}\right)^{-1} \text{ (ps/nm)} \tag{8}$$

where the chirp parameter $d\lambda_D/dz$ of the FBG is in units of nm/cm. The chirped FBG has a wider reflection bandwidth than the uniform FBG does because of its non-constant grating pitch. The chirped FBG is further apodized with a suitable index-change profile for an equalized performance. The FBG with a Sinc apodization function demonstrated the optimum performance for both the ideal Gaussian pulses and a direct modulated laser. (Pastor et al., 1996)

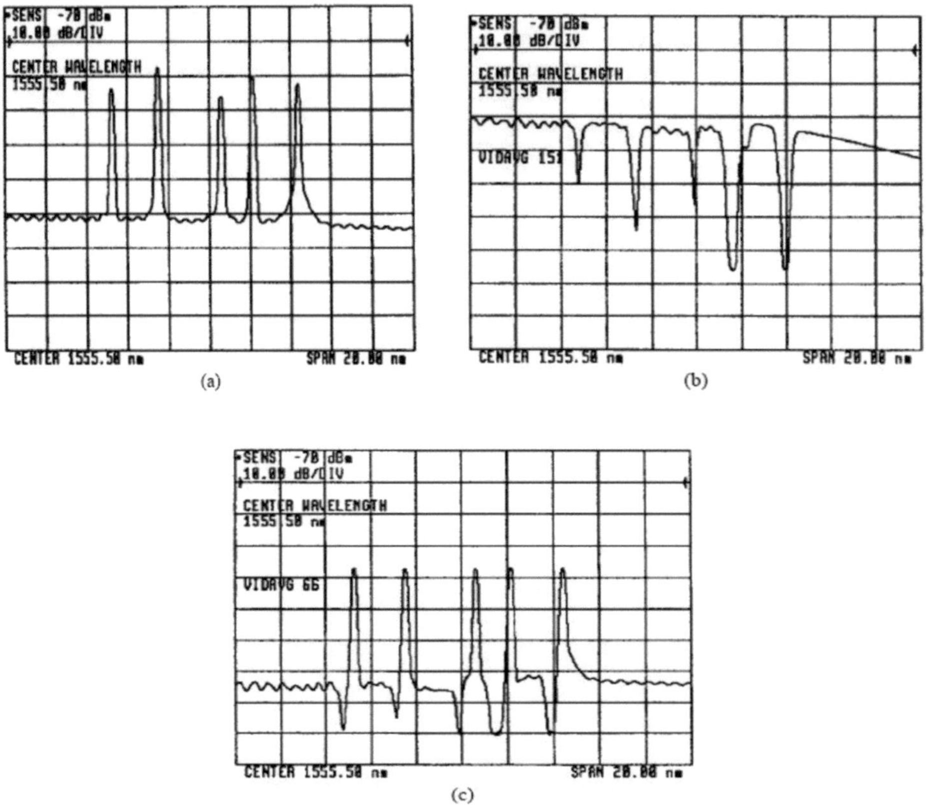

Fig. 7. Spectra of the 5-channel EDFA system (a) spectra of the 5 channels before the strain-tunable FBG chain (b) transmission spectra of the FBG chain. (c) equalized output signal channels after the FBG chain. (Liaw et al., 1999)

The single channel transmission over a 700 km distance for the 10 Gb/s signal was demonstrated by using a chirped FBG. (Loh et al., 1996) The chirp FBG length was as long as 10 cm for achieving dispersion as high as 5000 ~ 8000 ps/nm. Multi-channel transmission using the chirped FBGs for dispersion compensation is much more complicated. The disadvantage of the chirped FBG is its limited bandwidth. However, by increasing the FBG length up to meter range, the bandwidth is extended. The simultaneous dispersion compensation for multi-channels using the chirped FBG is possible. Transmission of the 16x10 Gb/s WDM system over 840 km single-mode fiber was demonstrated using the chirped FBGs. The chirped FBGs used were 1-m long with a nominal dispersion of -1330 ps/nm over a bandwidth as wide as 6.5 nm to compensate the dispersion of the 16 channels at the same time, as shown in Fig. 8 (Garrett et al., 1998). The chirped FBGs are packaged with the optical circulators in the dispersion compensation modules. The insertion loss of the module is 3-4 dB. The grating modules are inserted between stages of the 2-stage EDFAs. The signals were amplified and dispersion-compensated for every 80 km single-mode fiber span, as shown in Fig. 9. (Garrett et al., 1998)

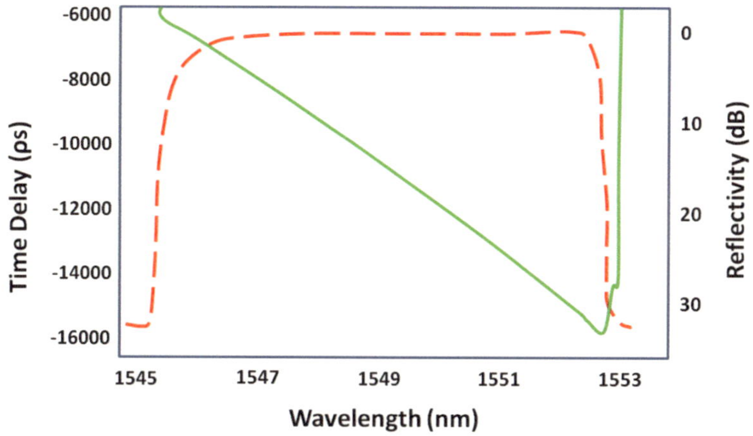

Fig. 8. Characteristics of the chirped FBG. (Garrett et al., 1998)

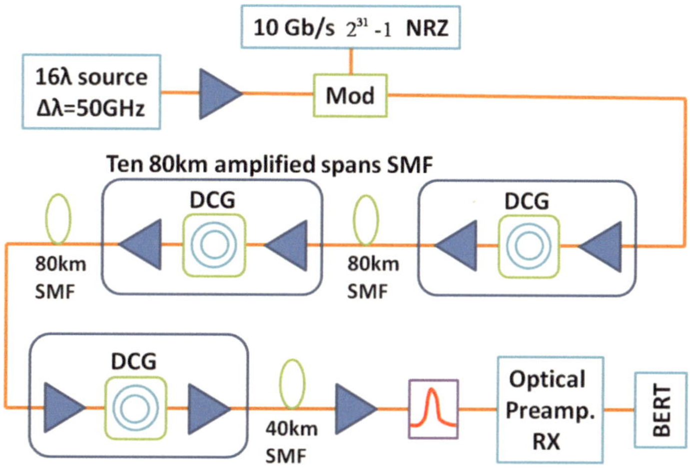

Fig. 9. Schematic of the transmission system. Total fiber length: 840 km. DCG: dispersion compensating grating. (Garrett et al., 1998)

3.3 Signal/pump reflection

Various optical fiber amplifier configurations have been proposed with improved performance utilizing filters or reflectors. The FBGs were used as the wavelength-selective reflectors in the fiber amplifier for the signals or pump to either increase the amplifier gain or recycle the residual pump power. The signal reflector is usually used with an optical circulator in the fiber amplifier. The signal travels through the gain fiber twice so the small-signal gain is nearly doubled and this amplifier configuration is called the double-pass structure. Also the ASE is suppressed by the FBG signal reflector since the FBG is a narrow-band reflector. The pump reflector is helpful to increase the amplifier output saturation power, and the small-signal gain with an increase of 1 ~ 3 dB. The enhancement is related to the amount of the residual pump power. Six configurations of reflected signal and pump in an EDFA are shown in Fig. 10 (a). The calculated gain of the EDFAs with the signal and pump reflectors is shown in Fig. 10 (b). (Giles, 1997)

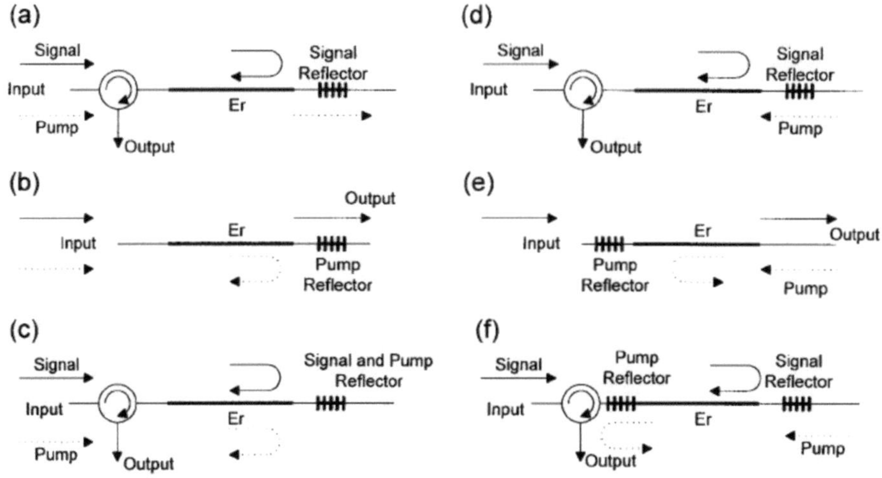

(a)

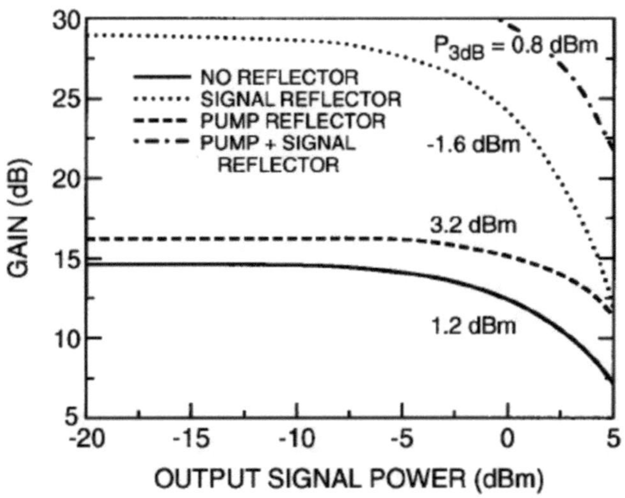

(b)

Fig. 10. (a) Six configurations of reflected signal and pump in an erbium fiber amplifier. (b) Calculated gain of an erbium-doped fiber amplifier with signal and pump reflectors. P_{3dB} is the 3-dB output saturation power. (Giles, 1997)

The pump power required in the Raman fiber amplifier is usually high. By using the double-pass Raman amplifier configuration, the required pump power is reduced nearly 50%. The configuration of the double-pass Raman fiber amplifier is shown in Fig. 11(a). (Tang, 2003) An FBG with R> 99% and a stop bandwidth of 0.2 nm was used as the signal reflector. A section of 3-km DCF was used as the Raman gain medium for its relatively larger Raman efficiency than the standard single-mode fiber. The experiment and simulation results of the Raman gain versus the pump power are shown in Fig. 11(b). (Tang, 2003) The pump power required for 20-dB gain was reduced from the original 29.6 dBm to 26.9 dBm with the use of the double-pass structure.

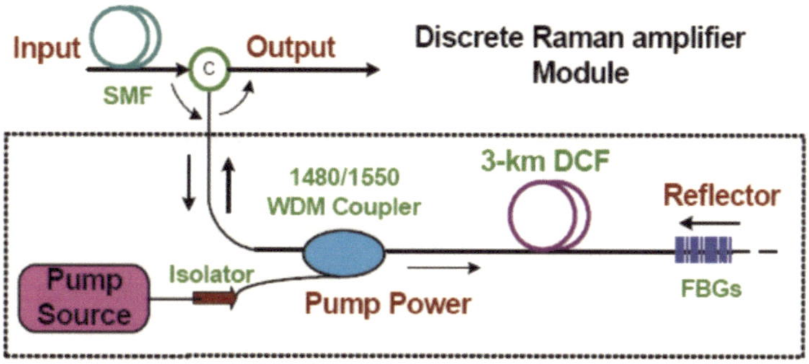

(a)

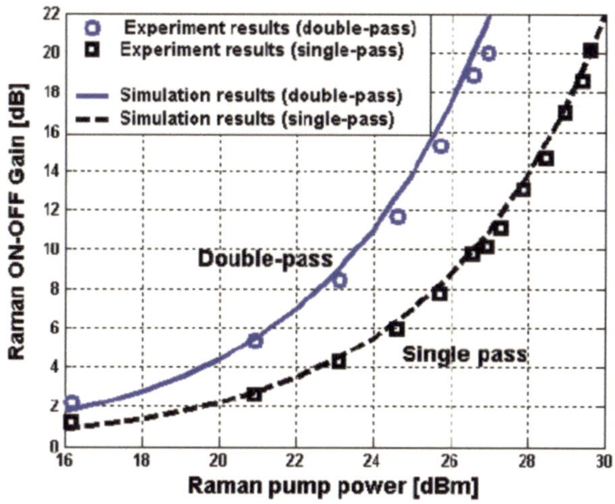

(b)

Fig. 11. (a) Double-pass discrete Raman amplifier configuration and (b) Raman gain versus pump power at the double- and single-pass configuration. (Tang, 2003)

4. FBGs play multiple-function roles in hybrid fiber amplifiers

4.1 Hybrid fiber amplifiers

Conventional erbium-doped fiber amplifiers (EDFAs) operating in the C-band wavelength-division-multiplexing (WDM) system is quite mature nowadays. For the L-band amplification, the Raman fiber amplifier (RFA) has a lower noise figure (NF) than the L-band EDFA and better performance in some circumstances (Jiang et al., 2007). Consequently, a hybrid amplifier is highly promising for terabit dense WDM (DWDM) systems. The hybrid Raman/Erbium-doped fiber amplifier designed for maximizing the span length and/or minimizing the impairments of fiber nonlinearities. It was also used to enlarge the EDFA gain-bandwidth (Curri, C.V. & Poggiolini, P, 2001). In this section, we discuss a serial type hybrid C+L band hybrid amplifier (Liaw et al., 2008), a parallel type hybrid C+L band hybrid amplifier (Liaw et al., 2009), a bridge type C+L band hybrid amplifier (Liaw et al., 2010) and a bidirectional C+L band hybrid amplifier (Liaw et al., 2010). All of them may simultaneously attain gain-flattening and dispersion management of the WDM channels. Figure 12 (Liaw et al., 2008) shows the common concept of four schemes using one high-power pump laser at 1480 nm for C-band EDFA and L-band RFA simultaneously. The C-band EDFA is based on Er^{3+} ions through population inversion amplification mechanism while the L-band RFA is based on Raman shift amplification. If the pump wavelength is 1495 nm, the corresponding gain peak could then be shifted according to the following equation (G. P. Agrawal, 1995).

$$\Delta\lambda = -\lambda \times \frac{\Delta f}{f} = -\lambda^2 \times \frac{\Delta f}{c} = 97nm \tag{9}$$

where Δf = -13THz and $\Delta\lambda$=97 nm are the total amount of detuning with respect to the pump frequency and wavelength, respectively. The maximum gain therefore occurs at around 1592 nm in the L band region. To discuss how critical the pump wavelength is to the amplification band, we compare two pump wavelengths of 1480 nm and 1495 nm, respectively. For the 1480 nm pump LD, its maximum Raman gain will occur at 1575 nm, which will lead to a rather low gain level for the longer RFA region. On the other hand, the 1495 nm pump LD will provide enough RFA gain for the 1595-1610 region. Although the longer pump wavelength will degrade the gain for the C-band EDFA, the gain is still at an acceptable level for the entire C band.

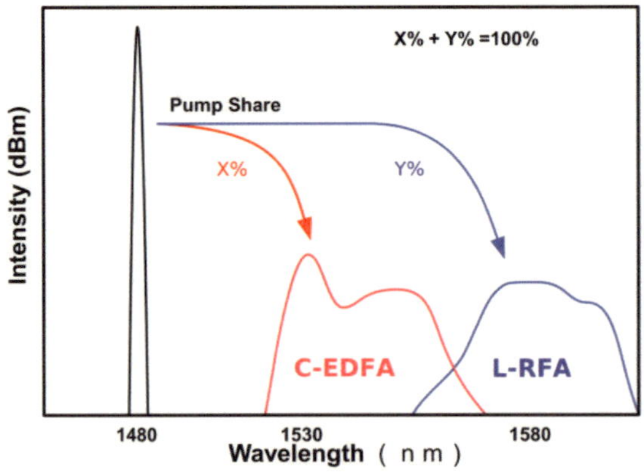

Fig. 12. Concept of using 1480 nm pump source(s) to amplify C-band EDFA and L-band RFA simultaneously. (Liaw et al., 2008)

The EDFA gain is defined as:

$$g = \frac{\sigma_{em}\tau}{h\nu_P}\frac{P_{abs}}{A_{eff}}\frac{F}{\eta_P} \tag{10}$$

where σ_{em} is the emission cross-section, τ is the upper-state lifetime, $h\nu_p$ is the pump photon energy, A_{eff} is the fiber core area, P_{abs} is the absorbed pump power, F is the overlapping integral between the pump and signal fields in the transverse dimensions, and η_p is the fractional pump energy. The gain of L-band RFA is defined as (Liaw et al., 2007).

$$G_A = \exp(\frac{g_R P_0 L_{eff}^p}{A_{eff}}) \tag{11}$$

where g_R is the Raman gain coefficient as the wavelength difference function between the signal and pump, P_0 is the pump power at the amplifier input, and L_{eff}^{p} is the effective pump length.

4.2 Serial type hybrid amplifier

Figure 13 (Liaw et al., 2008) shows the proposed serial type hybrid amplifier configuration. Both C+L-band signals are combined via a C/L WDM coupler. After that they will come into the OC1 and pass through a common segment of the dispersion compensation fiber (DCF). The L-band signals are amplified by an RFA and the residual pump power then comes into the EDF for C-band signal amplification. There is a C-band pump reflector at the far end to reflect the residual pump power for further EDF pumping. The DCF group is composed of several DCF segments in different lengths with one FBG for each section. Each FBGj has a central reflected wavelengths to match to a certain signal. Each signal travels through different DCF lengths before being reflected by its corresponding FBG for precise dispersion compensation. Note that the double-pass route may save 50% of the gain mediums (i.e. C-band EDF and L-band DCF). The Raman pump power travels through the entire DCF segment and the residual pump power is partially reflected by the L-band reflector located between the EDFA and the RFA.

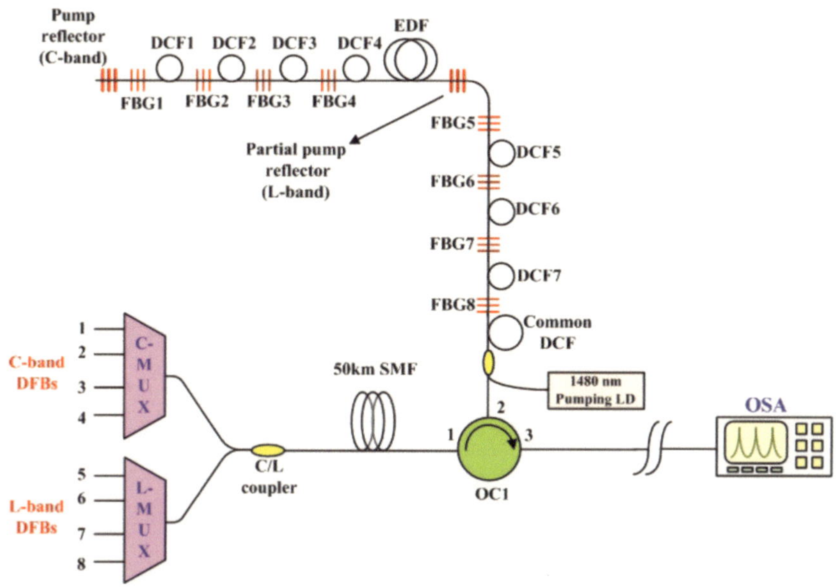

Fig. 13. Serial type hybrid amplifier. (Liaw et al., 2008)

For the pump sharing issue, the optimum pump sharing ratio is reached when 1550 nm and 1580 nm signals have the same gain. First, the total residual pump power is for the C band only by setting 0% reflectivity for the L band reflector. After that, we increase the reflectivity of L band reflector to reduce the residual pump power for C band until the gain of them are equal. Figure 14 (Liaw et al., 2008) shows the gain at 1550 nm (C band) and 1580 nm (L band) versus residual pump power for L band reflector. Where the horizontal axis means reflectivity of the L band reflector ($0 \leqq R \leqq 100$). The gain at 1550 nm with or without reflector means the C band reflector of 100% reflectivity is used or not. The gain for 1550 nm (C band) is 0- to 2.5 dB improved with the C band reflector. The suggested optimum reflectivity for L band reflector is 87.5% as the 1550 nm and 1580 nm have identical gain value. In this simulation, the total pump power is 545 mW and the residual pump power after passing through the L band RFA is 160 mW. So, the idea launched power for the C band is 20 (160×12.5%) mW.

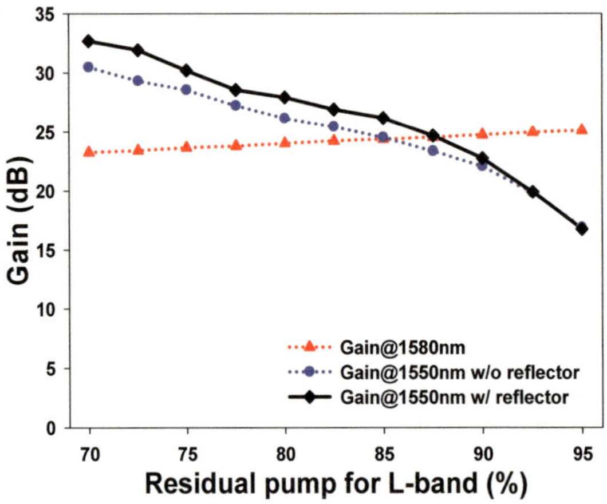

Fig. 14. Signals gain versus residual pump for L band (%) reusing. (Liaw et al., 2008)

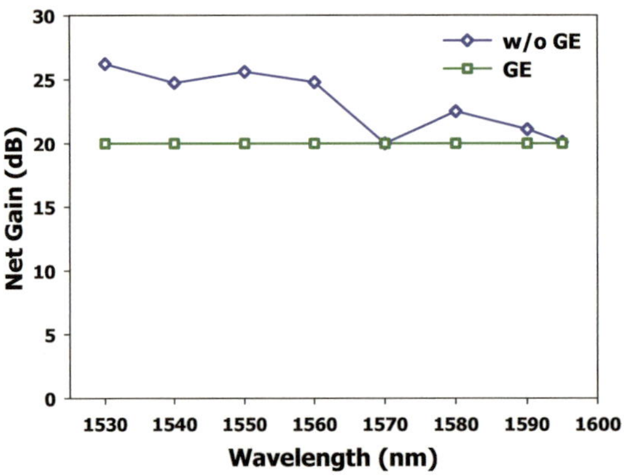

Fig. 15. The WDM signals gain equalization by adjusting the reflectivities of FBGs (Liaw et al., 2008)

For dispersion management, the travel length for each signal in the DCF is controlled by a corresponding FBG, the central wavelength of which is designed to match the signal wavelength. Each required DCF segment length can be predicted based on the following equation:

$$L_{DCF} = -\frac{L_{SMF} D_{SMF}(\lambda)}{2D_{DCF}(\lambda)} \quad (12)$$

where D_{SMF} and L_{SMF} are the SMF dispersion parameter and the length, respectively, and D_{DCF} is the DCF dispersion parameter. To minimize the residual WDM channel dispersion in a system with 50 km SMF, the required DCF length is 6987 m for the longest wavelength at 1595 nm. It has the minimum dispersion value using this common DCF. As the signal wavelength decreases, the extra dispersion value increases. The residual dispersion value may be precisely compensated by writing FBGs (FBG1, FBG2, . . . , FBG8) at different positions for individual signal channels. So, the residual dispersion can be eliminated exactly. To implement the gain equalization for C+L-band channels, the reflectivity of eight FBGs are set as 100% in the beginning before (without) gain equalisation (GE), as shown in Fig. 15. (Liaw et al., 2008) We adjusted the L-band pump reflector as 87.5%, as mentioned. Next, we equalizd the WDM channels by adjusting the reflectivities of the FBGs. The optimum FBGs' reflectivities are calculated to be 23, 37, 29 and 33% for the C band channels; and 100, 48, 67 and 98% for the L band channels, respectively. Thus, gain variation among eight WDM channels is less than 0.2 dB in the whole C + L region using one single-pump laser diode.

4.3 Parallel type hybrid amplifier

The proposed hybrid C+L band EDFA/RFA shares the same pump LD as shown in Fig. 16 (Liaw et al., 2009). At the input part, a C/L-band WDM coupler is used to separate the C-band and L-band signals. The C-band signals are then amplified using an EDFA and the L-band signals are amplified with a RFA. They share the same pump LD at 1480 nm. The EDFA and RFA gains are designed to equalize the signal level of C+L band channels. Both the C-band and L-band WDM signals travel through the optical circulator (OC) from port 1 and are then divided by the left-hand side C / L WDM coupler neighboring to OC. There are eight channel signals in the C band amplified by the forward pump EDFA. The C band pump reflector at the end of the EDF may reflect part (rather than 100%) of the residual pumping power to pass through the EDF again. Meanwhile, the L band signals will by-pass the EDF by going through a piece of standard fiber in-between the C/L band WDM coupler pair to avoid being absorbed by EDF. After that, the C+L band signals together with the Raman pumping power are fed into the dispersion compensation module (DCM). Each FBG is matched to a certain channel signal. Inside the DCM, different signals travel through different DCF lengths/segments. For example, signal L2 passes through DCF_{L2} and is then reflected by FBG_{L2}. Signal C2 passes through DCF_{L2}, DCF_{L3}...DCF_{C2} is then reflected by FBG_{C2} and so forth. The Raman pump travels through all DCF segments and those FBGs in between them, and its residual pumping power will go back to this DCM again according to this configuration. Thus, the pumping light also double-passes the gain medium of the DCF to increase the pumping efficiency. All WDM signals travel back to the DCM after being reflected by their corresponding FBGs and are then divided into C- and L bands signals again by the right-hand-side C/L band WDM coupler. The C band signals are amplified again in the EDF section and are then combined with the L band signals via the left-hand-side C/L band WDM coupler. The combined C+L band signals travel to the terminal coming out of port 3 of the OC.

After the C band signals are amplified twice, the gain and noise figure (NF) of the amplifier for the WDM signals may be calculated at port 3 of the OC. Fig. 17 (a) and (b) (Liaw et al., 2009) show the gain and NF characteristics with and without gain equalization. Note that from 1530 nm to 1545 nm, the gain is less than 20 dB when the FBG reflectivity is 99%, as shown in Fig. 17(a). It is interesting to find that the gain values increase to about 20 dB as the FGB reflectivity is reduced. The reason may be attributed to the EDFA homogeneous broadening characteristics (HBC). As the reflectivities of signals in the longer wavelength range of 1550-1565 nm drop lower than those in the 1530-1545 nm range, the shorter wavelengths will obtain extra gain thank to HBC even though reflectivities of the corresponding FBGs are reduced.

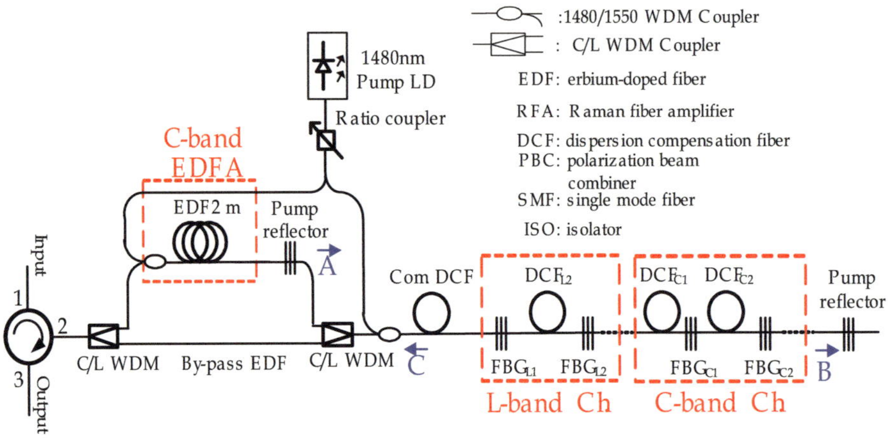

Fig. 16. Parallel type hybrid amplifier. (Liaw et al., 2009)

4.4 Bridge type hybrid amplifier

Figure 18 (Liaw et al., 2010) shows the bridge-type hybrid amplifier scheme using single-wavelength pump source. All C+L band channels reach the 3-port OC after 50 km SMF transmission and then divided into two groups using a C/L WDM coupler. The C band signals enter the C band branch at the upper side and are then amplified by a pump source. A pump reflector is positioned at the EDF end which reflects the residual pump power for further EDF segment pumping. The L band signals enter the dispersion compensation module (DCM) via the L band branch of the C+L WDM coupler and then pass through the same DCM in the opposite direction. They will then be reflected by the corresponding FBGs and then travel back along the same DCF. A commercial SPF (also named as C band filter) may be located after the pump reflector to suppress the L-band amplified spontaneous emission (ASE) noise. When the residual signals enter the opposite band of the C+L WDM coupler (e.g., L band signals enter C band branch or vice versa), most of the residual signals and noise will be suppressed. The DCM is composed of several DCF segments with a FBG for each. Note that the path is a round-trip scheme for all signals to save 50% of the DCF length. After passing through several DCF segments, the C band signal λ_{Cj} comes from the upper side of the DCM and travels along the fiber until it is reflected by FBG_{Cj} ($1 \leqq j \leqq M$). After that, λ_{Cj} will travel back along the same path and then be amplified by EDFA again. Meanwhile, the L band signal λ_{Lk} ($1 \leqq k \leqq N$) comes from the lower side of the DCM and is then reflected by the FBG_{Lk} after passing through several DCF segments. Note that the DCM functions as a double-pass gain medium for the L band RFA. Finally, the C+L band signals are merged via the C/L WDM coupler, and then travel to their destination after leaving the 3-port OC. Without loss of generality, we assume there are eight channels ranging from 1530-1565 nm for the C band and six channels ranging from 1570-1595 nm for the L band. The designed input power is -20 dBm per channel before being launched into

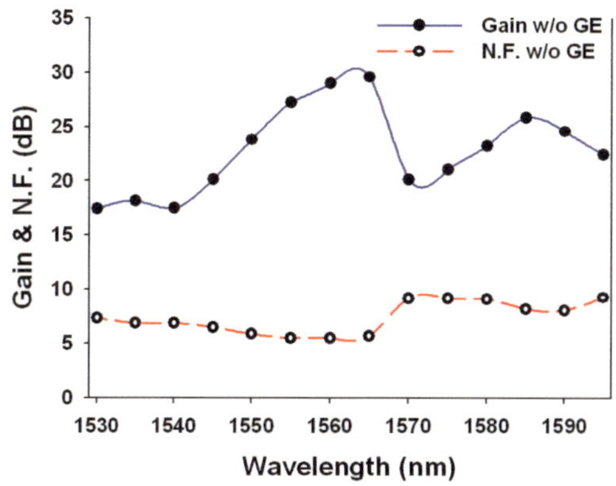

(a)

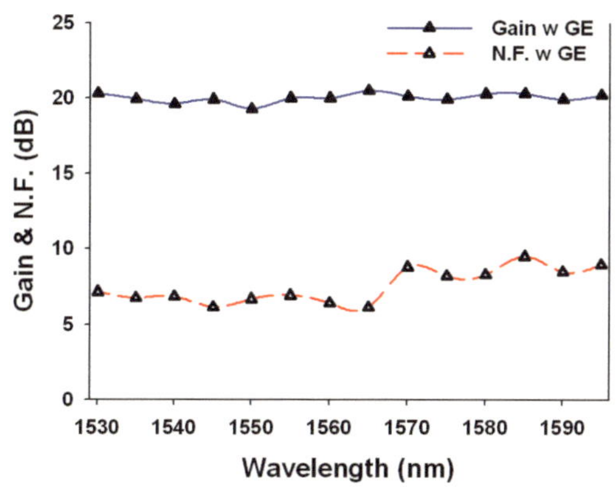

(b)

Fig. 17. The gain and noise figures for the hybrid amplifier (a) without gain equalization (b) with gain equalization. (Liaw et al., 2009)

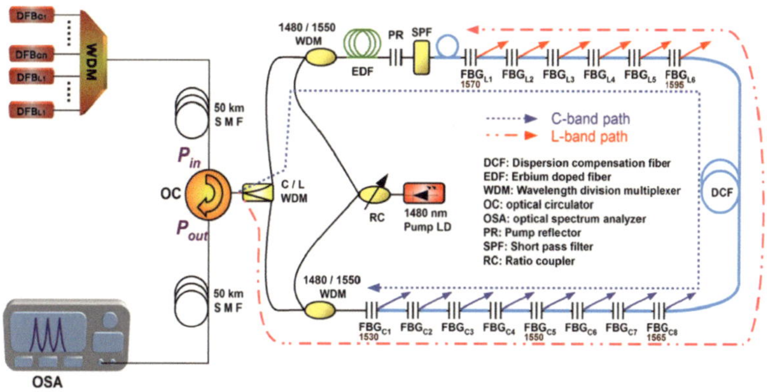

Fig. 18. The bridge type hybrid amplifier scheme. (Liaw et al., 2010)

the 3-port OC and the output power is 0 dBm after leaving port 3 of the OC, corresponding to a net gain of 20 dB for all C+L band channels. For dispersion compensation at 1595 nm in 100 km SMF transmission, a DCF length of 1772 m (i.e., 8860 m in double-pass scheme) is required. Note that 8860 m is the common DCF length required, then the extra DCF needed for compensating the other WDM channels' residual dispersion could be achieved by writing the corresponding FBGs at different positions.

To implement the gain equalization for C+L band channels, the reflectance of fourteen FBGs are set as 99% at the beginning before (w/o) gain equalization (GE), as shown in Fig. 19(a) (Liaw et al., 2010). The maximum gain is 27 dB at 1530 nm and the minimum gain is 20.5 dB at 1570 nm, respectively. Next, we equalize the WDM channels by reducing the FBG reflectance (except that of 1530 nm) to achieve 20.5 dB net gain for all WDM channels. As shown in Fig. 19(b) (Liaw et al., 2010), the required FBG reflectance is calculated to be 23%, 24%, 37%, 30%, 29%, 29%, 33%, and 68% for the C band channels, and 99%, 69%, 48%, 51%, 67%, and 98% for the L band channels, respectively, with a 5 nm channel spacing. We calculate both the forward and the backward noises in DCM until it is convergent. The NF values of the channels range from 6.8 to 7.3 dB for the C band channels and 5.4 to 6.6 for the L band channels, respectively.

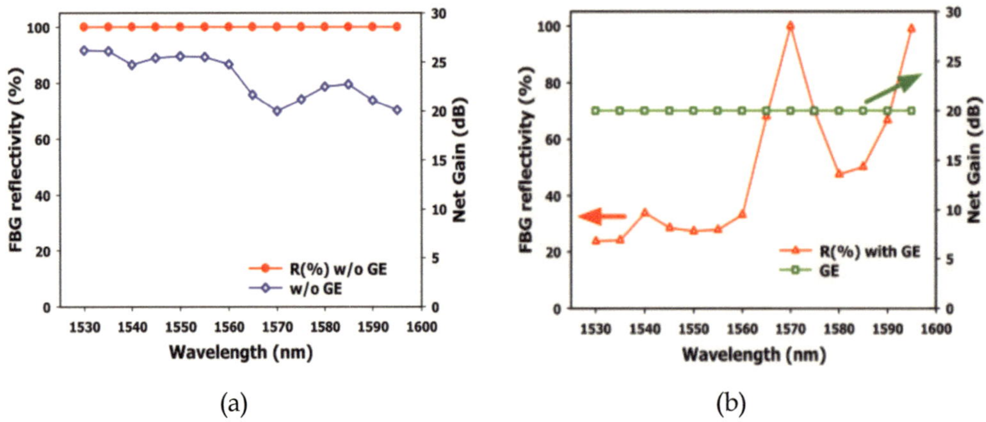

Fig. 19. The net gain spectra and FBGs reflectance for all WDM channels (a) before gain equalization and (b) after gain equalization. (Liaw et al., 2010)

4.5 Bidirectional hybrid amplifier

Figure 20 (Guo et al., 2010) presents the hybrid fiber amplifier located in between two OCs. The hybrid fiber amplifier is then applied in a 50-km bi-directional LAN. The C band signals carry downlink data and the L band signals carry uplink data. Bi-directional operation can be realized using a pair of OCs. The pump light propagates together with the L band signals first. Bi-directional pumping is used for the L band signals to reduce the RFA polarization dependent gain (PDG) and to increase its gain. The residual pumping power transmits partially through the pump reflector to provide backward pumping for the C band signals in EDF. This pumping direction may increase the EDFA gain. In this pumping direction and ratio distribution we may obtain optimum gain results for both the EDFA RFA, respectively. For the optimal dispersion compensation issue, the C-band FBG arrays (FBG_{C1}, FBG_{C2} ... and FBG_{Cn}) together with DCF segments may optimally compensate the chromatic dispersion for all downlink channels. Different downlink channels are reflected by the corresponding FBGs at different positions, thus, they undergo different DCF lengths. Meanwhile, the L-band FBG array (FBG_{L1}, FBG_{L2} ... and FBG_{Ln}) optimally compensate for the chromatic dispersion of all uplink channels with various DCF lengths. The principle is similar to that of the C-band channels. The pump laser is coupled into the bi-directional hybrid amplifier by a 1495/1550 nm WDM coupler and the C/L coupler separates up and down link data. To suppress the possible leftover C-band signal in the L-band receiver or vice versa, the C band and L band filters may be inserted in front of the L band receiver and C band receiver, respectively.

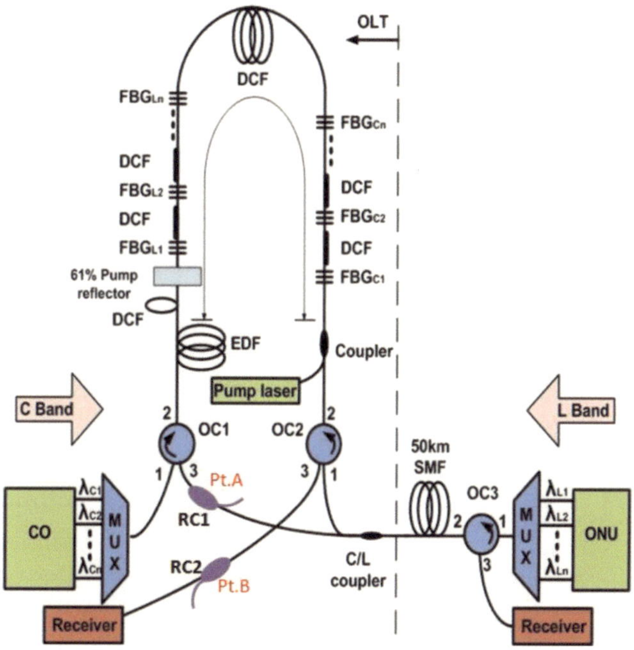

Fig. 20. Bidirectional hybrid amplifier (CO: central office, OC: optical circulator, FBG: fiber Bragg grating). (Guo et al., 2010)

We first performed a theoretical analysis on the proposed hybrid amplifier. All of the channels have an input power of -20 dBm/ch. We then used a common DCF segment with a length of 7.82 km for dispersion compensation to all channels. Thus, only the 1605 nm channel is optimally dispersion compensated while other channels still have residual dispersion. Without loss of generality, experimental work was also carried out for two C band wavelengths (1545 and 1553 nm) and two L band wavelengths (1582 and 1597 nm), respectively. Two 1495 nm pumping sources with a total power of 900 mW were combined via a polarization beam combiner (PBC). They were launched into the DCF via a 1495/1550 nm WDM coupler. Using simulation software, the reflectivities of all FBGs are set at 99% initially before gain equalization and shown as the black squares in Fig. 21 (a). Then we achieve gain equalization by adjusting FBGs' reflectivities with 99.9% for 1570 nm and 4.2% for 1605 nm channel individually as indicated by the red circles. On the other hand, the experimentally measured FBGs reflectivities for 1545, 1553, 1582 and 1597 nm are 9.5%, 5.2%, 20.3% and 9.7% respectively after gain equalization. In Fig. 21 (b), simulation results show that the gain variation is as large as 4 dB among channels, as indicated by black squares before gain equalization. Nevertheless, the variation is greatly reduced to 0.5 dB after gain equalization by adjusting individual FBG's reflectivities which is shown by red circles in Fig. 21 (b). Experiment results also confirm the gain values for channels at 1545, 1553, 1582 and 1597 nm are 4.6, 5.0, 4.8 and 5.1 dB, respectively. The measured NFs after residual dispersion compensation, using the polarization nulling method, for 1545-, 1553, 1582 and 1597 nm are 5.4, 5.0, 4.9 and 5.3 dB respectively. In this design, the L band signals neither transmit through nor interact with the EDF. We also find that the L band signals have no impact on the gain/power and NFs of the C-band signals.

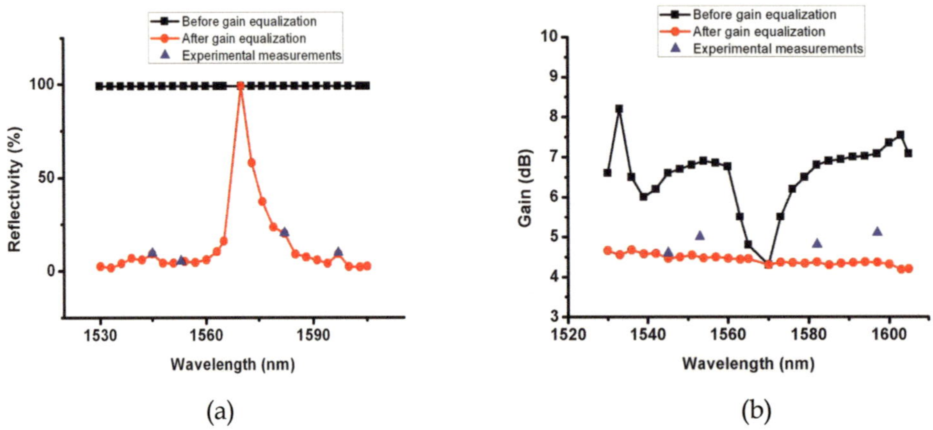

Fig. 21. (a) FBG reflectivity versus wavelength before and after gain equalization; (b) Gain versus wavelength of the hybrid fiber amplifier before and after equalization. (Guo et al., 2010).

4.6 Summary

Four hybrid fiber amplifier for simultaneously amplifying the C band EDFA and L band RFA were studied. These hybrid C+L band amplifiers have several common advantages: (1) The required DCF length for chromatic dispersion compensation is 50% saved. (2) Residual dispersion is exactly compensated for all C+L band channels by writing FBGs at appropriate locations individually. (3) The reduction of gain variation could be realized after optimizing the reflectivity of each FBG. (4) Pump efficiency is improved by recycling the residual pumping power using a pump reflector. (5) The ASE noise and residual signal power that arise from the opposite bands are greatly suppressed by WDM coupler or filter. With these merits, these proposed hybrid amplifiers may find vast applications in uni- or bidirectional WDM long-haul systems, local area network, and optical networks where dispersion management and power equalization are crucial issues.

5. FBG-based optical networks devices with built-in fiber amplifier

5.1 Introduction

Recently, a significant amount of effort has been devoted to the design of high-capacity, flexible, reliable and transparent multi-wavelength optical networks (Liaw et al., 1998). Among them, optical add-drop multiplexer (OADM) is one of the fundamental building blocks for a multi-wavelength optical network (Riziotis et al., 2002), such as the dense wavelength-division-multiplexing (WDM) system and optical networks. A reconfigurable OADM (ROADM) is capable of accessing all wavelengths to and from the WDM optical networks, providing the network flexibility, versatility, survivability (Liaw et al., 1998). Compared to unidirectional transmission, bidirectional transmission has the merit of 50% cost reduction for optical fibers and the required number of passive fiber components may therefore be reduced. In optical networks, an optical cross-connect (OXC) device is also an essential equipment for wavelength exchanging and routing. It allows the optical network to be reconfigured on a wavelength-by-wavelength basis to interchange and optimize traffic patterns, and provides the routing function, facilitate network growth, and enhance network survivability (Brackett et al., 1996).

FBGs have the advantages of better uniformity, high contrast ratio and low cost. Hence, FBG-based unidirectional ROADM, bidirectional ROADM were investigated, with a low-cost optical amplifier acts as a gain provider in both cases. In addition, an FBG-based OXC integrated with optical limiting amplifiers (OLA's) to provide a large dynamic range and self-equalization is also introduced in this section.

5.2 Amplifier in Unidirectional ROADM

Figure 22 (Liaw et al., 2007) shows a unidirectional ROADM which could add/drop one channel simultaneously. The ROADM consists of two three-port OCs, an l x N mechanical OSW pair and N pieces of FBGs. One TFBG is connected to port 1 of the OSW pair while the other (N-1) pieces of FBGs are temperature-compensated with fixed wavelengths. The central reflective wavelengths of the original $TFBG_1$ and the FBG_i, (i=2…4) are designed to match the WDM wavelengths in our experiment. In order to compensate the optical loss, a low-gain optical amplifier is constructed by sharing a low-cost 980 nm pump laser with two pieces of EDFs located between OC1 and OC2. A certain WDM channel can be dropped by adjusting the OSW pair to the proper port. For instance, when port 3 of the OSW pair is connected to the OC pair, the launched λ_3 is reflected by FBG_3 and then dropped from port 3 of the OC1. Meanwhile, the remaining three WDM signals pass through OC2 of the ROADM. Moreover, a new signal λ'_3 with the same wavelength of λ_3 is added into the ROADM via port l of the OC2. If there is no channel to be dropped, the OSW is switched to port 1 and the TFBG is tuned away from the original wavelength of λ_1. In another scenario, if the dropped channel does not match to any FBG wavelength at the beginning, port 1 of the OSW is selected and the TFBG is appropriately tuned to match that certain wavelength. This simple lxN mechanical OSW pair can fulfill these functions when the switching speed is within several msec. To avoid data loss, a faster switch such as an opto-electrical switch or magnetic switch could be used instead.

Figure 23 shows an experimental setup to verify the ROADM performance. Four distributed feedback (DFB) lasers with central reflective wavelengths matched to λ_1, λ_2, λ_3 and λ_4 are externally modulated with a 10 Gb/s 2^{31}-1 pseudo random bit sequence (PRBS), non-return-to-zero (NRZ) format. Two spools of 50 km SMF are used as the transmission link. Two dispersion compensation modules (DCMs) are adopted, with each of them can compensate for 50 km SMF. An optical band pass filter (OBPF) with 3-dB bandwidth of 1.2 nm and insertion loss of 2.0 dB is located at the receiving end to select the WDM channels for detection. An InGaAs avalanche photodiode (APD) of -19.4 dBm sensitivity is used as the receiver with low noise. A small power penalty of 1.2 dB @10^{-9} BER for the added channel is observed. This 1.2 dB power penalty may be due to the residual signal (λ_3) power that contaminates the added channel, and back reflection occurred at the connector interface. Hence, the BER performance confirms the feasibility of the ROADM.

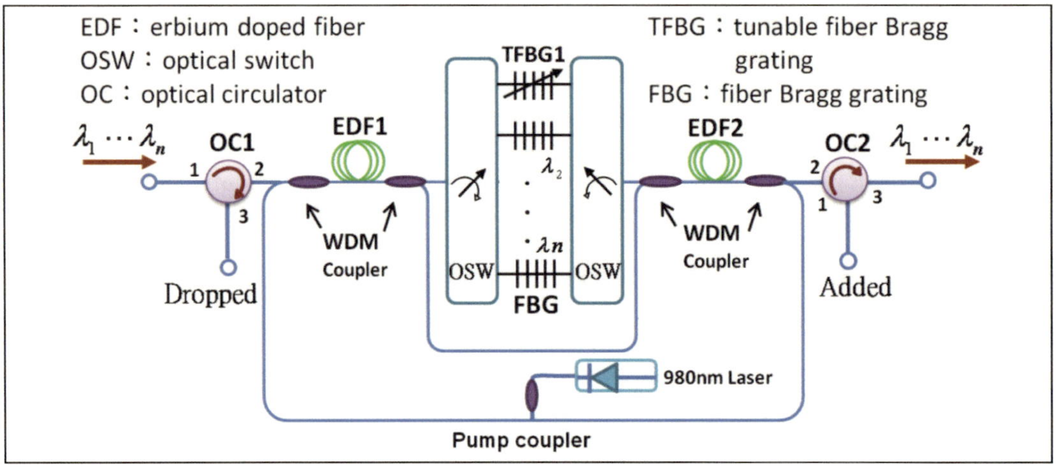

Fig. 22. Scheme of the unidirectional ROADM. (Liaw et al., 2007)

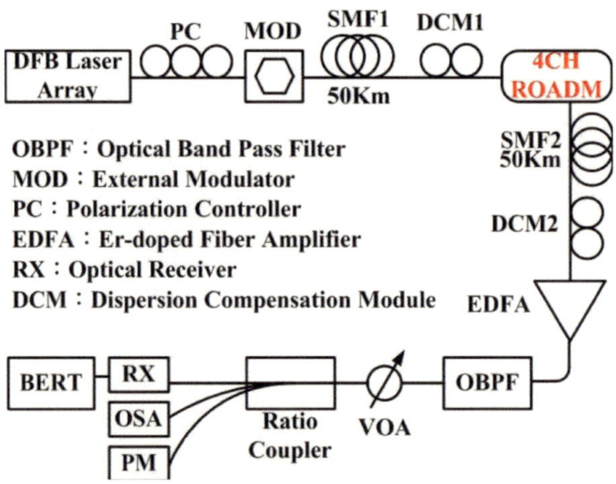

Fig. 23. Experimental set up to verify the BER performance of the ROADM. (Liaw et al., 2007)

This ROADM has four advantages; firstly, it could increase the pumping efficiency by using the residual pumping power to pump other piece of EDF. Either the added, dropped or pass-through channel are amplified twice. Secondly, all channels are pumped along the forward and then the backward paths. Such pump scheme could decrease the NF while increasing the gain of EDFA. Thirdly, the ROADM is cost effective with high reliability. And fourth, it provides a net gain with small gain variation.

5.3 Amplifier in bidirectional ROADM

A bidirectional ROADM (Bi-ROADM) is a promising candidate in Bi-WDM networks or ring networks because it could replace two ROADMs for each transmission direction. The proposed Bi-ROADM is shown in Fig. 24 (Liaw et al., 2008). The three-port OC1 is for adding and dropping downstream signals, while OC3 is for adding and dropping upstream signals, respectively. Here λ_1, λ_3... λ_{2n-1} are downstream signals and λ_2, λ_4... λ_{2n} are upstream signals. 2N pieces of TFBGs were written on EDF or spliced within the EDF segment are used, with whose original reflective wavelength set to match one of the individual channels $\lambda_1, \lambda_2, \ldots, \lambda_{2n}$. The FBG reflectivity should be as high as possible to avoid homodyne crosstalk. The wavelengths could be designated according to the ITU grid of 100 GHz channel spacing $\Delta\lambda$ or its multiple. To reduce the possible interband cross talk, two groups of bidirectional channels $\lambda_1, \lambda_3, \ldots, \lambda_{2n-1}$ and $\lambda_2, \lambda_4, \ldots, \lambda_{2n}$ could be interleaved. The FBG-based ROADM has a narrow 3-dB bandwidth, so the noise near the signal wavelength will be greatly suppressed after the wavelength signal is reflected (filtered). Therefore, there will be less cross talk induced from the adjacent channels compared with a WDM thin-film filter (TFF) or an arrayed waveguide grating (AWG). We define λ_3 as the passed-through signal because it does not need to be dropped midway. If TFBG3 is tuned by 1 nm away, λ_1 will be dropped via port 3 of the OC1. Meanwhile, another new wavelength λ'_1 will come into the Bi-ROADM via port 1 of the OC1 and will reach the four-port OC via its port 2. Finally, λ'_1 will reach the downstream terminal at the right-hand side of Bi-ROADM. Two WDM coupler are used in this Bi-ROADM to launch the pump power at 1480 nm for obtaining enough gain for power compensation.

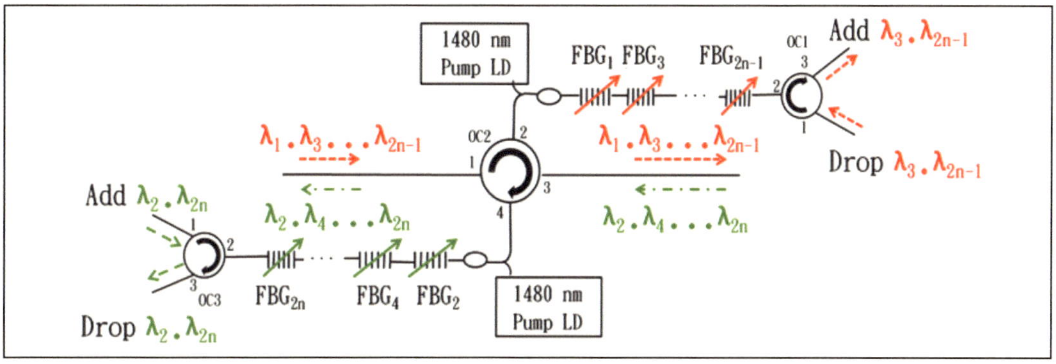

Fig. 24. Bidirectional ROADM. (Liaw et al., 2008)

Figure 25(a) (Liaw et al., 2008) shows variation in power level of the dropped channel λ_1 with or without using an EDFA for power compensation when $TFBG_1$ is tuned away from its original wavelength (λ_1 and λ_3 :downstream, λ_2 and λ_4 : upstream). For broadband and gain equalization among this ROADM channels, the designed channel spacing $\Delta\lambda$ is 2.4 nm for the bidirectional transmission signals in an interleaved wavelength allocation and 600 GHz for the unidirectional transmission. The optimal wavelength shift of the TFBG is 2.4 nm, and can be realized via strain or compression. In this ROADM, it is possible to simultaneously drop multiple wavelengths. Each FBG is individually controlled by a three-point bending device. Approximately 0.7% of the λ_3 signal power will penetrate $TFBG_3$ and then go to the dropped port. The power level difference between λ_1 and λ_3 is 24.79 dB after adding EDFA, which is an interband (heterodyne) cross talk to λ_1 and can be neglected (Ho et al., 1998) because the channel spacing is almost three times the 1.6 nm spacing. Intraband (homodyne) cross talk for λ_3 also occurs. The intend-to-add signal λ'_3 is reflected by $TFBG_3$ and goes to port 3 of the OC1. The power loss for the dropped channel could be compensated by the built-in EDFA. Figure 25(b) (Liaw et al., 2008) shows the superimposed optical spectra of the passed-through signal λ_3 observed at the downstream path output with or without an EDFA. Two weak upstream signals λ_2 and λ_4 will arise due to −30 dB Rayleigh backscattering along the fiber. A residual signal power for λ_1 still remains due to the connector reflection from port 2 of the OC2 and the Rayleigh backscattering of λ_1 when it travels along the path from OC2 to OC1. Figure 25(c) (Liaw et al., 2008) shows the added signal λ'_1 and the passed-through signal λ_3 observed at the output of the downstream path. In summary, the insertion loss of dropped, added, and passed-through signals have the same values to make the Bi-ROADM a symmetric module. The power level of amplified signals in Figs. 25(a)–25(c) are almost the same, making the Bi-ROADM a gain flattening and lossless module.

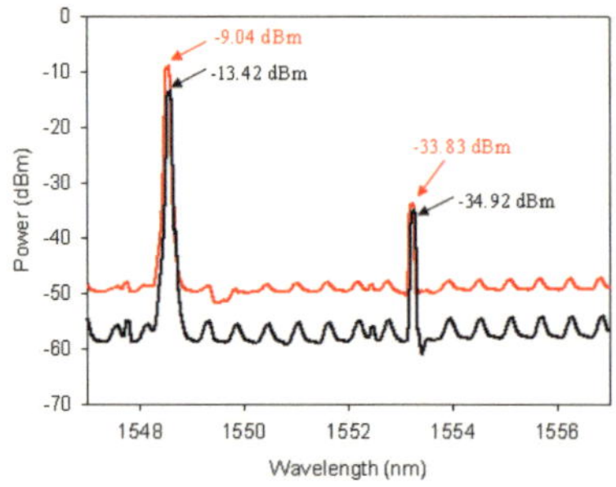

(a)

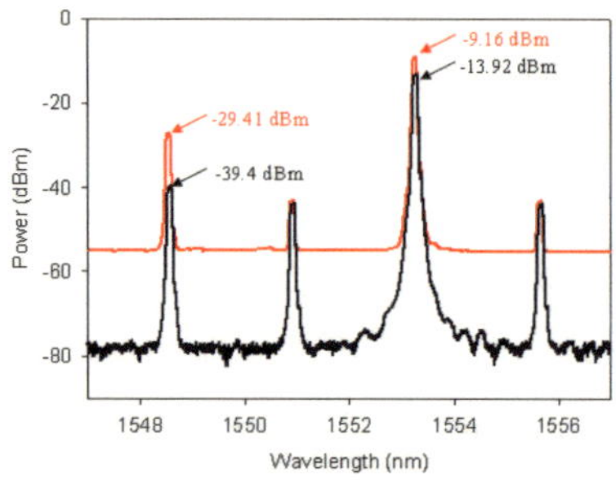

(b)

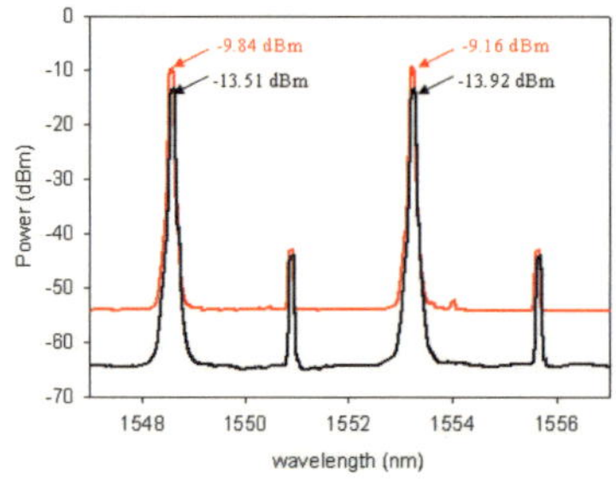

(c)

Fig. 25. (a) Dropped signal (λ_1) with and without use of an EDFA, (b) superposed optical spectra of the passed-through signals (λ_3) at the output terminal of the downstream path with or without use of an EDFA, and (c) added signal λ'_1 and the passed-through signal λ_3 at the downstream output port. Insertion loss of all wavelengths is included during measurement. (Liaw et al., 2008)

5.4 Amplifier in unidirectional optical cross connect

Figure 26 (Liaw et al.,1999) shows the schematic diagram of the unidirectional OXC. There are two input ports I_1 and I_2 and two output ports O_1 and O_2 in the OXC. The OXC also consists of numbers of XC units and two sets of bidirectional EDFAs (Bi-EDFAs). Each XC unit includes one OSW, one short piece of SMF and one FBG_i (i=1, 2, . . . , N). The FBG_i is designed to match to the WDM-channel signals of λ_i and λ'_i transmitted in the upper and lower fiber link. Without wavelength interchange, all signals are reflected by the FBGs due to bar-state status of all the OSW's and then travel back to port 3 of the corresponding OC1/OC2, thus all wavelength channels are the passing channels. If wavelength interchange is required, for example, the exchange of λ_1, λ_N with λ'_1, λ'_N, the corresponding OSW could be switched to the cross-state. Therefore, λ_1, λ'_1, λ_N, and λ'_N will pass through the FBG chain of XC units and exchange to another output port (I_1 to O_2, I_2 to O_1). The WDM channels λ_1, λ'_1, λ_N and λ'_N are reflected by the corresponding FBGs in the XC units and pass through via port I_1 to O_1 and port I_2 to O_2, respectively. Even if λ_1 and λ'_1, for example, appear simultaneously, because the interaction distance is very short and other reflections are small, Rayleigh back-scattering may induce negligible degradation. All WDM channels are amplified twice by the Bi-EDFA(s). The passing channels travel round-trip and are amplified twice by the same Bi-EDFA while the crossing channels are amplified once by each Bi-EDFA. For both cases, two Bi-EDFAs act as the optical limiting amplifiers (OLAs) to improve the input dynamic range and increase the link budget. The laser pump can be shared by two Bi-EDFAs for cost saving. In these implementation of the OXC, all passing channels are reflected by the corresponding FBGs. In another implementation, the label of O_1 and O_2 can be interchanged and all passing channels (i.e., signals from I_1 to new O_1 or from I_2 to new O_2) do not interact with the corresponding FBGs, but rather the crossing channels interact with the corresponding FBGs.

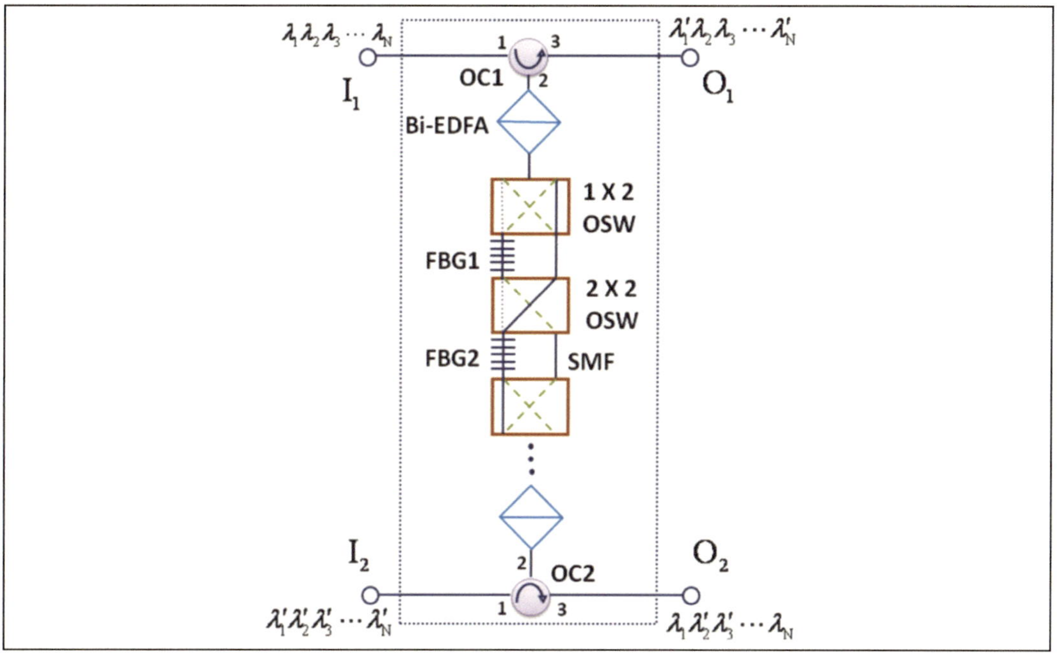

Fig. 26. Unidirectional optical cross connect. Bi-EDFA: bidirectional EDFA. (Liaw et al.,1999)

When TLS1 and TLS2 are launched from I1, Fig. 27(a) (Liaw et al., 1999) shows the passing signal of 1557.1 nm observed at O1 and Fig. 27(b) (Liaw et al., 1999) is the crossing signal of 1559.4 nm observed at O2. The insertion loss of the OXC for both the passing and crossing signals is about 2.5 dB. No Bi-EDFA is used during measure the insertion loss of the OXC.

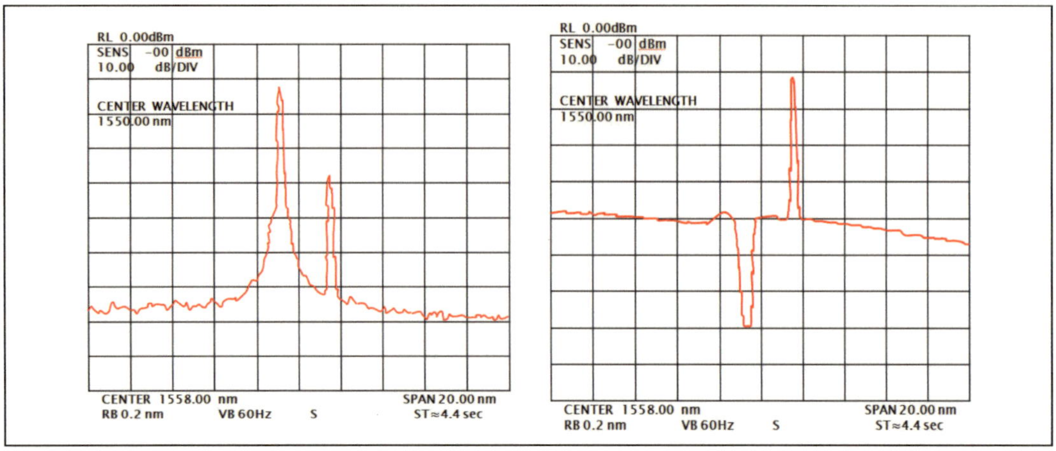

Fig. 27. Optical spectra of (a) the passing signal of 1557.1 nm at port O1 and (b) the crossing signal of 1559.4 nm at port O2. The insertion loss of the OXC for both the pass-through and cross-connect signals is about 2.5 dB. No Bi-EDFA was used during spectra measurement. (Liaw et al., 1999)

5.5 Conclusion

TFBGs with superior tuning ability, high reflectivity, and a high cross-talk rejection ratio were introduced inside the ROADM module. The power compensated ROADM based on TFBG/FBGs is low cost, low noise figure, equal gain among channels and with 8.0 dB net gain to compensate the insertion loss. The Bi-ROADM has benefits of low cost and equal gain among channels. It is symmetric characteristic, simpler, lower power penalties and smaller optical amplifier gain variation than a conventional ROADM system does. The 2×2 OXC based on FBG's has the advantages of low channel crosstalk, high dynamic range, uniform loss spectra for WDM channels, simple operation mechanism and low cost which could provide more reconfiguration flexibility and network survivability for WDM networks.

6. References

Kashyap, R. (2010). *Fiber Bragg gratings,* Academic Press, ISBN: 978-0-12-372579-0, United States of America

Hill, K. O. & Meltz, G. (1997). Fiber Bragg grating technology fundementals and overview. *J. Lightwave Technoogy,* Vol. 15, No. 8, 1263–1276, ISSN: 0773-8724

Kogelnik, H. & Shank, C. W. (1972). Coupled wave theory of distributed feedback lasers. *J. Appied. Physics,* Vol. 43, No. 5, 2327-2335, ISSN: 0021-8979

Yamada, M. & Sakuda, K. (1975). Optical waveguide filters: synthesis. *J. Optical Society of America,* Vol. 65, No. 7, 804-809, ISSN: 1084-7529

Lee, C. L. & Lai, Y. (2002). Evolutionary programming synthesis of optimal long-period fiber grating filters for EDFA gain flattening. *IEEE Photonics Technology Letter,* Vol. 14, No. 11, 1557-1559, ISSN: 1041-1135

Chuang, K. P.; Sheu, L. G. & Lai, Y. (2004). Complex fiber grating structures fabricated by sequential writing with polarization control. *Optics Letters*, Vol. 29, No. 4, 340-342, ISSN: 1539-4794

Hsu, K. C.; Sheu, L. G.; Chuang, K. P.; Chang, S. H. & Lai, Y. (2005). Fiber Bragg grating sequential UV-writing method with real-time interferometric side-diffraction position monitoring. *Optics Express*, Vol. 13, No. 10, 3795-3801, ISSN: 1094-4087

Qiu, T.; Suzuki, S.; Schülzgen, A.; Li, L.; Polynkin, A.; Temyanko, V.; Moloney, J. V. & Peyghambarian, N. (2005). Generation of watt-level single-longitudinal-mode output from cladding-pumped short fiber lasers. *Optics Letters*, Vol. 30, No. 20, 2748-2750, ISSN: 1539-4794

Archambault, J.-L. & Grubb, S. G. (1997). Fiber gratings in lasers and amplifiers. *J. Lightwave Technology*, Vol. 15, No. 8, 1378–1390, ISSN: 0933-8724

Liaw, S.-K. & Jhong, G. S. (2008). Tunable fiber laser using a broadband fiber mirror and a tunable FBG as laser-cavity ends. *IEEE J. Quantum Electronics*, Vol. 44, No. 6, 520-527, ISSN: 0018-9197

Vengsarkar, A. M.; Pedrazzani, J. R.; Judkins, J. B.; Lemaire, P. J.; Bergano, N. S. & Davidson, C. R. (1996). Long-period fiber-grating-based gain equalizers. *Optics Letters*, Vol. 21, 336–338, ISSN: 1539-4794

Shehadeh, F.; Vodhanel, R. S;. Krain, M.; Gibbons, C.; Wagner, R. E. & Ali, M. (1995). Gain-equalized, eight-wavelength WDM optical add-drop multiplexer with an 8-dB dynamic range. *IEEE Photonic Technology Letters*, Vol. 7, 1075–1077, ISSN: 1041-1135

Kim, H. S.; Yun, S. H.; Kim, H. K.; Park, N. & Kim, B. Y. (1998); Actively gain-flattened erbium-doped fiber amplifier over 35 nm by using all-fiber acoustooptic tunable filters. *IEEE Photonic Technology Letters*, Vol. 10, 790–792, ISSN: 1041-1135

Liaw, S.-K.; Ho, K.-P. & Chi, S. (1999). Dynamic power-equalized EDFA module based on strain tunable fiber Bragg gratings. *IEEE Photonic Technology Letters*, Vol. 11, 797–799, ISSN: 1041-1135

Hill, K. O.; Bilodeau, F., Malo, B.; Kitagawa, T.; Thériault, S.; Johnson, D. C.; Albert, J. & Takiguchi, K. (1994). Chirped in-fiber Bragg gratings for compensation of optical-fiber dispersion. *Optics Letters*, Vol. 19, 1314–1316, ISSN: 1539-4794

Pastor, D.; Capmany, J.; Ortega, D.; Tatay, V. & Marti, J. (1996). Design of apodized linearly chirped fiber gratings for dispersion compensation. *IEEE Journal of Lightwave Technology*, Vol. 14, 2581–2588, ISSN: 0933-8724

Loh, W. H.; Laming, R. I.; Ellis, A. D. & Atkinson, D. (1996). 10 Gb/s transmission over 700 km of standard single-mode fiber with 10-cm chirped fiber grating compensator and duobinary transmitter. *IEEE Photonic Technology Letters*, Vol. 8, 1258–1260, ISSN: 1041-1135

Garrett, L. D.; Gnauck, A. H.; Forghieri, F.; Gusmeroli, V. & Scarano, D. (1998). 16x10 Gb/s WDM transmission over 840 km SMF using eleven broadband chirped fiber gratings. *ECOC'98*, pp. 267–268, Madrid, Spain, September 1998.

Giles, C. R. (1997). Lightwave applications of fiber Bragg gratings. *IEEE Journal of Lightwave Technology, Vol.* 15, 1391–1404, ISSN: 1041-1135

Tang, M. & Shum, P. (2003). Design of double-pass discrete Raman amplifier and the impairments induced by Rayleigh backscattering. *Optics Express*, Vol. 11, 1887–1893, ISSN: 1094-4087

Jiang, S.; Bristiel, B.; Jaouën, Y.; Gallion , P.; Pincemin, E. & Capouilliet, S. (2007). Full characterization of modern transmission fibers for Raman amplified-based communication systems. *Optical Express,* Vol. 15, 4883-4892, ISSN: 1094-4087

Curri, C.V. & Poggiolini, P. (2001). On the optimization of the hybrid Raman/Erbium-doped fiber amplifiers. *IEEE Photonics Technology Letter.* Vol. 13, 1170-1172, ISSN: 1041-1135

Liaw, S.-K. & Huang, Y.-S. (2008). C +L-band hybrid amplifier using FBGs for dispersion compensation and power equalization. *Electronic Letters,* Vol. 44, 844-845, ISSN: 003-5194

Liaw, S.-K.; Dou, L.; Xu, A. & Huang, Y.-S. (2009). Optimally gain-flattened and dispersion-managed C + L-band hybrid amplifier using a single-wavelength pump laser. *Optics Communications,* Vol. 282, 4087-4090 ISSN: 0030-4018

Liaw, S.-K.; Huang, Y.-S.; Hung, H.-K.; Chen, N.-K.; Hsu, K.-C.; Yu, Y.-L.; Wang, T.; Manshina, A. & Tver'yanovic, Y. (2010). Dispersion management and gain flattened for a bridge-type hybrid amplifiers in a pumping recycling mechanism. *FOAN workshop, ICUMT 2010.*

Zhou, X., Lu; Shum, P.P. & Cheng, T.-H. (2001). A simplified model and optimal design of a multiwavelength backward-pumped fiber Raman amplifier. *IEEE Photonics Technology Letters,* Vol. 13, 945–947, ISSN: 1041-1135

Guo, M.-N.; Liaw, S.-K.; Shum, P.-P.; Chen, N.-K.; Hung, H.-K. & Lin, C. (2011). Single-wavelength-pump-based bi-directional hybrid fiber amplifier for bi-directional local area network application. Optics Communications, vol. 284, 573–578

Agrawal, G. P. (1995). *Nonlinear Fiber Optics,* 2nd ed. Academic, ISBN: 978-0-12-369516-1, New York

Liaw, S.-K.; Dou, L. & Xu, A. (2007). Fiber-Bragg-gratings–based dispersion-compensated and gain-flattened Raman fiber amplifier. *Optics Express,* Vol. 19, 12356-12361, ISSN: 1094-4087

Liaw, S.-K.; Ho, K.-P. & Chi, S. (1998). Multichannel add/drop and cross connect using fiber Bragg gratings and optical switches. *Electronics Letters,* Vol. 34, 1601–1602, ISSN: 003-5194

Riziotis, C. & Zervas, M.-N. (2002). Performance comparison of Bragg grating-based optical add-drop multiplexers in WDM transmission systems. Circuits, Devices and Systems, *IEE Proceedings,* Vol. 149, 179-186, ISSN: 1751-9683

Brackett, C.–A. (1996). Foreward— Is there an emerging consensus on WDM networking. *J. Lightwave Technoogy,* Vol. 14, 936-941, ISSN: 0773-8724

Liaw, S.-K.; Wang, C.-J.; Chen, S.-H. & Lin, Y.-T. (2007). Reconfigurable optical add/drop multiplexer with 8.0 dB net gain using dual-pass amplified scheme. *IEICE Transactions on Communications,* Vol. 8, 2016-2021, ISSN: 1745-1345

Liaw, S.-K.; Hsieh, Y.-S.; Cheng, W.-L.; Chang C.-L. & Ting, H.-F. (2008). Bi-directional reconfigurable optical add/drop multiplexer with power compensated using built-in optical amplifiers. *OSA Journal of Optical Networking,* Vol. 7, 622-671

Ho, K.-P.; Chan, C.-K.; Tong, F. & Chen, L.-K. (1998). Exact analysis of homodyne crosstalk induced penalty in WDM networks. *IEEE Photon Technology Letters,* Vol. 10, 457–458, ISSN: 1041-1135

18

Burst-mode Optical Amplifiers for Passive Optical Networks

Ken-Ichi Suzuki
NTT Access Service Systems laboratories, NTT Corporation
Japan

1. Introduction

Passive optical network (PON) systems are being used to create cost-effective optical access network systems (H. Shinohara, 2007) because the central office equipment and the optical transmission fibers can be shared by several customers. To offer PON-based optical access services more economically, the optical line terminal (OLT) installed in the central office must accommodate as many customers as possible and as efficiently as possible. In areas that have relatively lower service demand, it is important to enlarge the transmission distance between the OLT and optical network units (ONUs) in customer premises. In areas that have relatively higher densities, it is important to increase the splitting number of optical splitters. Both actions minimize the need to install additional OLTs in the central office. Enlarging the transmission distance enables remote customers to be serviced while retaining the cost benefits of the PON systems and increasing the splitting ratio enables a system to offer the services to many more urban customers.

To enlarge the transmission distance, we developed a PON repeater or a PON extender box with 3R (reshaping, re-timing, and regenerating) functions (K-I. Suzuki, et al., 2002). However, the 3R functions require a dedicated repeater to be developed for each PON system given that there are various transmission bit-rates and transmission protocols specified in ITU-T Recommendation G.984 series, IEEE 802.3ah, and so on. Therefore, we investigated the use of optical amplifiers as the PON repeaters because these can amplify optical signals regardless of the transmission bit-rate and/or protocol because their amplifications do not require any O/E (optical to electrical) conversion and/or E/O (electrical to optical) conversion. Therefore, we have intensively investigated long-reach PON systems based on optical amplifiers (K-I. Suzuki, et al., R. Davely, et. al., 2009, F. Saliou, et. al., 2009). Moreover, PON-extender boxes were standardized as ITU-T Recommendation G.984.6 to realize long-reach gigabit PON systems.

Figure 1 shows the long-reach PON system with the repeatered system configuration for single fiber WDM (Wavelength Division Multiplexing). In current PON systems, which have been widely installed in the world, upstream and downstream wavelengths lie in the 1.31 μm region and the 1.49 μm region, respectively. Different maximum OLT-ONU distances are specified in different standards but the range is, logically, from 10 to 60 km. Engineering-wise, however, the maximum OLT-ONU transmission distance depends on the splitting loss of the optical splitter and is as short as 7 km with the splitting number of 32 (H. Ueda, et. al., 1999). Moreover, the maximum number of branches continues to increase with the example being a G-PON with 64 branches (logically up to 128). Basing the PON repeater on optical amplifiers is a promising approach to achieving both longer transmission distances and higher splitting numbers.

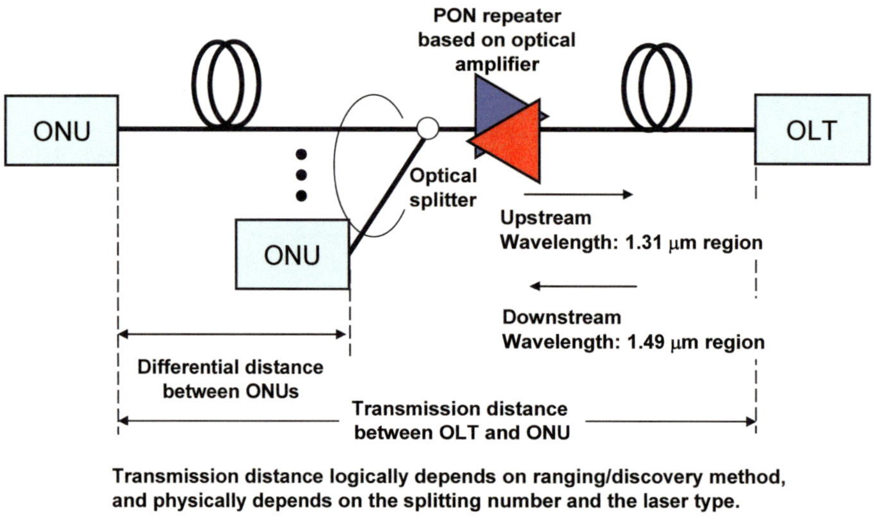

Fig. 1. Long-reach PON system with the repeatered system configuration for single fiber WDM

I next explain the transmission signals in PON systems. As the downstream optical signal is a continuous wave, we can employ an optical amplifier with conventional techniques. In the upstream direction, on the other hand, as the distances between the OLT and each ONU differ, the OLT must be able to receive optical burst signals with different intensities from the ONUs. It is clear that the PON repeater based on optical amplifiers must also be able to amplify these signals without any distortion. However, optical burst signal amplification leads to optical surges as shown in Fig.2, which may well cause failure of the optical receiver as well as interfering with the reception of normal signals at the OLT due to gain dynamics. So we have to use burst-mode optical amplifiers to suppress these optical surges and achieve gain stabilization.

In this chapter, I present burst-mode optical amplifiers for PON systems based on a couple of linear-gain control techniques, gain-clamping (GC) (G. Hoven, 2002, K-I. Suzuki, et. al., 2005), fast automatic gain controlling (fast AGC) (Y. Fukada, et. al., 2008, H. Nagaeda, et. al., 2008), and fast automatic level controlling (fast ALC) (K-I. Suzuki, et. al., 2009) for optical amplifiers. We also discuss the system design based on the signal to noise ratio (SNR) of the long-reach PON systems (K-I. Suzuki, et. al., 2006). On the other hand, 10 Gbit/s class high-speed PON systems have been receiving great attention, which were standardized in IEEE 802.3av and FSAN/ITU-T to cover the future demand, created by the rapid growth of Internet access and IP video delivery services, for high capacity communication. So, I also introduce 10 Gbit/s optical burst signal amplification based on GC optical fiber amplifiers (OFA's) (K-I. Suzuki, et. al., 2008) and optical automatic level control (ALC) techniques applied to fast AGC-OFA's (K-I. Suzuki, et. al., 2009) to ease the requirements of the receiver's dynamic range to confirm their feasibility.

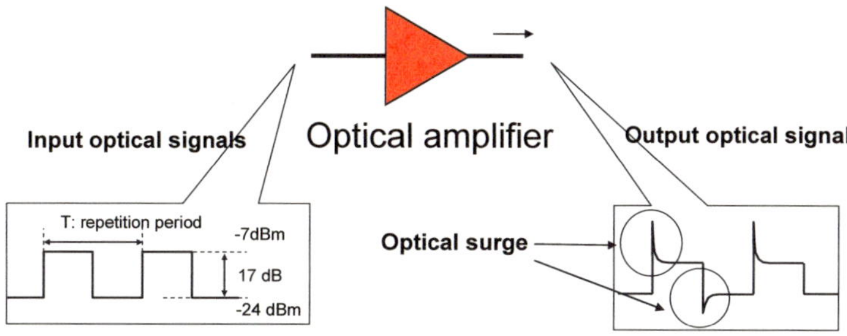

Fig. 2. Optical surge generation

2. Optical surges; Principle of gain-dynamics of optical amplifier

Figures 3 (a)-(b) show optical surges and Fig. 4 shows the numerical and experimental results of normalized optical surge intensity as a function of burst repetition rate (1/T) in the case of 1.3 μm fiber optical amplifier using PDFA (Praseodymium-doped fiber amplifier) using the experimental setup shown in Fig. 2. Mark and space levels are set to -7 dBm and -24 dBm, respectively. Gain characteristics are shown as those of PDFA without gain-clamping in Fig. 9 (a). We normalize optical surge intensity using the ratio of optical surge peak level to restored normal signal level in the case of a space to mark level (S-M) transition, see Fig. 3(a). In the case of a mark to space level (M-S) transition, shown in Fig. 3(b), we use the ratio of optical surge bottom level to restored normal signal level. Numerical results were calculated using measured gain relaxation time constants for both transitions. We estimated the gain relaxation time constants as 8.0 μs for S-M transition and 76 μs for M-S transition. Accordingly, we can calculate the gain dynamics *G(t)* by using the following simple equation.

$$G(t) = G_f + \Delta G \exp(-t/\tau) \quad (1)$$

where G_f is the initial gain value at the transition, ΔG is the difference between G_f and the intrinsic gain value for the input signals after the transition, and τ is the gain relaxation time constant. In this case, it takes several hundred micro seconds to restore gain to its normal

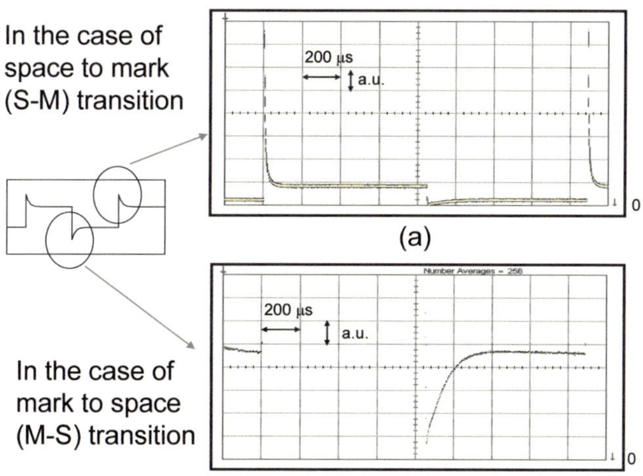

Fig. 3. Optical surges at (a) S-M transition and (b) M-S transition.

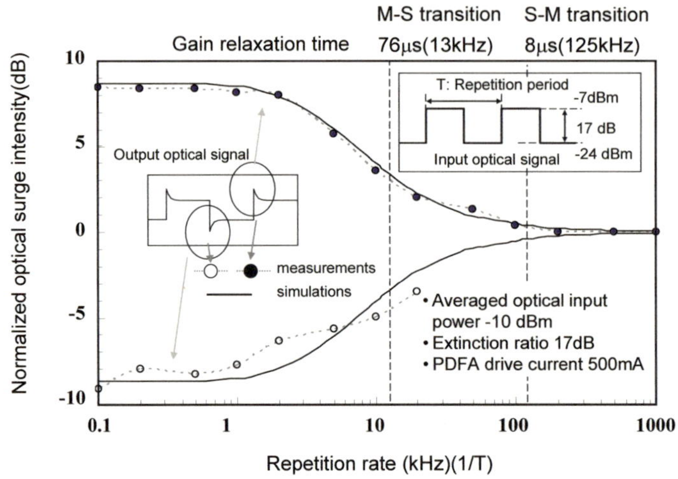

Fig. 4. Numerical and experimental results of normalized optical surge intensity as a function of burst repetition rate.

state because of its large gain relaxation time. Therefore, we adopt numerical values after the calculation becomes a steady state in excess of several hundred micro seconds to well handle repetition periods much smaller than the gain relaxation time.

3. Signal to noise ratio of Long-reach PON system based on optical amplifier

In this section, we confirm the validity of the optical-amplifier for PON systems by estimating the limits placed on the upstream transmission distance (ONU to OLT) in Long-reach PON systems. Figure 5 shows the schematic diagram for the signal to noise ratio (SNR) calculation of a Long-reach PON system consisting of an ONU, an OLT, an optical amplifier with optical splitter and optical fibres. The signal wavelength is 1.31 μm. We assume that the total loss of the 2^n way splitter is 3.5n + 0.5 dB (n is the number of stages in the multistage splitter) because of its 0.5n dB deviation of the splitting ratio (deviation is 0.5 dB/stage) and insertion loss in 0.5 dB in addition to its 2^n way splitting loss of 3.0n dB. For example, the total loss of a 32 way splitter is estimated to be 18 dB.

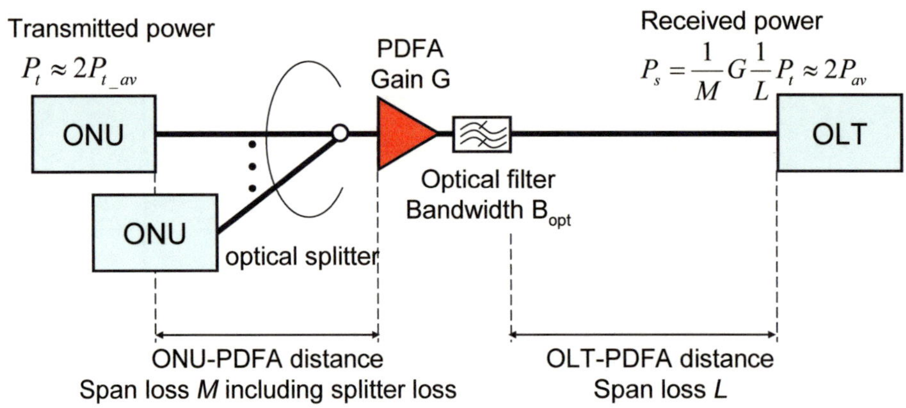

Fig. 5. Schematic diagram for SNR calculation of Long-reach PON system.

No.	Quantity	Symbol	Value	Unit
1	wavelength	λ	1.3	μm
2	Bit rate	B	1.25	Gbit/s
3	quantum efficiency	η	0.8	
4	Equivalent noise current	I_{eq}	5.2	pA/$\sqrt{\text{Hz}}$
5	inverted population parameter	n_{sp}	2	
6	Gain	G	17	dB
7	Extinction ratio	r	0.1	
8	Fiber loss		0.5	dB/km
9	2^n way optical splitter loss		$3.5n + 0.5$	dB
10	Transmitted power (Averaged transmitted power)	P_t (P_{t_av})	4 (1)	dBm
11	Filter bandwidth	B_{opt}	3	nm

Table 1. Calculation parameters

Equation (2) was used to calculate SNR of this Long-reach PON system. In a conventional Long-reach PON system based on optical amplifiers, signal-spontaneous beat noise is the dominant problem with regard to receiver sensitivity. However, we also need to consider the circuit noise of the optical receiver because the amplified spontaneous emission (ASE) is reduced due to the fibre loss of the following span in the mid-span repeatered configuration.

$$SNR = \frac{\left(\frac{e\eta}{h\nu}(1-r)P_s\right)^2}{\left(\sqrt{A}+\sqrt{C}\right)^2 \frac{B}{4}} \tag{2}$$

$$A = e\frac{e\eta}{h\nu}P_s + I_{eq}{}^2 + \left(\frac{e\eta}{h\nu}\right)^2 2P_s\frac{P_{sp}}{L} + 2\left(\frac{e\eta}{h\nu}\frac{P_{sp}}{L}\right)^2 B_{opt}$$

$$C = e\frac{e\eta}{h\nu}rP_s + I_{eq}{}^2 + \left(\frac{e\eta}{h\nu}\right)^2 2rP_s\frac{P_{sp}}{L} + 2\left(\frac{e\eta}{h\nu}\frac{P_{sp}}{L}\right)^2 B_{opt}$$

$$P_{sp} = h\nu(G-1)n_{sp}$$

where r is the extinction ratio between mark and space signals, Ps is the optical power of the mark signal (the average power is almost $P_s/2$), B is the bit-rate of the transmission signals, Psp is the frequency density of ASE power of one side of the polarization components, and Bopt is the bandwidth of the optical filter used for ASE elimination. G and nsp are the gain and the inverted population parameter of the optical amplifier, respectively. A and C show the noise elements related to the mark and space signals, respectively. The first term of each equation is the shot noise (related to the optical-electrical signal conversion), the second term is the frequency density noise power (related to the equivalent noise current of the receiver front-end circuit), the third term is the signal-spontaneous beat noise, and the forth term is the spontaneous-spontaneous beat noise since we must consider the ASE component

with orthogonal polarization to the optical signals as well as that with the same polarization as the optical signals.
Equation (3) shows the relationship between bit error rate (BER) and Q factor and SNR (N. A. Olsson, 1989, N. S. Bergano, et. al., 1993, T. Takahashi, et. al., 1995).

$$\mathrm{BER} = \frac{1}{\sqrt{\pi}} \frac{\exp\left(-\frac{Q^2}{2}\right)}{Q} \tag{3}$$

where Q is given by:

$$Q = \sqrt{\frac{SNR}{4}} \tag{4}$$

Therefore, we can obtain the BER using Eqs (2), (3) and (4).
Figures 6(a) and (b) show the SNR values calculated using Eq. (2) as a function of the transmission distance between an OLT and a PDFA with the parameter of ONU-PDFA distance for two filter bandwidths (1 nm and 20 nm). Table 1 shows the calculation parameters. The SNR with 20 nm bandwidth filter is degraded compared as that with the 1 nm filter because of its large spontaneous-spontaneous beat noise. However, we confirmed that over 40 km transmission can be achieved with the relatively wide band-pass filter in 32 way splitting PON systems at the bit-rate of 1.25 Gbit/s.
Figure 7 shows SNR as a function of the splitting number; the parameter is the ONU-PDFA distance and the OLT-PDFA distance is fixed at 0 km (i.e., the PDFA is employed as a pre amplifier and the OLT-ONU distance is varied.). As shown in Fig. 7, we find that splitting numbers above 470 are possible at the ONU-PDFA distance of 10 km. Moreover, the splitting number can exceed 630 at the ONU-PDFA distance of 7km.

4. Burst-mode optical amplifier using gain-clamping

4.1 Gain-clamped Praseodymium-doped fiber amplifier for burst-mode amplification

There are two major gain control methods for optical amplifiers. One is automatic gain control (AGC), which uses feedback/forward gain controls to realize constant gain operation. The other is gain-clamping, which offers constant gain operation using relatively high power control lights compared to optical signals. AGC is being used in the optical repeaters in many long haul transmission systems. However, AGC response time depends on the gain dynamics of the optical amplifiers as well as the speed of the control circuits, so it is impractical to use AGC without any technology progress to handle burst signals. Since gain-clamping is independent of gain dynamics and the control circuit speed (H. Masuda, et. al., 1997; L. L. Yi, et. al., 2003; Yung-Hsin Lu, et. al., 2003; D. A. Francis, et. al., 2001; G. Hoven, 2002), we adopted to realize burst mode amplification in our initial investigation.
In this section, I explain the gain dynamics of PDFAs for designing a burst mode optical amplifier and confirm the gain-clamp effect for burst mode amplification. Figure 8 shows the configuration of a gain-clamped PDFA that supports burst mode amplification. The gain-clamped PDFA consists of two PDFA gain stages pumped by 0.98 μm laser diodes (LD's). The first gain stage uses forward pumping and backward gain-clamping. The second stage uses backward pumping and backward gain-clamping. Forward and backward pumping

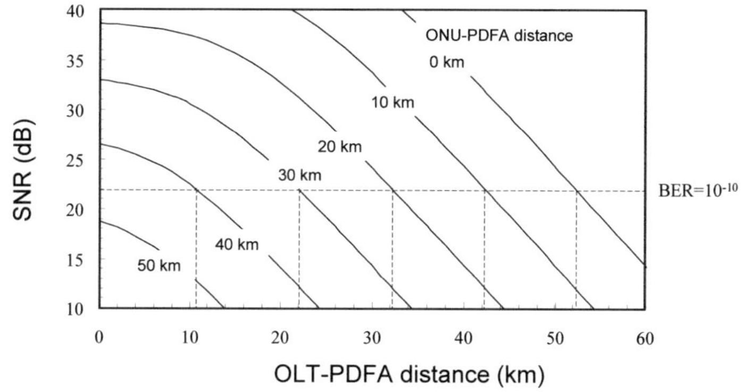

(a)

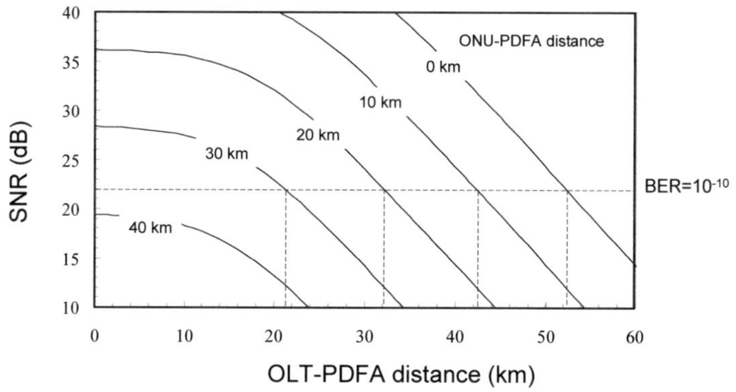

(b)

Fig. 6. SNRs as a function of transmission distance between OLT and PDFA with the parameter of ONU-PDFA distance with parameter of ONU-PDFA distance (a) B_{opt} is 1 nm, (b) B_{opt} is 20 nm.

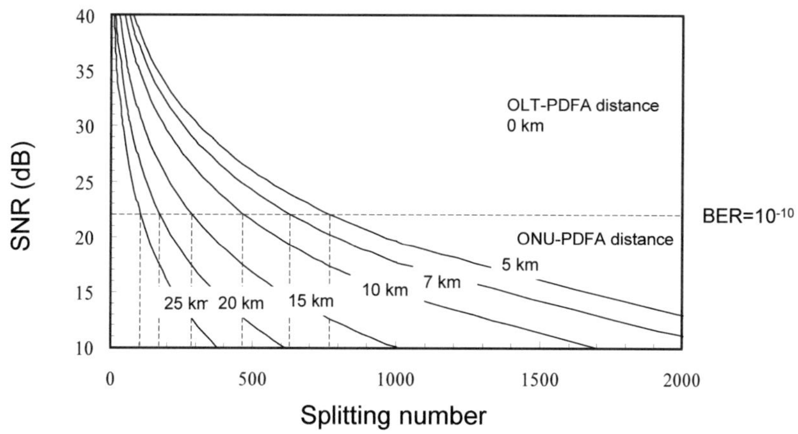

Fig. 7. SNRs as a function of splitting number with the parameter of ONU-PDFA distance and the fixed OLT-PDFA distance of 0 km. and B_{opt} of 20 nm.

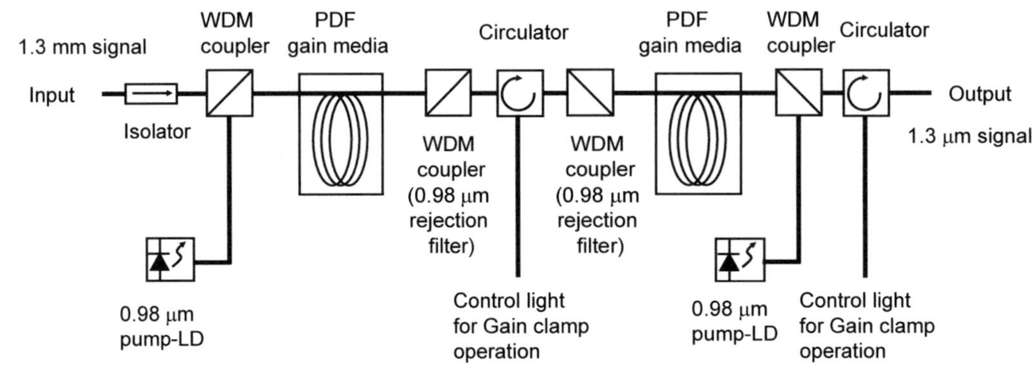

Fig. 8. Configuration of gain-clamped PDFA

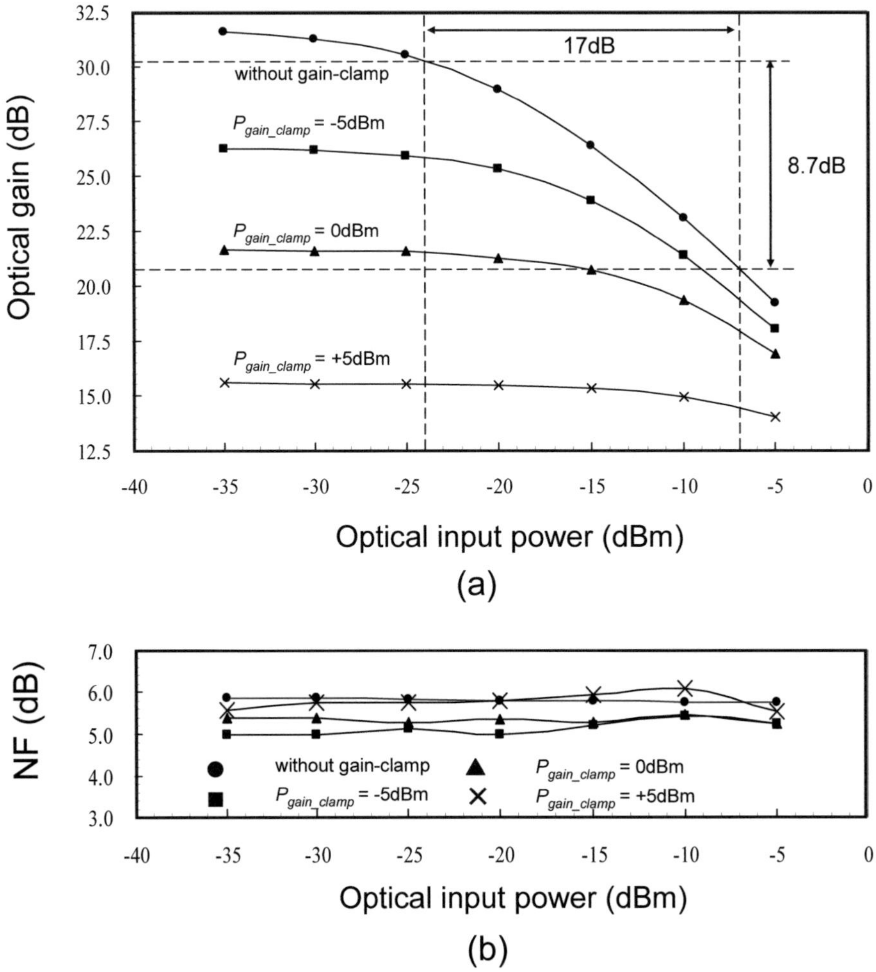

Fig. 9. (a) Optical gain (b) Noise figure as a function of optical input power; the parameter is the power of the gain-clamp light.

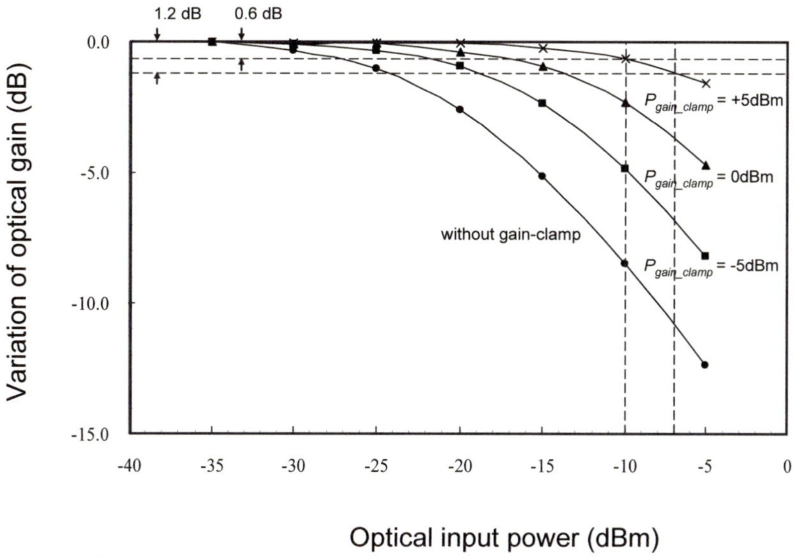

Fig. 10. Variation of optical gain as a function of optical input power; the parameter is the power of the gain-clamp light.

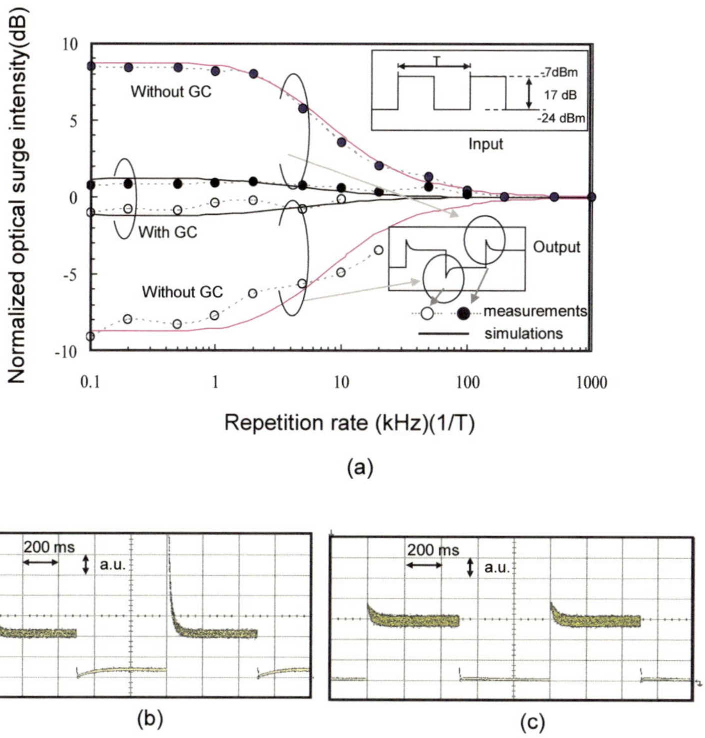

Fig. 11. (a) Numerical and measured results of normalized optical surge intensity as a function of burst repeatition frequency. Examples of optical signal traces at repetition rate of 1 kHz (b) without GC and (c) with GC.

are often used to realize low noise figure operation and to control the output power, respectively. In this case, we fix both pump powers and use the gain-clamping system for gain control. Backward gain-clamping is chosen because it simplifies the separation of the gain-clamp light from the optical signals.

Figure 9 (a)-(b) show the optical gain and the noise figure (NF) as a function of optical input power; the parameter is the power of the gain-clamp light. Figure 10 shows the variation of optical gain as a function of optical input power; the parameter is the power of the gain-clamp light. Although gain-clamping causes gain suppression as shown in Fig. 9, we find that gain-clamping drastically improves PDFA linearity as shown in Fig. 10. In particular, a 1.2 dB gain variation and 14.4 dB gain are achieved when the gain-clamp power is 5 dBm and the input power range is below -7 dBm. Moreover, we can improve the gain variation and the gain to 0.6 dB and 15 dB, respectively, at input powers below -10 dBm.

Figures 11 (a)-(c) show the numerical and measured results of normalized optical surge intensity as a function of burst repetition frequency and typical optical signal traces at the repetition rate of 1 kHz with/without GC. Although residual optical surges are observed because of 1-dB gain compression power of -7 dBm, gain-clamping does improve the gain dynamic properties and can suppress optical surges as shown in Figs. 11 (a)-(c). Note that gain-clamping does not work well if the input power to the gain-clamped PDFA exceeds -7 dBm. For example, the splitting number must be above 4 when the ONU-PDFA distance is 7 km.

5. 10 Gbit/s burst-signal amplification using gain-clamped optical amplifier

5.1 Experimental setup of 10 Gbit/s burst-signal amplification

In this section, I focuses on gain-clamp based burst-mode optical amplifiers (burst-AMP's) for 10 Gbit/s PON application to realize both long-reach and higher-speed PON systems. I then introduce the demonstration of 10 Gbit/s optical burst signal amplification to confirm their feasibility. Figure 12 shows the experimental setup for 10 Gbit/s optical burst-signal amplification; it consists of the burst-AMP under test, a burst-mode optical receiver (burst-Rx), a burst-mode optical transmitter 1 (burst-Tx1), and a burst-mode optical transmitter 2 (burst-Tx2). As the burst-AMP, we used our 0.98 μm pumped gain-clamped praseodymium-doped fiber amplifier (GC-PDFA) (K-I. Suzuki, et. al., 2007). This burst-AMP offers 17 dB gain and good gain linearity (the 1dB-gain compression power is -10 dBm). After the 3 nm optical band-pass filter (OBPF), optical gain is reduced to 16.2 dB by the 0.8 dB excess loss of the OBPF.

We used high-power with distributed feedback laser diodes (DFB-LD's) as burst-Tx's; they were directly modulated by 10.3125 Gbit/s signals with the pseudo random bit sequence (PRBS) 2^7-1 and 2^{23}-1, to generate burst signal sequences. The transmission timing of each burst-Tx was controlled by the enable signals from the timing pulse generator that was synchronized to the pulse pattern generator and the bit-error rate tester. The 99.3 nm guard time was used to separate burst signal sequences and the 74.5 ns preamble (continuous "101010" signal) was used to set the threshold level of received electrical signals. Basically, the measured burst signal sequence was modulated with PRBS 2^{23}-1 and the other burst signal sequence was modulated with PRBS 2^7-1 to distinguish them at the bit-error rate tester. So when we measured each burst signal sequence from the burst-Tx's, we replaced the PRBS patterns as required. The central wavelength, the averaged output power, and the extinction ratio of the burst Tx1 were 1303.0 nm, 5.6 dBm, and 9.6 dB, respectively. Those of the burst Tx2 were 1301.5 nm, 5.5 dBm, and 9.7 dB, respectively. Burst signal sequences

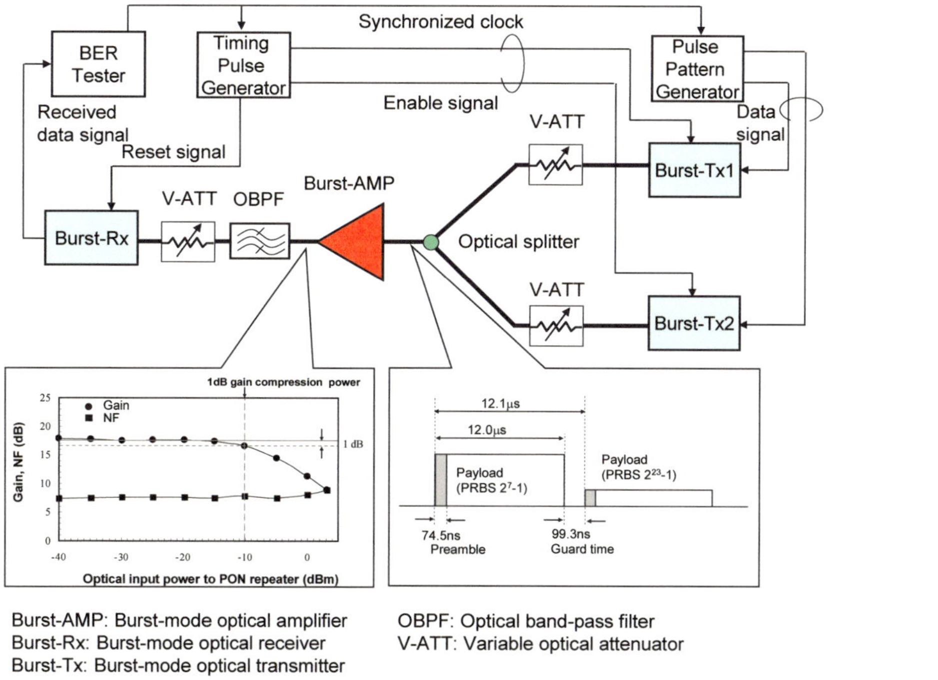

Fig. 12. Experimental setup for 10 Gbit/s burst signal amplification

with different intensities are generated by the variable attenuators (V-ATT) at the output of each burst-Tx. We set the link loss between the burst-AMP and burst-Tx1 and the burst Tx2 to 25.6 dB and 15.5 dB, so optical powers input to the burst-AMP were fixed at -20 dBm and -10 dBm, respectively. We varied the optical powers input to the burst-Rx to measure the allowable loss budget for 10 Gbit/s amplified PON systems.

The burst-Rx consisted of a burst-mode limiting amplifier (burst-LIM) and a burst-mode trans-impedance amplifier (burst-TIA) (S. Nishihara, et. al., 2007) followed by an avalanche photodiode APD (T. Nakanishi, et. al., 2007). Its sensitivity (BER=10^{-12}) and overload were estimated to be -22.2 dBm and -7.0 dBm, respectively, at the multiplication factor, M, of around 5.

5.2 Experimental results of 10 Gbit/s burst-signal amplification

Figure 13 shows the bit error rate (BER) as a function of averaged receive power to the burst-Rx. After the burst-AMP, we observed the power penalties, with and without the OBPF, of 1.2 dB and 2.2 dB, respectively, at BER=10^{-12}. On the other hand, at BER=10^{-4}, the power penalties are reduced to within 1.0 dB. By the way, as IEEE P802.3av requires a 29 dB loss budget considering the use of forward error correction (FEC), we should also consider the FEC effect. Assuming the use of the FEC based on RS(255,239), which can improve the BER from 10^{-4} to 10^{-12}, we expect that FEC can improve the receiver sensitivity to -28.0 dBm Note that IEEE 802.3av adopted the RS(255, 223) based FEC, which can improve the BER from 10^{-3} to 10^{-12}, after this experiment..

Figures 14 (a) and (b) show the signal traces for the burst-AMP output and the burst-TIA output. As optical inputs to the burst-AMP were -20 dBm and -10 dBm, we obtained the optical output of -3.8 dBm and +5.2 dBm. After optical burst signal amplification, we did not observe any significant waveform degradation such as optical surges, see Fig.6-3 (a). On the

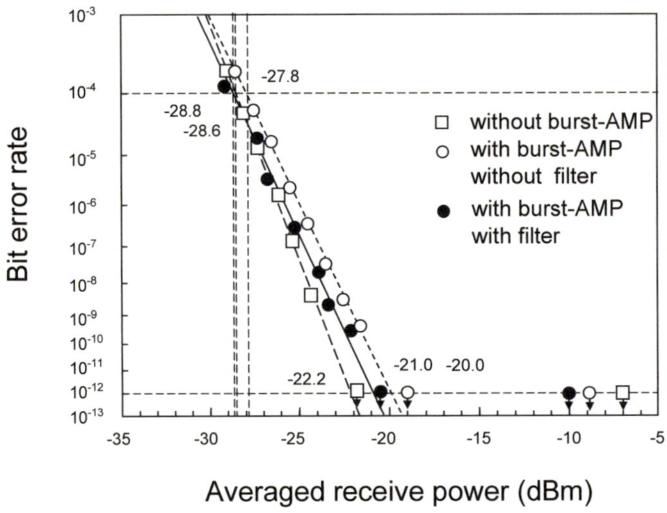

Fig. 13. BER as a function of averaged receive power to burst-Rx

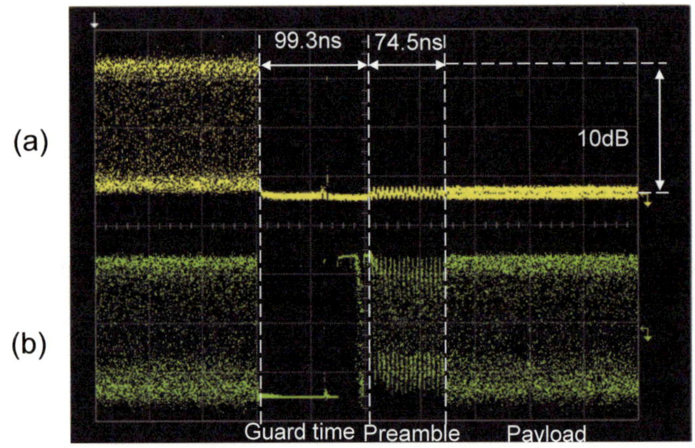

Fig. 14. Signal traces for (a) burst-AMP output and (b) burst-TIA output

other hand, received electrical signals were successfully regenerated within burst overhead time shown in Fig.14 (b).

Figure 15 shows the BER as a function of total loss. Not using the burst-AMP yields the estimated loss budget of 27.8 dB, which corresponds to 34.1 dB if FEC is used. On the other hand, when we use the burst-AMP, loss budget is improved to 42.6 dB, which corresponds to 50.4 dB when FEC is applied. Moreover, when we use the bust-AMP followed by the 3 nm OBPF, the loss budget is improved to 42.8 dB, which corresponds to 50.4 dB after FEC decoding. No significant difference is observed between cases of optical amplifier with/without the 3 nm OBPF, because the OBPF improves the receiver sensitivity but reduces the optical gain.

Therefore, we successfully achieved 15.0-dB improvement in the loss budget and a loss budget of 42.8 dB by using the burst-AMP. The use of FEC is expected to yield improvement values of 16.3 dB and 50.4 dB, respectively.

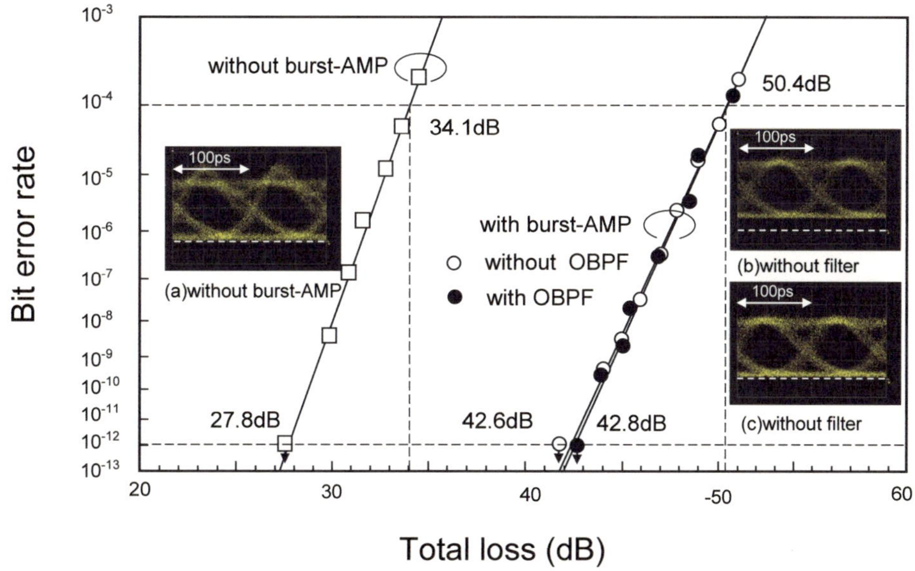

Fig. 15. Bit error rate as a function of total loss. Insets are eye patterns for optical signals (a) with burst AMP, (b) with burst-AMP without OBPF, and (c) with burst-AMP and OBPF

6. 10 Gbit/s burst-signal amplification using automatic level controlled burst-mode optical fiber amplifier

6.1 Automatic level controlled burst-mode optical fiber amplifier

To offer PON-based optical access services more effectively, we have investigated optically-amplified PON systems based on a couple of linear-gain control techniques, gain-clamping (GC) (K-I. Suzuki, et. al., 2007) and fast automatic gain controlling (fast AGC) for optical amplifiers (Y. Fukada, et. al., 2008, H. Nagaeda, et. al., 2008). In Section 5, we have introduced 10 Gbit/s optical burst signal amplification based on GC optical fiber amplifiers (OFA's) and confirmed their bit-rate independency. By the way, fast AGC techniques allow the linear gain region of OFA's to be expanded without gain suppression, GC techniques, on the other hand, are usually accompanied by considerable gain suppression. Accordingly, we expect AGC techniques will realize both higher linear gain and wider linear gain region (wider dynamic range). However, wide linear gain regions require that the burst-mode optical receivers (burst-Rx's) have wide dynamic range in order to take advantage of the fast AGC techniques.

In this section, I explain the optical automatic level control (ALC) techniques applied to fast AGC-OFA's to ease the requirements of the receiver's dynamic range. I then introduce the demonstration of 10 Gbit/s burst-RX based on the optical ALC techniques to confirm their feasibility.

Figure 16 shows the configuration of an ALC burst-mode OFA (burst-OFA); it consists of an AGC praseodymium-doped fiber amplifier (PDFA) block and an ALC circuit block. The AGC-PDFA employs a feed-forward (FF) controlled pump-LD for quick gain adjustment of the PDFA and offers constant gain of around 20 dB with optical input powers from -30 dBm to -10 dBm or above. The FF-controlled pump-LD quickly adjusts pump power according to the monitored signal power. The ALC circuit employs a FF-controlled variable optical attenuator (VOA) for quick adjustment of output optical level and the suppression of waveform distortion caused by the gain dynamics of the AGC-PDF. Using these techniques, the ALC burst-OFA realizes quick ALC response, of the order of sub-micro-seconds, and offers the constant output optical level of around -17 dBm with optical input powers from -30 dBm to -10 dBm or above.

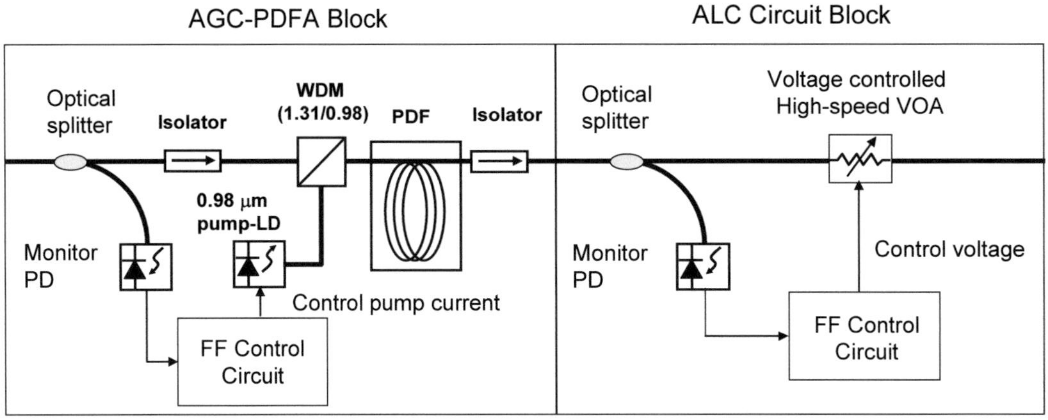

Fig. 16. Configuration of automatic level controlled burst-mode optical fiber amplifier

6.2 Experimental setup for bit-error rate measurement of 10 Gbit/s optically level controlled burst-signals

Figure 17 shows the experimental setup used to measure the bit-error rate (BER) of 10 Gbit/s optically level controlled burst-signals; the setup consists of a burst-Rx, a burst-mode optical transmitter 1 (burst-Tx1), and a burst-mode optical transmitter 2 (burst-Tx2). The burst-RX consists of the ALC-burst-OFA under test, a DC-coupled APD trans-impedance amplifier (APD-TIA), and a continuous-mode limiter amplifier (continuous LIM). We use a 3 nm band-pass filter (OBPF) for elimination of ASE noise from the ALC-burst-OFA. This reduces the optical output power of the ALC-burst-OFA to -18 dBm given the 1.0 dB excess loss of the OBPF.

We used DFB-LD's with quick response burst-mode LD drivers (H. Nakamura, et. al., 2008) as burst-TX's; they were directly modulated by 10.3125 Gbit/s signals with PRBS 27-1, to generate burst signal sequences. The transmission timing of each burst-Tx was controlled by the enable signals from the timing pulse generator that was synchronized tot the pulse pattern generator and the BER tester. The 198.5 ns guard time was used to separate burst signal sequences and a part of the 496.4 ns preamble (continuous "101010" signal) was used to set the optical output level of the ALC-burst-OFA (Normally, it is used to set the threshold level of the burst-mode LIM). The central wavelength, the averaged output power, and the extinction ratio of the burst Tx1 were 1307.5 nm, 3.6 dBm, and 6.1 dB, respectively. Those of the burst Tx2 were 1306.9 nm, 4.2 dBm, and 6.2 dB, respectively. Burst signal sequences with different intensities were generated by the VOA's at the output of each burst-Tx. We varied the optical powers input to the burst-Rx and measured the BER to confirm the allowable dynamic range of our proposed ALC-burst-OFA.

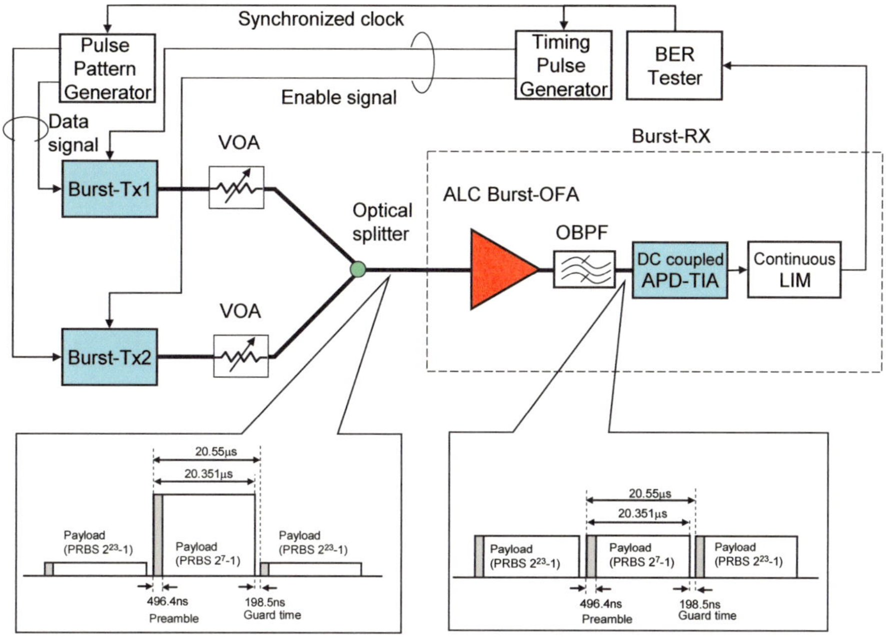

Fig. 17. Experimental setup for BER measurement of 10 Gbit/s burst signal

6.3 Experimental results of 10 Gbit/s burst-signal amplification

Figure 18 shows the signal traces for the ALC-burst-OFA inputs and outputs. We set the input power difference between burst-Tx's to 10.1 dB (Input powers from burst-Tx's were -20.3 dBm and -10.2 dBm), 19.7 dB (-29.9 dBm and -10.2 dBm), and 23.8 dB (-5.1 dBm and -28.9 dBm). As shown in Fig.18, the ALC-burst-OFA successfully achieved constant output level operation within the burst-overhead time; a small level difference was observed in Fig. 18(f).

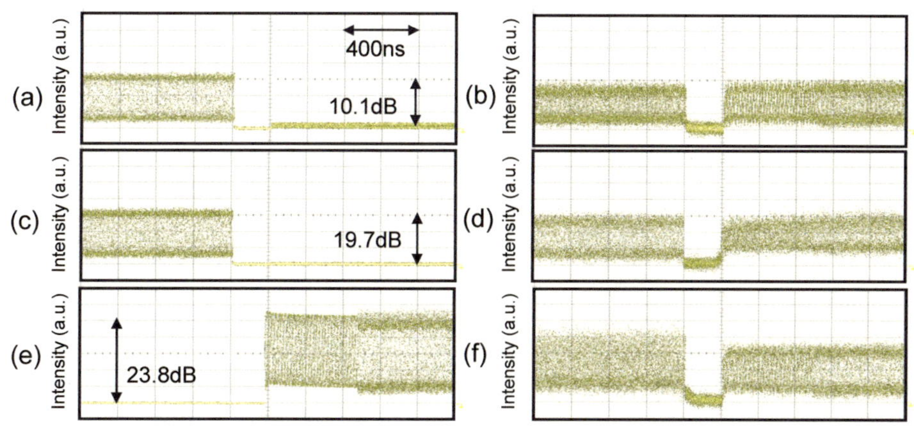

Fig. 18. Signal traces for ALC-burst-OFA inputs and outputs. (a)Input and (b) output signal traces at -20.3 dBm from burst-Tx1 and -10.2dBm from burst-Tx2. (c) Input and (d) output signal traces at -29.9 dBm and -10.2dBm. (e) Input and (f) output signal traces at -5.1 dBm and -28.9dBm.

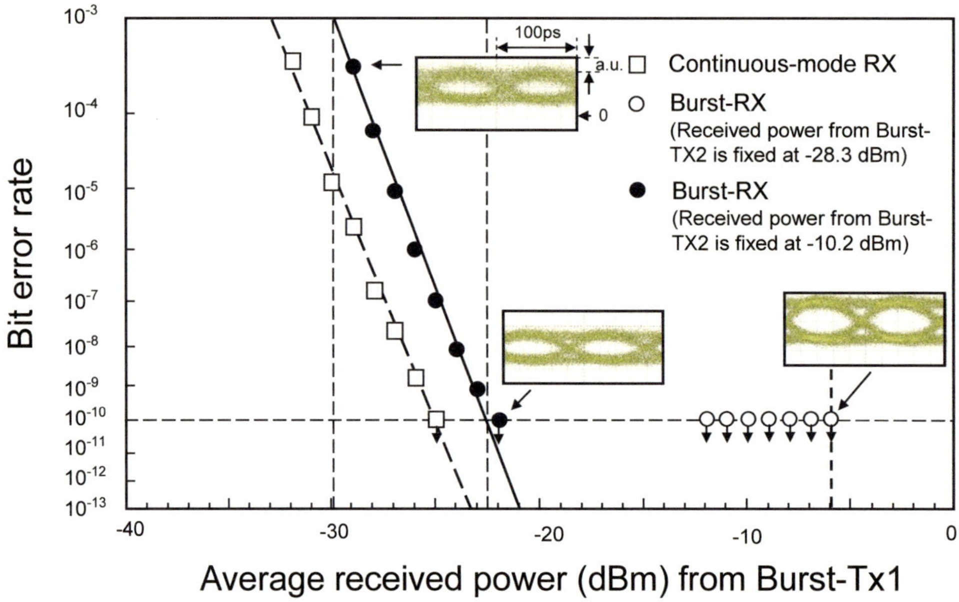

Fig. 19. Bit error rate as a function of averaged received power from Burst-TX1. Insets are eye patterns for optical signals.

Figure 19 shows the BER as a function of averaged received power to the burst-Rx at fixed received power from the burst-Tx2 (-28.3 dBm and -10.2 dBm). We also show the BER curve of the continuous-mode Rx for comparison; it consisted of a burst-mode (DC-coupled) APD-TIA and a continuous-mode LIM followed by continuous-mode PDFA. The receiver sensitivity was -22.5 dBm at BER=10^{-10}; the power penalty relative to the continuous-mode device was 2.6 dB. However, our ALC-burst-OFA successfully achieved a dynamic range of over 16.6 dB at BER=10^{-10}, thus confirming its feasibility in easing the receiver dynamic range requirement. By the way, as IEEE 802.3av requires a 29 dB loss budget considering the use of forward error correction (FEC), we should also consider the FEC effect. Assuming the use of FEC based on RS(255,223), which can improve the BER from 10^{-3} to 10^{-12}, we expect that FEC will improve the receiver sensitivity and the dynamic range to -29.7 dBm and 23.8 dB, respectively.

7. Conclusion

I introduced burst-mode optical amplifiers for PON systems based on a couple of linear-gain control techniques, gain-clamping (GC), fast automatic gain controlling (fast AGC), and fast automatic level controlling (fast ALC) for optical amplifiers. I also investigated the gain dynamics of a gain-clamped PDFA for achieving burst mode amplification and confirmed the ability of gain-clamping to suppress optical surges. I also investigated the validity of a gain-clamped PDFA for burst mode amplification by estimating the transmission distance and the allowable splitting number of Long-reach PON systems.

Moreover, I also introduced the demonstration of 10 Gbit/s optical burst signal amplification based on GC based optical fiber amplifiers (OFA's) and optical automatic level control (ALC) techniques applied to fast to ease the requirements of the receiver's dynamic

range to confirm their feasibility. Using GC based OFA's, we successfully achieved a 15.0 dB improvement in the loss budget, which stands at over 42.8 dB loss budget, by using the burst-mode optical amplifier based on gain-clamping. Moreover, the application of FEC can improve those results to 16.3 dB and 50.4 dB, respectively. No significant difference was observed between the cases of the GC based OFA with/without the optical band-pass filter. These results confirm the validity of the GC based OFA for 10 Gbit/s PON systems.
Using ALC based OFA's, we successfully received burst signals with 198.5 ns guard time and 496.4 ns preamble, and achieved a dynamic range of over 16.6 dB and receiver sensitivity of -22.5 dBm at BER=10^{-10} by combining our ALC based OFA, a DC-coupled APD trans-impedance amplifier, and a continuous-mode limiter amplifier; the results confirm the feasibility of our ALC based OFA in 10 Gbit/s PON systems.

8. References

H. Shinohara, "FTTH experiences in Japan", JON 6, pp. 616-623 (2007).

K-I. Suzuki, T. Furukawa, K. Saito, and H. Ueda, "Role and development of B-PON regenerator", OHAN2002 paper 4.4, pp.115-121(2002) or “B-PON repeater for enlarging transmission distance between OLT and ONT”, IEICE Transaction of Information and Communication Engineers B, vol.J86-B no.10, pp.2053-2064 (2003).

ITU-T Recommendation G.983, G.984, and G.987 series

IEEE Standard 802.3ah, 802.3av, IEEE P1904.1

K-I. Suzuki, Y. Fukada, D, Nesset, and R. Davey, "Amplified gigabit PON systems", JON 6, 422-433 (2007).

R. Davey, D, Grossman, M. Rasztovits-Wiech, D. Payne, D. Nesset, A. Kelly, A. Rafel, S. Appathurai, and S. Yang, "Long-reach passive optical networks", J. Lightwave Technol. Vol.27, No.3, pp.273-291(2009).

Z. Belfqih, P. Chanclou, F. Saliou, N. Genay and B. Landousies, “Enhanced Optical Budget System Performance of an Burst Extended PON at 10.7 Gbti/s over 60 km of Fiber”. ECOC’2008, paper Th.2.F.4, Brussel, Belgium, September, 2008.

F. Saliou, P. Chanclou, F. Laurent, N. Genay, J. Lazaro, F. Bonada, and J. Prat, " Reach Extension Strategies for Passive Optical Networks", J. Opt. Commun. Netw., Vol. 1, No. 4, C51-C-60(2009).

ITU-T Recommendation G.984.6, "Gigabit-capable passive optical networks (GPON): Reach extension", (2008).

H. Ueda and E. Maekawa, "Development of an ATM subscriber system (Model C)", NTT Review 11, 6, pp.27-32 (1999).

Gerlas van den Hoven, "Linear optical amplifier", LEOS2002 paper ThCC2.

K-I. Suzuki, Y. Fukada, D. Nesset, and R. Davey, "Amplified gigabit PON systems", J. Optical Networking, Vol. 6, no. 5, 422-433 (2007).

Y. Fukada, K-I. Suzuki, H. Nakamura, N. Yoshimoto, and M. Tsubokawa, "First demonstration of fast automatic-gain-control (AGC) PDFA for amplifying burst-mode PON upstream signal", ECOC 2008, papwer We.2.F.4.

H. Nagaeda, Y. Horiuchi, Y. Tanaka, N. Shiga, Y. Fukada, K-I. Suzuki, N. Yoshimoto, and M. Tsubokawa, "Burst-mode optical amplifier for signals with various burst length from 500 ns to 50 ms", ECOC 2008, paper P.1.13.

K-I. Suzuki, Y. Fukada, K. Saito, and Y. Maeda, “Long-reach PON System using Gain-clamped Optical Amplifier”, WTC2006, paper NT3 #2.

K-I. Suzuki, Y. Fukada, T. Nakanishi, N. Yoshimoto, and M. Tsubokawa, "Burst-mode optical amplifier for long-reach 10 Gbit/s PON application", OFC2008, paper OthL3.

K-I. Suzuki, Y. Fukada, N. Yoshimoto, K. Kumozaki, M. Tsubokawa, "Automatic Level Controlled Burst-Mode Optical Fiber Amplifier for 10 Gbit/s PON Application", OFC 2009, paper OTuH1.

N. A. Olsson, "Lightwave systems with optical amplifier", IEEE Journal of Lightwave Technol., vol. 7, no. 7, pp.1071-1082, July 1989.

N. S. Bergano, F. W. Kerfoot, and C. R. Davidson, "Margin measurement in optical amplifier system", IEEE Photon. Thecnol. Lett., vol. 5, no. 3, pp. 304-306, March 1993.

T. Takahashi, M. Aoyama, M. Murakami, and M. Amemiya, "Modeling of intersymbol interference effect on signal to noise ratio measurement in long haul optical amplifier systems", Electron. Lett., vol.31, pp. 2195-2197, 1995.

H. Masuda, K.-I. Suzuki, S. Kawai, and K. Aida, "Ultra-wideband optical amplification with 3dB bandwidth of 65 nm using a gain equalized two-stage erbium-doped fiber amplifier and Raman amplification", Electron. Lett., vol.33, no.9 pp.753-754, April 1997.

L. L. Yi, L. Zhan, Q. H. Ye, X. Hu, and Y. X. Xia, "Gain-clamped erbium-doped fiber-ring lasing amplifier with low noise figure by using an interleaver", IEEE Photon. Technol. Lett., vol.15, no.12, pp.1695-1697, December 2003.

Yung-Hsin Lu and Sien Chi, "Tow-stage L-band EDFA applying C/L-band wavelength-division multiplexer with the counterpropagating partial gain-clamping", IEEE Photon. Technol. Lett., vol.15, no.12, pp.1710-1712, December 2003.

D. A. Francis, S. P. DiJaili, and J.D. Wallker, "A single-chip linear optical amplifier", OFC2001 PDP13-1.

S. Nishihara, S. Kimura, T. Yoshida, M. Nakamura, J. Terada, K. Nishimura, K. Kishine, K. Kato, Y. Ohtomo, and T. Imai, "A 10.3125-Gbit/s SiGe BiCMOS Burst-Mode 3R Receiver for 10G-EPON Systems, " OFC2007 PDP8.

S. Nishihara, M. Nakamura, K. Nishimura, K. Kishine, S. Kimura, and K. Kato, " A fast-response and high-sensitivity PIN-TIA module with wide dynamic range for 10G burst-mode transmissions", ECOC2007 paper 4.4.1.

T. Nakanishi, K-I. Suzuki, Y. Fukada, N. Yoshimoto, M. Nakamura, K. Kato, K. Nishimura, Y. Ohtomo, and M. Tsubokawa, "High sensitivity APD burst-mode receiver for 10Gbit/s TDM PON system", IEICE Electronics Express 4, 588-592 (2007).

H. Nakamura, S. Kimura, K. Hara, N. Yoshimoto, M. Tsubokawa, M. Nakamura, K. Nishimura, A. Okada, and Y. Ohtomo, "500-ps response AC-coupled burst-mode Transmitter using baseline-wander common-mode rejection technique for 10-Gbit/s-class PON system", OFC2008, paper PDP26.

19

Cascaded Nonlinear Optical Mixing in a Noncollinear Optical Parametric Amplifier

Chao-Kuei Lee
Department of Photonics, National Sun Yat-Sen University
Taiwan, R.O.C

1. Introduction

Ultrafast optical science is a rapidly evolving multidisciplinary field: the ability to excite matter with femtosecond light pulses and probe its subsequent evolution on ultrashort time scales opens up completely new fields of research in physics, chemistry, and biology.[1] Furthermore, the high intensities that can be generated using femtosecond light pulses allow us to explore new regimes of light-matter interaction.[2] The implementation of more sophisticated spectroscopic techniques has been accompanied by improvements in laser sources. Considerable effort has been dedicated to the achievement of shorter light pulses,[3-5] to improve temporal resolution; other efforts have worked to expand the frequency tunability of the pulses, since this would make it possible to excite in resonance different materials, and to probe optical transitions occurring at different frequencies. Early sources of femtosecond optical pulses were based on dye laser technology; in that case, some frequency tunability could be achieved by simply changing the laser dye.[6] This flexibility, however, came at the expense of a complicated and time consuming reoptimization.

In recent years, femtosecond laser systems have become readily available. Rapid developments of widely tunable femtosecond laser radiation has also become power tools in the field of ultrafast spectroscopy, especially the transient ultrafast pump-probe spectroscopy on various from small organic compounds to complex enzymes, has opened up new research interests in femto-chemistry and femto-biology in the recent years.[7-9] The combination of high-intensity, ultrashort pulses from amplified solid-state lasers, for example, the regeneratively amplified Ti: sapphire laser systems running at 800-nm, with nonlinear optical techniques, such as optical parametric generator /amplifier (OPG/OPA), has made widely tunable femtosecond laser pulses routinely available in many research laboratories. Generally, the OPG process transfers energy from a high-power, fixed frequency pump beam to a low-power, variable frequency signal beam, thereby generating also a third idler beam. To be efficient, this process requires very high intensities of the order of tens of GW/cm^2; it is therefore eminently suited to femtosecond laser systems, which can easily achieve such intensities even with modest energies, of the order of a few microjoules. They have proved versatile as widely tunable coherent sources, especially for short pulses since they can offer both high gain and high-gain bandwidth. Gain bandwidths of several thousand wave numbers are possible,[10-15] corresponding to transform-limited pulse durations down to a few femtoseconds, and amplification of such short pulses has been demonstrated.[16,17] In addition, high powers have been generated over a wide range of wavelengths.[18-20]

The techniques are capable of generating widely tunable femtosecond light pulses from the near UV/VIS through the near-IR. [21-49] Among those reports, the seeded non-collinear optical parametric amplifier (NOPA) of whit-light super-continuum is commonly used for generating femtosecond widely tunable radiation.[37-49] The use of a seeder significantly reduces the pumping threshold required spontaneous parametric fluorescence and increases the output energy of OPA, and improves its pulse-to-pulse stability and spatial quality.[42] The use of NOPA has some unique advantages, it allows reduction of group velocity mismatch (GVM), increasing the interaction length between the pump pulse and the seeder and thus increasing the gain factor. However due to the limitation of phase-matching condition, the tuning range of a 400-nm pumped type-I β-BaB_2O_4 **(**BBO) NOPA covers only from 460nm to 720nm in the signal branch and from 900nm to 2.4μm in the idler branch. This is unfortunate since tunable femtosecond pulses in the blue-to-near UV region (<460nm) are most useful in many applications ranging from investigation of wide band-gap materials to probing biomolecules. In order to extend the tuning range, various nonlinear optical processes, such as sum frequency generation (SFG) and second harmonic generation (SHG), can be used to convert the output of NOPA to the desired short-wavelength radiation. Petrov *et al.* reported wavelength extension through cascaded 2nd-order nonlinear frequency-conversion process.[24]However, it requires an additional SHG or SFG device for the wavelength conversion. This complicates the entire tunable femtosecond laser system. Furthermore, conversion efficiency of SHG or SFG for NOPA is often limited.

Recently, several groups had demonstrated that femtosecond pulses near 400nm can be generated with a 267-nm-pumped NOPA.[49-52] Using a pump of 258nm from a frequency tripled femtosecond Ti:Sapphire laser system, Tzankvo's group had reported wavelength tunning from 346nm to 453nm. [53] These approaches to generating femtosecond deep UV pulses are typically inefficient and require special and limits conversion efficiency. We recently reported an alternative method of generating tunable femtosecond pulses from 380nm to 465nm near the degenerate point of a 405-nm pumped type-I BBO noncollinearly phase-matched optical parametric amplifier (NOPA).[54,55] This was tentatively attributed to cascading sum frequency generation (SFG) of the generated OPA pulses and the residual fundamental laser pulse at 810nm in the same crystal for NOPA. The tunable SFG output is readily available from the OPA stage without any additional frequency conversion device. In Section 2, we present a theoretically analysis of this much simpler scheme to generate femtosecond tunable radiation. In Section 3, we describe and experimentally demonstrate the arrangement and performance of this approach. Tuning range for various seeding angles, conversion efficiency and pulse profile are characterized in detail. With a pumping energy of 75 μJ at 405 nm, the optical conversion efficiency from the pump to the tunable SFG is more than 5%.

2. Theoretical treatment

Optical parametric amplification belongs to the three-wave frequency mixing processes. In order to achieve efficient frequency conversion, the conservation law of photon energies and momentums must be satisfied in this process [56]:

$$\begin{cases} \hbar\omega_p = \hbar\omega_s + \hbar\omega_i \\ \hbar\vec{k}_p = \hbar\vec{k}_s + \hbar\vec{k}_i \end{cases} \tag{1}$$

where ω_l and k_l denote the frequency and wave vector of the pump (l=p), signal (l=s) and idler (l=i) beam, respectively. Fig. 1 illustrates the phase matching scheme of the NOPA and cascading SFG to be employed in our setup. The signal beam of NOPA is injected into an appropriate nonlinear optical crystal with an angle α relative to the propagation direction of the pump beam. Here, the sign of α is defined to be positive when the direction of seeding signal beam is rotated counterclockwise from the pump beam. The idler beam is generated at an angle δ with respect to the signal beam. The cascading sum-frequency generation satisfies the following conservation law of energies and momentums:

$$\begin{cases} \hbar\omega_{SFG} = \hbar\omega_i + \hbar\omega_{810} \\ \hbar\vec{k}_{SFG}^{(e)} = \hbar\vec{k}_i^{(o)} + \hbar\vec{k}_{810}^{(o)} \end{cases} \tag{2}$$

where ω_{810}, k_{810}, ω_{SFG} and k_{SFG} denote the frequency and wave vector of the residual 810-nm pump beam and SFG, respectively. The resulting SFG is an e-ray for a type I BBO-NOPA and propagates with an angle δ' relative to the idler beam direction:

$$\begin{aligned} \delta' &= \tan^{-1}\left(\frac{\vec{k}_{810}^{(o)}\sin\delta}{\vec{k}_{810}^{(o)}\cos\delta + \vec{k}_i^{(o)}}\right) \\ \delta &= \theta_i - \theta_s \end{aligned} \tag{3}$$

where θ_s, θ_i are the angle of signal and idler with respect to the optic axis, respectively. The angle θ between the SFG beam and optic axis can be calculated from

$$\theta = \theta_i - \delta' \tag{4}$$

The index of refraction experienced by the SFG shall be

$$n_{SFG}^{(e)} = \frac{n_{SFG}^{(o)} n_{SFG}^{(e)}}{\sqrt{\left(n_{SFG}^{(o)}\sin\theta\right)^2 \left(n_{SFG}^{(e)}\cos\theta\right)^2}} \tag{5}$$

To properly depict the phase mismatching of the three-wave mixing processes in the NOPA and SFG, we express the summation of the wave vectors of optical pulses as:

$$\begin{cases} \Delta\vec{k} = \vec{k}_p - \vec{k}_s - \vec{k}_i \\ \Delta\vec{k} = \vec{k}_{SFG} - \vec{k}_{810} - \vec{k}_i \end{cases} \tag{6}$$

Upon the equations presented above, we can calculate the central wavelength and the tuning range of cascading SFG at various seeding angles by taking into account the phase-matching condition for both NOPA and SFG and the group velocity mismatch (GVM) among the optical pulses involved. Fig. 2 presents the tuning range of cascading SFG at various seeding angles α. When the seeding angle ranges from -3° to -18°, the phase-matching angle of the NOPA was found to overlap with that for SFG in a ~400nm-pumped type-I BBO-NOPA near its degenerate point of θ~29°. The result reveals that the central wavelength of the cascading SFG is adjustable from 410nm to 440nm by decreasing the seeding angle (see the filled squares for Δk=0). The tuning of the DFG can be achieved in two ways: one is to change the seeding angle between the white light supercontinuum (WLS) and the pump, the other is to scan the orientation of the crystal.

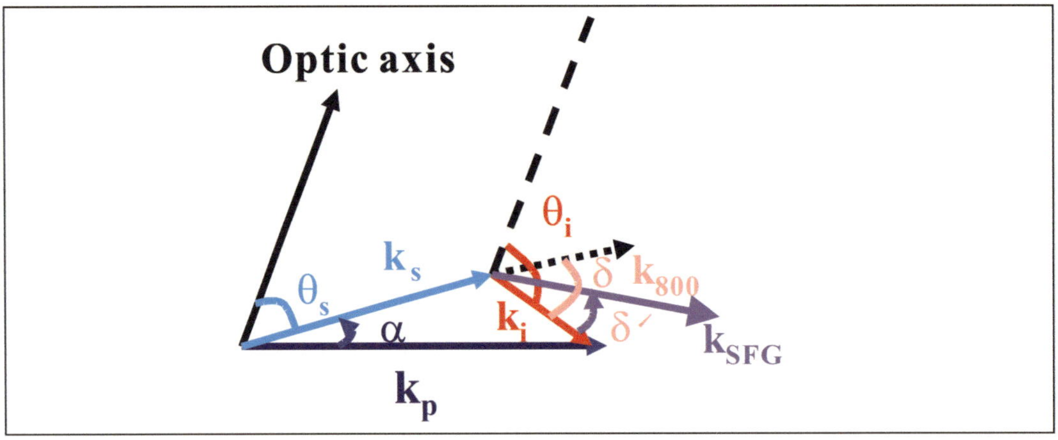

Fig. 1. Schematic showing the noncollinear phase matching condition for optical parametric amplifier (OPA) and the cascading SFG of OPA and residual 810nm pump beam.

As can be seen in Fig. 2 that at larger seeding angles, such as -15°, where the central wavelength with Δk=0 is near 410nm, small GVM and therefore large pulse interactive length are yielded for wavelength from 380nm to ~460nm. At seeding angle around -8°, the tuning range slightly increases and covers from ~390nm to ~500nm. At an even small angle such as -2° →-6°, the tuning range breaks into two separated regimes. But at such a small seeding angle, the GVM is large and the pulse interaction length becomes shorter. We need to apply higher pumping intensity in order to generate stable SFG. In addition to the

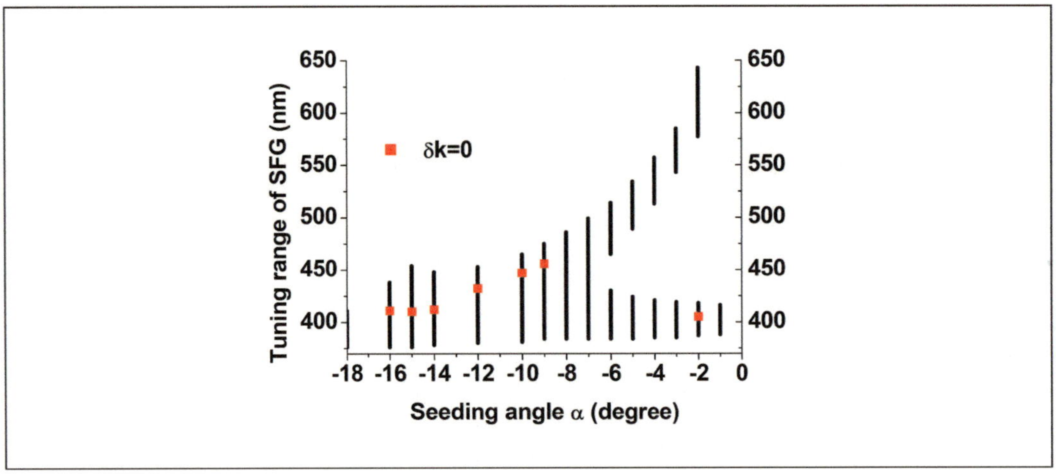

Fig. 2. Theoretical tuning curves of the cascading SFG at various seeding angles between OPA and residual 810-nm laser beam in a 405nm pumped type-I BBO-NOPA

drawback, the GVM from dispersion of WLS is also difficult to be compensated with the noncollinear phase matching scheme. At a given seeding angle, the SFG wavelength can also be tuned by changing the orientation of the NOPA crystal and thus continuous wavelength scan can be done.

3. Experimental setup

The experimental setup is shown in Fig. 3. An amplified femtosecond Ti: sapphire laser provides pulse energy of 1mJ at 810nm with pulse duration of ~90 fs and 1-kHz pulse repetition frequency. The output of the laser is split into two parts: 90% of the energy is frequency-doubled to 405nm with a 0.30-mm-thick BBO for pumping NOPA. The frequency-doubling pulse energy is adjustable from 50 to 120 μJ. The rest 10% of the Ti:sapphire laser beam is used to generate WLS with a 2-mm-thick CaF_2 plate. CaF_2 plate is employed to reduce the chirping in WLS and increase the seeding intensity so that the generated OPA pulses could be stronger and more stable. The generated WLS is collimated and then temporally and spatially overlaps with the SH beam in a 2-mm-long type-I BBO cut at 29°. The beam crossing angle between the seeding and the pump pulses is adjustable from -2° to -18°.

Around the degenerate point of the NOPA, a bright beam with wavelength tunable from near UV to the blue can readily be observed. The inset of Fig. 3 was taken with the output of the NOPA being projected onto a while paper. The ring is the parametric superfluorescence and the black spot at the center of the ring indicates the position of the residual pump beam at 405nm. We punched a hole on the screen to let the residual pump beam pass through the hole to avoid saturation of CCD camera by the scattered residual pump beam. The red spot on the right-hand side of the pump and near the superfluorescence ring is the output of OPA. On the left-hand side of the pump, the spot outside of the ring is originated from the tunable SHG of the OPA component and the bright spot inside of the ring is widely tunable from 380 nm to 460 nm.

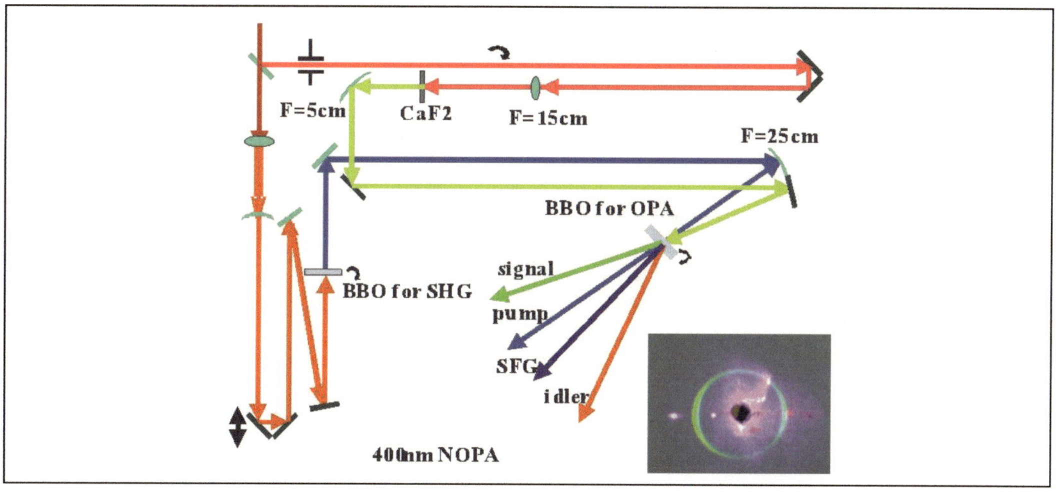

Fig. 3. Experimental setup of the noncollinear optical parametric amplifier and cascading SFG employed in this study. The inset shows the beam pattern of NOPA output projected onto a white screen. The image was taken with a seeding angle of -8°.

4. Results and discussions

We found experimentally that the bright and widely tunable spot at UV/blue can be attributed to the sum-frequency-generation (SFG) from the NOPA and the residual pump beam at 810 nm used in WLS generation. At seeding angle of -14°, the SFG was observable even below the pumping threshold of parametric superfluorescence. However when the seeding angle is decreased to -6°→-8°, we have to pump the NOPA crystal to a level at which a bright and stable parametric super fluorescence is obtained. To further reduce the seeding angle, the SFG becomes difficult to be generated. Based on the calculation of the GVM among SFG, OPA and 810-nm pulses, we found that seeding angle between -8.4° and -14° is most appropriate for the cascading SFG. At a small seeding angle of ~-1.5°, the large GVM leads to a short interactive length in the crystal and it weakens the cascading SFG. Note that at the small seeding angle we can still observe several blue spots, which could be attributed to the SHG of the OPA and cascading SFG. As shown in Fig. 4, at a small seeding angle of ~-2°, the phase matching curve of the cascading SFG crosses with that of the SHG of the idler with a wavelength near 800 nm. Therefore, it is possible to simultaneously observe the SHG of the NOPA idler wave and cascading SFG at small seeding angle. Although the SHG of NOPA allows extending the wavelength to 380-nm or lower, its output energy is significantly lower than the cascading SFG at large seeding angle.

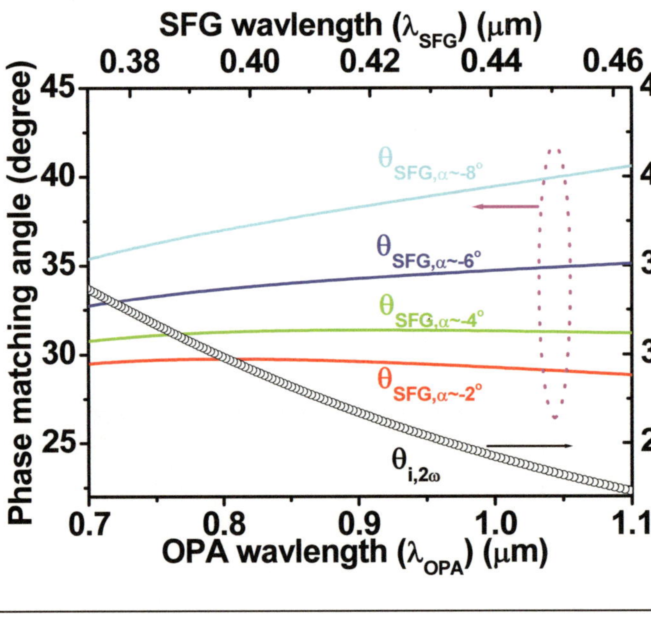

Fig. 4. Calculated curves of phase-matching angle of the cascading SFG (left y-axis) and the SHG of NOPA idler wave (right y-axis) as a function of optical wavelength of OPA (x-axis: bottom) and SFG (x-axis: top).

4.1 Tuning range of cascaded SFG and SHG

We can adjust the central wavelength of the cascading SFG by varying the orientation of the NOPA crystal with a given seeding angle. Fig. 5 shows the spectra of the cascading SFG with a seeding angle of -8.4° and -14°, respectively. As shown in Fig. 5(a) with a seeding angle of -8.4°, the tuning range covers from 395nm to 465nm and similarly with seeding angle of -14° the tuning range can extend from 380nm to 432nm. Figure 6 shows the tuning characteristics of the generated SFG and SHG, respectively, at a seeding angle of 8-degree. The SFG is tuned by rotating the BBO crystal. A tuning range from 389 nm to 425-nm had been observed as shown in the top figure. The maximum output of SFG is at 400-nm. With the same seeding angle of 8 degree, the tuning range of the cascaded SHG of OPA covers a very broad spectral range from 410-nm to 660-nm. The maximum output occurs at 439-nm. The bottom figure shows the spectra of the tunable SHG at various angles of the BBO.

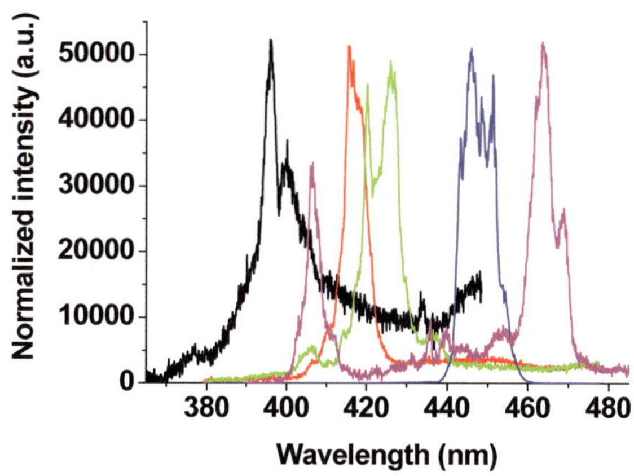

(a)

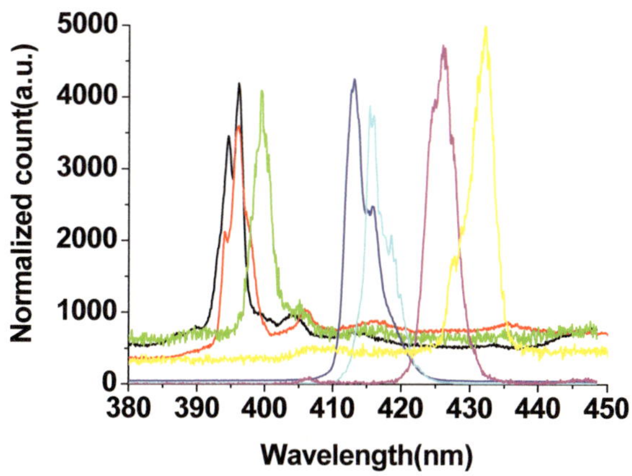

(b)

Fig. 5. Spectrum of the cascading SFG at various orientations of BBO. The seeding angle used is (a)~-8.4°; (b)~-14°

Figure 6 shows the tuning characteristics of the generated SFG and SHG, respectively, at a seeding angle of 8-degree. The SFG is tuned by rotating the BBO crystal. A tuning range from 389 nm to 425-nm had been observed as shown in the top figure. The maximum output of SFG is at 400-nm. With the same seeding angle of 8 degree, the tuning range of the cascaded SHG of OPA covers a very broad spectral range from 410-nm to 660-nm. The maximum output occurs at 439-nm. The bottom figure shows the spectra of the tunable SHG at various angles of the BBO. It is seen from Fig. 2 that the tuning range of the cascaded SFG/SHG fills the gap of the tuning range of a 400-nm pumped OPA in the wavelength range shorter than 460-nm. Experimentally, it is observed that the SFG and SHG can be generated simultaneously over a broad tuning range of the OPA.

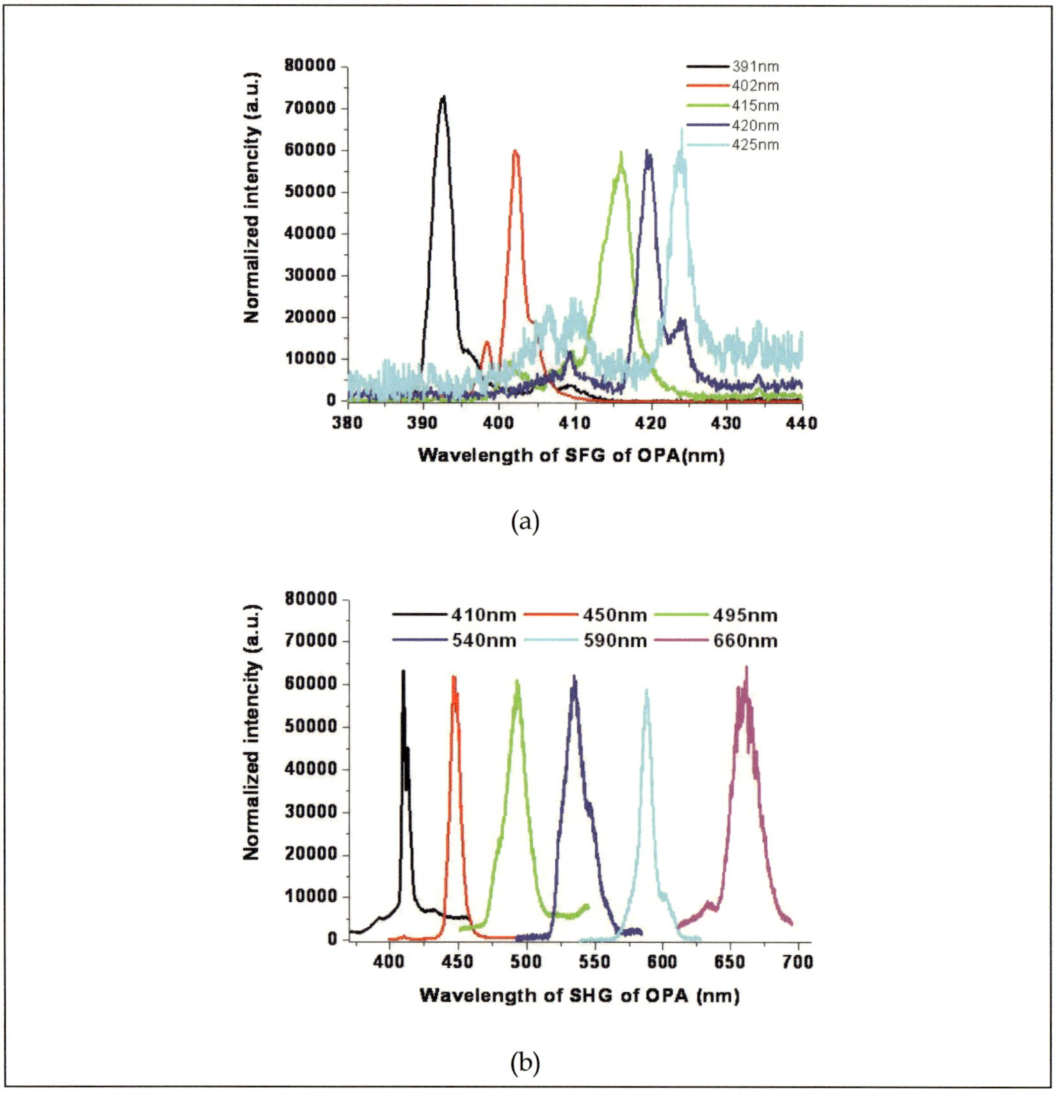

Fig. 6. Spectra of cascaded SFG (up-figure) and SHG (bottom-figure) of NOPA at various BBO orientations. The spectra were taken at a seeding angle of 8 degree.

In order to understand such a phenomenon, a theoretical calculation was done to find the phase-matching conditions for the OPA, the cascaded SFG and SHG. It can be shown that when the seeding angle lies between 3 to 18 degrees, the phase-matching angle of the cascaded SFG and SHG of OPA is coincidently overlapped with that of a 400-nm pumped type-I BBO NOPA at its degenerate point. Fig. 7 shows the theoretical tuning range and the SFG wavelength with $\Delta k=0$ for a variety of seeding angle α. The measured wavelengths with α= -8.4° and -14° are also included for comparison. It was found that the experimental results (open symbols) agree well with the theoretical calculation (solid lines). The measured SFG wavelength with maximum-gain (the star symbols) also coincides well with the calculated phase-matched SFG wavelength with $\Delta k=0$ (the filled squares). This firmly supports our concept about the origin of the generated tunable UV/blue radiation from NOPA.

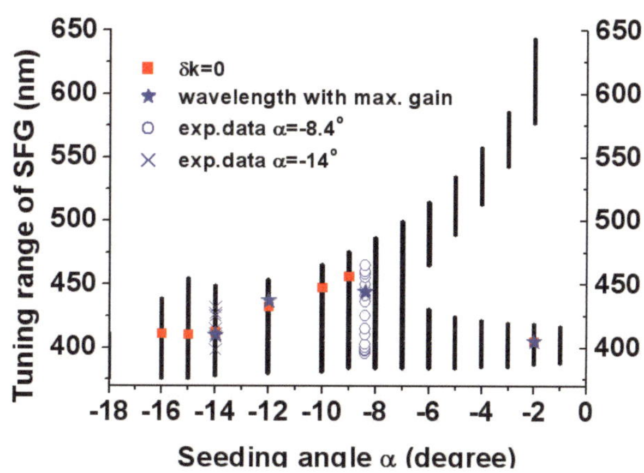

Fig. 7. Theoretical tuning range and the cascading SFG wavelength with maximum gain under various angles of α. Experimental data of SFG wavelength with seeding angle of -8.4° and -14° are also included for comparison.

4.2 Conversion Efficiency of Cascaded SFG-OPA

The cascading SFG yields fairly high optical conversion efficiency. With a pumping energy of 75 μJ per pulse, the output of the SFG was measured to be about 4μJ, which corresponds to an optical conversion efficiency of more than 5% from the pump to the SFG. The output of DFG can be increased by increasing the total energy used for WSL generation and thus the residual radiation at 810-nm is higher. Fig. 8 presents a plot of the SFG output power at a variety of residual pulse energy at 810nm along the WLS seeding beam. The observed linear dependence of SFG output power on residual 810nm beam provides extra supporting data for our notion of the generation mechanism of the cascading SFG. As noted above, we can also produce tunable near UV-VIS femtosecond pulses by frequency-doubling the output of the OPA or use SFG to combine the output of OPA with the fundamental laser at ~810nm. However, these schemes need extra NLO crystal and require synchronous tuning of at least two crystals. This makes the operation of these devices complicated.

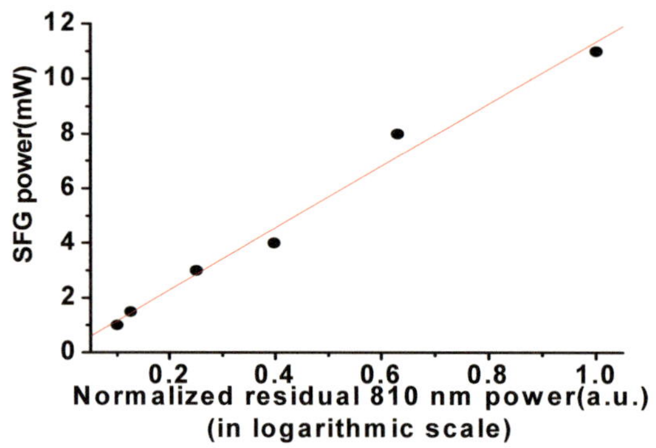

Fig. 8. Output power of the cascading SFG as a function of the averaged power of the residual 810nm pump beam

4.3 Pulse shape of SFG

By using a newly developed OPA-based frequency-resolved optical gating (OPA-FROG) [57], the output of the NOPA can be fully characterized. Fig. 9 shows the measured OPA-FROG trace and its retrieved amplitude and phase of the NOPA component at 571nm. The retrieved OPA-FROG trace shows that no significant chirping in the NOPA output pulse. The pulse duration of the NOPA is about 75-fs. We also employ a single-shot XFROG[58] to deduce the pulse profile of the cascading SFG and found the FWHM duration is about 100-fs. The value is longer than that of the corresponding NOPA but shorter than that of the residual 810-nm pulse. This result also agrees with the notion that the SFG component is a temporal convolution of the NOPA component and the residual fundamental beam. The OPA component is generally shorter than that of its pump at 405-nm, which is shorter than its fundamental pulse at 810 nm.

5. Summary

In conclusion, we have successfully demonstrated the generation of femtosecond laser pulses tunable from 380nm to 460nm via cascading SFG in a 405nm pumped type-I non-collinearly phase-matched OPA. Furthermore, we theoretically analyzed this much simpler scheme to generate femtosecond tunable radiation. Tuning range at various seeding angles, conversion efficiency and pulse profile are characterized in detail. The energy conversion efficiency of the SFG is more than 5%. The reported scheme provides a simple while effective way to extend the tuning range of a 405nm pumped type-I BBO-OPA from 460nm down to 380nm without any additional frequency up-conversion stage. It is also found that the beam quality and pulse shape of the SFG component is about the same as that of OPA.

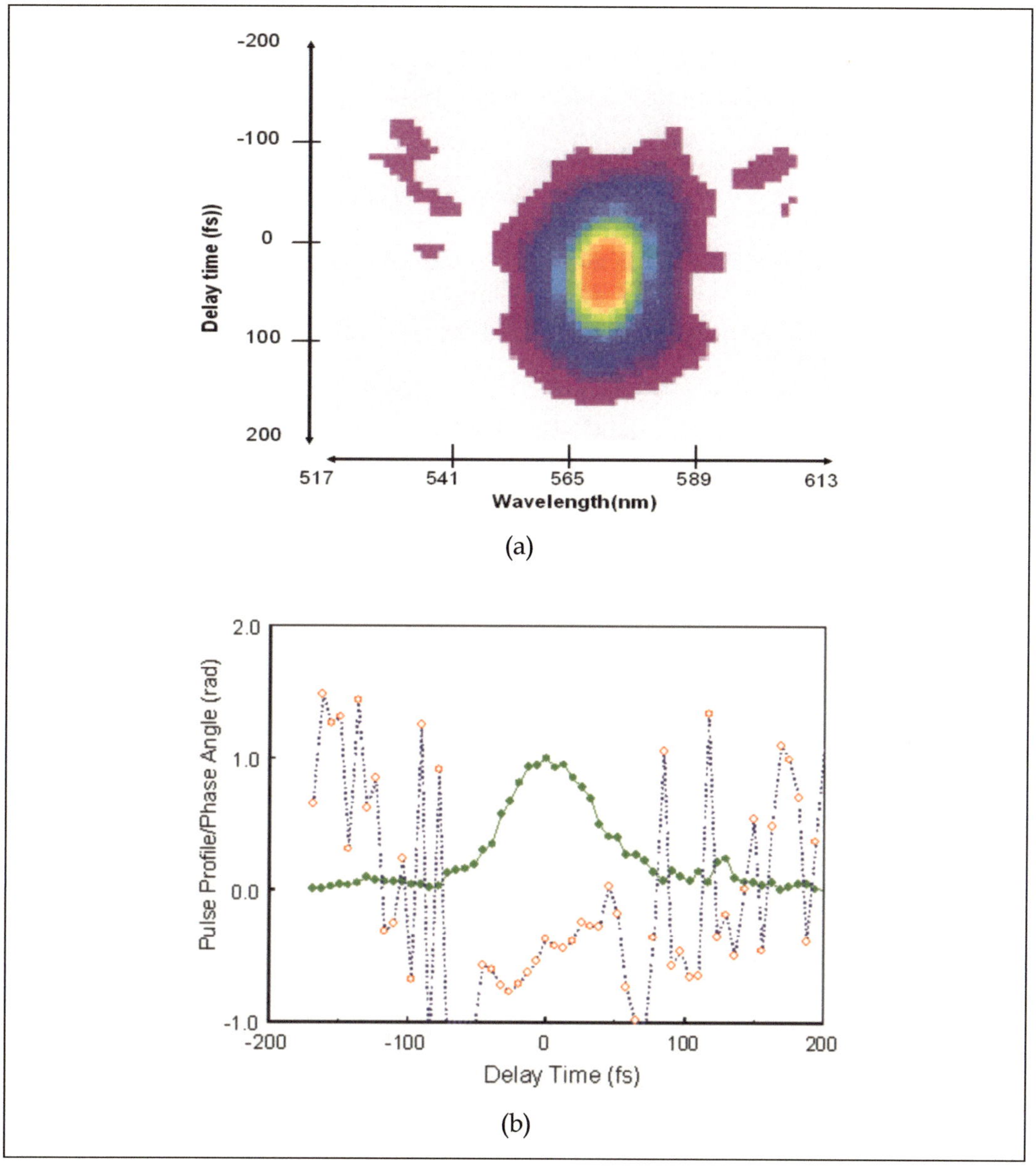

Fig. 9. (a) Measured OPA-FROG trace of NOPA output at 571 nm; (b) the retrieved pulse profile (solid line) and phase (dotted line) of the NOPA output.

6. References

[1] For a overview of the state of the art of ultrafast spectroscopy, see for example: Ultrafast Phenomena XII, Springer Verlag Series in Chemical Physics Vol. 66, edited by T. Elsaesser, S. Mukamel, M. M. Murnane, and N. F. Scherer ~Springer, Berlin, 2001.

[2] T. Brabec and F. Krausz, Rev. Mod. Phys. 72, 545 ~2000

[3] J. P. Zhou, G. Taft, C. P. Huang, M. M. Murnane, H. C. Kapteyn, and I. P. Christov, Opt. Lett. 19, 1149 ~1994

[4] A. Stingl, M. Lenzner, C. Spielmann, F. Krausz, and R. Szipocs, Opt. Lett. 20, 602 ~1995.

[5] G. Steinmeyer, D. H. Sutter, L. Gallmann, N. Matuschek, and U. Keller, Science 286, 1507 ~1999.

[6] R. L. Fork, B. I. Greene, and C. V. Shank, Appl. Phys. Lett. 38, 671 ~1981.

[7] A. H. Zewail, J. Phys. Chem. A 104, 5660 (2000)

[8] N. H. Damrauer, G. Cerullo, A. Yeh, T. R. Boussie, C. V. Shank, J. K. McCusker, Science 275, 54 (1997).

[9] T. Elsaesser, S. Mukamel, M. Murnane, N. F. Scherer, Ultrafast Phenomena XII, in: Proceedings of the 12th International Conference, Springer, New York, 2000.

[10] P. Di Trapani, A. Andreoni, C. Solcia, P. Foggi, R. Danielius, A. Dubietis, and A. Piskarskas, J. Opt. Soc. Am. B 12, 2237-2244 (1995).

[11] J. M. Liu, G. Zhou, and S. J. Pyo, J. Opt. Soc. Am. B 12, 2274-2287 (1995).

[12] R. Danielius, A. Piskarskas, A. Stabinis, G. P. Banfi, P. Di Trapani, and R. Righini, J. Opt. Soc. Am. B 10, 2222-2232 (1993).

[13] A. Piskarskas, A. Stabinis, and A. Yankauskas, 'Sov. Phys. Usp. 29, 969-979 (1986).

[14] R. Danelius, A. Piskarskas, V. Sirutkaitis, A. Stabinis, and A. Yankauskas, JETP Lett. 42, 122-124 (1985).

[15] J. Wang, M. H. Dunn, and C. F. Rae, Opt. Lett. 22, 763-765 (1997).

[16] A. Shirakawa, I. Sakane, H. Takasaka, and T. Kobayashi, 'Appl. Phs. Lett. 74, 2268-2270 (1999).

[17] G. Cerullo, M. Nisoli, S. Stagira, and S. De Silvestri, Opt. Lett. 23, 1283-1285 (1998).

[18] I. N. Ross, J. L. Collier, P. Matousek, C. N. Danson, D. Neely, R. M. Allott, D. A. Pepler, C. Hernandez-Gomez, and K. Osvay, Appl. Opt. 39, 2422-2427 (2000).

[19] F. Rotermund, V. Petrov, and F. Noack, Opt. Commun. 169, 183-188 (1999).

[20] K. Osvay, G. Kurdi, J. Klebniczki, M. Csatari, I. N. Ross, E. J. Divall, C. J. Hooker, and A. J. Langley, Appl. Phys. B 74, 163-169 (2002).

[21] J. Opt. Soc. Am. B 8 , 2087 (1991).

[22] R. Danielius , A. Piskarskas , A. Stabinis , G. P. Banfi , P. Di Trapani , and R. Righini , J. Opt. Soc. Am. B10 , 2222 (1993)

[23] F. Seifert, V. Petrov , and F. Noack , Opt. Lett. 19, 837 (1994).

[24] V. Petrov., F. Seifer, O. Kittlemann, J. Ringling, F. Noack, , J. Appl. Phys. 76, 7704 (1994)

[25] V. V. Yakovlev , B. Kohler , and K. R. Wilson , Opt. Lett. 19, 2000 (1994).

[26] M. Nisoli , S. De Silvestri , V. Magni , O. Svelto , R. Danielius , A. Piskarskas , G. Valulis, and A. Varanavicius , Opt. Lett. 19, 1973 (1994).

[27] V. Petrov, F. Seifert, and F. Noak, Appl. Phys. Lett. 65, 268 (1994).

[28] V. Petrov , F. Seifert , and F. Noak, Appl. Opt. 33 , 6988 (1994).

[29] M. K. Reed, M. K. Steiner-Shepard, and D. K. Negus, Opt. Lett.19, 1855-1857 (1994).
[30] S. R. Greenfield and M. R. Wasielewski, Appl. Opt. 34, 2688-2691 (1995).
[31] Opt. Lett. 20, 1394 (1995).
[32] R. Danielius , A. Piskarskas , P. Di Trapani , A. Andreoni , C. Solcia , and P. Foggi , Appl. Opt. 35 , 5336 (1996)
[33] P. Matousek , A. W. Parker , P. F. Taday , W. T. Toner , and M. Towrie, Opt. Commun. 127, 307-312 (1996).
[34] K. S. Wong, H. Wang, G. K. Wong and J. Y. Zhang, Appl. Opt. 36, 1889, (1997).
[35] P. Di. Trapani, A. Andreroni, C. Solicia, G. P. Banfi, R. Danielius, A. Piskarskas, P. Foggi, J. Opt. Soc. Am. B 14, 1245 (1997).
[36] J. Y. Zhang, Z. Y. Xu, Y. F. Kong , Chaowen Yu, and Yichen Wu, Appl. Opt. 37, 3299-3305 (1998).
[37] P. Di Trapani , A. Andreoni , P. Foggi , C. Solcia , R. Danielius , and A. Piskarskas , Opt. Commun. 119, 327-332 (1995).
[38] P. Di Trapani , A. Andreoni , G. P. Banfi , C. Solcia , R. Danielius , P. Foggi , M. Monguzzi , A. Piskarskas , and C. Sozzi , Phys. Rev. A 51, 3164 (1995).
[39] T. Wilhelm, J. Piel, and E. Riedle, Opt. Lett. 22, 1494-1496 (1997).
[40] G. Cerullo, M. Nisoli, and S. De Silvestri, Appl. Phys. Lett. 71, 3616-3618 (1997).
[41] G. Cerullo, M. Nisoli, S. Stagira, and S. De Silvestri, Opt. Lett. 23, 1283-1285 (1998).
[42] V. Krylov, O. Ollikainen, J. Gallus, U. Wild, A. Rebane, A. Kalintsev, Opt. Lett. 23, 100-102 (1998)
[43] A. Shirakawa and T. Kobayashi, Opt. Lett. 23, 1292-1294 (1998).
[44] F. Rotermund, V. Petrov and F. Noack, Optics Communs., 171, 183-188, (1999).
[45] E.. Riedle, M. Beutter, S. Lochbrunner, J. Piel, S. Schenkl, S. Sporlein, W. Zinth, Appl. Phys. B 71, 457 (2000).
[46] J. Piel, M. Beutter, E. Riedle, Opt. Lett., 25, 180-182 (2000).
[47] Howe-Siang Tan, Warren S. Warren, Elmar Schreiber, Opt. Lett. 26, 1812-1814 (2001)
[48] R. Huber, H. Satzger, W. Zinth, and J. Wachtveitl, Opt. Commun., 194, 443-448 (2001).
[49] P. Tzankov, I. Buchvarov, T. Fiebig, Opt. Commun., 203, 107-113 (2002).
[50] K. Osvay, G. Kurdi, J. Klebniczki, M. Csatari, and I. N. Ross, Appl. Phys. Lett. 80, 1704 ~2002!.
[51] K. Osvay, G. Kurdi, J. Klebniczki, M. Csatari, I. N. Ross, E. J. Divall, C. H. J. Hooker, and A. J. Langley, Appl. Phys. B: Lasers Opt. B74, S163~2002.
[52] K. Osvay, G. Kurdi, J. Klebniczki, M. Csatari, I. N. Ross, E. J. Divall, C.H. J. Hooker, and A. J. Langley, in Proceedings of the Conference on Lasers and Electro-optics, 18–22 June, Munich, Germany ~2001!.
[53] P. Tzankov, T. Fiebig, and I. Buchvarov, in Proceedings of the International Quantum Electronics Conference, 22–28 June, Moscow, Russia ~2002.
[54] Chao-Kuei Lee, Jing-Yuan Zhang+, J. Y. Huang and Ci-Ling Pan, Optics Express 11, 1702-1708 (2003).
[55] Chao-Kuei Lee, Jing-Yuan Zhang, J. Y. Huang and Ci-Ling Pan, *JOSA-B*. 21, 1494-1499 (2004).
[56] Y.R. Shen, "Principal of Nonlinear Optics", (John Wiely & Sons, 1984).

[57] J. Y. Zhang, Aparna Prasad Shreenath, Mark Kimmel, Erik Zeek and Rick Trebino, and Stephan Link, Optics Express, 11, 601-609 (2003).

[58] Jing-Yuan Zhang, Chao-Kuei Lee, Jung Y. Huang and Ci-Ling. Pan, *Optics Express* 12, 574-581 (2004)